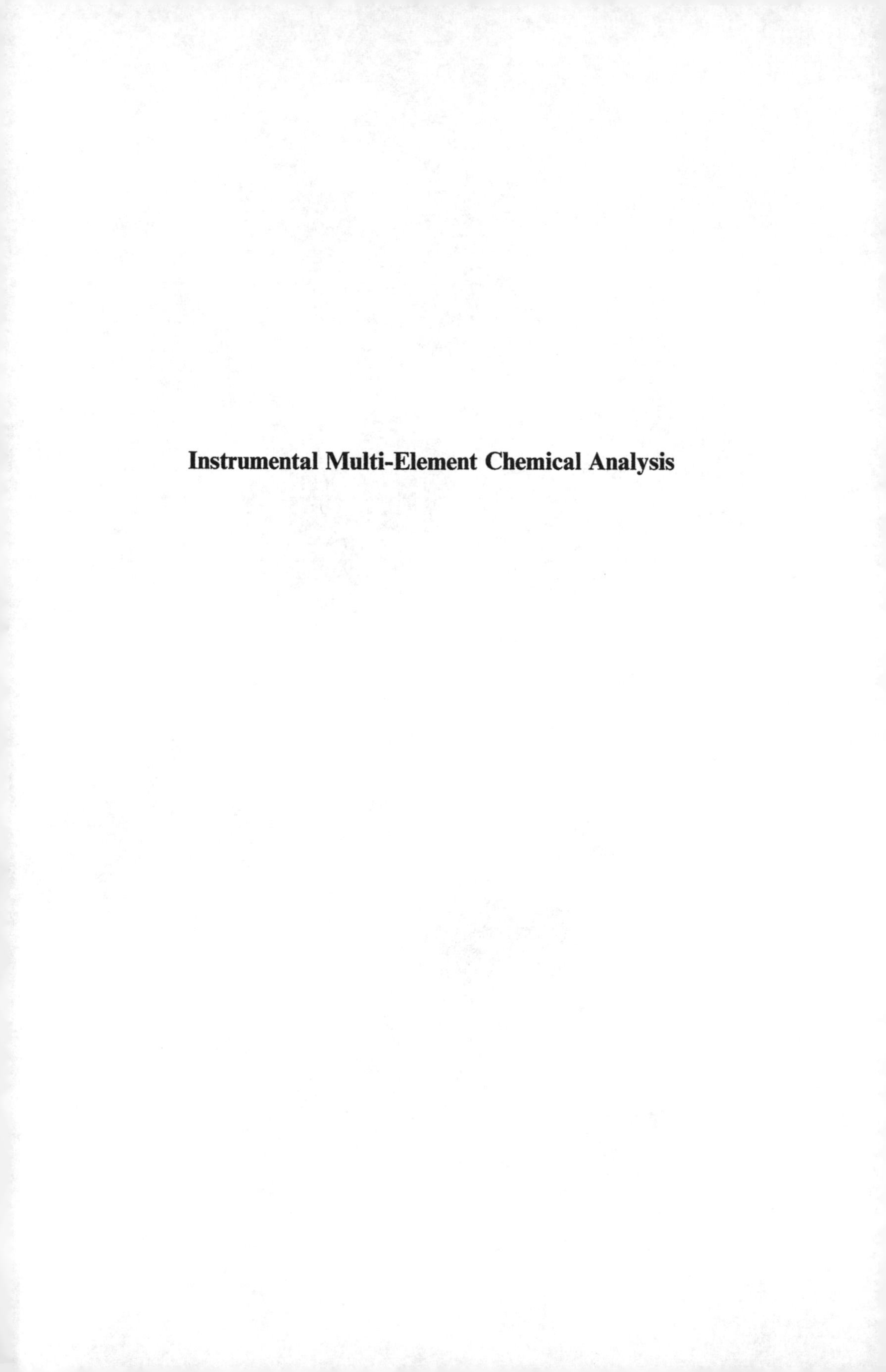

Instrumental Multi-Element Chemical Analysis

Instrumental Multi-Element Chemical Analysis

Edited by

Z.B. ALFASSI
Department of Nuclear Engineering
Ben-Gurion University of the Negev
Beer-Sheeva
Israel

SPRINGER-SCIENCE+BUSINESS MEDIA, B.V.

Library of Congress Catalog Card number: 98-66852

ISBN 978-0-7514-0427-2 ISBN 978-94-011-4952-5 (eBook)
DOI 10.1007/978-94-011-4952-5

Printed on acid-free paper

Originally published by Kluwer Academic Publishers in 1998

Typeset in 10/12pt Times by AFS Image Setters Ltd, Glasgow

to Sabina
with love

Contents

Contributors

Prof. Z.B. Alfassi — Department of Nuclear Engineering, Ben-Gurion University of the Negev, Beer-Sheeva 84102, Israel

Prof. N. De Silva — Department of Chemistry, Carleton University, Ottawa, Ontario, Canada K1S 5B6

S. Felix — Unit for Characterization of Materials, Institute for Applied Research, Ben-Gurion University of the Negev, Beer-Sheeva 84105, Israel

Dr M.D. Glascock — Research Reactor Center, University of Missouri, Columbia, MO 65211, USA

Dr D.C. Gregoire — Geological Survey of Canada, 601 Booth Street, Ottawa, Ontario, Canada K1A 0E8

Prof. P.C. Hauser — Department of Chemistry, University of Basel, Spitalstrasse 51, 5046 Basel, Switzerland

Dr P. Kregsamer — Atominstitut der Österreichischen Universitäten, Schüttelstrasse 115, A-1020 Wien, Austria

Dr S. Levin — Medtechnica, Efal St. 5 Kiriat Arie, Petach Tikva 49511, Israel

Prof. M. Mantler — Institut für Angewandte und Technische Physik, Technische Universtität Wien, Wiedner Hauptstrasse 8-10, A-1040 Wien, Austria

Prof. I.Z. Pelly — Department of Geological and Environmental Sciences, Ben-Gurion University of the Negev, Beer-Sheeva 84105, Israel

Prof. M. Polak — Department of Materials Engineering, Ben-Gurion University of the Negev, Beer-Sheeva 84105, Israel

Prof. E. Rauhala Accelerator Laboratory, Department of Physics, PO Box 9, University of Helsinki, 00014 Helsinki, Finland

Prof. P. Wobrauschek Atominstitut der Österreichischen Universitäten, Schüttelstrasse 115, A-1020 Vienna, Austria

Preface

Classical analytical methods such as gravimetry and absorption of special titrimetry complexes remain in use in many laboratories and are still widely taught in first-year analytical courses. These courses and particularly the laboratory experiments attached to them are excellent for training students in the work of an experimental chemist. These methods are also still the main ones when very high analysis of the major elements of a matrix is required. However, most analyses are now performed by instrumental analysis, and a rough estimate is that at least 99% of current analytical measurements are done by instrumental techniques such as emission and absorption spectrometry, electrochemical methods, mass spectrometry, various methods of gas and liquid chromatography or radiochemical methods. Several properties common to many of these methods are the reason for the shift from classical analytical chemistry to modern instrumental methods.

- *Sensitivity* Instrumental methods can reach detection limits undreamed of in classical methods. These higher sensitivities completely change the range of what is called trace elements. While not so long ago trace elements were those at a concentration of parts per thousand (permill, 10^{-3}), nowadays instrumental measurements of parts per million (ppm, 10^{-6}) concentration and even lower are very common.
- *Versatility and testing* The possibility of using various instrumental methods leads to verification of the analytical results. While using more than one method for routine analysis is less common, it is done when the required data are important for decision-making or for certification of reference materials.
- *Price and through-put* While the classical methods are labour-intensive and time-consuming, many of the instrumental methods are faster and hence allow cheaper analysis and a large through-put of samples.
- *Multi-element analysis* The classical methods require a separate experiment for the measurement of each element, while many of the modern instrumental methods allow determination of several elements in the same experiment. Frequently more than 10 elements and even 20 elements are determined simultaneously in inductively coupled plasma atomic emission or mass spectrometry or in neutron activation analysis or in EDXRF.
- *Computer interface and output* Most modern analytical instruments are PC-based: this allows fast, varied processing of the results (including statistical analysis) and the printing of the analysis results in various forms appropriate to the clients' requirement.

1 Preparation of samples

Z.B. ALFASSI and S. FELIX

1.1 Introduction

Analytical samples can be in various forms and in various phases – solid, liquid and gas. Most analytical methods cannot be used directly for solid samples. However, some instrumental methods like neutron activation analysis (NAA), sputtered neutral mass spectrometry (SNMS) and X-ray fluorescence spectrometry (XRFS) can be used to analyze solid samples without transferring them into solution. In these cases the only preparation step required is the cleaning (etching) of the surface to remove contamination which does not belong to the sample.

The advantage of this kind of analysis is the short time needed for the entire analysis, less contamination caused by the preparation steps and the economic aspect.

There are some disadvantages of the direct instrumental determination: (1) The reliability of the results decreases with decreasing concentration of the elements to be determined, since no preconcentration is done. (2) Systematic errors may occur as a result of spectral and non-spectral interference due to interference from the more abundant major elements. (3) There are not yet standards for all kinds of samples in the very trace concentration level.

For most other analytical methods a procedure of sample preparation, i.e. sample decomposition, separation and preconcentration of the trace elements, is required before the determination.

Preconcentration of the analytes will be dealt with in the next chapter and in this chapter we will discuss only the dissolution of matrices (Bajo 1992).

The decomposition of samples can be divided either according to the origin of the sample (geological, biological, etc.) or the type of the sample (elemental, inorganic, organic, etc.) or the type of the method of digestion (wet, combustion, fusion).

The proper choice among the various reagents and techniques for decomposing and dissolving analytical samples is crucial to the success of the analysis, particularly where refractory substances are involved or where the analyte is present in trace amounts.

1.2 Dissolution of geological and environmental inorganic samples

Samples should be completely dissolved to get reproducible and accurate results. The large variety in the composition of geological samples and the refractory properties of many elements make them very difficult to dissolve.

Most geological samples and environmental samples are either silicates or oxides or contain them as a major component. Silicates can be dissolved either in a wet process by dissolution with acids or in a dry process by fusion. The fusion is followed by dissolution in acids.

1.2.1 Acids

The use of acids in the dissolution process can be done in open or closed systems, depending on the elements to be analyzed.

The usual acid is hydrofluoric acid (HF) which readily dissolve silicates and aluminosilicates as well as oxides and many metals. The dissolving power of HF increases sharply with temperature. If PTFE-lined steel closed vessels are employed, the dissolution time can be decreased by the use of high pressure, which allows higher temperature than the boiling point. In many cases a mixture of HF and perchloric acid, as an oxidant, is used to speed up the dissolution. There is a risk that refractory oxides will again form if the $HF–HClO_4$ mixture is evaporated to dryness or near-dryness with too much heat. This can be avoided by using a water bath or an infrared heater. Even at normal pressure a 1:1 mixture of $HF:HClO_4$ at 95°C dissolves 16 of the 20 most common silicates in under 1 h (Langmyhr and Sween 1965). Of the four resistant minerals (beryl, kyanite, staurolite and topaz), the first three dissolve in 1 h with the same mixture in a closed vessel at 250°C. The rate of dissolution depends largely on the concentration of HF (French and Adams 1973). Some studies used aqua regia instead of $HClO_4$ for the purpose of oxidation, e.g. 5% HF + 0.2 ml aqua regia (French and Adams 1973). The aqua regia is added first to the sample and only then is HF added. If HF is added first the sample could be crusted.

It should be noted that some minerals such as chromite, corundum, zircon, rutile and quartz are not dissolved by acids.

The reaction between HF and silicates produces a milky suspension of insoluble fluorides (calcium and magnesium together with some mixed fluorides). To obtain a clear solution the fluorides are transformed into the soluble fluoroborates by the addition of boric acid (Bernas 1968, French and Adams 1973) or the fluorides are removed as volatile SiF_4 and HF by heating the solution to dryness in the presence of H_2SO_4 or $HClO_4$ (Langmyhr 1967), although many analysts are afraid of explosions in heating to dryness of a solution containing $HClO_4$.

Oxides are generally soluble in halogen acids. The dissolving power of the acid increases with the complexing ability of the halogen with the metal. As these acids have a low boiling temperature, they cannot dissolve fired oxides of several elements, e.g. aluminum, beryllium, chromium, titanium and zirconium. However, high fired magnesia and beryllia dissolved in halogen acids heated to high temperature in a closed vessel (200°C for 16 h). The sealed vessel can be of fused silica tube or stainless-steel PTFE-lined vessel or quartz or glassy carbon vessel.

Instead of the halogen acid in order to work at higher temperatures the higher-boiling phosphoric acid can be used. It dissolves even oxides of beryllium, thorium and protactinium.

Silicates can also be dissolved by hot (290°C) phosphoric acid, preferably together with $HClO_4$ to oxidize insoluble compounds as e.g. sulfides. The dissolution is done in a platinum crucible and was found to be efficient even for resistant minerals such as chromites (Hannaker and Qing-Lie 1984).

Geological samples are frequently dissolved by a selective procedure. Some sediments can be rich in a carbonate phase that is easily dissolved in cold or moderately warm dilute acid and carbonates can be separated from the rest of the sample. Another selective technique involves the use of moderately warm HCl and a reducing compound like hydroxylamine hydrochloride or citric acid. Such procedures preferentially dissolve metal hydroxide phases, whereas silicates are only slightly attacked.

1.2.2 Fluxes

The common flux for dissolving silicate is sodium carbonate, but alkali hydroxide and sodium peroxide have also been used. In Table 1.1 are listed the common fluxes used for fusion. When fused with silicate, these strong alkaline fluxes form acid-soluble alkali silicates and alkali salts of the present anionic elements together with carbonates and oxides of the cationic elements. The fusion is done at a temperature high enough to melt the flux and render it fluid. Fusion with Na_2CO_3 is the classic choice. Na_2O_2 is a better disintegrator but is difficult to obtain in sufficiently pure form. It is usually done in an electric furnace at 1000°C for 1 h or in the oxidizing flame of a Meker burner.

The cake obtained with these fluxes is readily soluble in acids, and usually HCl is employed. The cake can also be decomposed by water but not all the elements go into the solution. Cobalt, iron, scandium and rare-earth metals are not decomposed in this way. The solution has a high concentration of sodium. If sodium interferes it can be removed by an HAP (hydrated antimony pentoxide) column or preferably using K_2O_2 and/or KOH for flux. Potassium ions are removed by dissolving the cake in dilute perchloric acid, leaving potassium as the insoluble perchlorate salt, whereas most of the other perchlorates are very soluble.

Table 1.1 Common fluxes

Flux	Melting point (°C)	Type of crucible for fusion	Type of substance decomposed
Na_2CO_3	851	Pt	Silicates and silica-containing samples, alumina-containing samples, sparingly soluble phosphates and sulfates
Na_2CO_3 + an oxidizing agent, such as KNO_3, $KClO_3$, or Na_2O_2	–	Pt (not with Na_2O_2), Ni	Samples requiring an oxidizing environment; that is, samples containing S, As, Sb, Cr, etc.
$LiBO_2$	849	Pt, Au, glassy carbon	Powerful basic flux for silicates, most minerals, slags, ceramics
NaOH or KOH	318	Au, Ag, Ni	Powerful basic fluxes for silicates, silicon carbide and certain minerals (main limitation is purity of reagents)
Na_2O_2	decomposes	Fe, Ni	Powerful basic oxidizing flux for sulfides; acid-insoluble alloys of Fe, Ni, Cr, Mo, W and Li; platinum alloys; Cr, Sn, Zr minerals
$K_2S_2O_7$	300	Pt, porcelain	Acidic flux for slightly soluble oxides and oxide-containing samples
B_2O_3	577	Pt	Acidic flux for silicates and oxides where alkali metals are to be determined
$CaCO_3 + NH_4Cl$	–	Ni	Upon heating the flux, a mixture of CaO and $CaCl_2$ is produced; used to decompose silicates for the determination of the alkali metals

For analysis of silica-rich materials the digestion procedures require the use of lithium metaborate, which has the advantage of producing a glassy material rather than a cake. Techniques such as colorimetry and atomic absorption spectrometry (AAS) generally require a large excess of $LiBO_2$; Shur and Ingamells (1966) recommend a sample-to-flux ratio of 1:5 and Ingamells (1966) a ratio of 1:7 for complete dissolution of silica rocks. The sample is mixed with the flux in a crucible and fused over an oxidizing flame of a Meker burner for 5–10 min. The obtained glass is dissolved in 2% nitric acid (Ingamells 1970). Some sulfides present in the silicates were found to be only partially attacked.

Graphite crucibles are preferred by most workers because they are easier to use and cheaper. Walsh and Howie (1980) used a Pt crucible to fuse rocks with lithium metaborate. Govindaraju *et al.* (1976) used an automatic tunnel furnace with a capacity of 60 samples/h. The automated dissolution reduces the staff and the time of the analysis.

The fusion method is not sensitive enough for trace analysis down to the 0.5–1 ppm level, due to impurities in the flux materials.

1.3 Dissolution of biological (organic) material

Sample digestion is the least understood step in the analysis of biological materials (Mertz 1981). The big difference between biological and geological materials is the presence of organic compounds in a very great amount.

Biological material can be decomposed either by wet decomposition (called also wet-ashing) or by dry-ashing. Wet-ashing refers to the quantitative destruction of organic matter by oxidizing agents and the wet oxidation ash concentrate is then the sample for analysis. This can be done in open systems or in closed vessels, using conventional heating (gas burner, electric resistance heater, furnace) or by microwave heating. Dry-ashing comprises procedures carried out at elevated temperatures usually employing a muffle furnace, ashing procedures at lower temperatures using activated oxygen, and some specific combustion procedures.

1.3.1 Wet decomposition

Wet decomposition of biological materials can be done at atmospheric pressure in open systems or at higher pressure in closed systems. The reagents most commonly used for wet decomposition are oxidizing acids and partly also strong mineral acids and other oxidants like H_2O_2. The properties of most common decomposing agents are presented in Table 1.2 together with their main applications for various kinds of samples (not only biological). In many decomposition procedures, two or more of these reagents are applied in combination because of the advantages of each reagent; mixtures of acids or of acid + H_2O_2 are used, as e.g. 1 ml of concentrated HNO_3 + 3 ml of 30% H_2O_2 or HNO_3 + H_2SO_4 (5:1) or HNO_3 + $HClO_4$ + H_2SO_4 (10:5:3).

While the use of an open system is simpler experimentally it might cause systematic errors due to:

1. Losses of analyte due to volatilization.
2. Higher blank values due to the use of larger quantities of reagents than in a closed vessel. This problem can be solved by using an open system with a reflux condenser, the so-called Bethge apparatus (Figure 1.1).
3. Larger contamination from external sources.

Besides overcoming these problems the use of a closed system reduces

Table 1.2 Physical properties of selected acids and oxidizing agents used in wet-ashing

Compound	Molecular weight	Density ($kg\,l^{-1}$)	Boiling point (°C)	Concentration (M)	Main application
$HClO_4$	100.46	1.67	200	11.6	To destroy organic samples at fuming temperatures. Many times together with HNO_3. H_2SO_4 increases oxidizing power. Good solvent for sulfides
H_2SO_4	98.08	1.84	338	18.0	Used when high temperatures are required either to expel a volatile compound or to increase reaction rate. Both oxidizing and dehydrating agents added sometimes to reduce the danger of explosion due to $HClO_4$
HF	20.01	1.16	120	27.8	Used for silicates together with H_2SO_4 or $HClO_4$ and for many metals, carbides, nitrides and borides together with HNO_3
HCl HBr	36.46 80.92	1.12 1.50	110 126	7.7 8.7	Dissolves carbonates, phosphates and some oxides, but not organic samples
H_3PO_4	98.00	1.71	213	14.8	Used for dissolution of several oxides and also for some organic samples
HNO_3	63.01	1.40	121	14.4	Dissolve metals which are not dissolved by HCl. However noble metals are not dissolved. They can be dissolved by HCL + HNO_3, aqua regia. Used for decomposition of organic samples in mixtures with H_2O_2 or with H_2SO_4
H_2O_2	34.01	1.12	106	9.9	Used to increase the oxidation of the organic samples, usually together with HNO_3 or H_2SO_4. When the obtained solution is not clear, extra H_2O_2 is added, mainly in microwave application

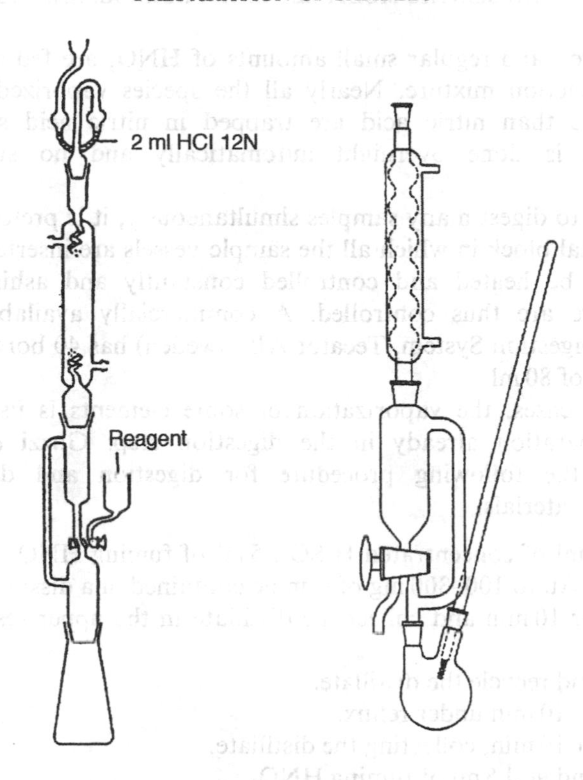

Figure 1.1 Two versions of Bethge apparatus for wet-ashing with refluxing. Generally this all-glass apparatus consists of three parts: a digestion flask, a flux reservoir and a reflux condenser. The apparatus on the left has the possibility of further additions of reagents during the ashing. On the right, the two-way stopcock allows the condensate to return to the flask or to be removed from the apparatus.

the reaction times by a factor of 3–10 and improves the decomposition due to the higher temperature used (above the boiling points of the reagents). The temperature that can be used depends on the vessel material. Polytetrafluoroethylene (PTFE, also called Teflon) and perfluoroalkoxyvinylether (PFA) can be used up to 200°C, but at high temperature the lifetime of the vessel is reduced. For temperatures higher than 200°C, vessels made of quartz are used. Borosilicate glasses are usually not used due to their high concentration of elements that are leached by acids, leading to high blank values.

Wet decomposition has the advantage of lower temperatures than dry-ashing and the result is less losses by reaction with the walls of the vessel, or adsorption on them. Bajo *et al.* (1983) described a simple apparatus for wet decomposition of any organic material by H_2SO_4/HNO_3 mixture. The digest material is treated with H_2SO_4 at a temperature of 300°C (such a high temperature cannot be used with HNO_3 owing to its lower boiling temperature), and regular small amounts of HNO_3 are fed automatically into the reaction mixture. Nearly all the species vaporized at a higher temperature than nitric acid are trapped in nitric acid solution. The dissolution is done overnight automatically and no supervision is required.

In order to digest many samples simultaneously, it is preferable to have a large metal block in which all the sample vessels are inserted. The metal block can be heated and controlled constantly and ashing time and temperature are thus controlled. A commercially available apparatus (Tecator Digestion System, Tecator AB, Sweden) has 40 borosilicate glass tubes each of 80 ml.

In some cases, the vaporization of some elements is used to obtain partial separation already in the digestion step. Guzzi *et al.* (1976) suggested the following procedure for digestion and distillation of biological materials.

1. Add 2 ml of concentrated H_2SO_4, 5 ml of fuming HNO_3 and 10 mg of carrier Au to 100–300 mg of sample contained in a dissolution flask.
2. Boil for 10 min and collect the distillate in the upper reservoir of the flask.
3. Cool and recycle the distillate.
4. Boil for 10 min under reflux.
5. Boil for 10 min, collecting the distillate.
6. Cool and add 5 ml of fuming HNO_3.
7. Repeat operations 2, 3, 4 and 5.
8. Cool and add 10 ml of H_2O_2 dropwise.
9. Distill until fumes of SO_3 appear, collecting the distillate.
10. Cool and add 5 ml of concentrated HBr.
11. Distill until fumes of SO_3 appear, collecting the distillate.
12. Repeat operations 10 and 11 twice.
13. Wash the cooling system and the distillation flask three or four times with 6 M HCl.

The dissolution–distillation can be carried out in an automatic system and Guzzi *et al.* used it for six samples. This procedure is a very long one and in most cases only dissolution without distillation is the routine process.

The use of microwave ovens for the wet decomposition of organic (and also inorganic) samples was first proposed in the mid-1970s and now it has become an important method for sample preparation, mainly in analytical chemistry specialized laboratories (Aysula *et al.* 1987, Kingston and Jassie 1988, Matusiewicz and Sturgeon 1989). The main advantage of microwave decompositions compared with the conventional methods of flame or hot plate is the shorter time required to complete the digestion. This is due to the different mechanism of energy transfer. In the conventional methods heat is transferred by conduction. The vessel materials are usually poor conductors and a long time is required to heat the vessel and then transfer the energy (heat) to the solution. Besides, owing to the poor conductivity of the solution there is a large difference of temperature in the solution and only a small fraction of the solution is maintained at the temperature of the vessel and thus at the boiling temperature of the liquid. In a microwave oven the energy of electromagnetic radiation (with wavelength of 10 cm, frequency of 2450 MHz) is absorbed by polar molecules (water, mineral acids, etc.). Thus the microwave energy is transferred directly and almost homogeneously to all molecules of the solvent. Boiling temperatures are reached throughout the entire solution very quickly. In the case of microwave digestion the vessels should be made of a good thermal insulator (so that less energy will be lost by conduction) and be transparent to microwaves (otherwise most of the energy will be absorbed by the vessel). PTFE and quartz fulfill both requirements. Although metals do not absorb microwaves, they reflect them and hence they will not reach the solution. Therefore most microwave digestions are performed in PTFE (Teflon) vessels. The melting point of this material is 300°C, and care should be taken not to exceed this temperature when using sulfuric and phosphoric acids, which have boiling points higher than 300°C. Special microwave ovens for chemical digestion are commercially available. They are much more expensive than domestic microwave ovens due to their higher power and the interior wall protection against corrosion.

Mainly because of vaporization of volatile elements, higher contamination and slower process, most analysts prefer to use wet digestion in closed systems than in open systems with Bethge apparatus (Figure 1.1). Most of the digestions in microwave ovens are done in closed systems. Several such systems are possible: (1) decomposition bombs; (2) low-pressure systems (up to 1.5 MPa); (3) high-pressure systems (up to 15 MPa).

Usual decomposition bombs are Teflon vessels contained in stainless-steel or nickel bodies, or most simply Teflon-covered stainless-steel vessels (Uhrberg 1982, Okamoto and Fuwa 1984). Sometimes for operating at higher temperatures a closed quartz tube is used (Faanhof and Das 1977). The digestion bombs are inserted in a furnace and kept at the required temperature for predetermined time.

In many laboratories the usual technique is to heat 100–500 mg of sample with an acid mixture overnight at about 140–170°C either in a high-pressure Teflon decomposition vessel (bomb) or in a reflux system.

Several devices for routine wet digestion are available commercially (Knapp 1988).

1. *Decomposition in closed quartz or glassy carbon vessel at high pressure (up to 20 MPa) with conventional heating.* An example is the High-

Pressure Asher (HPA), produced by the Anton Paar Company, Graz, Austria. The asher consists of a quartz digestion vessel mounted in an autoclave (Figure 1.2). The vessel is heated according to a preselected time–temperature program. In the autoclave a high pressure is kept in order that the inner pressure in the quartz vessel will not explode the vessel (what is crucial for breaking is the pressure difference rather than the inner pressure). A special pressure cap can be fitted onto the quartz vessel to measure the actual pressure inside the vessel and adjust to it the pressure in the autoclave. Depending upon the size of the digestion vessels, up to seven vessels can be placed inside the autoclave.

2. *Moderate-pressure digestion with microwave oven.* The vessels are made mainly from PTFE and have a safety relief valve. At pressure higher than the designed one the safety valve opens releasing the excess pressure and then the safety relief valve is again resealed. Sometimes when using sulfuric or phosphoric acids, which have higher boiling points than the softening temperature of PTFE, quartz or borosilicate glasses are used. The only medium pressure developed is due to the high boiling temperature of these acids. Figure 1.3(a) shows a PTFE moderate-pressure vessel for microwave digestion, produced by CEM

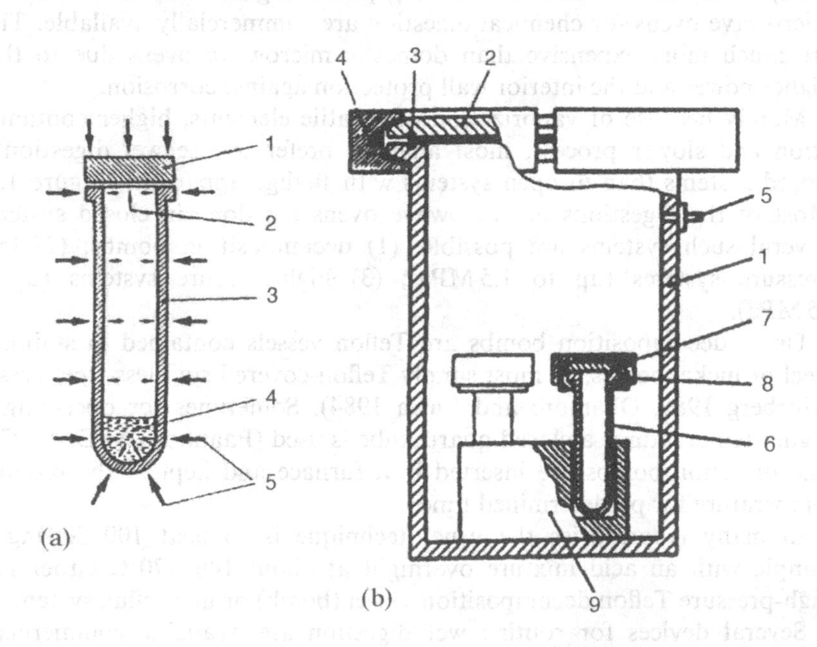

Figure 1.2 High-Pressure Asher. (a) Digestion vessel: 1, quartz lid; 2, pressure inside digestion vessel (<20 MPa); 3, quartz vessel; 4, sample and mixture of acids; 5, pressure outside decomposition vessel (20 MPa). (b) Autoclave together with digestion vessel: 1, pressure chamber; 2, lid of pressure chamber; 3, O-ring; 4, ring retainer; 5, pressure gas inlet; 6, quartz vessel (shown in a); 7, quartz lid; 8, steel screw; 9, heating block.

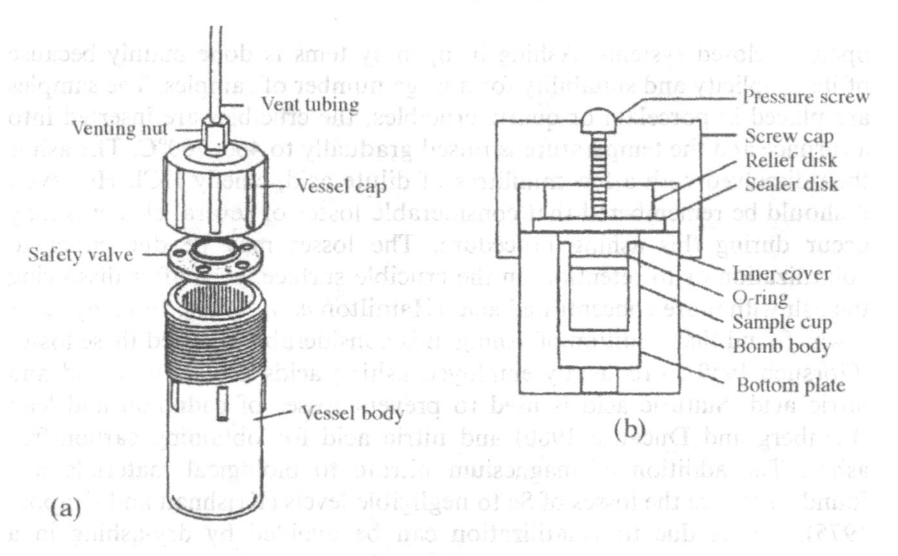

Figure 1.3 Closed vessels for microwave digestion: (a) medium-pressure vessel; (b) high-pressure vessel.

Corporation. The safety relief valve is designed to open at 0.85 MPa.

3. *Microwave digestion at high pressure*. The digestion is done up to 8–15 MPa. Figure 1.3(b) shows the vessel (bomb) produced by Paar Instrument Company. It is designed to operate at 8 MPa at up to 250°C. The sample with the acids is held in a PTFE cup supported in a heavy-wall bomb (vessel) made up of a polymeric material which is transparent to microwaves. The sample cup has on it a PTFE O-ring and on the top of it an inner cover. If the pressure exceeds 8 MPa the O-ring distorts and the excess pressure compresses the sealer disk; consequently the gases escape into the surroundings.

1.3.2 Dry-ashing

Dry-ashing of biological (organic) materials includes three types of ashing: at high temperature (400–800°C) and normal (air) pressure; at lower temperature (100–200°C) and reduced pressure of the oxidant (gas); and several combustion-like procedures. In analytical practice, ashing at high temperature is most frequently applied and is done by combustion in the presence of oxygen or air (Speziali *et al.* 1988). This method has the advantage that, since there is no addition of reagents, there are fewer sources for contamination. Most organic materials react with O_2 at temperatures ≥400°C to form CO_2 and H_2O, leaving most of the elements in the ash in the form of inorganic salts. Dry-ashing is performed in either open or closed systems. Ashing in open systems is done mainly because of its simplicity and suitability for a large number of samples. The samples are placed in porcelain or quartz crucibles, the crucibles are inserted into a furnace and the temperature is raised gradually to 400–700°C. The ash is then dissolved with a few millilitres of dilute acid, chiefly HCl. However, it should be remembered that considerable losses of several elements may occur during this ashing procedure. The losses may be due either to volatilization or to retention on the crucible surface, even after dissolving the ash with more concentrated acid (Hamilton *et al.* 1967). In many cases it was found that addition of ashing aids considerably reduced these losses (Gorsuch 1959). Frequently employed ashing acids are sulfuric acid and nitric acid. Sulfuric acid is used to prevent losses of cadmium and lead (Feinberg and Ducauze 1980) and nitric acid for obtaining carbon-free ashes. The addition of magnesium nitrate to biological materials was found to reduce the losses of Se to negligible levels (Krishnan and Crapper 1975). Losses due to volatilization can be avoided by dry-ashing in a

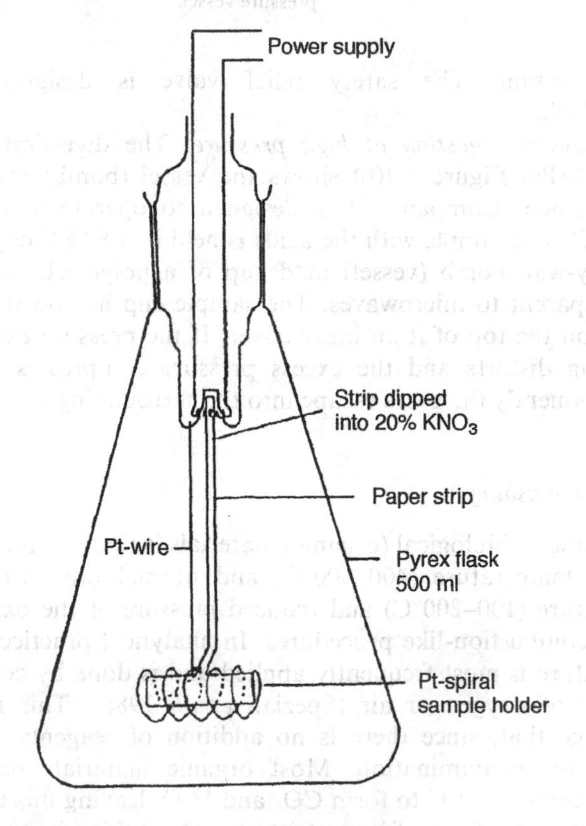

Figure 1.4 A typical Schoeniger system for dry-ashing in a closed system.

closed system and collecting the volatile compound on a cold finger (Knapp *et al.* 1981).

A common method for dry-ashing of biological samples in a closed system is to burn them in oxygen by the Schoeniger technique (also called the oxygen flask method) (Schoeniger 1961, Basset *et al.* 1978). A typical Schoeniger system is shown in Figure 1.4. This method is specially suitable for small samples. However, the method is not good for As, Pb and Bi, which react with the platinum holder (Gorsuch 1970) or even with the quartz holder (Desai *et al.* 1984). Desai *et al.* found also that Mn, Cu, Zn, Sb, Na and Hg were deposited on the holder and were not recovered completely.

Burning in oxygen in a static system suffers from the large surface of the apparatus, and hence from the risk of losing trace elements by adsorption. The disadvantage is overcome by a dynamic system of combustion in oxygen. The static system is restricted to small amounts of sample, owing to the small amount of oxygen in the vessel, whereas the dynamic method allows combustion of large quantities of sample, and consequently a larger ratio of the trace elements to be determined relative to the surface of the apparatus. The combustion can be done in a commercially available apparatus, the Trace-O-Mat (Anton Paar, Graz, Austria; H. Kurner, Neuberg, Germany). The description of the apparatus is given in Figure 1.5(a). The combustion chamber is made wholly of quartz, and it permits complete mineralization of 1 g of organic or biological solid sample. The combustion takes place in pure oxygen in a very small combustion chamber ($\approx 75\,cm^3$). The controlled incineration is started with an IR lamp. All volatile trace elements are condensed, together with the products of the combustion process, in a cold-finger cooling system filled with liquid nitrogen and mounted on top of the combustion chamber. Subsequent refluxing with a suitable acid in a quartz test tube mounted below the combustion chamber collects both the volatilized elements from the cooled areas and non-volatile elements in the ashing residue. The volume of acid is ≈ 2 ml. The decomposition process takes ≈ 50–60 min. This system was found to give recoveries $\geq 95\%$ for all elements studied in biological standard reference materials (Knapp *et al.* 1981) and also in oil and fats (Raptis *et al.* 1982). The method suffers from the disadvantage that only one sample can be digested each time whereas with dry-ashing in an oven several samples can be processed simultaneously. In the second generation of the apparatus the Trace-O-Mat II, higher sample throughput is available and it is possible to process 10 samples per hour (Knapp 1988). The various steps of operation are seen in Figure 1.5(b).

Another possibility for ashing organic materials with oxygen is to use an oxygen plasma which is excited by high frequency or microwaves (Raptis *et al.* 1983). The sample temperature is barely above 100°C and

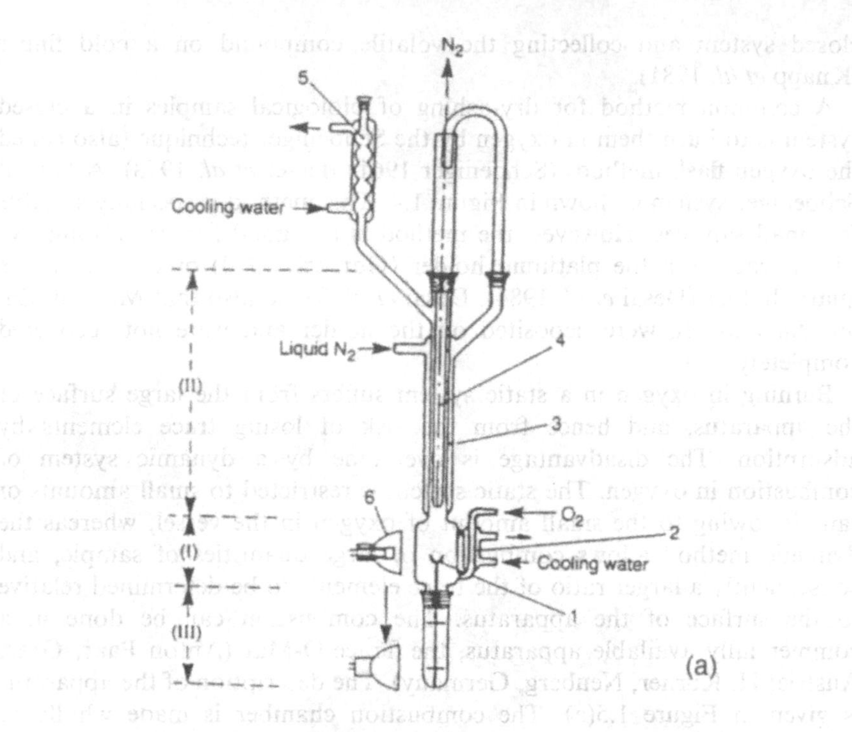

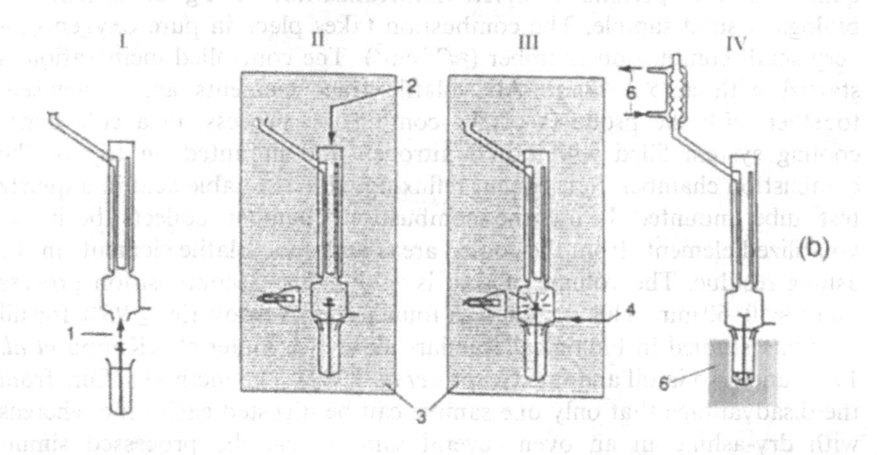

Figure 1.5 (a) Quartz combustion device of the Trace-O-Mat: (I) combustion chamber (ca. 75 ml capacity), (II) cooling system (liquid N_2), (III) quartz test tube; 1, sample holder; 2, oxygen inlet; 3, cooling mantle; 4, cold finger; 5, condenser; 6, IR lamps. (b) Trace-O-Mat II, flow diagram of the process: (I) sampling, (II) cooling, (III) burning, (IV) refluxing; 1, sample transfer into the burning chamber; 2, liquid nitrogen; 3, burning apparatus; 4, oxygen; 5, refluxing heater; 6, cooling water. (From Knapp 1988.)

hence the method is referred to as 'cold ashing'. The decomposition is performed in a closed quartz vessel equipped with a cooling finger. In spite of these low temperatures, As, Sb, Se and Hg will be partially volatilized, and the use of a cold finger prevents their loss. In order to reduce the time required for decomposition of organic samples of high mineral content, a magnetic stirrer is used to provide steady sample circulation during the decomposition process. All ashing residues and elements condensed on the surface of the decomposition vessel or cooling finger are collected in 1–2 ml of high-purity acid by refluxing. It was found that the system provides a universal apparatus for quantitative recovery of most elements and is characterized by low blanks. Even difficult-to-decompose organic materials like charcoal, graphite, all plastics (including PTFE), sugars, etc., can be ashed in this system without problems.

1.4 Contamination from reagents and equipment

Most reagents used in matrix decomposition are either acids or fluxes. When fluxes are used the final melts are dissolved in acids. The most soluble salts are the nitrates, perchlorates and chlorides and hence suitable acids are used. Every acid has a small concentration of impurities. The average concentrations of various trace elements in common acids are given in Table 1.3. Shimizu *et al.* (1988) suggested the use of tetramethyl-

Table 1.3 Average impurity concentration in doubly distilled acids (ng/kg)

Element	HCl	HNO_3	$HClO_4$	HF	H_2SO_4
Fe	2500	400	1200	2500	3700
Cr	150	50	3300	2200	100
Ca	100	100	600	1200	1000
K	300	150	200	800	2000
Mg	150	50	100	1000	2000
Al	1200	800	500	1000	1100
Na	800	600	1000	3800	6000
Se	40	60	500	200	–
Ni	125	100	350	450	125
Cu	80	60	100	250	150
Zn	120	60	200	200	250
Ag	225	60	60	200	300
Sr	40	30	30	40	150
Cd	40	30	60	60	170
Sn	120	30	140	150	130
Te	30	40	60	60	100
Ba	40	30	170	100	300
Tl	40	60	70	70	100
Pb	20	30	220	100	300

ammonium hydroxide for solubilization of biological material. This compound is a strong base equivalent to NaOH or KOH and is available in very high purity (ppb levels of contaminants). Zief and Mitchell (1976) summarized the existing techniques of purification of acids and water. A more recent review was done by Mitchell (1982) and Moody and Beary (1982). For water purification it was found that ion exchange columns are better than just distillation. Moody and Beary (1982) suggested the use of ion exchange followed by distillation from a fused quartz vessel (which has less contaminants than usual Pyrex glass). Nowadays the most used system for pure water is the Millipore Milli-Q system or its modification (Oehme and Lund 1979).

The most common method for purification of acids is sub-boiling distillation, which means distillation at temperatures below the boiling temperature of the acids. Figure 1.6 gives a schematic diagram of apparatus for sub-boiling distillation.

Another important aspect of contamination is the material used for containment and analysis. Polyethylene, polytetrafluoroethylene, Plexiglas tubing and synthetic quartz were found to be the purest materials (Robertson 1968). Other polyfluorocarbons, polychlorotrifluoroethylene, pyrolytic carbon, high-purity aluminum and platinum were found also to be clean materials (Zief and Mitchell 1976). The use of glassy carbon for simple laboratory-ware such as beakers was suggested (Tschoepel and Toelg 1982). Moody and Lindstrom (1977) found that the various

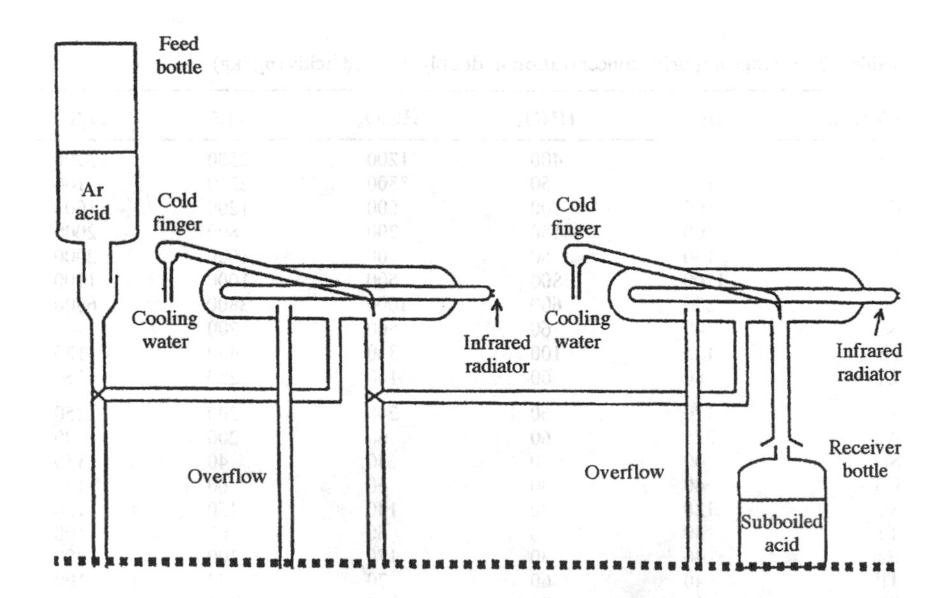

Figure 1.6 Schematic diagram of apparatus for sub-boiling distillation.

Table 1.4 Suggested method for cleaning plastic containers (Moody and Lindstrom 1977)

1. Fill the container with a 1 + 1 mixture of H_2O/HCl (AR grade)
2. Allow to stand for one week at room temperature; Teflon should be heated to 80°C
3. Empty the container and rinse with distilled water
4. Fill the container with a 1 + 1 mixture of H_2O/HNO_3 (AR grade)
5. Allow to stand for one week at room temperature; Teflon should be heated to 80°C
6. Empty the container and rinse with distilled water
7. Fill with the purest available water
8. Allow to stand for several weeks; change the water periodically to ensure continued cleaning
9. Rinse with the purest available water and allow to dry in a particle- and fume-poor environment

available polyfluorocarbons and conventional polyethylene are the least contaminating materials for storage, provided they have been cleaned properly. They suggested a quite long cleaning procedure using $HCl/HNO_3/H_2O$ as can be seen in Table 1.4. However, other workers found that for most purposes it is sufficient to soak the vessel in 10% HNO_3 for 48 h, followed by rinsing with H_2O (Laxon and Harrison 1981).

References

Aysula, P., Anderson, P. and Langford, C.H. (1987), *Anal. Chem.* **59**, 1582.
Bajo, S. (1992), in *Preconcentration Techniques for Trace Elements*, eds Z.B. Alfassi and C.H. Wai, CRC Press, Boca Raton, FL, p. 1.
Bajo, S., Suter, U. and Aschliman, B. (1983), *Anal. Chim. Acta* **149**, 321.
Basset, J., Denny, R.C., Jeffery, G.H. and Mendham, J. (1978), *Vogel's Textbook of Quantitative Inorganic Analysis*, 4th edn, Longman, London, p. 115.
Bernas, B. (1968), *Anal. Chem.* **40**, 1682.
Desai, H.B., Kayasth, R., Parthasarathy, R. and Das, M.S. (1984), *J. Radioanal. Nucl. Chem.* **84**, 123.
Faanhof, A. and Das, H.A. (1977), *Radiochem. Radioanal. Lett.* **30**, 405.
Feinberg, M. and Ducauze, C. (1980), *Anal. Chem.* **52**, 207.
French, W.J. and Adams, S.J. (1973), *Anal. Chim. Acta* **62**, 324.
Gorsuch, T.T. (1959), *Analyst* **84**, 135.
Gorsuch, T.T. (1970), *International Series of Monograph in Analytical Chemistry*, Pergamon Press, Oxford, vol. 39, p. 212.
Govindaraju, K., Mevelle, G. and Chourad, C. (1976), *Anal. Chem.* **48**, 1325.
Guzzi, G., Pietra, R. and Sabioni, E. (1976), *J. Radioanal. Chem.* **34**, 35.
Hamilton, E.I., Minski, M.J. and Cleary, J.J. (1967), *Analyst* **92**, 257.
Hannaker, P. and Qing-Lie, H. (1984), *Talanta* **31**, 1153.
Ingamells, C.O. (1966), *Anal. Chem.* **38**, 1228.
Ingamells, C.O. (1970), *Anal. Chim. Acta* **52**, 323.
Kingston, H.M. and Jassie, L.B. (1988), *Introduction to Microwave Sample Preparation. Theory and Practice*, American Chemical Society, Washington, DC.
Knapp, G. (1988), in *Trace Elements Analytical Chemistry and Biology*, eds P. Braetter and P. Schramel, vol. 5, Walter de Gruyter, Berlin, p. 63.
Knapp, G., Raptis, S.E., Kaiser, G., Toelg, G., Schramel, P. and Schreiber, B. (1981), *Fresenius Z. Anal. Chem.* **308**, 97.
Krishnan, S.S. and Crapper, D.R. (1975), *Radiochem. Radioanal. Lett.* **20**, 287.
Langmyhr, F.J. (1967), *Anal. Chim. Acta* **39**, 516.

Langmyhr, F.J. and Sween, S. (1965), *Anal. Chim. Acta* **32**, 1.
Laxon, D.O.H. and Harrison, R.M. (1981), *Anal. Chem.* **53**, 345.
Matusiewicz, H. and Sturgeon, R.E. (1989), *Prog. Anal. Spectrosc.* **12**, 21.
Mitchell, J.W. (1982), *J. Radioanal. Chem.* **69**, 47.
Mertz, W.M. (1981), in *Developments in Atomic Plasma Spectrochemical Analysis*, ed. R.M. Barnes, Heyden, Philadelphia, p. 635.
Moody, J.R. and Beary, E.S. (1982), *Talanta* **29**, 1003.
Moody, J.R. and Lindstrom, R.M., (1977), *Anal. Chem.* **49**, 2264.
Oehme, M. and Lund, W. (1979), *Fresenius Z. Anal. Chem.* **298**, 260.
Okamoto, K. and Fuwa, K. (1984), *Anal. Chem.* **56**, 1758.
Raptis, S.E., Kaiser, G. and Toelg, G. (1982), *Anal. Chim. Acta* **138**, 93.
Raptis, S.E., Knapp, G. and Schalk, A.P. (1983), *Fresenius Z. Anal. Chem.* **316**, 482.
Robertson, D.E. (1968), *Anal. Chem.* **40**, 1067.
Schoeniger, W. (1961), *Fresenius Z. Anal. Chem.* **181**, 28.
Shimizu, S., Che, T. and Murakami, Y. (1988), in *Trace Elements Analytical Chemistry in Medicine and Biology*, eds P. Braetter and P. Schramel, vol. 5, Walter de Gruyter, Berlin, p. 72.
Shur, N.H. and Ingamells, C.O. (1966), *Anal. Chem.* **38**, 730.
Speziali, M., de Casa, M. and Orvini, E. (1988), *Biol. Trace Elem. Res.* **17**, 271.
Tschoepel, P. and Toelg, G. (1982), *Trace Microprobe Tech.* **1**, 1.
Uhrberg, R. (1982), *Anal. Chem.* **54**, 1906.
Walsh, J.N. and Howie, R.A., (1980), *Mineral. Mag.* **43**, 967.
Zief, M. and Mitchell, J.W. (1976), *Contamination Control in Trace Element Analysis*, Wiley, New York.

2 Separation and preconcentration of trace inorganic elements

Z.B. ALFASSI

2.1 Introduction

The best method of chemical analysis for each matrix will be that (or those) method(s) which can determine the constituents of the matrix, both qualitatively and quantitatively, without any chemical separation. It is relatively easily done for the major constituents of the matrix, but for the determination of the trace elements in the matrix sometimes separation procedures are prerequisites. Some matrices allow the determination of most of their trace elements without any chemical treatment, as e.g. the measurement of the concentration of contaminants in pure silicon by instrumental neutron activation analysis (Alfassi 1990). Several books deal extensively with the separation and preconcentration of trace inorganic elements (Minczewski *et al.* 1982, Mizuike 1983, Alfassi and Wai 1992), including all the major methods but also those less frequently used as e.g. fire assay, isotachophoresis and separation on filter papers. However, owing to lack of space only the major preconcentration techniques will be dealt with in this chapter. The lowest limit of detection might be obtained if the sample is completely separated into its constituents; however, this is not always possible and when possible it is time-consuming and prone to errors and contaminations. Since most modern instrumental analytical methods are multi-element, it is sufficient to separate the sample into several fractions; each fraction consists of elements which can be determined simultaneously. The preconcentration step not only usually involves the increase of the concentration of the species of interest, but also eliminates part or all of the interferences (usually high-concentration elements of the matrix). Both aspects of preconcentration and interference removal must be examined for every preconcentration procedure. Usually the usefulness of a separation procedure is judged by four criteria:

1. Preconcentration factor P
2. Recovery yield R
3. Interference removal, called the decontamination factor D
4. Specificity

If the original sample weight m_0 grams, which includes w_0 grams of the element of interest, and after the separation the weights are m_f and w_f respectively, then

$$R = \frac{100w_f}{w_0} \qquad \text{(per cent)} \qquad (2.1)$$

$$P = \frac{w_f}{w_0}\frac{m_0}{m_f} \qquad (2.2)$$

If the main interference is from the matrix element, then

$$D = \frac{m_0}{m_f} \qquad \Rightarrow \qquad P = \frac{RD}{100} \qquad (2.3)$$

Das (1982) proposed the classification of preconcentration techniques by a relation between the preconcentration and decontamination factors. He used two other criteria to judge the combination of a separation technique with analytical measurement. One criterion is concerned with the minimal representative sample weight, which depends on the uniformity of the sample: the higher the uniformity, the smaller the sample that accurately represents the material. The second criterion was derived from the maximal acceptable sample weight. In the case of analysis involving radionuclides an additional factor is the maximal acceptable radiation dose.

2.2 Precipitation

2.2.1 *Matrix precipitation*

Precipitation is used mainly to concentrate various elements from dilute aqueous solutions, drinking water, waste water and sea water. Several elements (e.g. Cd, Co, Cu, Pb, Ni and Mo) were determined in water by inductively coupled plasma–atomic emission spectroscopy (ICP–AES) after preconcentration by precipitation (Xiao-qan *et al.* 1988). However in most cases precipitation is used to determine the major elements, as the accuracy of gravimetry is usually very high. For the determination of trace elements, the solubility with many common precipitating agents is too high to allow quantitative precipitation of the ions of interest. In these cases quantitative precipitation may be done by the addition of a coprecipitant, which will not only be precipitated but will also lead to the coprecipitation of the element of interest. This method will be dealt with in the next subsection. Precipitation is used sometimes to remove matrix elements by the formation of a sparingly soluble compound. Thus for example the trace elements in lead were determined by removal of Pb as $PbCl_2$ or $PbSO_4$ (Jackwerth 1979) and copper by precipitation of CuS (Jackwerth and Willmer 1976). In this case care should be taken that the elements of interest will not be removed with the matrix by coprecipitation. In order not to add another ion by the addition of the precipitating agent, it is advisable to use organic reagents to supply the desired anion. The use of the acid itself is not recommended in many cases, because the variation of pH changes the solubility. Thus for precipitation of sulfide, thioacetamide is added to the solution. The hydrolysis of thioacetamide leads to liberation of S^{2-} anions. This *in situ* slow formation of the precipitant also keeps low and uniform the concentration of the precipitant, leading to larger and more perfect crystals of the precipitate, which are easier to filter and which incorporate less impurities. Another example of slow precipitation from homogeneous solution is by precipitation of hydroxides with acetamide or urea. In those cases of producing the actual precipitating agent by hydrolysis, heating the sample leads to faster hydrolysis, and consequently precipitation. This is mostly important in the case of urea, for which the room temperature hydrolysis is very slow.

2.2.2 Coprecipitation

The incorporation into a precipitate of materials that under the experimental conditions are soluble (concentration below the solubility product) is usually called coprecipitation. In the preparation of pure materials, coprecipitation is a large problem; however for separation of trace elements it is one of the major procedures. It should be mentioned that actual coprecipitation includes also cases in which the element of interest exceeds the solubility product, but the amount of the precipitate is too little to be collected and coprecipitation is really a process of adding a carrier. Most of the processes of coprecipitation deal with reagents that form insoluble materials with the element of interest, although not always in these low concentrations.

There are two general types of mechanisms of coprecipitation: incorporation of the impurity into the crystal lattice of the host material, and adsorption of the trace elements on the surface of the precipitate of the host material. The two mechanisms differ considerably in the effect of the precipitating conditions on the ratio of the fractions of the trace elements and the host element precipitated. For the first mechanism the ratio of the fractions precipitated is independent of precipitating conditions such as excess of precipitating agent or the presence of highly charged ions. On the other hand adsorption is influenced considerably by the precipitating conditions.

Crystal formation. Coprecipitation via crystal formation was divided by Hahn (1936) into two categories, depending on whether or not crystals of the host precipitate can form mixed crystals with crystals of the impurities.

In the case of formation of mixed crystals the precipitate is a mixed crystal of both elements. Hahn refers to this type as **isomorphous mixed-crystal formation**. However, elements are incorporated into the crystal lattice even when macro-amounts of the trace and major compounds do not have similar crystallographic properties, yet it was found that the trace compound is distributed continuously in the precipitate. This case can be looked upon as a case of limited miscibility in the solid phase, whereas isomorphous mixed crystals are cases of complete miscibility in the solid phase. Hahn named this process of coprecipitation **anomalous mixed-crystal formation.**

The distribution of trace elements in **mixed crystals** was found to behave according to two limiting laws.

1. *Homogeneous distribution: Bertholt–Nernst distribution law.* This law applies when equilibrium is established between the ions in the interior of the crystals and those in the solution. In this case the distribution of the trace element in the precipitate is given by the Bertholt–Nernst distribution law:

$$\left(\frac{\text{trace elements}}{\text{carrier}}\right)_{\text{precipitate}} = D\left(\frac{\text{trace elements}}{\text{carrier}}\right)_{\text{solution}}$$

or

$$\frac{T}{C} = D\,\frac{T_0 - T}{C_0 - C} \tag{2.4}$$

where T_0 and C_0 are the initial amounts of the trace and carrier elements, and T and C are the respective amounts in the precipitate. The constant D is called the **distribution coefficient**. The larger the value of D, the higher is the enrichment of the trace element in the precipitate, leading to larger recovery of the trace element.

2. *Logarithmic distribution: Doerner and Hoskins law.* When the ions in the solution cannot equilibrate with the interior of the crystal, they are equilibrated only with the ions on an infinitesimally thin surface layer of the crystal. Since the concentration of trace ions in the solution is decreased as the coprecipitation proceeds, the concentration of the trace ions decreases with the distance from the inside of the crystal. This leads to a logarithmic distribution law:

$$\log\left(\frac{T_0 - T}{T_0}\right) = \lambda \log\left(\frac{C_0 - C}{C_0}\right) \tag{2.5}$$

Usually, actual coprecipitation may fall somewhere between the above two limiting distributions. In some cases fast coprecipitation leads to logarithmic distribution, but prolonged stay (aging) of the precipitate in the solution leads to a change toward homogeneous distribution.

Adsorption. Molecules and ions inside the crystal are surrounded on all sides by neighboring ions. On the other hand those on the surface interact with neighbors only from one direction – from inside the crystal – and hence the species on the surface attracts ions or molecules from the solution. Adsorption can be done either after the completion of precipitation or during the growth of the precipitate or both. Hahn (1936) divided the mechanism of adsorption into **internal adsorption** when the trace elements are adsorbed on surfaces of growing crystals, and **adsorption** when the trace elements are adsorbed on the surface of the final precipitate. Paneth–Fajan–Hahn's rule (Hahn 1936) states the following for coprecipitation by adsorption on ionic precipitates: 'Those ions whose compounds with the oppositely charged constituents of the lattice are (only) slightly soluble in the solution in question are well adsorbed on the ionic crystals.' In other words, adsorption increases as the solubility of the compound with the trace element decreases.

However, solubility is only one factor affecting adsorption. Two other main factors are the dissociation constant of the compound of the trace element and the deformation ability of the ions. The adsorptivity increases with decreasing dissociation constant and with increasing deformability of the adsorbed electrolyte. The deformability increases with increasing size of the ion; anions of large aromatic acids have specially large deformability. It was found that mathematically the amount coprecipitated, or rather the concentration coprecipitated C, by adsorption vs. the original concentration C_0 is given by either Freundlich or Langmuir adsorption isotherms:

$$\text{Freundlich} \qquad C = \alpha C_0^{\beta} \tag{2.6}$$

$$\text{Langmuir} \qquad C = \frac{C_\infty K C_0}{1 + K C_0} \tag{2.7}$$

α, β, K and C_0 are constants characteristic to any given system with usually $\beta < 1$. The Langmuir adsorption isotherm applies when the precipitate has only a limited number of surface sites, and C_∞ is the limiting value of C, i.e. monolayer capacity with any available site occupied.

2.2.3 *Coprecipitation of trace elements with inorganic precipitates*

The most common coprecipitants and the most-studied ones are the hydroxides and hydrous oxides, such as $Fe(OH)_3$, $Al(OH)_3$, $La(OH)_3$, $Zr(OH)_4$ and MnO_2; the most commonly used were $Fe(OH)_3$ and MnO_2. More than 100 references of coprecipitation with hydrous oxides (hydroxides) were reviewed (Alfassi 1992). Since most metals form insoluble oxides, Paneth–Fajan–Hahn's rule indicates that most of them are likely to be coprecipitated with hydroxide–oxide precipitate. Owing to the relatively low selectivity of coprecipitation with hydroxides, it is more common to use it for preconcentration from natural dilute solutions (sea water, fresh water, treated waste solutions) or from biological solutions in which the major constituents (Na, Cl) will not coprecipitate to a significant extent, while many of the trace elements will be coprecipitated quantitatively, than in order to separate between groups of elements. However, the most unselective is $Fe(OH)_3$, while the selectivity of $Al(OH)_3$ is a little higher, whereas MnO_2 coprecipitates less than half of the ions that are precipitated by $Fe(OH)_3$. $Bi(OH)_3$ and $Sn(OH)_4$ were found to coprecipitate only a few elements. Table 2.1 summarizes the elements collected by coprecipitation of hydrated oxides–hydroxides. The fact that the table does not show coprecipitation of an element by a coprecipitant does not mean in all cases that it is really not coprecipitated. It might be that its coprecipitation was not studied.

Coprecipitation is very pH-dependent, and separation between various trace elements can be done by choosing appropriate pH. Coprecipitation is done many times with NaOH. However, since usually not very high pH is required, less contamination can be caused by precipitation with NH_4OH, or more accurately by bubbling NH_3 through the solution. Usually NaOH is added to either concentrated NH_4OH or NH_4Cl solution and the solution is warmed to release NH_3. This NH_3 is bubbled through the studied solution. The common explanation for the coprecipitation with hydrated oxides (hydroxides) is by ion exchanging equilibrium with

Table 2.1 Elements coprecipitated on precipitate of hydroxides

Precipitate	Elements
$Fe(OH)_3$	Ag, Al, As, Ba, Be, Bi, Cd, Ce, Co, Cr, Cu, Ga, Ge, In, Ir, La, all the lanthanides and the actinides, Mg, Mn, Mo, Nb, Ni, Pb, Pd, Pt, Rh, Ru, Sb, Sc, Se, Sr, Ta, Tc, Te, Ti, Tl, V, W, Y, Zn, Zr
$Al(OH)_3$	Be, Bi, Cd, Ce, Co, Cr, Cu, Fe, Ge, Ir, lanthanides and actinides, Mg, Mn, Mo, Nb, Ni, Pb, Pd, Pt, Rh, Ru, Sc, Se, Sn, Ti, Tl, V, W, Y, Zn, Zr
MnO_2	Al, As, Au, Bi, Cr, Cu, Fe, Ga, In, Mo, Nb, Pa, Pb, So, Sb, Se, Sn, Te, Th, Tl, V
$Zr(OH)_4$	Al, As, Bi, Ca, Cd, Co, Cr, Cu, Fe, La, Mg, Mn, Mo, Ni, Pb, Ru, Sb, Sn, Tc, Ti, V, Y, Zn
$La(OH)_3$	Al, Ag, As, Au, Bi, Cd, Co, Cr, Cu, Fe, Ga, Mn, Mo, Ni, Pb, Rh, Sb, Se, Te, Ti, V, W, Y, Zn, Zr
$Ti(OH)_4$	Cd, Co, Cr, Cu, Fe, Hg, Ni, Zn
$Mg(OH)_2$	Cd, Cr, Mo, Pb, actinides
$Ga(OH)_3$	Al, Co, Cr, Cu, Fe, La, Mn, Ni, Pb, Ti, V, Y, Zn
$Sn(OH)_4$	Cd, Co, Cu, Fe, Zn
$Bi(OH)_3$	Cd, Co, Cu, Fe, Mn, Pb, Zn

protons. For simplicity let us write the iron hydroxide as FeH, then the adsorption of the trace element T proceeds through the equilibrium (Kurbatov *et al.* 1951, Morgan and Strumm 1964, Gadde and Laitinen 1974, Brunix 1975, James *et al.* 1975, Zhukova and Rachinskii 1978):

$$(FeH)_n + T \rightleftharpoons Fe_nT + nH^+ \tag{2.8}$$

The distribution coefficient, D, is defined as the ratio of the amounts of trace element in the precipitate, Fe_nT, and in the solution T:

$$D = [Fe_nT]/[T] \tag{2.9}$$

If K is the equilibrium constant of reaction (2.8) then

$$K = \frac{[Fe_nT][H^+]^n}{[T][FeH]^n} \quad \Rightarrow \quad K = \frac{[H^+]^n}{[FeH]^n} D \tag{2.10}$$

$$\log_{10} D = n\,\mathrm{pH} + \log_{10} K + n \log_{10}[FeH] \tag{2.11}$$

These equations explain the dependence of the coprecipitation on the pH, although not always a linear dependence of $\log D$ on the pH was observed. Although equation (2.11) does not show a dependence of D on the amount of the coprecipitated trace, usually D decreases with increasing amount of the coprecipitated trace, due to the limited number of coprecipitation sites. However, usually the data fitted Freundlich adsorption isotherms better than the Langmuir ones.

Very little effect of aging was found for coprecipitation with hydroxides, even after several weeks. This is in contrast to what was observed for mixed-crystal coprecipitation where the initial logarithmic distribution is shifted toward the homogeneous distribution on aging.

Other inorganic coprecipitants are sulfides (PbS, CuS, CdS, HgS, AgS, As_2S_3, ZnS, MoS_3, Bi_2S_3 and Sb_2S_3), sulfates ($PbSO_4$, $BaSO_4$, $SrSO_4$, $Th(SO_4)_2$), phosphates (Pb, Al, Ca, Sr, Bi), fluorides (CaF_2, ThF_4, LaF_3, NiF_2) and calcium oxalate. Some trace elements are coprecipitated with pure elements. The precipitate is formed in the solution by reduction of an oxidized form of the element. Elements used were Te, S, As, Se, Pb, Pd and H_2.

2.2.4 *Coprecipitation with organic collectors*

Several organic compounds when precipitated in aqueous solution have good selectivity in coprecipitation and they can concentrate some trace elements very effectively even from dilute solution with low concentration of 10^{-12} M. Organic collectors have also the advantage that if necessary they can be removed by simple combustion (most organic compounds are converted to CO_2, H_2O and NO_x when heated to 450–500°C in the presence of air) or by dissolution in an organic solvent.

In many coprecipitations by organic collectors, two compounds are added to the aqueous solution with the trace elements. One reagent is used in order to form complexes with the desired elements, the complexing agent being an organic chelating agent or an inorganic anion (SCN^-, Br^-, Cl^-) which forms anionic complexes with the metal to be separated. The second reagent is an organic compound that is sparingly soluble in aqueous solution, and it is added to the aqueous solution in solution of water-miscible organic solvent (methanol, ethanol, acetone, tetrahydrofuran, etc.). The second reagent is precipitated when the organic solution is added to the aqueous solution carrying with it the trace elements. Minczewski *et al.* (1982) described six types of mechanisms of coprecipitation of trace elements with organic compounds:

1. The sparingly soluble organic compound is a bulky organic cation which forms an ion-pair with the anionic complex. In this case one reagent is an inorganic ion and the other is an organic compound. An example is the use of SCN^- salts with antipyrine dye salts.

Table 2.2 Examples of trace elements coprecipitation with organic precipitates*

Organic coprecipitant	Coprecipitant elements	Conditions
Aluminum oxinate + thionalide	Cu, Fe, Pb, Mn, Ni, Sn, Zn	pH = 9, heating to 80–85°C for 1 h
Indium oxinate + tannic acid + thionalide	Ag, Al, Bi, Cd, Co, Cr, Cu, Fe, Ga, Ge, Mo, Ni, Pb, Sn, Ti, V, Zn	pH = 5.2, overnight aging
Magnesium oxinate	Co, Cd, Cu, Mn, Pb, Zn	Recoveries depend on pH; optimum pH = 8–9. Heating accelerates high recoveries, otherwise overnight recovery required
Ni + dimethylglyoxime + PAN	Cd, Cu, Mn, Pb, Zn	pH = 10, aging 1 h at 60°C
Diethyldithiocarbamic acid	As, Bi, Cd, Co, Cu, Fe, Hg, In, Mn, Ni, Os, Pb, Pd, Sb, Se, Sn, Te, Zn	Recovery depends on pH; optimal pH is different for different elements. Some elements do not coprecipitate with the acid but coprecipitate with its salts
PAN	Cd, Co, Cr, Cu, Eu, Fe, Hg, Mn, Ni, lanthanides, U, Zn	pH < 9 for decrease of coprecipitation of Ca. The solid floats on the solution and is precipitated either with a little amount of surfactant or aging at 60–80°C for a few minutes
Thionalide (thioglycolic-β-aminonaphthalide)	Ag, Au, Cd, Co, Cr, Cu, Fe, Hf, Hg, In, Ir, Mn, Os, Ru, Sb, Se, Sn, Ta, Tl, W, Zn	pH-dependent. Thionalide is added in acetone solution. Aging at 80°C

* More details and original references can be found in Alfassi (1992).

2. The organic reagent is an acid which forms insoluble salts with the metal cations. The small amount of the salt is coprecipitated with the excess of the acid, which is also insoluble in water. In this case only one reagent is used. An example is oxine (= 8-quinolinol) or thionalide.
3. A chelate complex is formed between the metal and acidic reagent. This chelate is coprecipitated with a precipitate formed by the excess of the acid reagent and a bulky cationic organic reagent. In this case two organic reagents are added.
4. Coprecipitation of an inner complex, of the ions to be separated with an organic reagent, with the large excess of sparingly soluble organic compound, e.g. PAN (1-(2-pyridylazo)-2-naphthol).
5. A chelate of the metal ion with an organic reagent is adsorbed–coprecipitated on a water-insoluble organic compound. This is actually a solid version of solvent extraction. An example is dithizone as the chelating agent and phenolphthalein as the insoluble organic compound.
6. 'Colloidal–chemical' sorption of the metal ions on a mixture of insoluble organic reagents.

Table 2.2 gives some examples of coprecipitation with organic collectors.

2.2.5 *Flotation of precipitates (coprecipitates)*

In order to simplify the separation of the coprecipitate from the solution we saw in Table 2.2 that long aging or heating plus aging is frequently used. Another method to simplify the separation was developed by Sebba (1962). By employment of flotation techniques the coprecipitate is readily separated from the mother liquid in a few minutes by floating the coprecipitate to the solution surface using surfactants with the aid of a rising stream of gas bubbles. The flotation technique eliminates also the difficulties associated with separation of colloidal precipitates or with the use of very large volumes. Several factors affect the flotation process:

1. *Chain length of the surfactant.* Surfactants are molecules with a polar end, to enable solubility in water, and a hydrocarbon chain, to enable the molecules to form hydrophobic bonds between the molecules. Thus they can be raised to the surface by the gas bubbles, together with the precipitate adsorbed onto them. Longer hydrocarbon chains increase the hydrophobic bonds but decrease the solubility of the surfactant in water. Chain lengths of 12–18 carbon atoms are most commonly used; the most frequently used surfactants are the anionic surfactants sodium oleate and sodium dodecyl sulfate.
2. *Surfactant concentration.* It was found that the separation efficiency

vs. the concentration of the surfactant is a function with a maximum rather than a monotonically increasing function. Several explanations were given for the inhibition effect of excess surfactant, such as the inability of the bubbles to levitate the large micelles.

3. *pH of the solution.* The sign and the magnitude of the electric charge on a variety of inorganic particulates influences their adsorption onto the surfactant micelles.

Other factors affecting flotation are ionic strength, presence of activators influencing the charge of the precipitate, presence of ethanol, gas flow rate, bubble size, foam stability and temperature.

2.2.6 *An example*

Lavi and Alfassi (1989) and Lan *et al.* (1990) coprecipitated from natural waters trace amounts of As, Cd, Co, Cr, Cu, Fe, Hg, In, La, Mn, Mo, Ni, Sb, Se, Ti, U, V and Zn using Pb and Bi salts of pyrrolidine dithiocarbamate. The efficiency of the coprecipitation was studied using radiotracers. The choice of Pb or Bi salt as the coprecipitation agent was done since the method of determination after coprecipitation was instrumental neutron activation analysis (INAA) for which Pb and Bi do not interfere. The pyrrolidine dithiocarbamate anion was added by addition of aqueous solution of ammonium pyrrolidine dithiocarbamate (APDC) followed by aqueous solution of $Pb(NO_3)_2$ or $Bi(NO_3)_3$. The pH of the solution was adjusted with HCl and NH_4OH. It was found that at pH = 3.5 most of the elements studied coprecipitated quantitatively, except Mo from which only 58% coprecipitated. It was found that Pb^{2+} + APDC can be replaced by a mixture of thionalide and oxine. Thionalide + oxine does not need always the Pb^{2+} ions as calcium and magnesium oxinate are also insoluble. However for some natural water samples the Ca and Mg concentrations were found to be insufficient. APDC + Pb has the advantage that Ca and Mg do not precipitate and do not interfere with measurement of short-lived radionuclides produced in the INAA. Besides, since Ca and Mg precipitated with oxine, the amount of oxine added should depend on their concentrations (which is not always traces). Oxine was dissolved in acetic acid + HCl and thionalide was dissolved in acetone, as both do not dissolve well in water.

To 2 L of natural water were added aqueous solutions containing 100 mg APDC and 2 mg $Pb(NO_3)_2$. Figure 2.1 gives the γ-ray spectra measured by INAA of the preconcentrated sea water. The four panels are for different irradiation times.

The same method was applied also to biological fluids, such as milk and blood, by Lavi and Alfassi (1989) and Lavi *et al.* (1990).

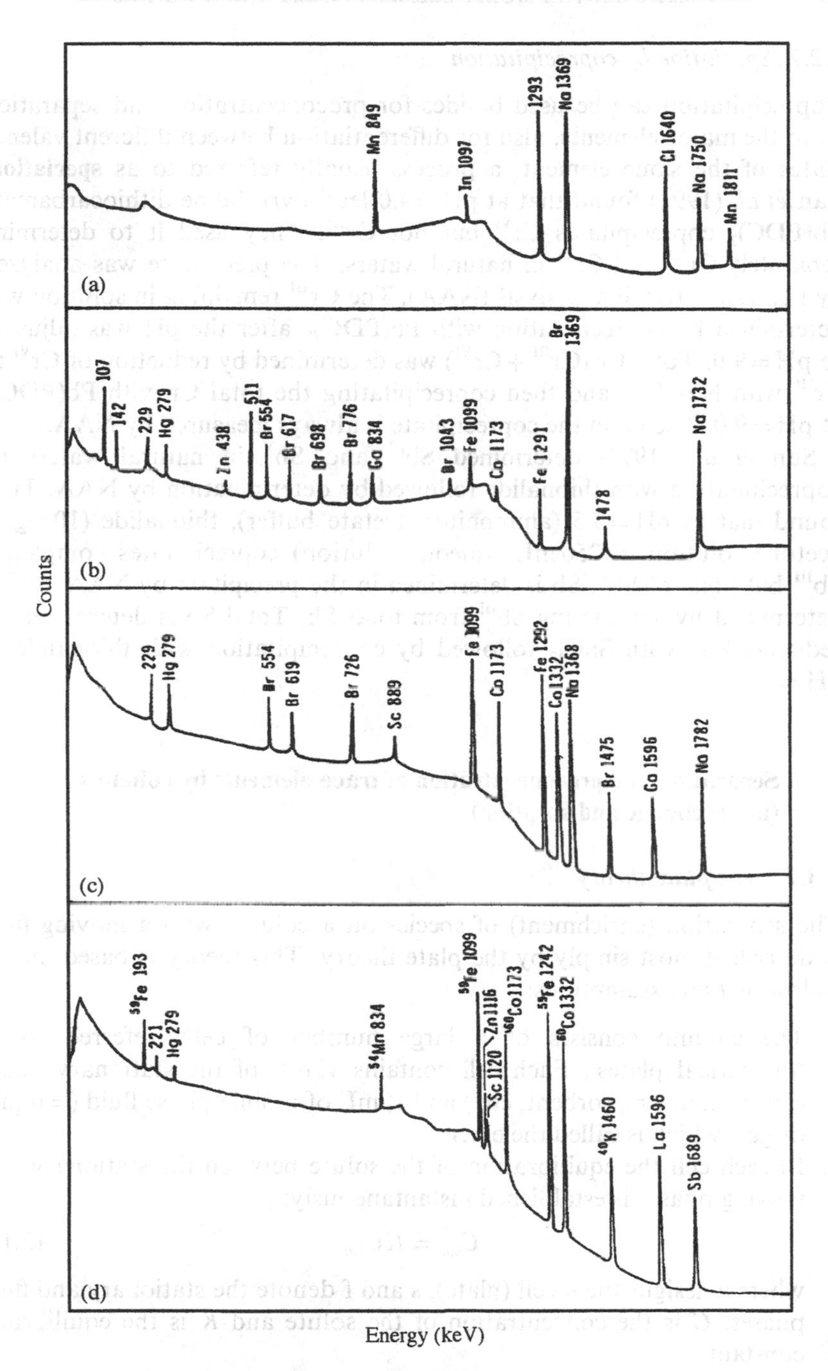

Figure 2.1 Determination of trace elements in natural water by instrumental neutron activation analysis after coprecipitation with lead pyrrolidine dithiocarbamate. The four panels are the gamma-ray spectra for different irradiation, decay and counting times.

2.2.7 Speciation by coprecipitation

Coprecipitation can be used besides for preconcentration and separation from the major elements, also for differentiation between different valence states of the same element, a process usually referred to as **speciation**. Lan *et al.* (1991) found that at pH = 4.0 lead pyrrolidine dithiocarbamate $Pb(PDC)_2$ coprecipitates Cr^{VI} but not Cr^{III}. They used it to determine separately Cr^{III} and Cr^{VI} in natural waters. The precipitate was analyzed by neutron activation analysis (NAA). The Cr^{III} remaining in solution was determined by coprecipitation with $Pb(PDC)_2$ after the pH was adjusted to pH = 9.0. Total Cr ($Cr^{III} + Cr^{VI}$) was determined by reduction of Cr^{VI} to Cr^{III} with $NaHSO_3$ and then coprecipitating the total Cr with $Pb(PDC)_2$ at pH = 9.0. The Cr in the coprecipitate is always measured by NAA.

Sun *et al.* (1993) determined Sb^{III} and Sb^{V} in natural waters by coprecipitation with thionalide followed by determination by NAA. They found that at pH = 5.5 (ammonium acetate buffer), thionalide (10 mg in acetone solution in 250 mL aqueous solution) coprecipitates completely Sb^{III} but none of Sb^{V}. Sb is determined in the precipitate by NAA. Sb^{V} is determined by subtracting Sb^{III} from total Sb. Total Sb is determined by reducing Sb^{V} with $SnCl_2$ followed by coprecipitation with thionalide at pH = 5.5.

2.3 Separation and preconcentration of trace elements by columns (ion exchange and sorption)

2.3.1 The plate theory

The separation (enrichment) of species on a column with a moving fluid is described most simply by the plate theory. This theory is based on the following two assumptions:

1. The column consists of a large number of cells referred to as 'theoretical plates'. Each cell contains V_s mL of the stationary phase (ion exchangers, sorbent, etc.) and V_f mL of mobile-phase fluid (= liquid or gas) which is called the eluent.
2. In each cell the equilibration of the solute between the stationary and moving phases is established instantaneously:

$$C_{s,n} = KC_{f,n} \tag{2.12}$$

where n assigns the n cell (plate), s and f denote the stationary and fluid phases. C is the concentration of the solute and K is the equilibrium constant.

When dV mL of eluent passes through the column, there is a passage of dV mL from each plate to the next one. The conservation of mass in the

nth cell requires that the difference between the amount of solute going in and out is equal to the change of concentration in both phases in this cell:

$$(C_{f,n-1} - C_{f,n})\,dV = V_f\,dC_{f,n} + V_s\,dC_{s,n} \tag{2.13}$$

The two differentials on the right are connected by the previous equation $C_{s,n} = KC_{f,n}$ leading to

$$dC_{s,n} = K\,dC_{f,n}$$

and thus

$$\frac{dC_{f,n}}{dV} = \frac{C_{f,n-1} - C_{f,n}}{V_f + KV_s} \tag{2.14}$$

In the beginning ($V = 0$) the solute is present only in the first plate – and its concentration there is C_0. The solution of the last equation for total fluid volume V yields

$$C_{f,n} = \frac{\alpha^n e^{-\alpha}}{n!} \qquad \text{where} \qquad \alpha = \frac{V}{V_f + KV_s} \tag{2.15}$$

The last equation is the definition of a Poisson distribution of C vs. n in the fluid phase. For sufficiently large values of n and $n \gg |\alpha - n|$, this distribution can be approximated by the Gaussian normal distribution

$$C_{f,n} = \frac{C_0}{\sqrt{2\pi n}} \exp[-(\alpha - n)^2/2n] \tag{2.16}$$

The retention volume – the volume required to elute the solute (see Figure 2.2) – is given by

$$V_R = I + KS \tag{2.17}$$

where I is the interstitial volume of the column (the volume of solvent initially in the column) and S is the volume of stationary phase in the column.

The width of the elution peak W is given by

$$W = \sqrt{2/N}\,V_R \tag{2.18}$$

where N is the number of theoretical plates in the column. According to the last equation a large number of theoretical plates are required to obtain narrow elution peaks, which will enable separation of various species. Larger number of theoretical plates will lead to less overlapping of close peaks. In general the number of theoretical plates increases with the column length and temperature, and decreases with the particle size of the column material, the column diameter, the flow rate and the viscosity of the eluent.

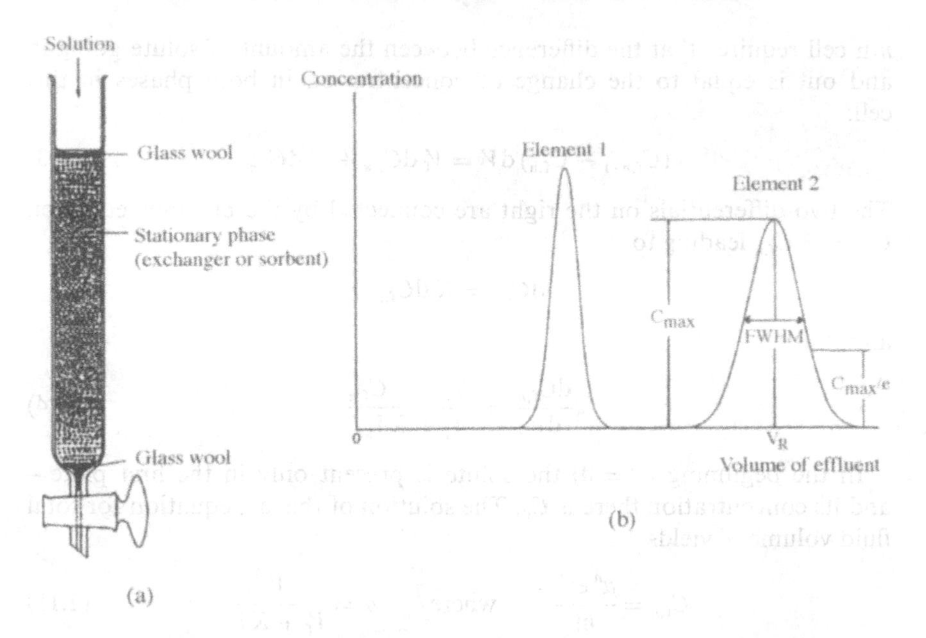

Figure 2.2 Concentration and separation by columns: (a) a scheme of a column; (b) an ideal chromatogram (concentration of the material in the effluent as a function of the volume of the effluent).

2.3.2 Ion exchangers

Solid ion exchangers are insoluble frameworks (a polymer) that contain ionic groups. The electric charge of this framework is balanced by mobile ions of opposite charge to these ionic groups. These mobile ions, the counter-ions, can be exchanged with ions of similar charge, when placed in a solution containing these ions. The exchange is reversible and difference in affinities of ions to the exchanger can lead to useful separation of ions, by some ions exchanging with the original counter-ions while others do not. Experimentally the simplest way to use ion exchangers is to pack them into a column and flow the solution containing the ions to be separated through the column by gravitation. The ions can be separated into two fractions, those exchanged and those remaining in the solution, or into several fractions by using different eluting solutions (eluents) to extract from the column the ions exchanged from the original solution.

Ion exchange resins are mainly copolymers of styrene and divinylbenzene (DVB) or copolymers of methacrylic acid with DVB containing ionizable functional groups. Strongly acidic exchangers (cation exchangers) are produced by sulfonation (the $-SO_3H$ group), while weakly acidic cation exchangers contain carboxylic groups. Weakly basic exchangers (anion exchangers) contain primary, secondary and tertiary amine groups while strong basic exchangers contain quarternary ammonium. The divinylbenzene, which contains two polymerizable groups (two olefinic bonds), is used to crosslink the polymer network. As the degree of crosslinking (i.e. percentage of DVB) increases, the wet volume capacity and the resistance to shrinking and swelling increase, while the rate of ion exchange and the permeability to large molecules decrease. Most used ion exchangers have 2–12% DVB.

Some ion exchangers also have functional groups that can form chelates. The most popular are the iminodiacetate $-CH_2N(CH_2COOH)_2$ or dithiocarbamate $\rangle$N–C(S)SH groups, but also other nitrogen- and sulfur-based resins are used, as e.g. chelating resins containing various substituted oximes. Besides the use of chelating resins prepared by chemical modification of various polymers, chelating ion exchange is done also by impregnating anion exchange or macroporous resins with chelating compounds.

The properties of various ion exchange resins are summarized in Table 2.3.

The exchange of an ion between the solution and the column (resin) can be represented by the equilibrium

$$M^{n+} + n(H^+)_c \rightleftharpoons (M^{n+})_c + nH^+ \tag{2.19}$$

where the subscript c denotes the species bonded to the resin in the column. The equilibrium constant of this reversible reaction is also called **selectivity coefficient**.

$$K_H^M = \frac{[H^+]^n[M^{n+}]_c}{[M^{n+}][H^+]_c^n} \tag{2.20}$$

K_H^M is a measure to compare the selectivity of this exchanger for the ion M under the given experimental conditions. In the case of dilute solution the usual measure for exchangeability is the **distribution ratio**, defined as

$$D_M = \frac{[M^{n+}]_c}{[M^{n+}]} = \frac{\text{amount of } M^{n+} \text{ adsorbed (mmol g}^{-1})}{\text{amount of } M^{n+} \text{ remaining in solution (mmol cm}^{-3})} \tag{2.21}$$

K_H^M is given by D_M/D_H. Similarly anion exchange, where the usual counter-ions are Cl^-, is described by the reversible reaction

$$A^{n-} + n(Cl^-)_c \rightleftharpoons (A^{n-})_c + nCl^- \tag{2.22}$$

with selectivity coefficient and distribution ratio

$$K_{Cl}^A = \frac{[Cl^-]^n[A^{n-}]_c}{[A^{n-}][Cl^-]_c^n} \quad \text{and} \quad D_A = \frac{[A^{n-}]_c}{[A^{n-}]} \tag{2.23}$$

Extensive measurements of the selectivity coefficients have been done for many ions and enabled the formation of a quantitative selectivity scale for common univalent and divalent ions (Bonner and Smith 1957). Larger selectivity was found for anion exchange than for cation exchange (Gregor

Table 2.3 Properties of various ion exchange resins

Type	Functional groups	pH range	Exchange capacity (meq g^{-1})	Trade names
Cation exchangers				
Strong acid	$—SO_3H$	0–14	4.3–5.2	Dowex 50,50 W Amberlite IR-112, 120 Zeokarb 215, 225 Diaion SK 1
Weak acid	—COOH	6–14	10–10.3	Dowex CCR-1 Amberlite IRC-50 Zeokarb 226 WK 10, 11
Anion exchangers				
Strong base	$—CH_2N(CH_3)_3Cl$	0–14	3.0–4.2	Dowex 1 Bio Rad AG1 Amberlite 401, 402 Zeokarb FF Duolite A-42 Diaion SA 10, 11
	$—CH_2CHOH$ (with $—CH_2—N(CH_3)_2Cl$ attached to CH)	0–14	3.0–3.3	Dowex 2 Bio Rad AG2 Amberlite IRA 410 Diaion SA 20
Weak base	$—CH_2NH(CH_3)_2OH$ $—CH_2NH_2CH_3OH$	0–7	5.0–6.0	Dowex 3 Bio Rad AG3 Amberlite IR-45 Zeokarb H
Chelating resin				
Imidoacetate	$—CH_2N(CH_2COONa)_2$	6–14	>0.5	Dowex A-1 Chelex-100 Amberlite XE-318 Diaion CR-10 Uniselec UR-10, 40
Dithiocarbamates	$>N—C(=S)—S$	1.0–6.0		Sumichelate Q-10 ALM-125

et al. 1955). Many selectivity coefficients are reported in the literature (Riley and Taylor 1968). For strongly acidic or strongly basic ion exchangers the trend of the exchangeabilities of cations and anions were found to obey the following rules.

1. *Effect of charge.* The distribution ratio (selectivity coefficient) is larger for higher charge ions. Thus the order for distribution ratios is

$$U^{4+} > Al^{3+} > Mg^{2+} > Na^{+} \quad \text{and} \quad SO_4^{2-} > Cl^{-}$$

Effect of ionic radius. For ions of equal charge the distribution ratio will be larger for the smaller ion (the radius of the hydrated ion). For monovalent cations:

$$Tl^+ > Ag^+ > Cs^+ > Rb^+ > K^+ > Na^+ > H_3O^+ > Li^+$$

For divalent cations:

$$Ra^{2+} > Ba^{2+} > Pb^{2+} > Sr^{2+} > Ca^{2+} > Ni^{2+} \geq Cd^{2+} \geq Co^{2+} \geq Zn^{2+} \geq Mg^{2+} \geq UO^{2+} \geq Be^{2+}$$

For trivalent cations:

$$Ac^{3+} > La^{3+} \geq \text{rare-earth ions (light to heavy)} \geq Y^{3+} > Lu^{3+} \geq Sc^{3+} \geq Al^{3+}$$

For monovalent anions:

$$I^- > NO_3^- > Br^- > Cl^- > OH^- > F^-$$

These rules apply only to strongly basic and acidic ion exchangers. On the other hand weakly acidic and weakly basic exchangers strongly adsorb H_3O^+ and OH^- respectively.

Selectivity for both types of resin can be increased further by combining ion exchange with complex formation. Very high selectivity effects are seen with the anion exchange of chloride and nitrate complexes (Bunney *et al.* 1959). Thus passing a solution on a Cl^- anion exchanger will remove most of the anions. Adding concentrated HCl to the eluted solution will transfer many of the cations to the anionic chloro-complexes

$$M^{n+} + qCl^- \rightarrow MCl_q^{-(q-n)} \tag{2.24}$$

$$\frac{[MCl_q^{-(q-n)}]}{[M^{n+}][Cl^-]^q} = K_{complex} \tag{2.25}$$

Flowing the solution through an anion or a cation exchanger will separate those ions which form complexes (adsorbed on an anion exchanger) from those which do not form a complex and hence do not adsorb on an anion exchanger. They can be separated also on cation exchangers (adsorbing only those which do not form anionic complexes). However, the use of an anion exchanger allows stepwise separation of the cations which form anionic chloro-complexes. Lowering the Cl^- concentration will lead to dissociation of the anionic chloro-complex to form the cation which will be eluted from the column. The Cl^- concentration in which the metal will be eluted from anion exchangers depends on $K_{complex}$, and hence varies from one metal to another.

Eluting from the column with aqueous solutions of varying concentrations of HCl will lead to stepwise separations of the ions which form the anionic chloro-complexes (Figure 2.3).

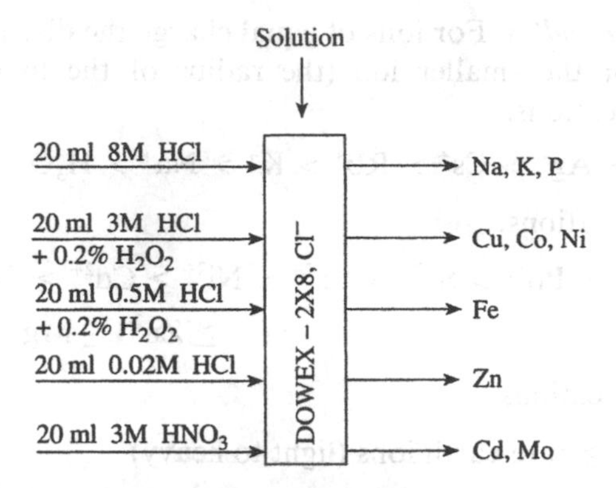

Figure 2.3 Stepwise separation of cations on anion exchangers using the formation of anionic complexes.

2.3.3 Preconcentration

Preconcentration by ion exchange techniques is applied mainly to natural water samples such as sea water, river and lake waters.

The most widely used group of resins for preconcentration of ions from natural water samples are the chelating organic ion exchange resins, since they have high selectivities for the transition elements but low ones for alkali and alkaline-earth metals. Thus for example a typical chelating resin Chelex-100 was found to adsorb selectively many transition metals from saline media (Riley and Taylor 1968). Chelex-100 is the most popular resin from various resins produced by various commercial producers in which the iminoacetate functional group, which is an analogue of EDTA, is covalently bonded to a styrene–divinylbenzene polymer that is only slightly crosslinked (1.5–2.0%). Kingston *et al.* (1978) suggested the following procedure for preconcentration of transition metals from sea water. They adjusted the pH to 5.0–5.5 with ammonium acetate buffer and passed the solution through a Chelex-100 resin in the ammonium form (i.e. the resin is conditioned with the same ammonium acetate buffer solution before adding the sea water solution). Selective elution of Na, K, Ca and Mg was achieved by the addition of 1.0 M ammonium acetate. The column was washed with distilled water before eluting the transition metals with 2.5 M HNO_3. Many studies used this procedure and the literature was summarized by van Berkel *et al.* (1988), who concluded that this procedure is not optimal and suggested to adjust 100 mL of sea water to pH = 5.7, passed through 0.4 g of resin and then washed with 30 mL of 0.4 M ammonium acetate or nitrate, without the washing by distilled water. In the same year Pai (1988) suggested that the optimal preconcentration conditions are pH = 6.5, 500 mL of sea water per

2 g of resin in the magnesium form at a flow rate of 4 mL min^{-1}. The following metals were found to be preconcentrated: Al, Ba, Be, Cd, Co, Cr, Cu, Fe, Hg, La, Mn, Ni, Pb, Ti, V and Zn. The elution of these metals is done mainly by 2.0–3.0 M HNO_3. Heithmar *et al.* (1990) and Siriraks *et al.* (1990) summarized the limitations suffered by the iminodiacetate resins. The first disadvantage is the large changes in volume with changes in the ionic form. This volume change causes large variations in the flow rates. The second drawback concerns the optimal pH for retention, 5.0–5.8, since in this pH range a number of trace elements, like Fe^{3+} and Cu, hydrolyze and precipitate. The change in volume can be made smaller by using a highly crosslinked macroporous resin. The precipitation can be overcome if the pH adjustment is done in a flow–injection system, which limits the hydrolysis and ascertains that the precipitate is transferred to the column. In this case some of the metals will remain on the column as a precipitate, but will in any case be eluted by the HNO_3 solution.

Chelating resins are not limited either to styrene–DVB copolymer matrices or to the iminodiacetate groups. Other resins used are chelating agents immobilized on glass (Ryan and Weber 1985), bonded to porous phenol–formaldehyde polymers (Kaczvinsky *et al.* 1985) or bonded to cellulose (Horvath *et al.* 1985). The second most popular group of chelating agents after the iminodiacetate group used for preconcentration are the polydithiocarbamate resins. In the previous subsection we saw that the same group is used for coprecipitation and in a later section we will see that the same group is used in solvent extraction. Kantipuli *et al.* (1990) reviewed the various groups capable of forming complexes and the possible methods of introducing them into a polymer matrix by chemical modification of the matrix.

2.3.4 Ion chromatography

Ion chromatoraphy is one of the more important fields of application of ion exchangers. It involves the separation of two or more ionic constituents by the difference in their distribution between a mobile phase and a stationary phase (the ion exchanger) which are in intimate contact. The ion exchange column is pretreated with the same electrolyte used later for elution. The sample is added in a small volume of the electrolyte to the top of the column and eluted down the column by the electrolyte (eluent). Two different ions which have different affinity for the resins travel down the column at different rates and are separated into separate peaks. If the sample is small compared to the capacity of the column each separate peak has a Gaussian shape (concentration vs. volume). The volume necessary to elute the maxima of these peaks, V_{max}, is given by the simple relationship:

$$V_{max} = D_w W + V_{void} \tag{2.26}$$

where W is the mass of resin in the column, V_{void} is the volume of free solution between the beads and D_w is the weight distribution coefficient defined by the expression:

$$D_w = \frac{\text{quantity of ion per unit weight of dry resin}}{\text{quantity of ion per unit volume of solution}} \tag{2.27}$$

The relationship between D_w and D, the distribution coefficient, is given by the equation

$$D_w = \frac{ID}{\gamma} \tag{2.28}$$

where I is the fractional interstitial volume and γ is the resin bed density. Two theories have been applied to ion chromatography: the simple plate theory (Beukenka *et al.* 1954) and the complex rate theory (Gleuckauf 1955). The simple plate theory is a mathematical oversimplification of the description of moving systems, yet it has large advantages in interpretation of chromatographic data and permits some useful predictions High-resolution ion exchange chromatography requires the use of small-particle resins, which are resistant to flow and hence cannot rely on gravity and need the application of high pressure to ensure sufficient flow rates. This is discussed in detail in Chapter 9 on HPLC.

2.3.5 *Practical column operation*

Many commercial-grade resins contain various kinds of impurities, such as soluble organic compounds, iron and calcium derived from the production processes, and they should be purified. Purification is usually done by repeated washing with acid, alkali and methanol. Some producers also sell prepurified resins, for which only washing in distilled water before use is required. The resin should be left in the distilled water for sufficient time for swelling of the resin. The purified resins are packed in a column by pouring as a water slurry. Packing should be as uniform as possible so a narrow sieve range of particle size is important. The resin in the column must be kept wet all the time, in order to prevent formation of channeling which considerably reduces the exchange efficiency.

In some cases the gravitational flow of the eluting solution takes too long. In order to speed the separation a vacuum line (or alternatively a pressurizing pump) can be connected to the columns as can be seen in Figure 2.4.

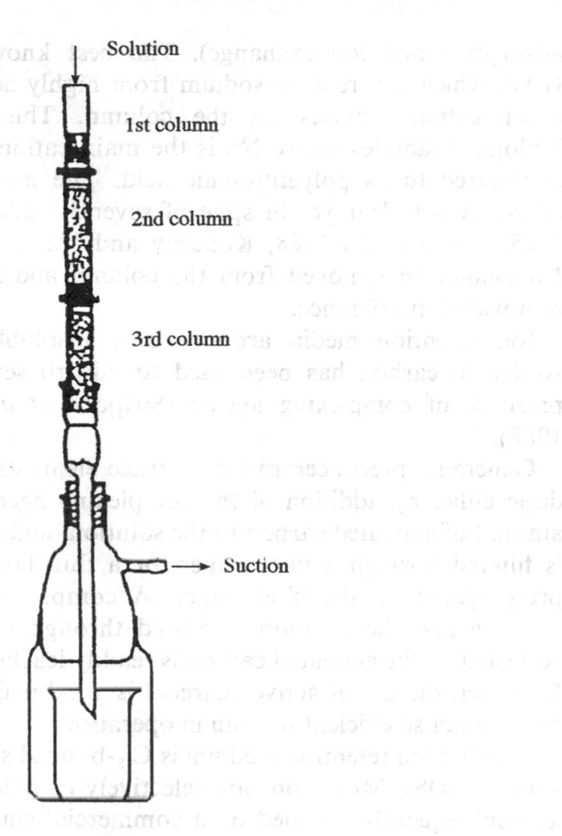

Figure 2.4 Increasing the rate of flow of the moving phase by suction. This system exemplifies the use of several columns in series.

2.3.6 Ion retention columns

Several inorganic materials were used also as packing materials for columns for separation of cations. Examples of these are acidic aluminum oxide (AAO), hydrated manganese dioxide (HMD), tin dioxide (TDO), hydrated antimony pentoxide (HAP), cadmium oxide (CDO), copper sulfide (CUS) and copper chloride (CUC). Pietra *et al.* (1986) devised many separation programs for trace cations in environmental, toxicological and biomedical samples using various combinations of these columns (see next section). Pietra *et al.* refer to these inorganic-packed columns as **inorganic ion exchangers**. However, this classification is questioned since often the process is irreversible and the cations cannot be eluted from the columns, and there are discrepancies between batch and column experiments. Fardy and Tan (1988) named these materials as **ion retention media**. The term describes materials which can remove elements from solution but which give rise to more than one process (e.g.

adsorption and ion exchange). The best known of these materials is HAP, which can remove sodium from highly acidic solution whereas no other cation remains on the column. This is very important for biological samples where Na is the main cation. Sometimes the material is referred to as polyantimonic acid. The mechanism of the retention of Na is not clear yet in spite of several studies (Girardi and Sabbioni 1968, Abe and Ito 1968, Konecny and Hartl 1975, Kyarku 1984). The Na cannot be removed from the column and it is usually used for the removal of interference.

Ion retention media are not only insoluble inorganic compounds. Activated carbon has been used to adsorb several trace metals in the presence of complexing agents (Siripone *et al.* 1983, Fardy and Tan 1988).

Generally, preconcentration of trace elements with activated carbon is done either by addition of the complexing agent together with sufficient amount of activated carbon to the solution, and after shaking, the solution is filtered through a filter paper, or a thin layer of activated carbon is pre-prepared on the filter paper. A complexing agent is added to the solution and the solution is passed through this thin layer. The metal collected on the activated carbon is readily leached out with hot nitric acid. However, the use of active charcoal is good mainly in batch experiments, but it is not so efficient in column operation.

Another ion retention medium is C_{18}-bonded silica (also called octadecyl silica – ODS). Metal ions are selectively complexed by a chelating agent and subsequently retained on a commercial small column of C_{18}-bonded silica. The advantage of these columns over those of activated carbon is the reversibility of the sorption on C_{18}-bonded silica (Fardy 1989). Sturgeon *et al.* (1982) used oxine (8-quinolinol) as the chelating agent and obtained more than 50-fold preconcentration for Cd, Co, Cu, Fe, Mn, Ni and Zn in sea water. The metals were eluted with methanol. Watanabe *et al.* (1981) added 5 mL 1% oxine to 1 L of sea water, adjusted the pH to 8.9 and passed through a 10 mm diameter ODS column. The metals were eluted with 5 mL ethanol.

2.3.7 *Some examples*

Figures 2.5 and 2.6 give examples of separation of mixtures of many elements into several groups, using columns of ion exchange and ion retention media. Since these methods were used to measure radionuclides, they also used columns on which the sorption is irreversible (mainly in Figure 2.6), since the radionuclides can be measured also on the columns. However, most of these elements can be eluted by highly acidic solutions.

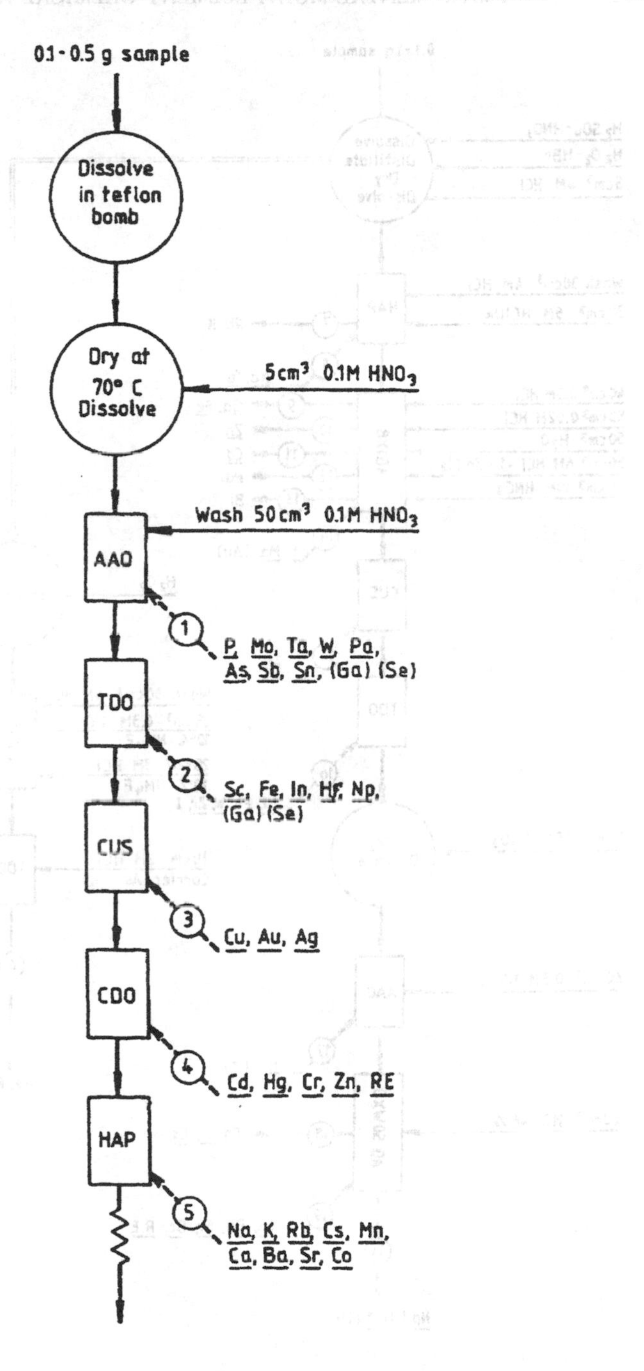

Figure 2.5 Separation and concentration using ion retention media columns (reproduced from Pietra 1986).

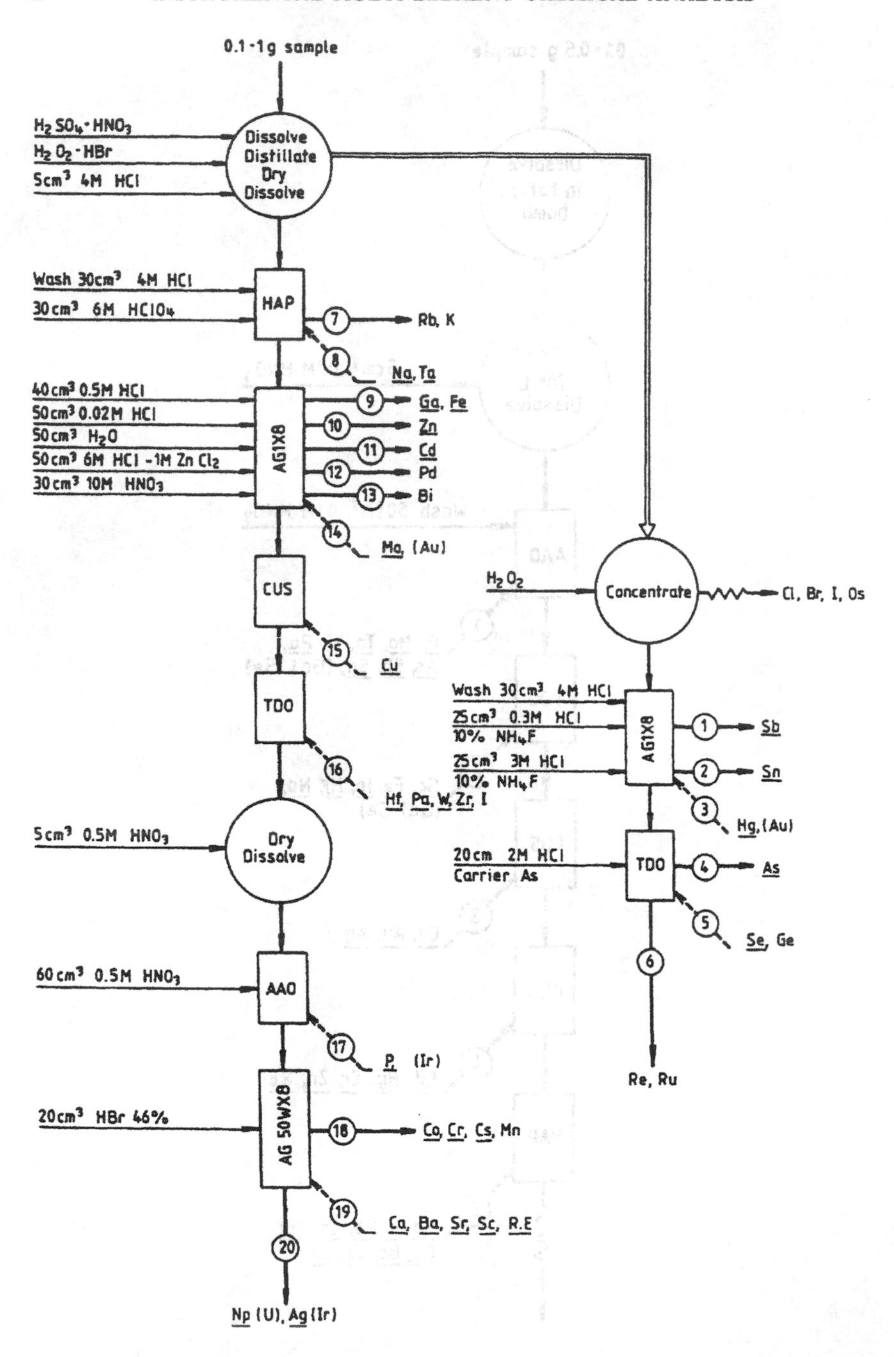

Figure 2.6 Separation and concentration using columns of ion exchange and ion retention media (reproduced from Pietra 1986).

2.4 Preconcentration of trace elements by solvent extraction

2.4.1 *General principles*

The distribution of a solute between two immiscible phases is given theoretically by the Nernst distribution law

$$K = \frac{C_2}{C_1} = \exp\left(-\frac{\mu_2 - \mu_1}{RT}\right) = \frac{S_2}{S_1} \tag{2.29}$$

where K is the distribution constant, C, μ and S are the concentration of the solute after the separation of the two liquid phases, the chemical potential and the solubility of the solute in the pure solvent. Actually this equation applies only for dilute solutions where the activity coefficients of the solute in the two solvents are close to unity and the solute does not affect the miscibility of one solvent in another. Another reason for the inconsistency of this equation is the real situation that the solute exists in more than one form, and the ratio of the various forms depends on the concentration of the solute, leading to variation of K with the concentration of the solvent. Since there is more than one form of the solute, it is more useful to relate to the total concentration of the respective element in the two solvents. The ratio of the total concentration is named the **distribution coefficient** D:

$$D = \sum C_2 / \sum C_1$$

Usually we are interested in extraction from an aqueous solution into an immiscible organic solvent, thus 2 refers to organic phase (or) and 1 to water (w):

$$D = \sum C_{\text{or}} / \sum C_{\text{w}} \tag{2.30}$$

$\sum C_{\text{or}}$ is the total concentration of the solute in various forms in the organic phase, and $\sum C_{\text{w}}$ is for the aqueous solution. However from now on we will skip the $\sum$ sign and write only C_{or} and C_{w}. The **efficiency of extraction**, E, expressed in percent, gives the percentage of the total amount of solute transferred from the aqueous phase to the organic phase. Using V_{or} mL of organic solvent to extract a solute from V_{w} mL of aqueous phase with concentration C_{w0} leads to

$$C_{\text{or}} = \frac{DV_{\text{w}}C_{\text{w0}}}{DV_{\text{or}} + V_{\text{w}}} \quad \Rightarrow \quad E = \frac{DV_0}{DV_{\text{or}} + V_{\text{w}}} \tag{2.31}$$

The concentration of the solute remaining in the aqueous phase is given by

$$C_{\text{w}} = \frac{V_{\text{w}}}{DV_{\text{or}} + V_{\text{w}}} C_{\text{w0}} \tag{2.32}$$

If the total amount of the organic extractant is divided into n fractions of equal volume (each V_{or}/n), and the extraction is done repetitively with a new fraction of the extractant, the concentration left in the aqueous phase after the nth extraction is

$$C_w = \left(1 + \frac{DV_{or}}{V_w}\frac{1}{n}\right)^{-n} C_{w0} \tag{2.33}$$

Equation (2.33) shows that the concentration left in the aqueous phase will decrease with the increase of n. Consequently the extraction process is more efficient if it is done with several small portions of the organic solvent, rather than with the whole volume of the solvent at once. However, repeating extractions is more time-consuming, and an optimal value of n has to be chosen, according to convenience and efficiency required.

The distribution coefficient as seen in equations (2.30)–(2.33) is an important parameter in determining the efficiency of preconcentration by liquid extraction. Separation of two trace elements by liquid extraction is described by the **separation factor** α, which is the ratio of the distribution coefficients for the two metals:

$$\alpha_{2,1} = D_2/D_1 \tag{2.34}$$

Organic solvents dissolve well only uncharged species (e.g. I_2, Cl_2 and $HgCl_2$). However, in aqueous solution most metals are in ionic form and cannot be extracted into organic solvents. In order to extract metal ions into organic solvents, they must be transferred into neutral form. The metal ions are converted into a neutral species by reaction with neutral organic ligands – HR which can replace H atoms with metal ions. The overall reaction is

$$M^{n+} + n(HL)_{or} \rightleftharpoons (ML_n)_{or} + nH^+ \tag{2.35}$$

The equilibrium gives the extraction constant K_{ex} as

$$K_{ex} = \frac{[ML_n]_{or}[H^+]^n}{[M^{n+}][HL]^n_{or}} \quad \Rightarrow \quad D = \frac{[ML_n]_{or}}{[M^{n+}]} = \frac{K_{ex}[HL]^n_{or}}{[H^+]^n} \tag{2.36}$$

The subscript 'or' means the organic phase, whereas the absence of a subscript means the aqueous solution.

The extraction constant can be given as a product of several equilibrium constants of the various steps involved in the extraction:

1. Distribution of the organic ligand between the organic and aqueous phase

$$HL_{(aq)} \rightleftharpoons HL_{(or)} \qquad D_H = [HL]_{or}/[HL] \tag{2.37}$$

2. Dissociation of the organic ligand in the aqueous phase

$$HL_{(aq)} \rightleftharpoons H^+ + L^-_{(aq)} \qquad K_H = [H^+][L^-]/[HL] \tag{2.38}$$

3. Association of the metal cation with the ligand anions

$$M^{n+}_{(aq)} + nL^-_{(aq)} \rightleftharpoons ML_{n(aq)} \qquad K_M = [ML_n]/[M^{n+}][L^-]^n \tag{2.39}$$

4. Distribution of the complex with the metal between the two solvents

$$ML_{n(aq)} \rightleftharpoons ML_{n(or)} \qquad D_M = [ML_n]_{or}/[ML_n] \tag{2.40}$$

Thus

$$K_{ex} = D_M D_H^{-n} K_M K_H^n \tag{2.41}$$

Equation (2.41) shows that the distribution coefficient for the metal, the distribution between the metal in the organic phase (as ML_n) and in the aqueous phase (as M^{n+}), depends on the ligand concentration and on the pH:

$$\log D = \log K_{ex} + n\log[HL] + n\,pH \tag{2.42}$$

Log D should be an increasing linear function of the pH, and the slope is equal to n. In practice, the observed slope is usually smaller than n due to side reactions in the systems. Formation of hydroxide species in aqueous solution would generally reduce the D value of an element at high pH. Another reason is that sometimes the extracted chemical species is not ML_n but includes also some molecules of HL.

The value of n is found from the slope of $\log D$ vs. log[HL] at constant pH, or vs. pH with constant HL concentration. K_M, called the **stability constant** of the complex of that element, is an important factor determining if a metal can be extracted with a specific combination of solvent-chelating agent. The larger K_M, the better is the extraction.

In addition to direct extraction of metal ions with a chelating agent–solvent, extraction can also be done with a combination of metal chelate–solvent. In this case only metals with larger stability constants than the metal added as chelate can be extracted:

$$n_1 M_2^{n_2+} + n_2 M_1 L_{n_1} \rightleftharpoons n_2 M_1^{n_1+} + n_1 M_2 L_{n_2} \tag{2.43}$$

The equilibrium constant for the exchange reaction is determined by the relative stability of the two metal chelates involved:

$$K = K_{M_2}^{n_1} / K_{M_1}^{n_2} \quad \text{or} \quad \frac{1}{n_1 n_2}\log K = \frac{\log K_{M_2}}{n_2} - \frac{\log K_{M_1}}{n_1} \tag{2.44}$$

Thus this method can be used to extract only a few metal ions while the others remain in the aqueous phase. Lo *et al.* (1977, 1982) used the diethyl dithiocarbamate chelate of lead, $Pb(DDC)_2$, in chloroform for concentration of Cu, Ag and Hg from sea water. The high stability constant of $Pb(DDC)_2$ eliminates the extraction of most of the common metal ions present in sea water.

2.4.2 Chelating agents

The formation of metal complexes is often described by the principle of hardness of acids and bases, stating that hard bases prefer to associate with hard acids and soft bases prefer to associate with soft acids (Kolthoff 1979).

Hard acids are metal ions characterized by one of the following properties: small size, high positive oxidation state and low polarizability. They include alkali metals, akaline-earth metals, trivalent rare-earth ions and lighter transition metals in higher oxidation states. Hard bases are oxygen- and nitrogen-containing ligands, such as alcohols, ethers, carboxylates, phosphates and amines. Heavier transition metals such as Ag^+, Hg^{2+}, Au^{3+} and transition metals in their lower oxidation states are soft acids and associate with soft bases which are sulfur- and phosphorus(III)-containing ligands such as thiols (RSH), sulfide (R_2S) and phosphines (R_3P).

Derivatives of dithiocarbamic acid (DTC). Dithiocarbamic acid is a compound where both oxygen atoms of the carbamic acid R_1R_2NCOOH are replaced by sulfur atoms:

$$(R_1)(R_2)N{-}C({=}S){-}S{-}H$$

The metal ion replaces the hydrogen of the SH and is bonded coordinatively to the second sulfur atom:

$$(R_1)(R_2)N{-}C(S)(S)M/n$$

The symbol M/n means that this complex formation bonds only one of the charges of the metal cation, or in other words n molecules of the ligands are required to complex the M^{n+}.

Several derivatives were used but two derivatives of dithiocarbamic acid are the common ones used. The most common one is diethyl-DTC in which $R_1 = R_2 = C_2H_5$. It is written usually as DDC. The second derivative is named pyrrolidine dithiocarbamate and abbreviated to PDC although the more accurate name is tetramethylene-DTC. In this case R_1 and R_2 form a ring with the nitrogen atom:

$$(CH_2{-}CH_2{-}CH_2{-}CH_2)N{-}C(S)(S)M/n$$

The dithiocarbamate ligands tend to form complexes with metal ions having partly filled d orbitals, or ions with filled d orbitals but low positive charge and ions with an 18 + 2 electronic structure. Thus they form complexes with the elements from Ti to Se (except Ge), Nb to Te and W to Bi. The dithiocarbamate complexes are generally insoluble in water (and hence used for coprecipitation as described earlier) but dissolve readily in organic solvents such as chloroform or methyl isobutyl ketone (MIBK). Extraction wih dithiocarbamates removes the alkali metals, alkaline-earth metals, halogens and rare-earth elements, which do not form complexes and hence are not extracted.

The chelating agents used are not the acids, which are not stable, but the sodium salt in the case of DDC (NaDDC) and the ammonium salt in the case of PDC (APDC), which are highly soluble in water (> 0.1 M). APDC is more stable than NaDDC in acid solution and therefore APDC is generally used for the extraction of trace metals from acidic solutions.

Bis(trifluoroethyl) dithiocarbamate was found to form volatile complexes with a number of metal ions which are suitable for gas chromatographic and supercritical fluid chromatographic analysis (Yu and Wai 1991).

The stability constants of the DDC complexes are given in Table 2.4. The data are averages from several measurements given in various studies. A more detailed table is given by Wai (1992).

The stability constants of $M(PDC)_n$ are usually lower than those of $M(DDC)_n$ but follow the same order for different metal ions (Cheng *et al.* 1982).

The factors determining the relative stabilities of metal–DDC complexes are largely unknown. However, quite good inverse correlation was found between $\log K_M$ and the pK of the solubility product of the metal sulfide. So, the solubility product can be used for estimation/prediction of K_M.

Dithiocarbamate preconcentration of trace elements was used for various aqueous solutions such as natural waters (Sugiyama *et al.* 1986), industrial waste water (Bone and Hilbert 1979), urine (Burguera *et al.* 1986), mine water (Mok and Wai 1989) and phosphoric acid (Havezov *et al.* 1987).

Table 2.4 Log stability constants of $M(DDC)_n$

M^{n+}	Ag^{+}	Tl^{+}	Cd^{2+}	Co^{2+}	Cu^{2+}	Fe^{2+}
$\log K_M$	18.7	6.6	9.1	7.4	12.1	6.7
M^{n+}	Hg^{2+}	Mn^{2+}	Ni^{2+}	Pb^{2+}	Pd^{2+}	Zn^{2+}
$\log K_M$	22.3	5.3	9.2	9.7	32.5	7.40
M^{n+}	As^{3+}	Au^{3+}	Bi^{3+}	Ga^{3+}	In^{3+}	Sb^{3+}
$\log K_M$	8.1	23.0	12.5	7.4	9.8	8.8

Extraction with dithiocarbamates can also be used for speciation separation. As^{III} can be extracted with APDC into chloroform from aqueous solutions with pH range 0–6, while As^{V} is not extracted at all (Mok *et al.* 1986). Sb^{III} is extracted in the same pH range while Sb^{V} is extracted only for pH < 3. Thus at pH 3.5–6.0 only Sb^{III} is extracted (Mok and Wai 1987). Se^{IV} is extractable by DDC, whereas Se^{VI} is not (Wyttenbach and Bajo 1975). In all these cases the lower-valent species is extracted. The opposite case exists for Cr (Gilbert and Clay 1973). One method of Cr speciation recommended by the US Environmental Protection Agency (EPA) involves the extraction of Cr^{III} with APCD into methyl isobutyl ketone (MIBK). The possible explanation for the inability to extract Cr^{III} is the very slow process of displacing its water of coordination. Wai *et al.* (1987) used a two-step extraction method (direct extraction and extraction after oxidation) to determine both Cr^{VI} and Cr^{III} in natural waters by extraction with NaDDC.

Crown ethers. Crown ethers (macrocyclic polyethers) are a new generation of selective chelating agents which form stable complexes with metal ions based on the ionic radius–cavity size concept. These chelating agents are used mainly for extraction of alkali and alkaline-earth metal ions (Pedersen 1988, Bartsch 1989). The cavity diameters of crown ethers vary, examples being 0.12–0.15 nm for 14-crown-4 (a selective chelant for Li^+ with ionic diameter 0.1366 nm), 0.18–0.21 nm for 15-crown-5, 0.26–0.32 nm for 18-crown-6 (selective for K^+ with ionic diameter 0.266 nm) and 0.34–0.42 nm for 21-crown-7.

Crown ethers with no substituents on the ring are usually soluble in water. The benzo-substituted crown ethers are slightly soluble in water, fairly soluble in aromatic solvents and very soluble in methylene chloride and chloroform. The complex formed by a neutral crown ether with a metal cation is positively charged and cannot be extracted efficiently into an organic phase. Lipophilic anions, such as picrate, thiocyanate, perchlorate and iodide, are often used to enhance the extraction of the complexes into organic solvents.

2.4.3 *Special extractions*

Masking. When two metal elements form extractable chelates with a particular chelating agent they can sometimes be separated if K_{ex} is sufficiently different so that, at some pH, D will be large for one metal while small for the other one. If K_{ex} is similar for both metals they can be separated by the use of a masking agent L which can change one D more than the other. The most commonly used masking agents include cyanide, tartrate, citrate, fluoride and EDTA. These chemicals form strong water-soluble complexes with many metal ions thus reducing D for the extraction of the metal into the organic solvent:

$$\log D = \log K_{ex} + n \log[HR]_0 + n\,pH - \log \alpha_M(L) \qquad (2.45)$$

where $\alpha_M(L) = 1 + K_M[L]$; $K_M[L]$ is the formation constant of the complex with the masking agent. Since HL is a weak acid the pH dependence of $\log D$ in the presence of a masking agent is generally not simple and sometimes a maximum and/or a minimum appears in the $\log D$ vs. pH curve.

Three-phase extraction. In some separations a mixture of two miscible organic solvents is used. In most cases the two organic solvents are still miscible in the presence of water but in some cases two organic phases and one inorganic phase is obtained at equilibrium. For example when a 7:3 mixture of benzene:chloroform containing diantipyrylmethane is mixed with an aqueous solution of metal ions, three phases appear. Most of the metal elements are in the bottom phase, which has the lowest volume.

Homogeneous extraction. In some cases the equilibration between the two solvents is quite low owing to the slow interaction between the metal ions in the aqueous phase and the chelating agent in the organic phase. The equilibrium can be speeded up by warming the mixture due to higher mutual solubility of the water and the organic solvent.

In extreme cases the mixture is heated up to form one phase as was done for water–propylene carbonate (80°C). In this case the equilibration is done in one homogeneous phase. The separation into two phases is done by cooling (Murata 1972).

Another case of homogeneous equilibration is done by using an organic solvent that is miscible with water. The separation of the two solvents into two phases is done by adding a large amount of inert salt to cause salting out. The usual salting-out agent is nitrate as it does not precipitate the metal ions (Kawamoto and Akaiwa 1973).

2.5 Preconcentration by formation of volatile compounds

Some elements can be preconcentrated by selective formation of volatile compounds (Story and Caruso 1992), which can be removed from the original mixture either by pump sucking or by flushing with an inert gas. The most common compounds are the hydrides but also chlorides, oxides and organic compounds were used. Formation of a volatile compound has additional advantages, besides the preconcentration, in cases where the analysis is done in a plasma (inductively coupled plasma, ICP) or a flame (atomic absorption spectrometry, AAS). In these cases the efficiency of the atomization and the excitation is much higher than for liquid samples since the usual introduction of liquids through pneumatic nebulizers has low efficiency.

Several review papers were written on preconcentration by hydride generation (Robinson and Caruso 1979, Nakahara 1983, 1990). The elements most commonly preconcentrated in this method are As, Bi, Ge, Pb, Sb, Sn and Te. Two main reactions are used for the formation of the hydrides. The first technique used the hydrogen emanating from the reaction of zinc metal with HCl

$$Zn + 2HCl \rightarrow ZnCl_2 + 2H$$

$$n\mathrm{H} + As \rightarrow AsH_3 \qquad (\text{or } SbH_3 \text{ or } SeH_4)$$

This method works well only with the three metals above (As, Sb and Se) and only in their lower valence state As^{III}, Sb^{III} and Se^{IV}. If the elements are in their higher valence state (5, 5 and 6 respectively) they must be reduced, usually by addition of KI (or KBr) and $SnCl_2$ to the acidified solution. This difference between the valence states can be used to preconcentrate only one of the valences, by forming first hydrides of the low valence state, and after reduction generating hydrides again. This method of hydride generation suffers from the disadvantages of relatively slow reaction and being difficult to automate, in addition to the limited number of elements.

The other method for hydride generation used $NaBH_4$ (sodium borohydride or sodium tetrahydroborate) instead of Zn:

$$NaBH_4 + HCl + 3H_2O \rightarrow NaCl + H_3BO_3 + 8H$$

With $NaBH_4$ all the above-mentioned elements form hydrides. Sodium borohydride solutions are not stable and are better prepared freshly. However they can be stabilized by 0.1–2% NaOH or KOH. The optimum concentration of $NaBH_4$ for efficient formation of the hydrides is usually around 2%. The HCl solution is usually 1–5 M. The formation of the hydride can be done in batch mode, where a pellet or solution of $NaBH_4$ is added to the sample, or in continuous mode, in which sample and $NaBH_4$ solutions are pumped and mixed continuously.

The continuous mode suffers from excess of H_2, which can be removed in the batch mode by trapping the hydrides in a cold trap while the hydrogen gas is removed. Hydrogen interferes more with AAS and there mainly the batch mode is used. Most ICP systems use the continuous mode, and various commercial systems for hydride generation are available. The systems employ a peristaltic pump connected to a mixing tee, into which both the sample solution and $NaBH_4$ solutions are pumped. From the mixing tee the solution is transferred to a gas–liquid separator and the gas is drawn into the ICP system.

Also other compounds besides hydrides were used in order to volatilize special elements. The disadvantge (and sometimes advantage) is that usually each element has its own method (if at all). Thus, mercury is volatilized as the element itself (Erguecyener *et al.* 1988, Welz and Schubert-Jacobs 1988). Osmium is preconcentrated as its volatile tetraoxide OsO_4 (Russ *et al.* 1987, Gregoire 1990). Nickel was separated as $Ni(CO)_4$ (Sturgeon 1989, Drews *et al.* 1990). Several elements form volatile chlorides – Bi, Cd, Ge, Mo, Pb, Sn, Tl, Zn and As (Skogerboe *et al.* 1975) – or fluorides – W, Mo, U, V, Re and Ge (Rigin 1989). Some metals form volatile organometallic compounds mainly with β-diketonates (Castillo *et al.* 1988).

2.6 Electrochemical preconcentration

Electrochemical methods are used for the preconcentration and in some cases also for the following analysis of the concentrate (von Wandruszka 1992). There are two distinct accumulation mechanisms in electrochemical preconcentration. The first one, and the oldest one, involves electron transfer from the ions, i.e. cathodic deposition of cations or deposition on the anode of anions. This mechanism applies only to ions. More recently another method of accumulation was developed for organic compounds and for organometallic complexes. This mechanism is based on adsorption of the analyte on the charged electrode in a non-faradic process (no electrons are transferred).

For both processes the accumulated analytes can be stripped in a faradic process by a reverse current. The stripping process can be used either for preconcentration or for analysis. More information on the analytic step can be found in Chapter 6 on analysis by electrochemical methods.

The separation or preconcentration of only a few of the metal ions present in the solution can be done in a controlled potential electrolysis, where only those ions having lower potentials than those applied are deposited. When the potential of the working electrodes exceeds the deposition potential by about 0.2 V, deposition yields larger than 99.9% are generally obtained.

Most concentration processes are those of cathodic deposition of metal ions

$$M^{n+} + ne \rightarrow M^0$$

The current at time t is given by the equation

$$i = nFAD\,\frac{C_s - C_e}{\delta}$$

where F is the Faraday constant, A is the electrode surface area, D is the diffusion constant of the metal ions, C is the concentration of the metal ion (the subscripts s and e stand for in solution and at the electrode) and δ is the diffusion layer thickness. The electrolysis diminishes the concentration near the electrodes and in order to speed up the homogenization the electrode is rotated or the solution stirred. The faradic current can be expressed in terms of the experimental parameters by the Levich equation

$$i = knFAD^{2/3}\omega^{1/2}\nu^{1/6}C_s$$

where k is a constant, ω is the rate of electrode rotation or solution stirring and ν is the kinematic viscosity of the solution.

When the cathodic deposition potential is large, any cation reaching the electrode will be immediately reduced leading C_e to approach zero, leading to a maximal (limiting) electrolysis current i_L

$$i_L = nFAD\frac{C_s}{\delta}$$

using the experimental parameters the following semi-empirical equation was obtained for a mercury-drop electrode:

$$i_L = 4\pi r_0 FDC_s + k\pi r_0^2 D^{2/3} C_s \omega^{1/2}$$

where r_0 is the radius of the mercury-drop electrode.

For long electrodeposition the change of current with time is exponential

$$i(t) = \alpha \mathrm{e}^{-\beta t}$$

leading to

$$Q(t) = Q_0(i - \mathrm{e}^{-\beta t})$$

where Q is the deposited amount, and α and β depend on i_L, ε, D, the electrode geometry, etc.

Materials used for working electrodes include platinum, platinum–iridium alloys, silver, copper, tungsten and carbon (in various forms, e.g. graphite, glassy carbon, etc.). Counter-electrodes (anodes) are usually platinum or platinum–iridium alloys. It should be kept in mind that the platinum anode dissolves slightly in acidic or ammoniacal electrolytes.

Sometimes several reagents are added to the analyzed solution in order to improve the electrical deposition. These reagents have the following functions:

1. *Depolarizers.* Anodic depolarizers (e.g. hydrazine or hydroxylamine) and cathodic depolarizers (HNO_3 and Cu^{2+}) prevent undesired electrode reactions, in cathodic and anodic depositions respectively, due to their preferential oxidation at the anode or reduction at the cathode, leading to less positive anode potential or less negative cathode potential. Thus hydrazine or hydroxylamine prevents the

deposition of cobalt oxide on the anode or the dissolution of platinum anode.

2. *Complexing agents.* To shift the deposition potential or improve the adhesion of the deposit to the electrode surface.
3. *Surfactants.* To smooth the metal deposits.

References

Abe, M. and Ito, T. (1968), *Bull. Chem. Soc. Jpn.* **41**, 333.
Alfassi, Z.B. (1990), in *Activation Analysis*, ed. Z. Alfassi, CRC Press, Boca Raton, FL.
Alfassi, Z.B. (1992), in Alfassi and Wai (1992), p. 33.
Alfassi, Z.B. and Wai C.M. (1992) *Preconcentration Techniques for Trace Elements*, CRC Press, Boca Raton, FL.
Bartsch, R.A. (1989), *Solv. Extr. Ion Exch.* **7**, 829.
Beukenka, J., Rieman, W. and Lindenbaum, S. (1954), *Anal. Chem.* **26**, 505.
Bone, K.M. and Hilbert, W.D. (1979), *Anal. Chim. Acta* **107**, 219.
Bonner, O.D. and Smith, L.L. (1957), *J. Phys. Chem.* **61**, 326.
Brunix, E. (1975), *Philips Res. Rep.* **30**, 177.
Bunney, R.L., Ballou, N.E., Pascual, J. and Foti, S. (1959), *Anal. Chem.* **31**, 324.
Burguera, J.L., Burguera, M., Cruzo, L.L. and Naranjo, O.R. (1986), *Anal. Chim. Acta* **186**, 273.
Castillo, J.R., Mir, J.M., Bendicho, C. and Laborda, F. (1988), *Fresenius Z. Anal. Chem.* **332**, 37.
Cheng, K.L., Ueno, K. and Imamura, T. (1982), *Handbook of Organic Analytical Reagents*, CRC Press, Boca Raton, FL.
Das, H.A. (1982), *Pure Appl. Chem.* **54**, 755.
Drews, W., Weber, G. and Toelg, G. (1990), *Anal. Chim. Acta* **231**, 265.
Erguecyener, A.S., Ataman, D.Y. and Temizer, A.J. (1988), *Anal. At. Spectrom.* **3**, 177.
Fardy, J.J. and Tan, M. (1988), *J. Radioanal. Nucl. Chem.* **123**, 573.
Gadde, R. and Laitinen, H. (1974), *Anal. Chem.* **46**, 2022.
Gilbert, T.R. and Clay, A.M. (1973), *Anal. Chim. Acta* **67**, 289.
Girardi, F. and Sabbioni, E. (1968), *J. Radioanal. Nucl. Chem.* **1**, 169.
Gleuckauf, E. (1955), *Trans. Faraday Soc.* **51**, 34; and *Ion Exchange and its Application*, Society of Chemical Industry, London, p. 34.
Gregoire, D.C. (1990), *Anal. Chem.* **62**, 141.
Gregor, H.P., Belle, J. and Marcus, R.A. (1955), *J. Am. Chem. Soc.* **77**, 2713.
Hahn, O. (1936), *Applied Radiochemistry*, Cornell University Press, Ithaca, NY.
Havezov, I., Ivanova, E., Koehler, P., Jordanov, M., Matchat, R., Reiher, K., Emrich, G. and Licht, K. (1987), *Fresenius Z. Anal. Chem.* **326**, 536.
Heithmar, E.M., Hinners, T.A., Rowan, J.T. and Riviello, J.M. (1990), *Anal. Chem.* **62**, 1185.
Horvath, Z.S., Barnes, R.M. and Murty, P.S. (1985), *Anal. Chim. Acta* **173**, 305.
Jackwerth, E. (1979), *Pure Appl. Chem.* **51**, 1149.
Jackwerth, E. and Willmer, P.G. (1976), *Fresenius Z. Anal. Chem.* **279**, 23.
James, R.D., Stiglich, P.J. and Healey, T.W. (1975), *Discuss. Faraday Soc.* **59**, 142.
Kaczvinsky, J.R., Fritz, J.S., Walker, D.D. and Ebra, M.A. (1985), *J. Radioanal. Nucl. Chem.* **91**, 349.
Kantipuli, C., Katragadda, S., Chow, A. and Gesser, H.D. (1990), *Talanta* **37**, 491.
Kawamoto, H. and Akaiwa, H. (1973), *Chem. Lett.* 259.
Kingston, H.M., Barnes I.L., Brady, T.J., Raines, T.C. and Champ, M.A. (1978), *Anal. Chem.* **50**, 2064.
Kolthoff, I.M. (1979), in *Treatise on Analytical Chemistry*, part 1, vol. 2, 2nd edn, eds I.M. Kolthoff and P.J. Elving, John Wiley, New York, p. 531.
Konecny, C. and Hartl, I. (1975), *Z. Phys. Chem. Leipzig* **256**, 17.

Kurbatov, M.H., Wood, G.B. and Kurbatov, J.D. (1951), *J. Chem. Phys.* **19**, 258.
Kyarku, S.K. (1984), *Anal. Lett.* **17**, 2213.
Lan, C.R., Sun, Y.C., Chao, J.H., Chung, C., Yang, M.H., Lavi, N. and Alfassi, Z.B. (1990), *Radiochim. Acta* **50**, 225.
Lan, C.R., Tseng, C.L., Yang, M.H. and Alfassi, Z.B. (1991), *Analyst* **116**, 35.
Lavi, N. and Alfassi, Z.B. (1989), *J. Radioanal. Nucl. Chem.* **130**, 71.
Lavi, N., Mantel, M. and Alfassi, Z.B. (1990), *Analyst* **115**, 817.
Lo, J.M., Wei, J.C. and Yeh, S.J. (1977), *Anal. Chem.* **49**, 1146.
Lo, J.M., Yu, J.C., Hutchinson, F.I. and Wai, C.M. (1982), *Anal. Chem.* **54**, 2536.
Minczewski, J., Chwastowska, J. and Dybcynski, R. (1982), *Separation and Preconcentration Methods in Inorganic Trace Analysis*, Ellis Horwood, Chichester.
Mizuike, A. (1983), *Enrichment Techniques for Inorganic Trace Analysis*, Springer-Verlag, Berlin.
Mok, W.M. and Wai, C.M. (1987), *Anal. Chem.* **59**, 233.
Mok, W.M. and Wai, C.M. (1989), *Water Res.* **23**, 7.
Mok, W.M., Shan, N.K. and Wai, C.M. (1986), *Anal. Chem.* **58**, 110.
Morgan, J.J. and Strumm, W. (1964), *J. Colloid Sci.* **19**, 347.
Murata, K. (1972), *Anal. Chem.* **44**, 805.
Nakahara, T. (1983), *Prog. Anal. At. Spectrosc.* **6**, 163.
Nakahara, T. (1990), in *Simple Introduction to Atomic Spectroscopy*, Analytical Spectroscopy Library, vol. 4, ed. J. Sneddon, Elsevier, Amsterdam, ch. 10.
Pai, S.C. (1988), *Anal. Chim. Acta* **211**, 271
Pedersen, C.J. (1988), *Science* **241**, 536.
Pietra, R., Sabbioni, E., Gallorini, M. and Orvini, E. (1986), *J. Radioanal. Nucl. Chem.* **102**, 69.
Rigin, V.I. (1989), *Fresenius Z. Anal. Chem.* **335**, 15.
Riley, J.P. and Taylor, D. (1968), *Anal. Chim. Acta* **40**, 479.
Robinson, W.B. and Caruso, J.A. (1979), *Anal. Chem.* **51**, 889A.
Russ, G.P., Bazan, J.M. and Date, A.R. (1987), *Anal. Chem.* **59**, 984.
Ryan, D.K. and Weber, J.H. (1985), *Talanta* **32**, 985.
Sebba, F. (1962), *Ion Flotation*, Elsevier, Amsterdam.
Siripone, C., Wals, G.D. and Das, H.A. (1983), *J. Radioanal. Chem.* **79**, 35.
Siriraks, A., Kingston, H.M. and Riviello, J.M. (1990), *Anal. Chem.* **62**, 1185.
Skogerboe, R.K., Dick, D.L., Pavlica, D.A. and Lichte, F.E. (1975), *Anal. Chem.* **47**, 568.
Smits, J., Nellisen, J. and Van Grieken, R.E. (1979), *Anal. Chim. Acta* **111**, 215.
Story, W.C. and Caruso, J.A. (1992), in Alfassi and Wai (1992), p. 333.
Sturgeon, R.E., Berman, S.S. and Willie, S.N. (1982), *Talanta* **29**, 167.
Sugiyama, M., Fujino, O., Kihara, S. and Matsue, M. (1986), *Anal. Chim. Acta* **181**, 159.
Sun, Y.C., Yang, J.Y., Lin, Y.F., Yang, M.H. and Alfassi, Z.B. (1993), *Anal. Chim. Acta* **276**, 33.
van Berkel, W.W., Overbosch, A.W., Feenstra, G. and Maessen, A.O. (1988), *J. Anal. At. Spectrom.* **3**, 249.
von Wandruszka, R. (1992), in Alfassi and Wai (1992), p. 133.
Wai, C.M., Tsay, L.M. and Yu, J.C. (1987), *Mikrochim. Acta* **2**, 73.
Watanabe, H., Goto, K., Taguchi, S., McLaren, J.W., Berman, S.S. and Russel, D.S. (1981), *Anal. Chem.* **53**, 738.
Welz, B. and Schubert-Jacobs, M. (1988), *Fresenius Z. Anal. Chem.* **331**, 324.
Wyttenbach, A. and Bajo, S. (1975), *Anal. Chem.* **47**, 1813.
Xiao-qan, S., Jun, T. and Guang-guo, X. (1988), *J. Anal. At. Spectrom.* **3**, 259.
Yu, J.J. and Wai, C.M. (1991), *Anal. Chem.* **63**, 842.
Zhukova, L.A. and Rachinskii, V.V. (1978), *Radiokhimia* **20**, 485.

3 Quality assurance, control and assessment

Z.B. ALFASSI

3.1 Introduction

The appraisal of quality has a considerable impact on analytical laboratories. Laboratories have to manage the quality of their services and to convince clients that the advocated level of quality is attained and maintained. Increasingly, accreditation is demanded or used as evidence of reliability. At present there are American and European Standards (ISO 25 and EN 45001) that describe how a laboratory ought to be organized in order to manage the quality of its results. These Standards form the basis for accreditation of analytical laboratories. Terms used frequently are 'quality assurance' and 'quality control'. Quality assurance is a wider term, which includes both quality control and quality assessment.

Quality control of analytical data (QCAD) was defined by the ISO Committee as: 'The set of procedures undertaken by the laboratory for continuous monitoring of operations and results in order to decide whether the results are reliable enough to be released.' QCAD primarily monitors the batchwise accuracy of results on quality control materials, and precision on independent replicate analysis of test 'materials'. **Quality assessment** was defined (Taylor 1987) as 'those procedures and activities utilized to verify that the quality control system is operating within acceptable limits and to evaluate the data'.

The Standards of **quality assurance** (American, ISO 25; European, EN 45001) were written for laboratories that do analyses of a routine nature and give criteria for the implementation of a quality system which ensures an output with performance characteristics stated by the laboratory. An important aspect of the quality assurance system is the full documentation of the whole analysis process. It is essential to have well designed and clear work-sheets. On the work-sheets both the raw data and the calculated results of the analyses should be written. Proper work-sheets reduce the chances of computing error and enable reconstruction of the test if it appears that a problem has occurred. The quality assurance system (or Standard) also treats the problems of personnel, equipment, materials and chemicals. The most important item is the methodology of the analysis. Quality control is not meaningful unless the methodology used has been validated properly. Validation of a methodology means the proof of suitability of this methodology to provide useful analytical data. A method is validated when the performance characteristics of the method are adequate and when it has been established that the measurement is under statistical control and produces accurate results.

Statistical control was defined as: 'A phenomenon will be said to be "statistically controlled" when, through the use of past experience, we can predict, at least within limits, how the phenomenon may be expected to vary in the future. Here it is understood that prediction means that we can state at least approximately, the probability that the observed phenomenon will fall within the given limits'.

The quality assurance systems required for accreditation of analytical laboratories are very important and are dealt with in several recent books (Funk *et al.* 1995, Kateman and Buydens 1993, Taylor 1987, Guenzler 1994, Pritchard 1997). However, these systems are well beyond the scope of this chapter, which will be devoted mainly to quality assessment of analytical data.

3.2 Quality assessment

The quality of chemical analysis is usually evaluated on the basis of its uncertainty compared to the requirements of the users of the analysis. If the analytical results are consistent and have small uncertainty compared to the requirements, e.g. minimum or maximum concentration of special elements in the sample and its tolerances, the analytical data are considered to be of adequate quality. When the results are excessively variable or the uncertainty is larger than the needs, the analytical results are of low or inadequate quality. Thus the evaluation of the quality of analysis results is a relative determination. What is high quality for one sample could be unacceptable for another.

A quantitative measurement is always an estimate of the real value of the measure and involves some level of uncertainty. The limits of the uncertainty must be known within a stated probability, otherwise no use can be made of the measurement. Measurement must be done in such a way that it can provide this statistical predictability.

Statistics is an integral part of quality assessment of analytical results, e.g. to calculate the precision of the measurements and to find if two sets of measurements are equivalent or not (in other words if two different methods gave the same result for one sample).

3.3 Statistical methods

3.3.1 *Mean and standard deviation*

The basic parameters that characterize a population of samples are the mean μ and the standard deviation σ. In order to determine μ and σ the entire population should be measured, which is usually impossible to do. In practical measurement of several samples (or replicate measurements on a given sample), estimates of the mean and the standard deviation are calculated and denoted as $\bar{x}$ and s. The values of $\bar{x}$ and s are used to calculate confidence intervals, comparison of precisions and significance of apparent discrepancies. The **mean** $\bar{x}$ and the **standard deviation** s of the values $x_1, x_2, \ldots, x_n$ obtained at n replicate measurements is given by the equations

$$\bar{x} = \frac{\sum_{i=1}^{n} x_i}{n} \tag{3.1}$$

$$s = \sqrt{\frac{\sum_{i=1}^{n}(x_i - \bar{x})^2}{n-1}} = \sqrt{\frac{\sum_{i=1}^{n} x_i^2}{n-1} - \frac{\left(\sum_{i=1}^{n} x_i\right)^2}{n(n-1)}} \tag{3.2}$$

The values of $\bar{x}$ and s can be calculated using a computer program or a calculator. It is important to note that all scientific calculators have two keys, one depicted as σ_{xn} and the other one as $\sigma_{xn\text{-}1}$. Equation (3.2) fits the key $\sigma_{xn\text{-}1}$. The other key uses n instead of $n-1$ in equation (3.2). The key σ_{xn} gives the standard deviation of our sample, but not that of the whole population, which can be obtained only by doing an infinite number of repeated measurements. For a small number of repetitions, the equation with $n-1$ gives a better estimate of the true σ, which is unknown. The mean $\bar{x}$ is a better estimate for the true value than one measurement alone and the standard deviation σ (or its estimate s) represents the dispersion of the measured values around the mean. The standard deviation has the same units as that of the measured values x_i. Often analysts prefer to use a dimensionless quantity to describe the dispersion of the results. In this case they use the **relative standard deviation** as a fraction (called also **coefficient of variation**, CV) or as a percentage (RSD):

$$\mathrm{CV} = s/\bar{x} \tag{3.3}$$

$$\mathrm{RSD} = \mathrm{CV} \times 100 \tag{3.4}$$

When calculating small absolute standard deviations by calculators, sometimes considerable errors are caused by rounding errors due to the limited number of digits used. In order to overcome this problem, and in order to simplify the punching on the calculator, it is worth subtracting a constant number from all the data points, so that x_i will be of the same order of magnitude as the differences between themselves. The standard deviation will have the same value as without the subtraction and the subtracted constant should be added to the mean. In other words, if we have n data points $x_1, \ldots, x_n$ that are large numbers, we will do better to key into the calculator $x_1 - c, x_2 - c, \ldots, x_n - c$ such that $x_i - c$ are no longer large numbers. The real mean of x_i is $\bar{x}_i = c + \overline{x_i - c}$ and the standard deviation remains the same, $s(x_i) = s(x_i - c)$. Thus for calculating the mean and standard deviation of 50.81, 50.85, 50.92, 50.96, 50.83, we can key 0.01, 0.05, 0.12, 0.16, 0.03 and obtain $\bar{x} = 0.074$ and $s = 0.06348$. The real mean is $50.8 + 0.074 = 50.874$ and s is the same, 0.06348.

Frequency tables. When large numbers of measurements are made (on the same sample if the sample is not consumed by the measurement or on different aliquots of the same sample or on different samples), some values are obtained more than once. Sometimes, instead of a discrete value, a range of values is chosen. In this case it is simpler to concentrate the data in a **frequency table** – a table that gives the number of times each value was obtained. For example, the concentration of salt in drinking water was measured each day during a whole year. The results are given in the following table (given to two significant figures):

Concentration (mg/L) x_i	Number of days f_i
3.5	18
3.6	22
3.7	25
3.8	35
3.9	46
4.0	55
4.1	45
4.2	40
4.3	32
4.4	27
4.5	20

In this case the mean and the standard deviation are calculated by the equations

$$\bar{x} = \frac{\sum_{i=1}^{n} f_i x_i}{\sum_{i=1}^{n} f_i} \tag{3.5}$$

$$s = \sqrt{\frac{\sum_{i=1}^{n} f_i (x_i - \bar{x})^2}{\left(\sum_{i=1}^{n} f_i\right) - 1}} \tag{3.6}$$

The summation is carried out over all the various values of x_i (n different values) and the total number of measurements is

$$\sum_{i=1}^{n} f_i$$

Most scientific calculators can calculate the mean value and the standard deviation from frequency tables. In our example the following results will be obtained: $\bar{x} = 4.0164$, $s = 0.2687$ (remember to use the n-1 key). The units of $\bar{x}$ and s are the same as those of x_i, i.e. mg/L.

3.3.2 *Distribution of the errors*

The standard deviation gives a measure of the spread of the results around the mean value. However, it does not indicate the shape of the spread. Frequency tables and, even more so, drawing them as a rod diagram or as a histogram give a clearer picture of the spread of the measurements. A histogram describes the real situation better since the real values are not of only two significant digits, and 3.7 mg/L stands, for example, for the range 3.65001 to 3.75000. If we arranged the table not according to two but to three significant digits, we would have many more columns in the histogram. Increasing the number of measurements and the number of significant figures will lead to a continuous distribution. The usual mathematical function for a random distribution is the **normal** or **Gaussian** distribution, which is described by the equation

$$p = \text{probability} = \frac{f_i}{\sum f_i} = \frac{\exp[-(x_i - \mu)^2/2\sigma^2]}{\sigma\sqrt{2\pi}} \tag{3.7}$$

where μ is the mean value and σ the standard deviation. The curve of this function is symmetrical about μ. The larger is σ/μ, the wider is the curve (larger spread of the data). The area under the curve is equal to 1, as it is the total probability. The probability that the measured value in an experiment will be equal to or higher than a value α is given by the integral

$$\int_{\alpha}^{\infty} p \, \mathrm{d}x$$

This integral cannot be calculated analytically but has been calculated numerically and is given in tables. The tables are given not for the actual measurements x_i but for the measurement in units of standard deviation, u_i, which is defined by the equation:

$$u_i = \frac{x_i - \mu}{\sigma} \tag{3.8}$$

The probability that one measurement will be in the range between x_1 and x_2 is given by the subtraction of the value in the cumulative normal distribution table for x_1 from the value obtained for x_2. Approximately 68% of the population in a normal distribution lies within 1σ of the mean (i.e. in the range $\mu - \sigma \leq u \leq \mu + \sigma$). Also, 95% of the values lie in the range $\mu \pm 2\sigma$ and 99.7% lie within $\pm 3\sigma$ of the mean.

It is usually assumed that repeated measurements of a single analytical quantity are distributed normally, and this is the case most of the time for measurement of different samples from the same kind, such as, for example, the concentration of glucose in blood measured for different healthy people.

Sometimes different frequency functions are observed. For example, the size of droplets formed by nebulizers used in atomic absorption or inductively coupled plasma spectrometry follows the log-normal distribution function. However, if the frequency is plotted against the logarithm of the size of the droplets, then the distribution curve is a normal one.

3.3.3 Confidence limits of the mean

The mean of a sample of measurements $\bar{x}$ provides an estimate of the true value, μ, of the quantity we are trying to measure. However, it is quite unlikely that $\bar{x}$ is exactly equal to μ, and an important question is to find a range of values in which we are certain that the true value lies. This range depends on the measured mean but also on the number of measurements done and on the question of how certain we want to be. The more certain we want to be, the larger is the range we have to take. The larger the number of experiments done, the closer is $\bar{x}$ to μ, and a smaller range has to be taken for the same percentage of certainty. Usually tables refer not to the number of repeated experiments but to the number of **degrees of freedom** (usually given the symbol ν or d.f.). In the calculation of the standard deviation the number of degrees of freedom is $n-1$, where n is the number of repeated measurements. The number of degrees of freedom refers to the number of independent variables. Calculating the standard deviation we use n terms of $(x_i - \bar{x})^2$, but only $n-1$ terms are independent, since the nth term can be calculated from the equation

$$\sum_{i=1}^{n}(x_i - \bar{x}) = 0$$

We mentioned before that 95% of the population in a normal distribution lies approximately in $\mu \pm 2\sigma$, or more correctly in $\mu \pm 1.96\sigma$. The same holds also for the mean value of a number of repeating experiments, each one consisting of n separate measurements with a mean $\bar{x}$. In this case we will say that 95% of the mean values are within $\pm\sigma/\sqrt{n}$ of the true value:

$$\mu - 1.96(\sigma/\sqrt{n}) < \bar{x} < \mu + 1.96(\sigma/\sqrt{n}) \tag{3.9}$$

In other words we can say that within **95% confidence** the true value μ is in the range

$$\bar{x} - 1.96(\sigma/\sqrt{n}) < \mu < \bar{x} + 1.96(\sigma\sqrt{n}) \tag{3.10}$$

For 99% confidence the 1.96 factor will be replaced by 2.58, and for

99.7% by 2.97. However, one problem remains, as we do not know σ, only its estimate, s. The larger n, the closer is s to σ, but it is not equal. To correct for this uncertainty, an additional factor was introduced. This factor was combined together with the factor due to the percentage of confidence to give the following equation for the **confidence limit**:

$$\mu = \bar{x} \pm t(s/\sqrt{n}) \tag{3.11}$$

Here t is a factor dependent on both the number of degrees of freedom (d.f. $= n - 1$) and the percentage of confidence, as can be seen in Table 3.1.

No quantitative experimental result is of value if an estimate of the errors involved in its measurement is not given. The usual practice is to give $\bar{x} \pm s$ or $\bar{x} \pm 2s$. It should be stated clearly if the cited error is one or two standard deviations. Any confidence limit can then be calculated from equation (3.11).

It should be emphasized that in many analytic papers results are

Table 3.1 Values for the two-sided Student's t-distribution. The percentages are for confidence intervals and the fractions for level of significance, α

d.f.	80% $t_{0.90}$	90% $t_{0.95}$	95% $t_{0.975}$	98% $t_{0.99}$	99% $t_{0.995}$
1	3.078	6.314	12.706	31.821	63.657
2	1.886	2.920	4.303	6.965	9.925
3	1.638	2.353	3.182	4.541	5.841
4	1.533	2.132	2.776	3.747	4.604
5	1.476	2.015	2.571	3.365	4.032
6	1.440	1.943	2.447	3.143	3.707
7	1.415	1.895	2.365	2.998	3.499
8	1.397	1.860	2.306	2.896	3.355
9	1.383	1.833	2.262	2.821	3.250
10	1.372	1.812	2.228	2.764	3.169
11	1.363	1.796	2.201	2.718	3.106
12	1.356	1.782	2.179	2.681	3.055
13	1.350	1.771	2.160	2.650	3.012
14	1.345	1.761	2.145	2.624	2.977
15	1.341	1.753	2.131	2.602	2.947
16	1.337	1.746	2.120	2.583	2.921
17	1.333	1.740	2.110	2.567	2.898
18	1.330	1.734	2.101	2.552	2.878
19	1.328	1.729	2.093	2.539	2.861
20	1.325	1.725	2.086	2.528	2.845
25	1.316	1.708	2.060	2.485	2.787
30	1.310	1.697	2.042	2.457	2.750
40	1.303	1.684	2.021	2.423	2.704
60	1.296	1.671	2.000	2.390	2.660
∞	1.282	1.645	1.960	2.326	2.576

reported as $\bar{x} \pm 2\sigma$ and stated with 95% confidence. Since many of these studies do not have more than 6–8 experiments, the factor for 95% confidence should be larger than 2.

The factor t in equation (3.11) is taken from a distribution called the t-distribution or Student's t-distribution. The source of this strange name is due to the fact that the statistician who first developed it (W.S. Gosset) worked for an employer that forbade publication of research work by its employees. Gosset published his first work under the pseudonym 'Student', leading to the distribution being known as Student's t-distribution. The t-distribution obeys the equation

$$p = \frac{p_0}{[1 + t^2/(n-1)]^{n/2}} \tag{3.12}$$

p_0 is a constant normalizing factor whose value depends on n in such a way that the area under the curve is a unit area. For large n, p as a function of t looks very similar to the normal distribution. However, as n decreases, p decreases in the mean and increases in the tail.

The confidence limit can be used as a test for systematic errors in the measurement, by assuming that, if the observed value is not within the confidence limits, there is a systematic error in the measurement. Another method for studying systematic errors will be given in the next section.

3.4 Significance tests

3.4.1 Introduction

As we saw in the previous section, the measured mean value is not exactly the true value but a value close to it within the confidence limit. So how can we know if an analytical method is a good one, since the mean is not the same as the expected true value of the standard used for quality control of the method? For example, let us assume that we prepared a standard with concentration of 0.5 mg/mL and measured it repeatedly by a new analytical method 10 times and observed $\bar{x} = 0.49$ with $s = 0.02$. Does this mean that this analytical method is inappropriate or is this difference acceptable with the required certainty due to the randomness of the sampling? A similar question can arise when comparing two materials. The first material was measured n_1 times and yielded a mean $\bar{x}_1$ with a standard deviation s_1; the appropriate values for the second material are n_2, $\bar{x}_2$ and s_2. The means $\bar{x}_1$ and $\bar{x}_2$ are different from each other, but a question that remains is if this difference is significant or only due to the randomness of the associated errors. It should be remembered that always when we ask if a difference is significant it has to be said at what level of certainty we want to know it. The testing of whether differences can be accounted for by the random errors is done by statistical tests known as **significance tests**. In making a significance test, we are testing the truth of a hypothesis, called the **null hypothesis**. This hypothesis assumes that there is no difference (null difference) between the two values we are comparing. Assuming that the null hypothesis is true, we can use statistical methods to calculate the probability that the observed difference arises solely as a result of random variations. There are two kinds of errors in these statistical decisions of rejecting or accepting the null hypothesis. If we reject a hypothesis when it should be accepted, that is an **error of type I**. An **error of type II** is made when we accept a hypothesis when it should be rejected. In testing a hypothesis, the maximum probability with which we would be willing to risk a type I error is called the **level of significance** of the test. In practice, 0.05 is taken as the common level of significance, so there is about 5% probability (5 chances in 100) that we would reject the hypothesis when it should be accepted. In other words there is 95% confidence that we have made the right decision. In this case the common language is to say that the hypothesis has been rejected at the 0.05 level of significance. The statistical meaning of level of significance α, with its critical value for the statistic (for example the t statistic) t_c, means that in the range $-t_c < t < t_c$ there is $100(1-\alpha)$% of the population. When comparing two values each of which can be larger than the other, so that the measured values can be on both sides of the mean, we call it a **two-tailed test** or a **two-sided test**. However, sometimes we are interested only in extreme values to one side of the mean, i.e. only in one 'tail' of the distribution. Such a test is called a **one-tailed test** or **one-sided test**. When using statistical tables for critical values, attention should be paid to whether the table is for a one- or two-tailed test. If the table is for a one-tailed test, a level of 2α (α level of significance) should be used for a two-tailed test; and if a table for a two-tailed test is used for a one-tailed test, then the critical value for $\alpha/2$ should be taken.

3.4.2 Comparison of an experimental mean with an expected value (standard)

One of the important steps in the development of a new analytical method (for example, a new method of wet chemistry for separation and enrichment or a new instrumental method) is the analysis of certified reference materials. There are certified reference materials of different kinds of matrices, e.g. geological, biological (either botanical or from animals), etc., prepared in large amounts, homogenized well to ensure homogeneity and analyzed by several laboratories for their constituents. These standard materials can be purchased from several sources, e.g. The National Institute of Standards and Technology in the USA. Repeating the measurement of the reference material by the new method (instrument)

n times yields a mean $\bar{x}$ and a standard deviation s. The certified value, assumed to be true, is μ. How do we know if the difference $|\bar{x} - \mu|$ is significant, which means a systematic error in the new method, or insignificant and the new method is a reliable one?

In order to decide whether the difference between μ and $\bar{x}$ is significant, we use equation (3.11) and rewrite it:

$$\mu = \bar{x} \pm ts/\sqrt{n} \quad \Rightarrow \quad t = |(\bar{x} - \mu)|\sqrt{n}/s \qquad (3.13)$$

From the experimental results t is calculated. If the value of t exceeds a certain value, found in tables (Table 3.1), then the null hypothesis is rejected, which means that $\bar{x}$ and μ are really different. If t is less than the value in the table, then the null hypothesis that $\bar{x}$ and μ are the same within randomness is accepted.

Example 3.1

A new method for measuring aluminum in biological samples was tested on a certified standard material of plant origin with a certified value of 1.37 ppm (μg/g). The five measurements gave a mean value of 1.25 ppm with a standard deviation of 0.12 ppm. Is the new method a good one, i.e. is there no systematic error in it? Calculating t according to equation (3.13) we obtain

$$t = |1.37 - 1.25|\sqrt{5}/0.12 = 2.24$$

From Table 3.1 it is seen that for d.f. = 4 (d.f. = $n - 1$) the critical value of t for 0.05 significance level is 2.78. Hence the null hypothesis is retained at this level of significance and the new method was not demonstrated to have a systematic error at 0.05 level of significance. We should note that if we want the level of significance to be 0.1 instead of 0.05, then $t_{\text{critical}} = 2.13$ and the data already indicate systematic errors. The same will be true if $\bar{x}$ and s were obtained from 15 experiments, since t_{critical} at the 0.05 level for d.f. = 14 is 2.14. Actually, equation (3.13) did not have to be used, and many authors suggest to use directly equation (3.11). Thus from equation (3.11) we get for the 95% confidence limit

$$\mu = 1.25 \pm 2.78 \times 0.12/\sqrt{5} \quad \Rightarrow \quad \mu = 1.25 \pm 0.149$$

Since the expected value is within the confidence interval (between the confidence limits), the experimental method is not demonstrated to have systematic errors. However, the use of equation (3.13) and calculation of t has the advantage that it shows up to what probability the null hypothesis is accepted by comparing the calculated t with the values in Table 3.1.

3.4.3 *Comparison of two samples*

This test will be used to check if two different methods give the same results within our level of significance, by comparing their means and standard deviations (in the previous section the expected value was assumed to be accurate with zero standard deviation and it actually can be considered as a special case of the more general one treated in this section) or to compare if two samples measured by the same method are equal within our significance level.

If the first sample (or method) has n_1 repeated experiments with a mean $\bar{x}_1$ and standard deviation s_1 and the second sample has the same with subscript 2, then equation (3.13) for t is replaced by

$$s^2 = s_1^2/n_1 + s_2^2/n_2 \qquad (3.14)$$

$$t = (\bar{x}_1 - \bar{x}_2)/s \qquad (3.15)$$

and the number of degrees of freedom is given by rounding the next result to the nearest integer

$$\text{d.f.} = \frac{s^4}{s_1^4/[n_1^2(n_1 + 1)] + s_2^4[n_2^2(n_2 + 1)]} - 2 \qquad (3.16)$$

In the case in which s_1 and s_2 can be assumed to be close to each other, as e.g. in two samples measured by the same method in the same laboratory, simpler equations can be used:

$$s^2 = [(n - 1)s_1^2 + (n_2 - 1)s_2^2]/(n_1 + n_2 - 2) \qquad (3.17)$$

$$\text{d.f.} = n_1 + n_2 - 2 \qquad (3.18)$$

$$t = (\bar{x}_1 - \bar{x}_2)/s\sqrt{1/n_1 + 1/n_2} \qquad (3.19)$$

Equation (3.17) is obtained by adding the squares of the differences from the means of the two samples:

$$s^2 = \frac{\sum (x_i - x)^2}{n - 1} \quad \Rightarrow \quad \sum (x_i - x)^2 = (n - 1)s^2$$

The degrees of freedom (number of independent variables) is the sum of the degrees of freedom of the two samples.

The testing of t with the values in Table 3.1 is done in the same fashion as in the previous section.

The cooperative standard deviation calculated by equation (3.16) is called the **estimated pooled standard deviation** and it is used to estimate the total standard deviation from several samples (not limited to two as in equation (3.16)), each containing n_i results.

Example 3.2
Two methods have been used to measure the concentration of fluoride in water collected in a well. The following results, all in ppm (μg/L), were obtained:

Method I	4.44	4.24	4.17	4.35	4.02	3.89	4.11
Method II	3.95	3.89	4.20	4.05	3.97	3.75	4.21

Do the two methods give results that differ significantly?

Let us calculate for each method the mean and the standard deviation: $\bar{x}_1 = 4.17$, $s_1 = 0.19$ and $\bar{x}_2 = 4.00$, $s_2 = 0.17$ (n_1 and n_2 are equal in this case, but they can be different). Using equations (3.14)–(3.16) gives: $s^2 = 0.009286$, $t = 1.76$, d.f. $= 14$. If we use equation (3.17)–(3.19) we will get $s^2 = 0.0325$, $t = 1.76$, d.f. $= 12$. So the two sets of equations give the same value of t but different degrees of freedom. Using Table 3.1 for d.f. $= 14$ we obtain that the two methods give the same results not only with significance level of 0.05, but almost at significance level of 0.1.

Sometimes two methods are tested not several times on the same sample (or aliquots of the same sample) but several different samples are measured by each of the analytical methods. In this case there is no meaning for the mean or the standard deviation in each method since the samples (the repeated measurements) are different and each has its own value. In this case we will apply the paired t-test described in the following section.

3.4.4 *Paired t-test*

In this case equal numbers of measurements must be done in both methods, since each sample is measured in both methods. In this case we calculate the mean of the differences between the two methods for each of the samples, $\bar{x}_d$, and the standard deviation of the differences, s_d. The statistic t is calculated by the equation

$$t = \bar{x}_d \sqrt{n}/s \tag{3.20}$$

where n is the number of pairs (samples measured by both methods) and the degrees of freedom is d.f. $= n - 1$.

Example 3.3
Let us look at the concentration of chlorides (in ppm) in rain water falling at different days in January, each day measured once by each method:

Method I	4.44	3.84	3.14	4.08	3.92	3.14
Method II	5.00	3.90	3.32	4.26	4.14	3.36

Do the results obtained by the two methods differ significantly?

If we use the methods of section 3.4.3 we will find: $\bar{x}_1 = 3.76$, $s_1 = 0.52$ and $\bar{x}_2 = 4.00$, $s_2 = 0.63$. Equations (3.17)–(3.19) will give $s^2 = 0.33$, d.f. = 10, $t = 0.72$. Equations (3.14)–(3.16) will give $s^2 = 0.11$, $t = 0.72$, d.f. = 11. (So the same t was obtained by both calculations but the d.f. differ by 1.) Referring to Table 3.1 we can see that the results are the same even for a level of significance of 0.1, but there are no meanings to the value of $\bar{x}$ and s as the samples are different and the concentration can be different each day. The differences between the methods for the various days are: 0.56, 0.06, 0.18, 0.18, 0.22 and 0.22, which gives $\bar{x}_d = 0.24$ and $s_d = 0.17$, yielding according to equation (3.20) $t = 3.46$ with d.f. = 5. According to Table 3.1, the difference will not be significant only at the level of 0.01. As the common requirement is a level of significance of 0.05, we must conclude that the two methods give different results.

It should be emphasized that this method of paired t-test assumes that any errors, either random or systematic, are independent of concentration. Over a wide range of concentration this assumption may no longer be true, and hence this method should not be used over too large ranges of concentrations. When a new method is tested it should be tested over the whole range for which it is applicable. In order to perform this comparison, several samples with various concentrations should be prepared, and for each one several repeated measurements should be performed by each method, applying the usual t-test for means (section 3.4.3) for each concentration.

Another method that is the preferred one in analyzing over a wide range of concentration is via linear regression, which will be treated in section 3.5.

3.4.5 Comparing two variances – the F *test*

The previous tests examined the significance of variation of two means using their standard deviations. These tests assumed that the two samples are distributed normally with the same σ (population standard deviation). The t-test was used rather than the normal distribution because of the small sizes of the samples that are usual in analytical chemistry. These tests for comparison of the means assuming the same σ are used for detecting systematic errors. Yet, sometimes different methods have different precisions – i.e. different standard deviation or, as actually used, squares of the standard deviation s^2, called the **variance**. The comparison of two variances takes two forms depending on whether we ask if method A is better than method B (neglecting the possibility that B is better than

A), i.e. a one-tailed test, or whether we allow also the probability that method B is better and ask only if the two methods differ in their precision, i.e. a two-tailed test. Thus if we want to test whether a new analytical method is preferable to a standard one, we use a one-tailed test, because unless we prove that the new method is a better one we will stick to the old method. When we are comparing two new methods or two standard methods, a two-tailed test is the right one.

In this test we calculate a quantity called the *F*-statistic, which is given by the ratio of the two variances:

$$F = s_1^2/s_2^2 \tag{3.21}$$

The indices 1 and 2 are chosen so that F is always equal to or greater than 1 ($s_1^2 \geq s_2^2$). The calculated F are checked against the values in Table 3.2. If the calculated value is larger than that found in the table, the null hypothesis that the two methods are not different in their precision is rejected.

Example 3.4

A standard method for measuring copper ions in drinking water was repeated eight times and found to have a mean of 4.52 μg/mL with standard deviation of 0.253 μg/mL. Seven measurements of a new method yield the same mean but with standard deviation of 0.207 μg/mL. On a first consideration the new method seems to be better, more precise, since the standard deviation is lower but is this difference significant or it is just a random one?

For testing this we calculate the *F*-statistic. Since only one method is a standard one, a one-tailed test should be used. The degrees of freedom of the larger s is 7 (= 8 − 1) and that of the smaller s is 6, so we write this F as $F_{7,6}$. The first number is the d.f. for the numerator and the second for the denominator,

$$F_{7,6} = \frac{0.253^2}{0.207^2} = 1.4938$$

From Table 3.2 we get $F_{7,6} = 4.207$, for 0.05 level of significance. Since the calculated value is lower than the critical value, the null hypothesis cannot be rejected and hence it was not proven that the new method is better than the standard one. The lower standard deviation can be, at this level of certainty that we might be wrong in 5% (1 from 20) cases, due completely to a random origin. However, if the new method had a standard deviation of 0.113, then we calculate $F_{7,6} = 5.01$, and since this value is larger than the critical value, we can say that the new method is more precise than the standard one.

Table 3.2 Critical values of the F-statistic for a significance level of $p = 0.05$

(a) One-tailed test

	d.f.$_{1}$												
d.f.$_{2}$	1	2	3	4	5	6	7	8	9	10	12	15	20
1	161.4	199.5	215.7	224.6	230.2	234.0	236.8	238.9	240.5	241.9	243.9	245.9	248.0
2	18.51	19.00	19.16	19.25	19.30	19.33	19.35	19.37	19.38	19.40	19.41	19.43	19.45
3	10.13	9.552	9.277	9.117	9.013	8.941	8.887	8.845	8.812	8.786	8.745	8.703	8.660
4	7.709	6.944	6.591	6.388	6.256	6.163	6.094	6.041	5.999	5.964	5.912	5.858	5.803
5	6.608	5.786	5.409	5.192	5.050	4.950	4.876	4.818	4.772	4.735	4.678	4.619	4.558
6	5.987	5.143	4.757	4.534	4.387	4.284	4.207	4.147	4.099	4.060	4.000	3.938	3.874
7	5.591	4.737	4.347	4.120	3.972	3.866	3.787	3.726	3.677	3.637	3.575	3.511	3.445
8	5.318	4.459	4.066	3.838	3.687	3.581	3.500	3.438	3.388	3.347	3.284	3.218	3.150
9	5.117	4.256	3.863	3.633	3.482	3.374	3.293	3.230	3.179	3.137	3.073	3.006	2.936
10	4.965	4.103	3.708	3.478	3.326	3.217	3.135	3.072	3.020	2.978	2.913	2.845	2.774
11	4.844	3.982	3.587	3.357	3.204	3.095	3.012	2.948	2.896	2.854	2.788	2.719	2.646
12	4.747	3.885	3.490	3.259	3.106	2.996	2.913	2.849	2.796	2.753	2.687	2.617	2.544
13	4.667	3.806	3.411	3.179	3.025	2.915	2.832	2.767	2.714	2.671	2.604	2.533	2.459
14	4.600	3.739	3.344	3.112	2.958	2.848	2.764	2.699	2.646	2.602	2.534	2.463	2.388
15	4.543	3.682	3.287	3.056	2.901	2.790	2.707	2.641	2.588	2.544	2.475	2.403	2.328
16	4.494	3.634	3.239	3.007	2.852	2.741	2.657	2.591	2.538	2.494	2.425	2.352	2.276
17	4.451	3.592	3.197	2.965	2.810	2.699	3.614	2.548	2.494	2.450	2.381	2.308	2.230
18	4.414	3.555	3.160	2.928	2.773	2.661	2.577	2.510	2.456	2.412	2.342	2.269	2.191
19	4.381	3.522	3.127	2.895	2.740	2.628	2.544	2.477	2.423	2.378	2.308	2.234	2.155
20	4.351	3.493	3.098	2.866	2.711	2.599	2.514	2.447	2.393	2.348	2.278	2.203	2.124

(b) Two-tailed test

	d.f.$_{1}$												
d.f.$_{2}$	1	2	3	4	5	6	7	8	9	10	12	15	20
1	647.8	799.5	864.2	899.6	921.8	937.1	948.2	956.7	963.3	968.6	976.7	984.9	993.1
2	38.51	39.00	39.17	39.25	39.30	39.33	39.36	39.37	39.39	39.40	39.41	39.43	39.45
3	17.44	16.04	15.44	15.10	14.88	14.73	14.62	14.54	14.47	14.42	14.34	14.25	14.17
4	12.22	10.65	9.979	9.605	9.364	9.197	9.074	8.980	8.905	8.844	8.751	8.657	8.560
5	10.01	8.434	7.764	7.388	7.146	6.978	6.853	6.757	6.681	6.619	6.525	6.428	6.329
6	8.813	7.260	6.599	6.227	5.988	5.820	5.695	5.600	5.523	5.461	5.366	5.269	5.168
7	8.073	6.542	5.890	5.523	5.285	5.119	4.995	4.899	4.823	4.761	4.666	4.568	4.467
8	7.571	6.059	5.416	5.053	4.817	4.652	4.529	4.433	4.357	4.295	4.200	4.101	3.999
9	7.209	5.715	5.078	4.718	4.484	4.320	4.197	4.102	4.026	3.964	3.868	3.769	3.667
10	6.937	5.456	4.826	4.468	4.236	4.072	3.950	3.855	3.779	3.717	3.621	3.522	3.419
11	6.724	5.256	4.630	4.275	4.044	3.881	3.759	3.664	3.588	3.526	3.430	3.330	3.226
12	6.554	5.096	4.474	4.121	3.891	3.728	3.607	3.512	3.436	3.374	3.277	3.177	3.073
13	6.414	4.965	4.347	3.996	3.767	3.604	3.483	3.388	3.312	3.250	3.153	3.053	2.948
14	6.298	4.857	4.242	3.892	3.663	3.501	3.380	3.285	2.209	3.147	3.050	2.949	2.844
15	6.200	4.765	4.153	3.804	3.576	3.415	3.293	3.199	3.123	3.060	2.963	2.862	2.756
16	6.115	4.687	4.077	3.729	3.502	3.341	3.219	3.125	3.049	2.986	2.889	2.788	2.681
17	6.042	4.619	4.011	3.665	3.438	3.277	3.156	3.061	2.985	2.922	2.825	2.723	2.616
18	5.978	4.560	3.954	3.608	3.382	3.221	3.100	3.005	2.929	2.866	2.769	2.667	2.559
19	5.922	4.508	3.903	3.559	3.333	3.172	3.051	2.956	2.880	2.817	2.720	2.617	2.509
20	5.871	4.461	3.859	3.515	3.289	3.128	3.007	2.913	2.837	2.774	2.676	2.573	2.464

When using two methods of the same level – either both standards or both new – we will use a two-tailed test.

3.4.6 *Comparison of several means*

If, for example, we want to measure the effect of various conditions of storage (e.g. effect of temperature, kind of material from which the containers are made, ratio of surface area to volume of the containers), then we will take from each container n samples and analyze them to have mean ($\bar{x}_i$) and standard deviation (s_i), $i = 1, 2, \ldots, k$, where k is the number of different containers. In order to see if the containment conditions influence the chemical in question, we should compare the data obtained. This is done in the following steps:

1. Calculation of the *within-sample* estimate of the variance, V_{ws}: we have

$$\text{averaged variance} = \sum_{i=1}^{k} s_i^2/k \quad \Rightarrow \quad V_{ws} = n\sum_{i=1}^{k} s_i^2/k \tag{3.22}$$

   i.e. V_{ws} is the product of the size of each sample in the arithmetic average of the k values of s_i^2. The degrees of freedom is the sum for all the k samples, i.e. d.f. $= k(n-1)$.
2. Calculation of the *between-sample* estimate of the variance, V_{bs}: The total mean is calculated as the arithmetic average of the individual means,

$$\bar{x} = \sum_{i=1}^{k} \bar{x}_i/k$$

   The variance of the different individual means is calculated by the equation:

$$\text{variance of the means} = \sum_{i=1}^{k} (\bar{x}_i - \bar{x})^2/(k-1)$$

   V_{bs} is calculated by multiplying the variance of the means by the number of measurements in each sample, n. The d.f. is $k - 1$. So

$$V_{bs} = n\sum_{i=1}^{k} (\bar{x}_i - \bar{x})^2/(k-1) \tag{3.23}$$

3. These two variances are used to calculate the F-statistic

$$F_{k-1,\,k(n-1)} = \frac{V_{ws}}{V_{bs}} \tag{3.24}$$

4. If the calculated F is larger than the critical value (one-tailed test) then the means are significantly different and the containment conditions

are important. If the calculated value is lower than that from the table, the means are not different at that level of significance.

Example 3.5

Six containments are tested. From each, five samples were measured and the data are given below:

$\bar{x}$	100	104	96	106	98	103
s	2.5	4	3.5	5.1	3	5

Thus we find that:

averaged variance $= (2.5^2 + 4^2 + 3.5^2 + 5.1^2 + 3^2 + 5^2)/6 = 15.75$

$$V_{ws} = 5 \times 15.75 = 78.75$$

averaged mean $= 101.17$

variance of the mean $= 3.817$

$$V_{bs} = 5 \times 3.817 = 19.08$$

$$F_{5,24} = \frac{78.75}{19.08} = 4.127$$

We do not have $F_{5,25}$ in the table but it has to be lower than $F_{5,20}$, which is 2.711. Since the calculated F is larger than the critical one, the means are different from each other.

3.4.7 *Separation of sources of variances by analysis of variances (ANOVA)*

In the previous section we saw that **analysis of variances** (usually called by the acronym ANOVA) can show if several means differ significantly or not. The same analysis can be used also to separate the variance into its sources, into the part that is due to the random error of measurement and the part that is due to difference between the samples. Before continuing, it is desirable to show how the calculation of ANOVA (as in the previous section) can be simplified.

Let us remember that we have k samples, each consisting of n measurements with a mean $\bar{x}_i$ and standard deviation s. We calculated two kinds of variances, between-sample and within-sample. Let us assign the total number of measurements $k \times n$ by N, i.e. $N = kn$. The sum of all measurements in sample i is

$$T_i = \sum_{j=1}^{n} x_{ij}$$

and the sum of all measurements is

$$T = \sum_{i=1}^{k} T_i = \sum_{i=1}^{k} \sum_{j=1}^{n} x_{ij}$$

It can be shown that

$$V_{bs} = (1/n)\sum_{i=1}^{k} T_i^2 - T^2/N \qquad \text{(d.f.} = k-1) \qquad (3.25)$$

$$V_{ws} = \sum_{i=1}^{k}\sum_{j=1}^{n} x_{ij}^2 - (1/n)\sum_{i=1}^{k} T_i^2 \qquad \text{(d.f.} = N-k) \qquad (3.26)$$

Example 3.6

Let us now look at five containers with powdered coal ash. From each container, five samples are taken and measured for their iron content (in %):

Container A	Container B	Container C	Container D	Container E
1.78	1.81	1.78	1.84	1.76
1.76	1.80	1.80	1.80	1.79
1.75	1.79	1.76	1.83	1.74
1.76	1.83	1.77	1.79	1.73
1.80	1.82	1.79	1.82	1.78
Mean 1.77	1.81	1.78	1.82	1.76

We will assume that the random error of measurement is normally distributed with a variance σ_0^2 and that the concentration between the containers is also normally distributed with a variance σ_1^2. The means cannot tell us about σ_1^2 since they involve both variances. However, V_{ws} can yield σ_0^2 directly. It can be shown that $V_{bs} = \sigma_0^2 + \sigma_1^2$. If it can be shown that V_{ws} and V_{bs} do not differ significantly, then we can assume σ_1 to be zero. Otherwise we will get

$$\sigma = V_{ws} \qquad \sigma_1^2 = (1/n)(V_{bs} - V_{ws}) \qquad (3.27)$$

V_{bs} and V_{ws} can be calculated from equations (3.25) and (3.26), but it is easier not to treat the original x_{ij} but rather after subtracting a constant number we choose from all of them. Thus if we choose the constant subtracted number as 1.80 the values are:

						T_i	T_i^2	$\sum x_{ij}^2$
A	−0.02	−0.04	−0.05	−0.04	0	−0.15	0.0225	0.0061
B	0.01	0	−0.01	0.03	0.02	0.05	0.0025	0.0015
C	−0.02	0	−0.04	−0.03	−0.01	−0.1	0.01	0.0030
D	0.04	0	0.03	−0.01	0.02	0.08	0.0064	0.0030
E	−0.04	−0.01	−0.06	−0.07	−0.02	−0.20	0.0400	0.0106

$$T = \sum T_i = -0.32$$
$$\sum T_i^2 = 0.0814$$
$$\sum x_{ij}^2 = 0.0242$$
$$n = 5 \qquad k = 5 \qquad N = 5 \times 5 = 25$$

Thus we obtain

$$V_{bs} = \frac{1}{5} \times 0.0814 - \frac{1}{25} \times 0.1024 = 0.01218 \qquad (\text{d.f.} = 5 - 1 = 4)$$

$$V_{ws} = 0.0242 - \frac{1}{5} \times 0.0814 = 0.00792 \qquad (\text{d.f.} = 25 - 5 = 20)$$

$$F_{4,20} = \frac{0.01218}{0.00792} = 1.533$$

From Table 3.2 the critical value at the level of $p = 0.05$ is 3.056. Since the calculated F is less than the tabulated one, then $\sigma_1 \neq 0$.

3.4.8 *The chi-squared (χ^2) test*

This test is concerned with the frequency with which various results are obtained. Thus it can be used to test if obtained data are distributed normally or not. The quantity χ^2 is calculated according to the equation

$$\chi^2 = \sum_{i=1}^{n} \frac{(O_i - E_i)^2}{E_i} \tag{3.28}$$

where O_i is the observed frequency and E_i is the expected frequency, as for example those expected from a normal distribution or other expectation, and n is the number of different values. The null hypothesis is that there is no difference between the expected and observed frequencies. If the calculated χ^2 is larger than the critical value, given in Table 3.3, then

Table 3.3 Critical values of χ^2 for level of significance 0.05

Number of degrees of freedom	Critical value
1	3.84
2	5.99
3	7.81
4	9.49
5	11.07
6	12.59
7	14.07
8	15.51
9	16.92
10	18.31

the null hypothesis is rejected. The critical value depends as in other cases on the significance level of the test and the number of degrees of freedom $= n - 1$.

3.4.9 Testing for normal distribution – probability paper

In the previous section we saw that the χ^2 test can be used to test if the measurements are distributed normally by comparing the expected frequency from a normal distribution to the observed frequency for each value (or interval). Usually the χ^2 test should not be applied for samples smaller than 50. Another way to test for normal distribution is by plotting on **probability paper**. This is, in principle, the same as semi-logarithmic paper. On semi-logarithmic paper the x-axis is linear while the y-axis is logarithmic, i.e. the distance between 1 and 2 (ratio 2) is equal to the distance between 3 and 6 or between 4 and 8, or the distance between 1 and 10 is equal to the distance between 10 and 100. On probability paper the distance is plotted according to the area under the normal curve.

The data are organized according to their values in increasing order starting from the lower value. Then their cumulative frequency is calculated by addition. If we plot the cumulative frequency vs. the value on ordinary graph paper, we will obtain a graph with an S shape. However, if we plot it on probability paper, a straight line should be obtained. So the check for the linearity of this line is a check for its normal distribution.

3.4.10 Outliers

An important question every experimentalist meets quite often concerns one or two data points that deviate greatly from the others and hence influence the standard deviation quite strongly although not so much the mean. The tendency is usually to say that some exceptional error was involved in the measurement and to discard such points. However, is this justified? Perhaps the deviation is just due to random error. We refer to measurements like these as **outliers** and this section will discuss the question of whether we can disregard these outliers. Outliers can result, for example, from blunders or malfunctions of the methodology or from unusual losses or contaminations. Assume that we measure aliquots from the same sample and obtain the values for manganese content as 5.65, 5.85, 5.94, 5.73, 6.52, 5.80. At first glance we will tend to reject the fifth result and use only the other five in order to calculate the mean and the standard deviation. Is this rejection justified? We will test this rejection assuming that the whole population is normally distributed. There are several tests for outliers. The most common one is Dixon's Q test.

Table 3.4 Values for use in Dixon's Q test for outliers

Number of observations, n	Risk of false rejection			
	0.5%	1%	5%	10%
3	0.994	0.988	0.941	0.886
4	0.926	0.889	0.765	0.679
5	0.821	0.780	0.642	0.557
6	0.740	0.698	0.560	0.482
7	0.680	0.637	0.507	0.434
8	0.725	0.683	0.554	0.479
9	0.677	0.635	0.512	0.441
10	0.639	0.597	0.477	0.409
11	0.713	0.679	0.576	0.517
12	0.675	0.642	0.546	0.490
13	0.649	0.615	0.521	0.467
14	0.674	0.641	0.546	0.492
15	0.647	0.616	0.525	0.472
16	0.624	0.595	0.507	0.454
17	0.605	0.577	0.490	0.438
18	0.589	0.561	0.475	0.424
19	0.575	0.547	0.462	0.412
20	0.562	0.535	0.450	0.401

Dixon's Q test. The test statistic Q is calculated from the equation

$$Q = |\text{suspected value} - \text{nearest value}|/(\text{largest value} - \text{smallest value}) \tag{3.29}$$

i.e. in our case

$$Q = (6.52 - 5.94)/(6.52 - 5.65) = 0.666$$

Table 3.4 gives the value of the critical Q. For a sample size of 6, the critical value of Q is 0.621, so it means that our Q exceeds the critical value and the value of 6.52 should be rejected. If we did not perform the last experiment, we will have the same Q, but since the sample size is 5, the critical value is 0.717, and thus $Q < Q_{\text{critical}}$ and hence the 6.52 measurement should be retained.

Actually the Q calculated by equation (3.29) is correct only for sample size n in the range $3 \leq n \leq 7$. For larger sizes of sample, other ratios are calculated. If we arrange the x_i in either increasing or decreasing order, such that the outlier is the last one x_n, the following ratios are calculated:

$$3 \leq n \leq 7 \qquad Q = |(x_n - x_{n-1})/(x_n - x_1)|$$

$$8 \leq n \leq 10 \qquad Q = |(x_n - x_{n-1})/(x_n - x_2)|$$

$$11 \leq n \leq 13 \qquad Q = |(x_n - x_{n-2})/(x_n - x_2)|$$

$$14 \leq n \leq 25 \qquad Q = |(x_n - x_{n-2})/(x_n - x_3)|$$

However, n is rarely larger than 7, and equation (3.29) is commonly used.

Grubb's test for outliers. The following procedure should be followed:

1. Rank the data in either ascending or descending order such that x_n is the outlier, i.e. if the outlier is the largest then $x_n > x_{n-1} > \ldots > x_1$, but if the smallest is the possible outlier then $x_n < x_{n-1} < \ldots < x_2 < x_1$.
2. Calculate the mean and the standard deviation of all the data points.
3. Calculate the statistic T from the equation

$$T = |x_n - \bar{x}|/s \tag{3.30}$$

4. Compare the calculated T with the critical values given in Table 3.5.

For the case we treat in the previous test, we will have $\bar{x} = 5.915$, $s = 0.313$ and hence

$$T = \frac{6.52 - 5.915}{0.313} = 1.933$$

For $p = 0.05$, this value is larger than the critical value (1.822) in Table 3.5 and hence should be rejected.

Youden test for outlying laboratories. Youden (1982) developed a statistical test applicable to a group of laboratories analyzing the same samples, as e.g. for the purpose of certifying the contents of standard reference materials, in order to reject the data of laboratories that are outliers. Each laboratory gets several different samples to analyze. For each sample the laboratories are ranked such that the one obtaining the

Table 3.5 Values for Grubb's T test for outliers

Number of data points, n	Risk of false rejection				
	0.1%	0.5%	1%	5%	10%
3	1.155	1.155	1.155	1.153	1.148
4	1.496	1.496	1.492	1.463	1.425
5	1.780	1.764	1.749	1.672	1.602
6	2.011	1.973	1.944	1.822	1.729
7	2.201	2.139	2.097	1.938	1.828
8	2.358	2.274	2.221	2.032	1.909
9	2.492	2.387	2.323	2.110	1.977
10	2.606	2.482	2.410	2.176	2.036
15	2.997	2.806	2.705	2.409	2.247
20	3.230	3.001	2.884	2.557	2.385
25	3.389	3.135	3.009	2.663	2.486
50	3.789	3.483	3.336	2.956	2.768
100	4.084	3.754	3.600	3.207	3.017

Table 3.6 Values for Youden's range to identify outlying laboratories (5% two-tailed limits)

Number of participants	Number of materials												
	3	4	5	6	7	8	9	10	11	12	13	14	15
3		4	5	7	8	10	12	13	15	17	19	20	22
		12	15	17	20	22	24	27	29	31	33	36	38
4		4	6	8	10	12	14	16	18	20	22	24	26
		16	19	22	25	28	31	34	37	40	43	46	49
5		5	7	9	11	13	16	18	21	23	26	28	31
		19	23	27	31	35	38	42	45	49	52	56	59
6	3	5	7	10	12	15	18	21	23	26	29	32	35
	18	23	28	32	37	41	45	49	54	58	62	66	70
7	3	5	8	11	14	17	20	23	26	29	32	36	39
	21	27	32	37	42	47	52	57	62	67	72	76	81
8	3	6	9	12	15	18	22	25	29	32	36	39	43
	24	30	36	42	48	54	59	65	70	76	81	87	92
9	3	6	9	13	16	20	24	27	31	35	39	43	47
	27	34	41	47	54	60	66	73	79	85	91	97	103
10	4	7	10	14	17	21	26	30	34	38	43	47	51
	29	37	45	52	60	67	73	80	87	94	100	107	114
11	4	7	11	15	19	23	27	32	36	41	46	51	55
	32	41	49	57	65	73	81	88	96	103	110	117	125
12	4	7	11	15	20	24	29	34	39	44	49	54	59
	35	45	54	63	71	80	88	96	104	112	120	128	136
13	4	8	12	16	21	26	31	36	42	47	52	58	63
	38	48	58	68	77	86	95	104	112	121	130	138	147
14	4	8	12	17	22	27	33	38	44	50	56	61	67
	41	52	63	73	83	93	102	112	121	130	139	149	158
15	4	8	13	18	23	29	35	41	47	53	59	65	71
	44	56	67	78	89	99	109	119	129	139	149	159	169

highest value is ranked 1, and the second highest 2, etc. At the end the different rankings of each laboratory are summed up. The expected ranges for the cumulative scores, for a random distribution, at 0.05 level of significance (95% confidence), are given in Table 3.6. Any laboratory that is outside of this range is treated as an outlier.

Example 3.7
Let us look at the case of seven laboratories each analyzing five different samples of ores for their copper content (in %), with the results shown below:

Laboratory	Measured results					Score					Sum of scores
	1	2	3	4	5	1	2	3	4	5	
A	9.8	15.8	21.3	18.6	24.9	4	2	3	4	5	18
B	10.0	16.1	20.5	18.7	25.3	2	4	7	3	1	17
C	10.2	17.2	29.7	18.8	25.2	1	1	1	2	2	7
D	9.7	15.9	21.5	18.9	25.0	5	3	2	1	4	15
E	9.6	16.8	20.7	18.3	25.1	6	6	5	5	3	25
F	9.9	16.3	20.8	18.1	24.8	3	5	4	7	6	25
G	9.3	15.4	20.6	18.2	24.0	7	7	6	6	7	33

Table 3.6 shows that with 95% confidence limit the range is expected to be within 8 to 32. Accordingly laboratory C is considered to provide higher results and laboratory G lower results than the other laboratories. In calculating the certified value of a reference standard material, the results of these laboratories will be rejected, at least for Cu.

3.4.11 *Non-parametric analysis*

All the previously mentioned statistical methods have one basic assumption, that the population (or at least the means of small samples) are normally distributed (Gaussian distribution). However, for some cases it was proven that this assumption is wrong. Besides, for cases where the sample size is small ($n = 3$–4) this assumption is not justified. In these cases a different kind of statistical test should be used. Statistical tests that make no assumption about the shape of the distribution of the population from which the data are taken are referred to as **non-parametric tests**. Two other aspects that lead to the popular use of these methods in the social sciences are their simplicity and very small amount of calculation, compared to those involved in the calculation of standard deviations in parametric tests. However, now that statistical programs are available on every spread-sheet program on almost all PCs and scientific calculators can easily calculate the mean and the standard deviation of a sample of data, this point of speedier analysis has lost most of its significance. Just in order to see the simplicity of these methods we will show the use of two of these methods, the sign test and the Wald–Wolfowitz runs test.

While parametric analysis tests the mean of the sample, non-parametric analysis involves the median. While the value of the mean involves some calculation $\bar{x} = \sum x_i/n$, the finding of the median does not require any calculation, only ranking of the data points according to either decreasing or increasing order. The **median** is defined as the number for which half of the data points have larger (or smaller) or equal value. Thus if η is the median, then its definition is $p(x_i \leq \eta) = 0.5$. For n data points ranked in a monotonic direction, the median is the value of the $0.5(n + 1)$th observation if n is odd and the average of the 0.5nth and $(0.5n+1)$th observations for even n.

The sign test. Assume that there are eight observations which are (arranged by descending values) 6, 5, 4, 3, 3, 3, 3, 2. Since there are eight observations, the median is the average of the fourth and fifth observations and hence the median is 3. Now we subtract the median from all values and check the number of plus signs and minus signs (neglecting all differences that are equal to zero, i.e. all the values that are equal to the median). Thus we have three plus signs and one minus sign. To test the significance of this observation (three plus signs out of four – as we neglect those equal to the median in calculation of the signs but not in the calculation of the median) we use the binomial distribution. From this we have Bernoulli's equation for the probability of k successes in n trials where the probability of a success in a single experiment is p, H_n^k being this probability:

$$H_n^k = C_{n}^kp^k(1-p)^n \tag{3.31}$$

$$\text{the binomial coefficient } C_n^k = \frac{n!}{k!(n-k)!} \tag{3.32}$$

There are equal chances of a plus sign as of a minus sign and hence $p = 0.5$. Thus the probability of obtaining three plus signs out of four is

$$H_4^3 = C_4^3 \times 0.5^3 \times 0.5^1 = 4 \times 0.0625 = 0.25$$

When we check the significance of the difference, we are checking the tail of the distribution, i.e. we check the probability of obtaining three or more equal signs. So we must calculate also the probability of obtaining four plus signs out of four:

$$H_4^4 = C_4^4 \times 0.5^4 \times 0.5^0 = 0.0625$$

Thus the probability of obtaining three or more plus signs is $0.25 + 0.0625 = 0.3125$. However, the test should be two-tailed, i.e. what are the probabilities of obtaining three or more equal signs (either plus or minus) out of four. Since the probability of plus is the same as of minus, the total probability is $2 \times 0.3125 = 0.625$. This probability is much higher than our common level of significance, $p = 0.05$, and hence we cannot reject our null hypothesis, i.e. that the data come from a population with median = 3. The binomial probability does not have to be calculated, as was done here, and it can be taken from Table 3.7.

Parametric analysis will treat this by the t-test. The eight data points will yield $\bar{x} = 3.625$ and $s = 1.302$. So

$$t = \frac{|3.625 - 3|\sqrt{8}}{1.302} = 1.3572$$

Table 3.7 The one-tailed binomial probability for r successes in n trials (for a two-tailed test multiply by 2)

n	$r = 0$	1	2	3	4	5	6	7
4	0.063	0.313	0.688					
5	0.031	0.188	0.500					
6	0.016	0.109	0.344	0.656				
7	0.008	0.063	0.227	0.500				
8	0.004	0.035	0.144	0.363	0.637			
9	0.002	0.020	0.090	0.254	0.500			
10	0.001	0.011	0.055	0.172	0.377	0.623		
11	0.001	0.006	0.033	0.113	0.274	0.500		
12	0.000	0.003	0.019	0.073	0.194	0.387	0.613	
13	0.000	0.002	0.011	0.046	0.133	0.290	0.500	
14	0.000	0.001	0.006	0.029	0.090	0.212	0.395	0.605
15	0.000	0.000	0.004	0.018	0.059	0.151	0.304	0.500

For d.f. $= 7$ the critical value of t for 95% confidence is 2.36. Since the calculated t is less than the critical one, we cannot reject the null hypothesis that the true mean is 3 (the median value).

The Wald–Wolfowitz runs test. This test can be used to check if the operation of an instrument is changed during the day, or if the performance of the analyst changes during the day. Running replicate analyses of the same sample, we write + any time the last measurement is larger than the previous one or − any time it is smaller. From n measurements we will have $n - 1$ signs. Let us denote the number of plus signs by P and the minus signs by M. We write the signs in the order of their appearance (i.e. the sign between the first and second experiment will be the first, while that between the second and third experiment will be the second, etc.). Every time a sign is changed (from plus to minus or vice versa) we called it a **run**. Let us denote the number of runs by R. We want the number of runs to be random to show that the analysis is independent of the time of the measurement. Table 3.8 shows the range of the number of runs for each P and M (note that they are interchangeable so $P = 5$ and $M = 6$ is the same as $P = 6$ and $M = 5$, and hence the table gives only $M > P$), which can be regarded as due to a random distribution. If R is out of this range it is not random and must be due to machine changes or human behavior change. For example, in one day of experiments the following results were obtained: +, +, +, −, −, −, −, −, −, +, +, +. We have six pluses so $P = 6$, and six minuses so $M = 6$. There are only two runs (changing of signs) and since the table shows that R must be larger than 4, we must conclude that the changes are not random, and the reason for it must be looked for.

Table 3.8 The Wald–Wolfowitz runs test

N	M	At $P = 0.05$, the number of runs is significant if it is	
		less than	greater than
2	12–20	3	na
3	6–14	3	na
3	15–20	4	na
4	5–6	3	8
4	7	3	na
4	8–15	4	na
4	16–20	5	na
5	5	3	9
5	6	4	9
5	7–8	4	10
5	9–10	4	na
5	11–17	5	na
6	6	4	10
6	7–8	4	11
6	9–12	5	12
6	13–18	6	na
7	7	4	12
7	8	5	12
7	9	5	13
7	10–12	6	13
8	8	5	13
8	9	6	13
8	10–11	6	14
8	12–15	7	15

3.5 Errors in instrumental analysis – calibration lines

3.5.1 *Instrumental analysis vs. classical 'wet chemistry' methods*

Many instrumental methods also require wet chemistry, since methods like atomic absorption spectrometry or plasma emission spectrometry require the measurement of solutions. Instrumental methods differ from the classical methods in calibration. The classical 'wet chemistry' methods like gravimetry and titration are absolute ones and completely linear, and hence do not require the use of standards and calibration curves. However, instrumental analysis has many advantages over the classical methods, such as sensitivity, versatility, price, large throughput, multi-element analysis and computer interfacing. Very few of the instrumental methods are absolute and usually the instrument signals that are used for calculation of the amount of the analyte in the analyzed sample depend on several of the instrument's parameters. The quantitative determination of the analyzed sample is done by comparison to standards. In most cases more than one concentration of standards is used and a calibration curve is plotted, a curve that gives the concentration vs. the electronic signal measured by the instrument. These calibration curves also give us the certainty (confidence limits) of our measurement and the range of concentrations for which the method can be applied accurately.

3.5.2 Standards for calibration graphs

Standards can be purchased commercially or prepared by the analyst. There are commercially available standards for single elements or multi-element standards in various concentrations; however, many times the analyst prepares standards or part of them. Great care should be taken in the preparation of standards used for calibration purposes. Only chemicals with exact known stoichiometry should be used, and one has to be sure that the amount of water of hydration is known exactly. It is preferable to use chemicals without any water of hydration. The standard should be dried, at sufficiently high temperature, till constant weight is reached. A standard solution is prepared by weighing a known mass of this chemical into a known volume. Care should be taken that all other solutions of lower concentrations will *not* be prepared by dilution of one standard solution, since in the case of an error in preparation of the first solution it will propagate to all other standard solutions. It is recommended that three or four different solutions will be prepared by weighing and these solutions will be diluted. Since weighing is usually more accurate than pipetting, it is recommended to use weighing whenever possible. If a standard solution is prepared by dilution of another standard solution, it is recommended to use a buret and use at least 25 mL of the original solution. If two chemicals with exact known stoichiometry are available for the same element, it is recommended to use standards of each one of them, thus verifying that both stoichiometries are really known. It should always be remembered that the accuracy of our results cannot be better than the accuracy of our standards.

3.5.3 Calculation of an equation for the calibration graphs

Calibration graphs and/or functions are obtained by measuring several standard solutions with our instrument and thus forming a set of data points each consisting of the concentration of the solution (x_i) and the corresponding instrument signal (y_i). We can plot the data on millimeter paper and connect the points by either a broken line, a straight line that seems to us to be the best or a curve done with a curved ruler, and then extract the concentration x of the unknown solution from the instrument's signal y. However, this way takes a long time in routine work and may also lead to larger errors and subjective errors depending on the analyst. The best and most objective way is to use the set of coupled values to calculate a function $y(x)$ and to use it to calculate the concentration of the analyzed samples according to their signal y. Many analysts today are not bothered how this function is calculated, since, in contrast to even 10–20 years ago when the calibration function was calculated by the analyst, today instruments (all of which are PC-based) are supplied with programs to calculate the calibration function. If not calculated by the instrument's program, the functions are usually calculated by one of the spread-sheet programs. However, to understand how this is done on the computer, we will give the principles here. Several scientific calculators also have the possibility to calculate this function assuming it is linear.

Finding the best functional equation that fits the experimental data is referred to as **regression**. Most calibration graphs are linear and hence the search for the best linear function is called **linear regression**.

How do we choose the best line? The basic assumption is that the values of x are exact (although actually this is not true since there will be at least a random error in the volumes of the standard solutions taken for calibration) while the values of y are subject to a normal distribution. In linear regression we are interested in finding a function $y = ax + b$ that will fit best to our data. What is the criterion for best fit? We cannot ask for the sum of deviations between our observed values (y_{obs}) and our calculated values from the fitted equation ($y_{cal} = ax + b$) to be minimum (minimal value of $\sum_{i=1}^{n} y_i - (ax_i + b)$, where n is the number of observations (x_i, y_i) used to calculate the calibration equation), since in this case negative deviations will cancel positive deviations and minimal values of the sum do not warrant small deviations. In order to cancel the effect of the sign, instead of summing of the deviations, the summing of either the absolute values of the deviations or their squares can be done. These deviations are referred to as **residuals**, so we have two alternatives when looking for a minimum, either the sum of the absolute residuals (SAR) or the sum of the squares of residuals (SSR). Which of the two alternatives is the better one? Without referring to a better quality, the prevailing use is that of the minimum SSR. Erjavec (1978) presented an example to show that the SAR method is not unique and will lead to a worse fit than judged by eye. Nalimov (1963) claims that, since we used the mean (which is the value for minimum SSR) and not the median (which is the value for minimum SAR), we should also use SSR for curve fitting. The disadvantage of SSR is that it gives a larger influence to larger deviations, and often larger deviations occur in the limits of the possibility of measurement by the instrument and hence have larger associated errors. The main advantage of the SSR is that the minimum values can be calculated by simple calculus methods, leading to simple equations for calculating a and b. This method of minimizing the SSR is called the **least-squares method**:

$$\mathrm{SSR} = \sum_{i=1}^{n} [y_i - (ax_i + b)]^2 \tag{3.33}$$

where n is the number of experimental pairs of values measured for the calibration.

Differentiating SSR with respect to a and b and equating the derivatives to zero leads to the following equations:

$$\begin{aligned} \sum x_i[y_i - (ax_i + b)] = 0 \quad &\Rightarrow \quad a\sum x_i^2 + b\sum x_i = \sum x_i y_i \\ \sum y_i - (ax_i + b) = 0 \quad &\Rightarrow \quad a\sum x_i + bn = \sum y_i \end{aligned} \tag{3.34}$$

Solving these two equations for a and b yields

$$a = \frac{n\sum x_i y_i - \sum x_i \sum y_i}{n\sum x_i^2 - \left(\sum x_i\right)^2} \qquad b = \frac{\sum x_i^2 \sum y_i - \sum x_i \sum x_i y_i}{n\sum x_i^2 - \left(\sum x_i\right)^2} \tag{3.35}$$

Another form in which to write the equations for a and b is by the use of the means of x_i and y_i:

$$a = \frac{\sum (x_i - \bar{x})(y_i - \bar{y})}{\sum (x_i - \bar{x})^2} \qquad b = \bar{y} - a\bar{x} \tag{3.36}$$

Using a calculator, it is usually simpler to use equation (3.35) since all scientific calculators give $\sum x_i$ and $\sum x_i^2$.

The variance of the calculated y is given by the SSR divided by the number of degrees of freedom. Since n data points were used to calculate the two constants a and b, the number of degrees of freedom (independent variables) is $n - 2$. So

$$s_o^2 = \frac{\sum_{i=1}^{n} (y_{\mathrm{obs},i} - y_{\mathrm{cal},i})^2}{n-2} \quad \Rightarrow \quad s_o^2 = \frac{\sum_{i=1}^{n} (y_i - ax_i - b)^2}{n-2} \tag{3.37}$$

where a and b are those calculated by equations (3.35) or (3.36). From the total variance s_o^2 given in equation (3.37), we can calculate the variance of the coefficients a and b:

$$s_a^2 = \frac{ns_o^2}{n\sum x_i^2 - \left(\sum x_i\right)^2} = \frac{s_o^2}{\sum (x_i - \bar{x})^2} = \frac{s_o^2}{\sum x_i^2 - n(\bar{x})^2} \tag{3.38}$$

$$s_b^2 = \sum x_i^2 s_a^2 \tag{3.39}$$

Confidence limits of the regression parameters. The confidence limits of the regression parameters are calculated as for the measurement of the mean by the Student's t-statistic with the appropriate level of significance (probability level) using $n-2$ degrees of freedom. Thus the confidence limits of the parameters are:

$$a \pm ts_a \quad \text{and} \quad b \pm ts_b \tag{3.40}$$

3.5.4 *Test for linearity*

In the previous section we assumed that the calibration curve is linear and calculated its parameters. But is this assumption a correct one? And how do we check it?

The most common method for testing for linear correlation is by the statistic R, called the **product-moment linear correlation coefficient**. Often it is called just the **correlation coefficient** since it is the most used one, although there are also other types of correlation coefficients. It is given by

$$R^2 = \frac{\left[\sum (x_i - \bar{x})(y_i - \bar{y})\right]^2}{\sum (x_i - \bar{x})^2 \sum (y_i - \bar{y})^2} = \frac{\left(n\sum x_i y_i - \sum x_i \sum y_i\right)^2}{\left[n\sum x_i^2 - \left(\sum x_i\right)^2\right]\left[n\sum y_i^2 - \left(\sum y_i\right)^2\right]} \tag{3.41}$$

For absolute linearity R^2 is equal to 1. The smaller is R^2, the larger is the deviation from linearity. However, quite large R^2 is obtained also for non-linear curves, as e.g. with one side of a parabola, and it is advisable also to check the obtained linearity by observation of the calculated line vs. the observed data points. Parametric testing of the significance of R can be done using the t-test. The statistic t is calculated from the equation

$$t = \left[\frac{R^2(n-2)}{\sqrt{1-R^2}}\right]^{1/2} \tag{3.42}$$

The calculated value of t is compared with the tabulated value (a two-tailed test as was done for the means) with $n-2$ degrees of freedom. The null hypothesis in this case is that there is no correlation between x and y. If the calculated value of t is larger than the critical tabulated value, then the null hypothesis is rejected, and we conclude that within this level of confidence there is a correlation.

Another parametric test for linearity is the F-test. In this test we compare the variance of the linear regression s_o^2 with the standard deviation for a single measurement, obtained by repeating measurements of the same standard solution and calculating the single measurement variance s_s^2. F is calculated from the ratio $F = s_o^2/s_s^2$. If the calculated F value is larger than the tabulated critical value, then the null hypothesis of linear fit is rejected. If F is less than the tabulated value, we retain the hypothesis of a linear correlation. But which of the single points of the calibration line to repeat? It is best to repeat several of them (or even all of them) and pool together the standard deviations of all of them. As we said before, the pooling together of the standard deviations is done by their weighted arithmetic average when the weights are their degrees of freedom:

$$s_s^2 = \frac{\sum_{i=1}^{n} \text{d.f.}_{\cdot i} \times s_i^2}{\sum_{i=1}^{n} \text{d.f.}_{\cdot i}}$$

where $\text{d.f.}_{\cdot i}$ and s_i are for each calibration point that was repeated more than once, and n is the number of calibration points (different concentrations).

A non-parametric test of the linearity is done by the Wald–Wolfowitz runs test, where the signs are taken to be those of $y_{obs} - y_{calc}$.

Another test of linearity is the comparison of the least-squares linear regression parameter of dependence on x with its standard deviation. The smaller is the ratio s_a/a, the better is the linearity fit.

All curve-fitting programs and spread-sheet programs calculate both parameters, their standard deviations and R^2, and actually nowadays there is no need to calculate these values ourselves. The equations are given only so that we can understand what are the meanings of the program's output.

In the calibration curve

$$\text{signal} = a \times \text{concentration} + b$$

the parameter b is called the 'blank' of the method, as this is the reading obtained for zero concentration. The parameter a is called the sensitivity of the method. As we will see in the next section, a influences the error of the determination, mainly for large concentrations. On the other hand b influences mainly the limit of determination and the errors at lower concentrations. Ideally b should be zero, but in many applications there is a non-zero signal even for zero concentration, the blank reading b, which could be due to impurities in the water or to instrumental reasons.

3.5.5 Calculation of a concentration

Once we have the calibration equation $y = ax + b$, which actually stands for the equation

$$\text{concentration} = (1/a) \times (\text{instrument's signal} - b) \tag{3.43}$$

then it is very simple to calculate the concentration from the instrument's signal. The calculation becomes more complicated when we want to state the confidence limit (interval) of the concentration. Remember that there are confidence limits for a and b and those uncertainties in their values are reflected by uncertainty in the concentration. In many cases an approximate equation is used for the calculation of the standard deviation of the calculated concentration from equation (3.43). If the experimental signal for the unknown solution is y_u and the linearization parameter a was calculated from the n points (x_i, y_i) then the variance s_c^2 of the concentration (the square of the standard deviation) is given by:

$$s_c^2 = \frac{s_o^2}{a^2}\left(1 + \frac{1}{n} + \frac{(y_u - \bar{y})^2}{a^2 \sum_i (x_i - \bar{x})^2}\right) \tag{3.44}$$

Instead of using $\sum (x_i - \bar{x})^2$ we can substitute it by $\sum x_i^2 - \left(\sum x_i/n\right)$. If m readings were done on the unknown solution and y_u is their mean, then the first term in the brackets is changed from 1 to $1/m$:

$$s_c^2 = \frac{s_o^2}{a^2}\left(\frac{1}{m} + \frac{1}{n} + \frac{(y_u - \bar{y})^2}{a^2 \sum_i (x_i - \bar{x})^2}\right) \tag{3.45}$$

If the concentration obtained by substituting y_u in equation (3.43) is c, then the confidence limits of the concentration are $c \pm ts_c$.

Another way of calculating the confidence limits for the concentration is by using the confidence limits for the parameters – equation (3.40). Treating a and b independently, we have four possibilities of extreme values of concentrations, i.e. using first $a + ts_a$ and $b + ts_b$ in equation (3.43) and then change to $a + ts_a$ and $b - ts_b$, and the same with $a - ts_a$. Calculation of these four extreme concentrations gives us the confidence limit around c, calculated from a and b. The disadvantage of the latter method is that it does not show how the confidence limits are changed if the signals for the unknown are read more than once. In this case for each reading the confidence limit is calculated and then the extreme values are averaged and the averaged confidence limit is divided by $\sqrt{m}$. However, this method requires a lot of calculations, although they can easily be added to the least-squares program or written separately with a, b, s_a and s_b as the input of the program.

The confidence limits can be narrowed (i.e. improved) by increasing either n, the number of points on the calibration line, or m, the number of repeating experiments on the unknown sample, or both.

3.5.6 Weighted least-squares linear regression

We mentioned before that sometimes the points at the end of the calibration range are more erroneous than the others since we are going to the limits of the instrument's capabilities. These points are influencing the fit more than the others since their deviations are larger in contrast to what should be obtained. In order to overcome the disadvantage that the usual least-squares linear regression gives similar importance to all points, more accurate and less accurate ones, a weighted regression line can be calculated. In this regression method each point gets its weight depending on its accuracy. How do we know the accuracy, or relative accuracy of each point? The common way to give a weight to a point is to measure each concentration point several times, and then the weight is taken to be inversely proportional to the variance of the various measurements of y for this x:

$$w_i = \frac{1}{s_i^2} \tag{3.46}$$

The weighted least-squares linear regression is obtained by minimizing the sum $\sum_i^n w_i(y_{obs} - y_{cal})^2$. Then each sum in equations (3.34) has in it also w_i and n is replaced by $\sum w_i$. Thus

$$\begin{aligned} a &= \frac{\sum w_i \sum x_i y_i - \sum w_i x_i \sum w_i y_i}{\sum w_i \sum w_i x_i^2 - \left(\sum w_i x_i\right)^2} \\ b &= \frac{\sum w_i x_i^2 \sum w_i y_i - \sum w_i x_i \sum w_i x_i \sum w_i x_i y_i}{\sum w_i \sum w_i x_i^2 - \left(\sum w_i x_i\right)^2} \end{aligned} \tag{3.47}$$

Some people prefer to normalize the weights so their sum is equal to n, such that in equations (3.47) $\sum w_i$ is replaced by n, similarly to equations (3.35). In this case w_i is defined by

$$w_i = \frac{n/s_i^2}{\sum_i (1/s_i^2)} \tag{3.48}$$

3.5.7. Non-linear calibration equations

Several analytical methods give non-linear calibration plots, or they are linear only in a limited range, and at higher concentration they become curved. Usually the curves are curved downward, reducing the sensitivity of the method in this range. Thus, if possible, it is recommended to dilute the analyzed solution to the linear range. However, there are cases where accuracy is sacrificed for shorter times of analysis, prohibiting any additional step of dilution. So we would like to be able also to fit our data to non-linear functions. There are computer programs that can fit the data to any function we choose. Most curved calibration lines are fitted to a quadratic equation

$$y = ax^2 + bx + c$$

or to a third-power equation

$$y = ax^3 + bx^2 + cx + d$$

The principle of least-squares fitting to these polynomials is the same as for the linear function. Thus for a quadratic regression we want to minimize the SSR:

$$\sum_i [y_i - (ax_i^2 + b_i x_i + c_i)]^2$$

By differentiating the SSR with respect to a, b and c and equating the derivatives to zero, we will get three equations with unknowns a, b and c. From these three equations we can derive formulas for a, b and c, similarly to equations (3.35). It is possible to do these polynomial fittings also with spread-sheets that do not have polynomial regression. Although they have only linear regression, they also have multilinear regression, which means that the regression assumes that y depends linearly not only on x but also on z, etc. For example, for bilinear regression the program minimizes the SSR of the function $y = az + bx + c$. However, if in the column of z_i we insert a column of x_i^2, we will perform least squares of the quadratic equation

$$y = ax^2 + bx + c.$$

3.5.8 *Limit of detection*

Almost the first question asked by the analyst about an analytic method or an instrument is about its **limit of detection**, which means the minimum amount that can be detected by the instrument, within the confidence level that this detection is not a false one. This does not tell us the confidence limits of the concentration calculated, just the confidence level that the element is really present in the sample. The limit of detection is not changed by repeated measurements, but the confidence limits of the measured concentration will be narrowed by repetition of the measurement. The limit of detection of the analyte is determined by the limit of detection of the signal. If the blank signal was zero with standard deviation of zero, any signal measured by the instrument would be significant, but actual measurements usually have blank signals and it is not an accurate number but has a distribution.

So how do we recognize if a signal is really due to the presence of the element and not just a blank signal? Different people use different criteria for this limit of detection. The suggestion made by the National Institute of Standards and Technology of the USA is to use the 95% level of confidence. So if the blank is distributed normally with standard deviation σ_0 then the limit of detection of the signal is

$$s_D = 3.29\sigma_0$$

This means that only if our signal is $b + 3.29\sigma_0$ can we say that we have a true signal of the analyte with false positive and false negative risks each of 5%. The 3.29 value is taken from $2z_{0.95}$ where z is the value of the probability of the normal distribution. The factor 2 comes from the fact that not only the blank has a normal distribution but also our measured signal has a normal distribution. If we neglect the factor 2 we will obtain what is called a **limit of decision**. It will guarantee (with 95% confidence) that we do not claim the presence of the analyte when it is actually absent, but it will not ensure that we will not claim that it is absent while actually present. Many people use the factor 3 instead of 3.29. This is rounding off the factor. Using the normal distribution table shows that this factor ($z = 1.5$) means a risk of 7% of either false positive and false negative answers. In most cases we do not know the real value of σ_0 and we estimate it from the calculation of s, obtained from repeated measurements of the blank. In this case the Student's t-distribution should be used rather than the normal one and

$$s_D = 2ts$$

where t is taken for d.f. $= n - 1$ where n is the number of repeated measurements.

If the blank value is obtained from a least-squares linear regression, then σ_0 is equal to $\sqrt{2}s_b$. Instead of this other people suggest using s_0 as σ_0.

The analyte detection limit c_D is related to the signal detection limit by the sensitivity of the measurement, the parameter that transforms signals to concentration (the parameter a in the calibration linear equation):

$$c_D = s_D/a$$

Limit of determination or limit of quantitation. These terms (synonyms) are sometimes used for the lower limit of a precise quantitative measurement in contrast to limit of detection, which deals with qualitative detection. A value of $10s_b$ was suggested as the limit of determination of the signal and $10s_b/a$ as the limit of determination of the concentration. It should be remembered that the accuracy of the measurement can be improved by repeated measurements of the sample. In any case very few reports include this term and usually only limit of determination is given. It should always be borne in mind that it is actually almost impossible to measure quantitatively samples with concentration close to c_D. We can only detect the presence of the element but its real value will be considerably in error (large confidence intervals).

3.5.9 The method of standard addition

When the matrix of the analyzed sample is considerably different from that of the standards and it is known, or the analyst suspects, that the matrix has an influence on the measurement, the usual calculation of the concentration via a calibration plot or equation cannot be done. Matrix effects can be found in many instrumental analytical methods; even for inductively coupled plasma atomic emission or mass spectrometry, which are relatively free from matrix effects, it is only 'relatively', and many matrices influence the measurements (the instrument's reading). A possible way to discover matrix effects, but not always to overcome them, is by diluting the solution (or the solid sample, but for solids mixing and homogenization is more problematic) by a factor of 2 and checking if the concentration calculated has also decreased to a half. However, this method cannot be used when the concentration of the analyte in the original solution is close to the limit of detection of the instruments.

In the case of matrix effects the concentration of the analyte is determined by the method of standard addition. This method applies mostly to instrumental methods that use solutions and not for those determining solid samples, because of the problem of mixing and homogenizing solid samples. In this method, to several aliquots of volume V_1 from the original solution a volume V_2 is added that contains a known amount of the analyzed element. Several standards are used so that the volume V_2 will have different amounts of the analyte. All these mixtures are measured by the instrument and the instrument readings are plotted vs. the added amount in the volume V_2. Assuming that the instrument reading is linear with the concentration, the extrapolation of the plot to the added amount axis will give the amount of the analyte in the analyzed solution (in absolute value, as it intersects the added amount axis in its negative part). If we analyze the data of the instrument readings (y_i) vs. the added amount (x_i) by linear regression of $y = ax + b$, then the concentration of the analyzed sample is $c = b/a$. The variance of the measured concentration is given by the equation

$$s_c^2 = \frac{s_0^2}{a^2}\left(\frac{1}{n} + \frac{y^2}{a^2 \sum_i (x_i - \bar{x})^2}\right)$$

where n is the number of different amounts added (number of points on the plot). The confidence limit is, as in the use of a calibration line, $c \pm ts_c$. Precision can be improved (narrower confidence limit) by the increase of n and most of the time also by the increase of the range of x as this increases $\sum (x_i - \bar{x})^2$, but care should be taken to use only a range where the signal is linear with the concentration.

This method has found wide application in activation analysis or X-ray fluorescence of liquid samples, in atomic absorption, emission spectrometry and various electroanalytical methods, but from the point of view of the analyst it means a lot of work and determination of several solutions in order to measure the concentration of one sample. Each sample needs its own standard addition line, unless the matrices are exactly the same, a quite rare event. The statistician sees the disadvantage of this method in that it uses extrapolation rather than the more accurate interpolation. However, when there are matrix effects that cannot be eliminated by dilution, the method of standard addition might be the sole solution.

References

Erjavec, J. (1978), in *Statistics*, ed. R.F. Hirsch, The Franklin Institute, Philadelphia. (A collection of papers on the use of statistics in analytical chemistry, presented in an analytical chemistry symposium.)

Funk, W., Damman, V. and Donnevert, G. (1995), *Quality Assurance in Analytical Chemistry*, VCH, Weinheim.

Guenzler, H. (1994) *Accreditation and Quality Assurance in Analytical Chemistry*, Springer, Berlin.

Kateman, G. and Buydens, L. (1993), *Quality Control in Analytical Chemistry*, John Wiley, New York.

McCormick, D. and Roack, A. (1987) *Measurement, Statistics and Computation*, John Wiley, Chichester.

Meier, P.C. and Zund, R. (1993), *Statistical Methods in Analytical Chemistry*, John Wiley, New York.

Miller, J.C. and Miller, J.N. (1984), *Statistics for Analytical Chemistry*, Ellis Horwood, Chichester.

Nalimov, V.V. (1963) *The Application of Mathematical Statistics to Chemical Analysis* Addison-Wesley, Reading, MA.

Pritchard, E. (1997), *Quality in the Analytical Chemistry Lab*, John Wiley, Chichester.

Taylor, J.K. (1987), *Quality Assurance of Chemical Measurements*, Lewis Chelsea, Michigan.

Youden, W.J. (1982), *J. Qual. Technol.*, **4**, 1.

4 Activation analysis

M.D. GLASCOCK

4.1 Introduction

Activation analysis is one of the most sensitive and versatile techniques possible for elemental analysis. The technique involves irradiation of a sample with neutrons, charged particles or photons to induce instability in some of the sample atoms. Measurement of the characteristic radiation emitted from the unstable atoms enables the analyst to establish an elemental fingerprint for the sample (i.e. amounts of different major, minor and trace elements present).

The history of activation analysis began in 1934 when Irene Joliot Curie and Frederic Joliot discovered that artificial radioactivity could be induced in foils made of aluminum, boron and magnesium when bombarded by naturally occurring alpha-particles. The notion that nuclear reactions might be used for quantitative elemental analysis first occurred to Georg Hevesy and Hilde Levy in 1936 when they exposed samples of rare-earth element (REE) salts to neutrons emitted from a Ra(Be) source. Hevesy and Levi discovered that the radiation emitted by the different elements decreased with different time constants. By measuring the different radiations, they were able to determine the elemental compositions in different mixtures of the REE compounds.

The construction of charged-particle accelerators and nuclear reactors in the 1940s, 1950s, and 1960s was followed by numerous studies of basic nuclear physics and chemistry. Thousands of experiments were performed to measure nuclear parameters (i.e. reaction cross-sections, half-lives, isotopic abundances, gamma-ray energies and branching ratios, etc.). Improvements in the sophistication and sensitivity of instrumentation used to make nuclear measurements took place throughout this period and continue to the present time. If it were not for the building of these machines, development of more sensitive instrumentation and determinations of nuclear parameters, use of activation analysis very likely would have remained only a passing curiosity.

The utilization of activation analysis is not necessarily limited to the field of analytical chemistry but also includes a broad range of related applications. Among these are measurement of beam and flux intensities, production and measurement of radioisotopes for medicine and industry, experiments involving the use of stable isotopes as tracers, and quality control monitoring of industrial process streams.

A good appreciation of activation analysis and its various aspects benefits from an understanding of the elementary principles of nuclear physics and chemistry. This chapter begins by introducing some of the basic nuclear properties (i.e. nuclear structure, radioactive reactions, nuclear decay) and continues with descriptions of irradiation sources, radiation detection and radionuclide production. The chapter concludes with descriptions of different activation analysis techniques and common applications.

4.2 Nuclear structure

This section presents a brief introduction to the principles of nuclear physics. More extensive reading can be found in several textbooks on nuclear physics and chemistry (Adloff and Guillaumont 1993, Ehmann and Vance 1991, Friedlander *et al.* 1981) and books describing activation analysis (Alfassi 1989, De Soete *et al.* 1972, Kruger 1971, Parry 1991).

4.2.1 The atomic nucleus

The nuclear model of the atom was proposed in 1911 by Rutherford to interpret the data from his alpha-particle scattering experiments. In Rutherford's model, the atom had all of its positive charge and most of the mass contained within a nucleus whose dimensions were on the order of 10^{-12} cm in diameter. In order to balance the charge of the atom, the nucleus was surrounded by a sufficient number of negatively charged electrons distributed over atomic dimensions of order 10^{-8} cm. The discovery of neutrons in 1932 by Chadwick led to a revised model of the nucleus where protons and electrically neutral neutrons represented the unit building blocks of nuclear matter known as **nucleons**. A strong but short-ranged nuclear force between the nucleons holds the nucleus together.

The number of protons in the nucleus, Z, is called the **atomic number** and the protons determine the chemical behavior of the atom. For any uncharged atom, the number of orbiting electrons is equal to the number of protons in the nucleus. The number of neutrons in the nucleus is represented by N. The atomic number of any nucleus, A, is equal to the total number of nucleons present in the nucleus:

$$A = Z + N$$

A **nuclide** is defined as any species of atom that exists for a measurable length of time. A shorthand notation commonly used to symbolize individual nuclides is ^{A}X, where X is replaced by the chemical symbol for the element. For example, the nuclide ^{12}C has six protons and six neutrons for a total of 12 nucleons.

Nuclides are sometimes classified according to their stability. Accordingly, three different groups of nuclides can be defined:

1. The **stable nuclides** include those not known to undergo radioactive decay. Altogether, 264 stable nuclides are known with half-lives longer than 10^{18} years, the current limit of detection for radioactive emission. Examples of stable nuclides include ^{12}C, ^{16}O, ^{27}Al and ^{197}Au.
2. The **naturally occurring radionuclides** include two types: **primordial** and **cosmogenic**. The primordial radionuclides include long-lived radionuclides and their shorter-lived daughters that have existed since the origins of the Earth. Cosmogenic radionuclides are produced continuously as a result of nuclear reactions occurring in the environment, such as those induced by the actions of cosmic rays and natural fission. About 64 naturally occurring radionuclides are known. Familiar examples of these include ^{238}U ($t_{1/2} = 4.47 \times 10^9$ y), ^{40}K ($t_{1/2} = 1.28 \times 10^9$ y), ^{226}Ra ($t_{1/2} = 1600$ y, a continually produced daughter of ^{238}U) and ^{14}C ($t_{1/2} = 5730$ y, produced by the interaction of cosmic-ray-produced neutrons on ^{14}N).
3. The **artificial radionuclides** are man-made and do not normally occur in Nature. Greater than 2000 artificial radionuclides are known, including ^{24}Na ($t_{1/2} = 15$ h), ^{60}Co ($t_{1/2} = 5.27$ y), ^{137}Cs ($t_{1/2} = 30.2$ y, a product of fission) and ^{198}Au ($t_{1/2} = 64.8$ h).

Certain elements have a number of **isotopes** that differ from one another by the number of neutrons present in the nucleus. The individual isotopes of each element are found in different amounts (i.e. **abundances**) in Nature. For example, the element hydrogen ($Z = 1$) has two naturally occurring isotopes, ^{1}H (99.985% natural abundance) and ^{2}H (deuterium, 0.015% natural abundance). In addition to the naturally occurring hydrogen isotopes, one radioactive isotope of hydrogen exists in the form of ^{3}H (tritium, $t_{1/2} = 12.3$ y). Another example occurs for the element chlorine ($Z = 17$), which has four frequently mentioned isotopes. The isotopes ^{35}Cl (75.77% natural abundance) and ^{37}Cl (24.23% natural abundance) are both stable, while ^{36}Cl ($t_{1/2} = 3 \times 10^5$ y) and ^{38}Cl ($t_{1/2} = 37.2$ min) are radioactive. Several shorter-lived isotopes of chlorine have also been identified. By multiplying the weights of all stable isotopes by their natural abundances, the **atomic weight** of the element can be determined.

The number of atoms of a particular isotope present in a sample depends on the mass, Avogadro's number and the isotopic abundance. The equation to calculate the number of atoms in a sample is

$$n = \frac{m}{M} N_A \theta \tag{4.1}$$

where m = mass of the sample (g), M = atomic weight of element (g), N_A = Avogadro's number (6.02×10^{23} molecules/mole) and θ = isotopic abundance.

4.2.2 *Excited states and nuclear disintegration*

The properties of different nuclei cannot be explained satisfactorily by a single nuclear model, although certain analogies between nuclei and atoms and molecules exist (i.e. atomic structure, rotational and vibrational motions in molecules). Nucleons, like the electrons in an atom, occupy different energy states depending upon their arrangement inside the nucleus. As a result, an individual nucleus can exist in different levels of excitation, either a ground level or in one of several discrete excited states, unique to the particular species of nuclide. When a nucleus undergoes a spontaneous transformation in its internal structure from a higher excited state to a lower one, an accompanying emission of particle(s) or gamma-ray(s) takes place. This process is called **nuclear disintegration** or **radioactive decay**.

During radioactive decay, mass is converted into energy as predicted by Einstein's theory of relativity such that the laws of conservation of energy, momentum, charge and mass number are obeyed. A nucleus will decay by one or more modes if the mass(es) of the product(s) are less than that of the original nucleus. Certain decay modes are favored based on the properties of both the original and product nucleus. In general, neutron-rich nuclei will transform by β^- decay, resulting in a decrease in the number of neutrons by one, while increasing the number of protons by one. Neutron-deficient nuclei decay by β^+ or electron capture decay, which reduce the number of protons by one while increasing the number of neutrons by one. The alpha decay mode is important for very heavy nuclides and results in a loss of a pair of neutrons and a pair of protons (i.e. four nucleons) from the original nucleus. The neutron decay mode is important for certain fission products and results in a loss of one neutron from the original nucleus.

Disintegrations by alpha decay, beta decay or neutron decay frequently produce a product nucleus in an excited state. The product nucleus continues the disintegration process by emitting one or more gamma-rays or particles until arriving at a ground state. A particular nuclide undergoing decay can produce more than one energy gamma-ray photon and the number of photons emitted is not necessarily equal to the number of nuclei undergoing decay. The number of gamma-rays of a specific energy emitted from a radionuclide per 100 decays is called the **intensity** of that gamma-ray, in per cent. The excited states populated by nuclear decay are often illustrated by energy level diagrams or **decay schemes**.

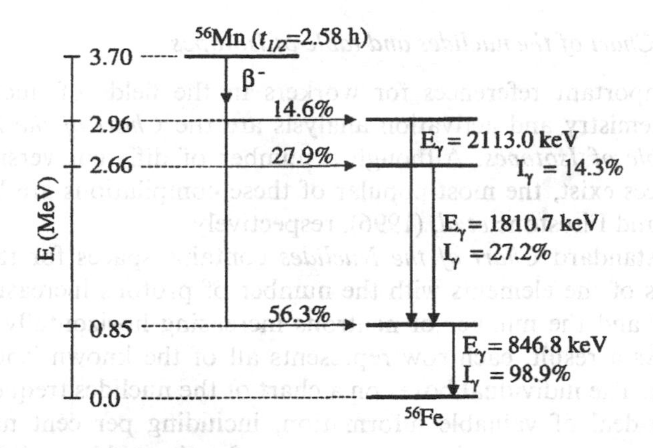

Figure 4.1 The radionuclide ^{56}Mn decays with a half-life of 2.58 h by β^- emission to the excited states of ^{56}Fe. Energy levels at 2.96, 2.66 and 0.85 MeV are populated by 14.6%, 27.9% and 56.3% of the beta decays, respectively. The remaining beta decays populate a number of higher-lying energy levels. The excited levels at 2.96 and 2.66 MeV decay to the 0.85 MeV level by emitting 2113.0 and 1810.7 keV gamma-rays, respectively. The 0.85 MeV level decays directly to the ground state by emitting an 846.8 keV gamma-ray. The gamma-rays at 846.8, 1810.7 and 2113.0 keV are observed with absolute intensities of 98.9%, 27.2% and 14.3%, respectively. Several minor gamma-rays account for the remaining intensity.

Decay schemes for several radionuclides are illustrated in Figures 4.1 and 4.2.

Although a majority of the excited states decay in less than 10^{-12} s, some excited states last long enough to be measured. Excited states with measurable lifetimes are commonly known as **metastable** or **isomeric** nuclear states. Metastable states usually de-excite by emitting gamma-rays, but some of the longer-lived metastable nuclides undergo beta decay.

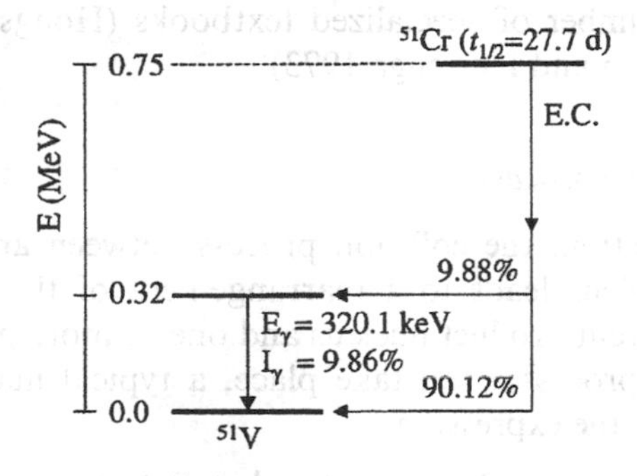

Figure 4.2 The radionuclide ^{51}Cr decays with a half-life of 27.7 d by electron capture to the 0.32 MeV and ground states of ^{51}V. The 0.32 MeV and ground states are populated by 9.88% and 90.12% of the decays, respectively. The 0.32 keV level is depopulated by a 320.1 keV (intensity = 9.86%) gamma-ray to the ground state.

4.2.3 Chart of the nuclides and table of isotopes

Two important references for workers in the fields of nuclear physics, radiochemistry and activation analysis are the *Chart of the Nuclides* and the *Table of Isotopes*. Although a number of different versions of these references exist, the most popular of these compilations are Walker *et al.* (1989) and Firestone *et al.* (1996), respectively.

The standard *Chart of the Nuclides* contains spaces for the individual isotopes of the elements with the number of protons increasing vertically upward and the number of neutrons increasing horizontally from left to right. As a result, each row represents all of the known isotopes for the element. The individual boxes on a chart of the nuclides frequently contain a great deal of valuable information, including per cent natural abundance, neutron absorption cross-sections for the stable nuclides, half-lives and modes of decay, energies of the most intense gamma-rays, and spins and parities of ground states.

The *Table of Isotopes* contains elemental and isotopic properties, including natural abundance, neutron absorption cross-sections, half-lives and modes of decay, charts of the energy states for each isotope, decay schemes for all known radioisotopes, and tabulations of the gamma-ray energies and branching intensities following radioactive decay. Further information in the *Table of Isotopes* includes fission yields, spins and parities of excited states, and other useful nuclear data.

4.3 Nuclear reactions

Nuclear reactions occupy a central role to all methods of activation analysis and it is, therefore, essential to present the aspects of nuclear reactions that are pertinent to activation analysis. Additional reading can be found in a number of specialized textbooks (Hodgson 1971, Satchler 1980, Vandenbosch and Huizenga 1973).

4.3.1 The reaction process

In a nuclear reaction, the collision process between an incident particle and a target nucleus leads to a rearrangement of the nucleons, usually producing a different product nucleus and one or more particles. Although several reaction processes can take place, a typical nuclear reaction can be represented by the expression

$$\mathrm{a} + \mathrm{A} \rightarrow [\mathrm{X}] \rightarrow \mathrm{b} + \mathrm{B} + Q$$

or in shorthand notation

$$\mathrm{A(a,b)B}$$

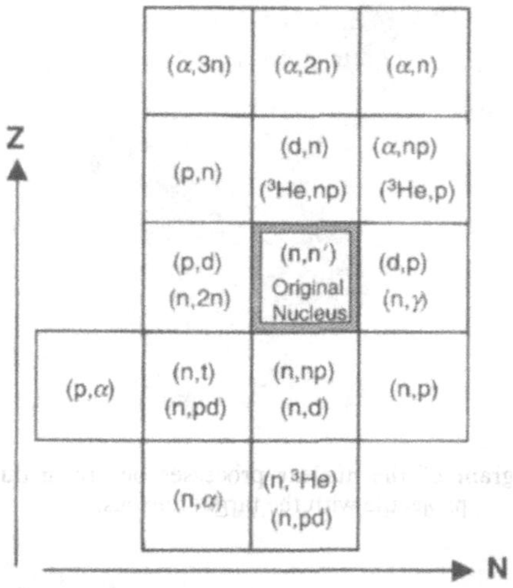

	(α,3n)	(α,2n)	(α,n)
	(p,n)	(d,n) (^{3}He,np)	(α,np) (^{3}He,p)
	(p,d) (n,2n)	(n,n') Original Nucleus	(d,p) (n,γ)
(p,α)	(n,t) (n,pd)	(n,np) (n,d)	(n,p)
	(n,α)	(n,^{3}He) (n,pd)	

Z
N

Figure 4.3 The relative locations on a chart of nuclides of products following selected particle-induced nuclear reactions.

where a represents the incoming particle or radiation, A is the target nuclide, [X] is the compound nucleus in a state of excitation, B is the product nuclide, b is the exiting particle or radiation, and Q is the total change in energy for the system. Both B and b are products of the reaction. If b is a particle, the only difference is that b is lighter than B. The diagram shown in Figure 4.3 shows the positions of products for selected nuclear reactions relative to the target nucleus on a chart of the nuclides.

Mass–energy, charge, nucleon number and momentum are each conserved quantities for reactions of interest to activation analysis. The value of Q, above, accounts for the amount of energy released or absorbed during the nuclear reaction. When Q is a positive value, the reaction is **exoergic**. When Q is negative, the reaction is **endoergic**. Most nuclear reactions are of the latter type and require the incoming particle to have a certain amount of kinetic energy in order for the reaction to occur.

Some of the incident particle's energy is used to conserve momentum for the particle-nucleus system, and some energy is also required to overcome the Coulomb repulsion forces between incident charged particles and the nucleus. Therefore, the incident particle must have a kinetic energy in excess of the Q value. The minimum amount of energy required for a particular reaction to occur is known as the **threshold energy**. For exoergic reactions involving neutrons, the required threshold energy is zero. Thus, the reaction can occur when the incident neutron has a kinetic energy of nearly zero (i.e. the situation for thermal neutron capture reactions). For incident particles with less than 30 MeV of energy, most nuclear reactions are classified as one of two broad types: **scattering reactions** or **compound**

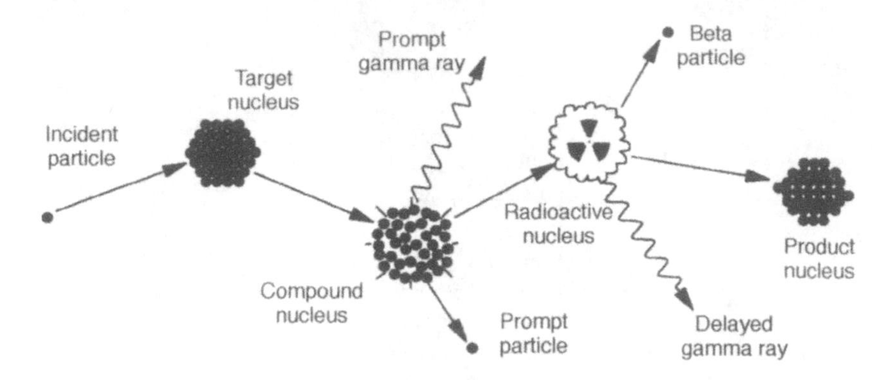

Figure 4.4 Schematic diagram of the nuclear processes occurring during interaction of a projectile with the target nucleus.

nucleus formation. Figure 4.4 illustrates the sequence of events following a collision between a projectile and target nucleus leading to compound nucleus formation.

4.3.2 Compound nucleus formation

Reactions leading to compound nucleus formation are usually called **direct reactions** by nuclear physicists. Direct reactions involve a two-step process of formation and evaporation. During the formation step, the incident particle merges with the target nucleus such that the total excitation energy (kinetic plus binding) is quickly distributed among all nucleons in the compound nucleus. Once a state of statistical equilibrium is reached, the incident particle(s) are no longer distinct from any of the other nucleons in the compound nucleus. Depending on the amount of excitation energy present, the compound nucleus is usually unstable with respect to particle emission. As a result of random collisions within the nucleus, a single nucleon (or a collection of nucleons, such as an alpha-particle) can escape. The most probable outcome is that only a portion of the excitation energy will be taken away by the exiting particle(s); thus, it is likely that the residual nucleus will have sufficient energy remaining for additional emission(s) of particles and/or gamma-rays to take place. This de-excitation process for the compound nucleus is commonly known as **nucleon evaporation**, because it resembles the way in which molecules in a drop of hot liquid escape into the vapor phase.

Nuclear fission is another type of compound nucleus reaction involving only the heaviest elements ($Z \geq 90$). Fission occurs when the compound nucleus splits into two large fragments. This splitting process generally releases several free neutrons at the same time. The escaping fission fragments typically have a large kinetic energy on the order of 200 MeV. The fragments themselves are generally unstable and undergo subsequent de-excitations by emitting other nuclear particles or gamma-rays. The fission-produced neutrons may induce additional nuclear reactions in other nuclei (including fission).

Neutron capture (n,γ) reactions are among the more common types of nuclear reactions that involve compound nucleus formation. In most situations involving low-energy neutrons, the compound nucleus has insufficient energy for particle emission; thus only gamma-ray emission is possible. At higher neutron energies, the compound nucleus has a greater likelihood of emitting charged particles, neutrons and/or gamma-rays.

In general, neutron capture (n,γ) reactions are more useful for activation analysis than charged-particle- or gamma-ray-induced reactions because they offer greater selectivity and sensitivity for elemental analysis. Activation analysis techniques based on the measurement of prompt and delayed radiation following the formation step are known as **prompt activation analysis** and **delayed activation analysis**, respectively.

4.3.3 Reaction cross-sections

The interaction rate for a given nuclear reaction depends on the number of incident particles, the number of target nuclei of the particular type available for interaction and the probability of occurrence for the reaction. To first approximation, the reaction probability for a single incident particle is similar in magnitude to the cross-sectional area of the target nucleus, which for most nuclei is on the order of 10^{-24} cm^2. For the sake of convenience, reaction cross-sections are usually expressed in units of barns (b), where 1 b = 10^{-24} cm^2. Actual reaction cross-sections vary widely with the energy and type of incident particle, the identity of the target nuclide and the specific type of reaction. In general, low-energy neutrons have larger cross-sections for radiative capture (n,γ) reactions while high-energy neutrons and charged particles have smaller cross-sections and favor reactions such as (n,p), (n,α), (p,n), (p,d), (d,p), (d,n), (d,^{3}He) and (d,α). As the incident particle energy is increased, the reactions tend to become even more complex.

Examples of nuclides showing the variability of thermal neutron (n,γ) reaction cross-sections are ^{1}H (0.33 b), ^{27}Al (0.233 b), ^{113}Cd (19 910 b), ^{157}Gd (254 000 b) and ^{208}Pb (0.5 mb). Thermal-neutron-induced fission reaction cross-sections are often quite large, such as those for the nuclides ^{233}U (531 b), ^{235}U (584 b), ^{239}Pu (742 b) and ^{241}Pu (1009 b). On the other hand, cross-sections for charged-particle-induced reactions range from about 0.5 to 10^{-12} b.

4.3.4 Measurement of cross-section

Reaction cross-sections are frequently used to measure neutron and charged-particle fluxes and spectra, to estimate the amounts of radioactivity that will be produced in samples and to perform certain activation analysis calculations. Experimental determination of a reaction cross-section requires measurement of the rate of production, the number of incoming particles and other sample-related parameters. The methods of calculation are different depending upon the relative dimensions of the sample (i.e. thickness) and the sample's effect on the energy and intensity of the irradiating particles.

Thin samples (or targets) are defined as those in which the energy and intensity of the incident particle beam are not significantly reduced when passing through the sample. The flux (number of incident particles $cm^{-2}\ s^{-1}$) is assumed to be constant throughout the sample. Examples of reactions with thin samples are well-defined beams of charged particles or neutrons incident on one surface of a thin foil or a sample undergoing bombardment in all directions by a uniform flux of reactor neutrons. In both cases, calculations are based on the following assumptions:

1. The product nuclide has a half-life much longer than the total irradiation time such that no appreciable decay occurs during irradiation.
2. The number of target nuclei is not depleted during irradiation. (For short and moderate length irradiations, this assumption is usually valid. For very long irradiation times and for targets with high reaction cross-sections, target **burnup** may be a concern.)
3. The cross-section is constant throughout the sample. (This assumption is valid as long as the incident particle energy is not significantly reduced while passing through the sample.)

For the example of a thin foil irradiated by a beam of particles, the equation describing the rate of production, R, for a particular nuclear reaction is

$$R = in'x\sigma \tag{4.2}$$

where i = flux of incident particles per second, n' = number of target nuclei per unit of volume (cm^{-3}), x = target thickness (cm) and σ = cross-section per incident particle (cm^2).

For the sample undergoing irradiation in a uniform neutron flux, the rate of production can be rewritten as follows:

$$R = n\phi\sigma \tag{4.3}$$

where n = number of target atoms belonging to the nuclide of interest in the sample, ϕ = neutron flux ($cm^{-2}\ s^{-1}$) and σ = cross-section (cm^2). For any irradiation time, T_i, the total number of product nuclei is calculated as

$$N_{\text{product}} = n\phi\sigma T_i \tag{4.4}$$

With thick targets or very large samples, the particle beam undergoes significant attenuation. As a result, the flux of particles exiting a thick target or penetrating a thick sample is less than the incident beam. For samples in a flux of neutrons, few samples need to be considered thick because neutrons have no charge and pass through most samples with little attenuation, unless the sample contains a significant amount of high cross-section elements. Since charged particles interact readily, most charged-particle reactions require calculations to correct for target thickness.

4.3.5 *Excitation functions*

The detailed variation between the reaction cross-section and bombarding particle energy is called an **excitation function**. In general, excitation functions for a particular reaction type cannot be calculated, but instead they must be measured over the energy range of interest. Extensive tabulations of excitation functions for a variety of reactions are available (Brune and Schmidt 1974, Mughabghab *et al.* 1984).

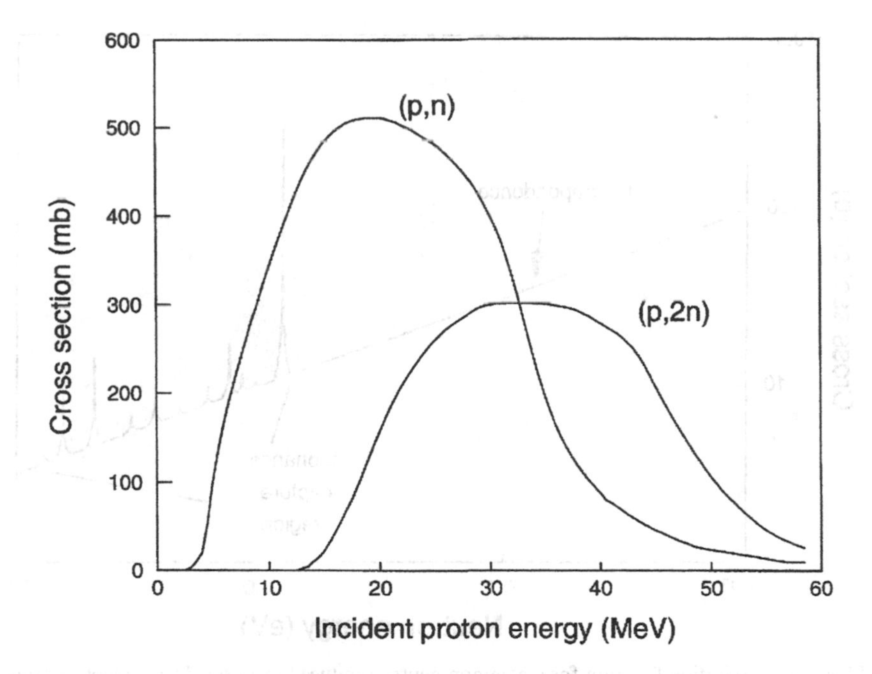

Figure 4.5 Excitation functions for common proton-induced reactions on a low-Z target nuclide.

With few exceptions, thermal neutrons will induce (n,γ) reactions only. Reactions involving higher-energy neutrons, charged particles and photons will induce more complex reactions but require kinetic energies in excess of the threshold energy before the reaction will take place. As the energies for neutrons and charged particles continue to increase, the probability of a given reaction usually increases from the threshold energy to some maximum value, approximately 10 MeV above the threshold, and then drops rapidly to a lower value. As illustrated in Figure 4.5, the observed drop is usually accompanied by a rise in the cross-sections of more complex reactions.

Frequently, excitation functions for reactions exhibit peaks when the excitation energy coincides with the energy of an excited state of the compound nucleus. The peaks are known as **resonances** and are well separated as long as the differences between excited states are large. At higher energies, the resonance peaks frequently overlap such that the cross-section becomes monotonic. When neutrons are used as the irradiating particles, resonance peaks are frequently observed in the range from 10 eV to about 0.5 MeV as shown in Figure 4.6. Below 10 eV, the cross-section usually varies smoothly with $E^{-1/2}$ or $1/v$, where v is the velocity of the neutron. This behavior is sometimes known as the **one-over-v law**.

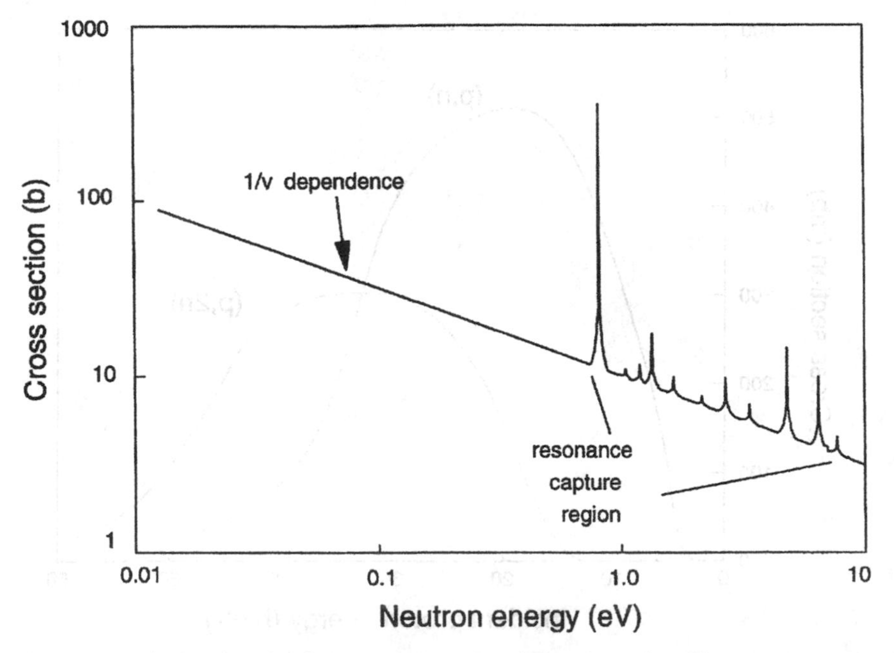

Figure 4.6 Excitation function for a common neutron-induced reaction. The neutron capture cross-section (σ) is plotted against low incident neutron energies with resonance peaks shown.

4.4 Decay rates

Radioactive decay is the spontaneous transformation of atoms by emission of particles or gamma-rays from the nucleus, or of X-rays after capture of shell electrons by the nucleus. In any application of radioactivity, the amount of radioactivity present and the length of time that the radioactivity will be present must be known. In this section, a mathematical treatment of the laws of radioactive decay will be presented, including the equations for decay and growth of multiple radioactive products in a decay chain. Finally, the formation and decay of a radioactive species following a nuclear reaction will be discussed.

4.4.1 Radioactive decay law

Radioactive decay is a statistical process. For a sample containing a large number of radioactive atoms, there is no way to predict which atom will be next to decay, but the decay characteristics of the whole sample can be described. The number of atoms of a particular radionuclide that decay per unit of time (i.e. rate of decay) is defined as the **activity**, A, and is defined by the fundamental law of radioactivity

$$A = -\frac{\mathrm{d}N}{\mathrm{d}t} = \lambda N \tag{4.5}$$

where N is the number of radioactive atoms and λ is defined as the **decay constant**.

A general expression describing the decay of a parent nuclide P with the decay constant λ into a stable daughter nuclide D is

$$\mathrm{P} \xrightarrow{\lambda} \mathrm{D}$$

The value of λ is different for each species of radionuclide. An equation describing the time dependence of the number of atoms of the radionuclide is

$$N(t) = N_0\,\mathrm{e}^{-\lambda t} \tag{4.6}$$

where N_0 is the number of radioactive atoms at time $t = 0$. Therefore, the process of radioactive decay is an exponential law, and the activity of the radioactive nuclide is controlled by a characteristic property known as the **half-life** (i.e. the period of time during which half of the original atoms of that nuclide disintegrate).

Since the activity is proportional to the number of radioactive atoms, it decreases exponentially with time as well

$$A(t) = A_0\,\mathrm{e}^{-\lambda t} \tag{4.7}$$

The exponential behavior of the law of radioactive decay, as defined by equations (4.6) and (4.7), is due to the fact that radioactive transmutations are random. Deviations from this behavior are increasingly smaller as larger numbers of radioactive atoms are monitored.

Half-life is related to the decay constant according to

$$t_{1/2} = \frac{\ln 2}{\lambda} = \frac{0.693}{\lambda} \tag{4.8}$$

The half-lives of radionuclides range from milliseconds to several times the age of the universe.

4.4.2 *Units of radioactivity*

The amount of radioactivity in a radioactive sample can be expressed in a variety of ways. The historical unit of activity has been the **curie** (Ci), defined as exactly 3.7×10^{10} disintegrations per second (dps), which owes its origins to the activity of 1 g of pure ^{226}Ra. A more modern SI unit of radioactivity known as the **becquerel** (Bq) is also frequently used. By definition, one becquerel is equal to one disintegration per second. The standard metric prefixes also apply to the curie; for example, millicurie (mCi) = 10^{-3} Ci, microcurie (μCi) = 10^{-6} Ci, nanocurie (nCi) = 10^{-9} Ci, and picocurie (pCi) = 10^{-12} Ci. Thus

$$1\ \mu\text{Ci} = 3.7 \times 10^4\ \text{Bq} = 3.7 \times 10^4\ \text{dps}$$
$$1\ \text{Bq} = 2.703 \times 10^{-11}\ \text{Ci} = 27.03\ \text{pCi}$$

The millicurie is a fairly common working level for radioactivity in medical procedures. Activation analysis procedures generally involve measurements of activities at the microcurie and nanocurie level.

The **specific activity**, A_{sp}, of a radioactive source is defined as the activity per unit of mass of the sample. If a sample consists of a single nuclear species, its specific activity can be calculated according to

$$A_{sp} = \frac{\lambda N}{NM/N_A} = \frac{\lambda N_A}{M} \tag{4.9}$$

In practice, most samples are diluted by larger concentrations of stable atoms from the same element. If the sample is not in pure elemental form, the specific activity is further diluted by the other elements.

In order to characterize a radioactive substance, the radionuclide(s) present, the level(s) of radioactivity and the time of measurement must be stated.

4.4.3 *Growth and decay of radioactive products in a chain*

Thus far, the only decay process we have considered involved a radioactive parent that produced a stable daughter nuclide. However, often the daughter nuclide, formed by the decay of the parent, is also radioactive and its decay is characterized by its own decay constant. A generalized example of a genetic chain of radionuclides is

$$(1) \xrightarrow{\lambda_1} (2) \xrightarrow{\lambda_2} (3) \xrightarrow{\lambda_3} \ldots (n) \xrightarrow{\lambda_n} \text{stable}$$

The nuclide (1) is called the parent nuclide and the others are called the first-, second-, etc., daughter nuclides. The number of atoms of the parent nuclide is not influenced by the instability of its daughter nuclides and is correctly described by the equations mentioned previously. However, the number of atoms and activities of the daughter nuclides vary with time in a more complicated manner.

In the case where the parent nuclide (1) and the first-daughter nuclide (2) are both radioactive, the parent nuclide decays at the rate $-\lambda_1 N_1$, to produce the first-daughter nuclide, while the first-daughter nuclide decays at the rate $-\lambda_2 N_2$ to produce the second-daughter nuclide (3). As a result, the overall change in the number of atoms of the first-daughter nuclide is given by the sum of its formation and decay:

$$\frac{\mathrm{d}N_2}{\mathrm{d}t} = \lambda_1 N_1 - \lambda_2 N_2 \qquad (4.10)$$

and we can write

$$N_1 = N_1^0 \,\mathrm{e}^{-\lambda_1 t} \qquad (4.11)$$

where N_1^0 is the number of parent nuclide atoms at time $t = 0$. After substituting equation (4.11) into equation (4.10), we arrive at the following differential equation:

$$\frac{\mathrm{d}N_2}{\mathrm{d}t} = \lambda_1 (N_1^0 \,\mathrm{e}^{-\lambda_1 t}) - \lambda_2 N_2 \qquad (4.12)$$

After rearrangement, the following first-order linear differential equation is obtained:

$$\frac{\mathrm{d}N_2}{\mathrm{d}t} + \lambda_2 N_2 - \lambda_1 N_1^0 \,\mathrm{e}^{-\lambda_1 t} = 0 \qquad (4.13)$$

The solution to this equation yields an expression that can be used to calculate the number of first-daughter atoms present as a function of time:

$$N_2 = \frac{\lambda_1}{\lambda_2 - \lambda_1} N_1^0 (\mathrm{e}^{-\lambda_1 t} - \mathrm{e}^{-\lambda_2 t}) + N_2^0 \,\mathrm{e}^{-\lambda_2 t} \qquad (4.14)$$

Multiplication of equation (4.14) by λ_2 gives the activity of first-daughter atoms as a function of time:

$$A_2 = \lambda_2 N_2 = A_1^0 \frac{\lambda_2}{\lambda_2 - \lambda_1} (\mathrm{e}^{-\lambda_1 t} - \mathrm{e}^{-\lambda_2 t}) + A_2^0 \,\mathrm{e}^{-\lambda_2 t} \qquad (4.15)$$

where $A_1^0 = \lambda_1 N_1^0$ and $A_2^0 = \lambda_2 N_2^0$ at time $t = 0$.

Using equations (4.14) and (4.15), we can consider three situations in which the ratios of decay constants of the parent and first-daughter nuclides differ:

1. *Situation 1.* If the half-life of the parent nuclide is at least 10 times greater than that of the first-daughter nuclide, then its activity and also the number of atoms will not vary significantly during the observation time. Under this condition, $\lambda_1 \ll \lambda_2$ and $e^{-\lambda_1 t} \approx 1$. Therefore, we can assume that $\lambda_2 - \lambda_1 \approx \lambda_2$ and $N_1 = N_1^0 = \text{constant}$. Equation (4.14) then simplifies to

$$\frac{N_2}{N_1} = \frac{\lambda_1}{\lambda_2} \quad \text{or} \quad \lambda_2 N_2 = \lambda_1 N_1 \tag{4.16}$$

   This means that after a decay period of 7–10 half-lives of the first-daughter nuclide, a steady-state condition called **secular equilibrium** is attained. In the steady-state condition the activities of the first-daughter and the parent nuclides are essentially equal (i.e. $A_2 = A_1$) such that after equilibrium is reached the first-daughter nuclide will appear to have the same half-life as its parent.
2. *Situation 2.* If the activity of the parent nuclide decays more slowly than the first-daughter but is measurable during the period of observation, a condition called **transient equilibrium** is achieved. In this case, the half-life of the parent nuclide is approximately 3–10 times longer than the first-daughter nuclide. This means that the condition $\lambda_1 < \lambda_2$ is satisfied. After an elapsed time $t >$ 5–10 times the half-life of the first-daughter has elapsed, the term $e^{-\lambda_2 t}$ in equation (4.14) becomes very small and simplifies to

$$N_2 = \frac{\lambda_1}{\lambda_2 - \lambda_1}(N_1^0 e^{-\lambda_1 t}) \tag{4.17}$$

   which is equivalent to

$$N_2 = \frac{\lambda_1}{\lambda_2 - \lambda_1} N_1 \tag{4.18}$$

   Rearrangement of equation (4.18) yields

$$\frac{N_2}{N_1} = \frac{\lambda_1}{\lambda_2 - \lambda_1} \tag{4.19}$$

   In the case of transient equilibrium, the ratio of first-daughter atoms to parent atoms becomes a constant after an elapsed time several times the half-life of the first-daughter nuclide has passed. It is also true that the ratio of the activities of the parent nuclide and first-daughter nuclide becomes a constant:

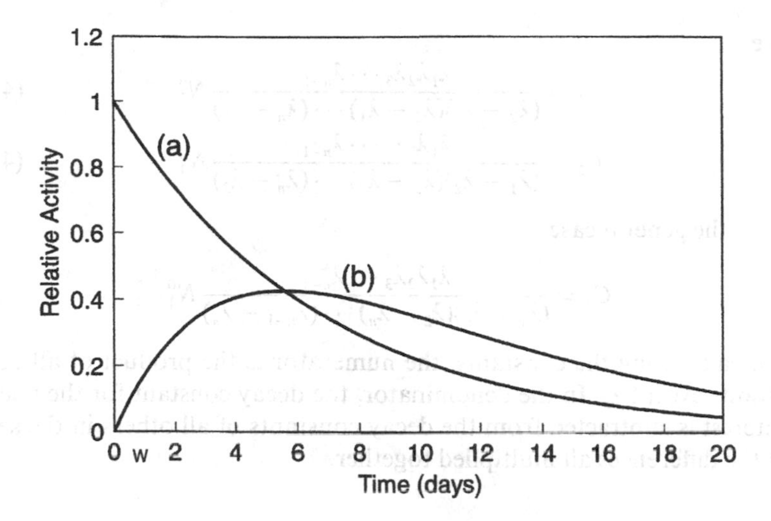

Figure 4.7 Parent (a) and daughter (b) activities as a function of decay time in the case of transient equilibrium.

$$\frac{A_2}{A_1} = \frac{\lambda_2}{\lambda_2 - \lambda_1} \tag{4.20}$$

The situation of transient equilibrium is illustrated in Figure 4.7.

3. *Situation 3.* When the half-life of the first-daughter nuclide is greater than that of the parent, no situation of equilibrium will ever occur. After a certain period of time the activity of the parent radionuclide will decrease to a negligible value and the activity of the first-daughter first rises and then declines. A constant ratio between the parent and first-daughter activities never occurs.

If daughter nuclides beyond the first-daughter are also radioactive, the situation becomes more complex as illustrated by the expression for the rate of change in second-daughter radionuclides:

$$\frac{dN_3}{dt} = -\lambda_3 N_3 + \lambda_2 \left(\frac{\lambda_1}{\lambda_2 - \lambda_1} N_1^0 (e^{-\lambda_1 t} - e^{-\lambda_2 t}) + N_2^0 e^{-\lambda_2 t} \right) \tag{4.21}$$

Expressions for third-daughter and fourth-daughter radionuclides would be still more complicated. The Bateman equation permits calculation of the activity for any member of a chain, provided that at time $t = 0$, the parent radionuclide is the only one present (i.e. $N_2^0 = N_3^0 = \ldots = N_n^0 = 0$). In this case, the number of atoms of the nth-daughter radionuclide present at any time can be calculated from

$$N_n = C_1 e^{-\lambda_1 t} + C_2 e^{-\lambda_2 t} + C_3 e^{-\lambda_3 t} + \cdots + C_n e^{-\lambda_n t} \tag{4.22}$$

where

$$C_1 = \frac{\lambda_1\lambda_2\lambda_3\cdots\lambda_{n-1}}{(\lambda_2-\lambda_1)(\lambda_3-\lambda_1)\cdots(\lambda_n-\lambda_1)}N_1^0 \tag{4.23}$$

$$C_2 = \frac{\lambda_1\lambda_2\lambda_3\cdots\lambda_{n-1}}{(\lambda_1-\lambda_2)(\lambda_3-\lambda_2)\cdots(\lambda_n-\lambda_2)}N_1^0 \tag{4.24}$$

and for the general case

$$C_n = \frac{\lambda_1\lambda_2\lambda_3\cdots\lambda_{n-1}}{(\lambda_1-\lambda_n)(\lambda_2-\lambda_n)\cdots(\lambda_{n-1}-\lambda_n)}N_1^0 \tag{4.25}$$

When calculating the constants, the numerator is the product of all decay constants except λ_n. In the denominator, the decay constant for the nuclide of interest is subtracted from the decay constants of all others in the series with the differences all multiplied together.

4.4.4 *Growth of radioactive products during nuclear irradiation*

In most activation analysis applications, the goal is to determine the amounts or concentrations of one or more elements present in a sample. The sensitivity of the analysis will depend on a number of factors, including the choice of analytical technique, the sample matrix and the experimental and nuclear parameters.

For prompt activation analysis, the total number of product nuclei created during the measurement period is used to determine the number of target nuclei present and analytical sensitivity. In this case, the rate of production and length of measurement are the key factors for prompt activation analysis and half-life is not important.

On the other hand, for delayed activation analysis the number of nuclei decaying during the post-irradiation measurement period is measured. The number of radioactive nuclei decaying depends on the total number of radioactive nuclei produced during irradiation less the number of radioactive nuclei that decayed both during the irradiation period and in the interval between the end of irradiation and the beginning of measurement. Thus, the rate of production, half-life, length of irradiation, length of decay and length of measurement are the key factors for delayed activation analysis.

Mathematically, the rate of change in the number of radioactive nuclei during an irradiation is the difference between the rate of production and the rate of decay, i.e.

$$\frac{dN^*}{dt} = R - \lambda N^* \tag{4.26}$$

where N^* is the number of atoms of the radionuclide present.

Assuming irradiation in a uniform nuclear reactor flux and substituting for the rate of production from equation (4.3) leads to the equation:

$$\frac{dN^*}{dt} = n\phi\sigma - \lambda N^* \quad (4.27)$$

where n is the number of target atoms. If equation (4.27) is integrated with the initial condition $N_0^* = 0$, then the number of radioactive nuclei present at the end of irradiation (EOI) is

$$N^*_{\mathrm{EOI}} = \frac{n\phi\sigma}{\lambda}(1 - e^{-\lambda T_i}) \quad (4.28)$$

where the irradiation time is given by T_i.

If irradiation times are short when compared to the half-life (i.e. $T_i \ll t_{1/2}$), decay during irradiation can be neglected such that the factor $(1 - e^{-\lambda T_i})$ is approximately equal to λT_i and equation (4.28) becomes $N^*_{\mathrm{EOI}} = n\phi\sigma T_i$. If the irradiation time is long when compared to the half-life (i.e. $T_i \gg t_{1/2}$), then the factor $(1 - e^{-\lambda T_i})$ is approximately 1. In the latter case, the rate of production is equal to the rate of decay and irradiation of the sample for times longer than about five times the half-life offers no significant advantage. The maximum number of radioactive nuclei that can be produced is called the **saturation value** N^*_{sat}:

$$N^*_{\mathrm{sat}} = \frac{n\phi\sigma}{\lambda} \quad (4.29)$$

For a particular reaction, it is possible to increase the saturation value by increasing the flux or the target mass, but not by lengthening the irradiation time.

4.4.5 Decay of radioactive products during measurement

In general, the measurement of radiation emitted from an activated sample does not begin immediately after irradiation, but after some decay time, T_d. Under normal conditions, short decay times are used when the radionuclide of interest has a short half-life and long decay times are used when the half-life is either medium or long. The long decay time also allows possible short-lived radionuclides with initially high activities time to become insignificant.

The equation describing the number of radioactive atoms present at the start of the counting period (SOC) is

$$N^*_{\mathrm{SOC}} = N^*_{\mathrm{EOI}}\, e^{-\lambda T_d} = N^*_{\mathrm{sat}}(1 - e^{-\lambda T_i})\, e^{-\lambda T_d} \quad (4.30)$$

Sample counting occurs during an interval of time called the counting time, T_c. It is a relatively easy matter to show that the number of atoms that decay during the counting period, ΔN, is

$$\Delta N = N^*_{\mathrm{sat}}(1 - e^{-\lambda T_i})\, e^{-\lambda T_d}(1 - e^{-\lambda T_c}) \quad (4.31)$$

4.5 Irradiation sources

Utilization of a particular reaction for activation analysis is highly dependent upon the availability of irradiation and measurement facilities. The selection of an irradiation source implies that a suitable nuclear reaction has been identified. When selecting the source, three factors must be considered:

1. The type of particle or radiation
2. The energy of the particle or radiation
3. The magnitude or intensity of the beam or flux

Some possible sources of nuclear particles for activation analysis are nuclear reactors, charged-particle accelerators and radioisotopic sources. The main properties of these irradiation sources are briefly described.

4.5.1 Nuclear reactors

Neutrons are the most widely employed particles for activation analysis, owing to their greater range of penetration into target materials, larger reaction cross-sections and higher possible fluxes. The neutron fluxes of present-day research reactors range from 10^{12} to 10^{15} neutrons $cm^{-2}\,s^{-1}$.

Nuclear reactors produce neutrons as by-products of thermal-neutron-induced fission of uranium and offer the capability of neutron activation analysis. In a reactor, the fissionable uranium isotope, ^{235}U, absorbs thermal neutrons and the fission process outputs additional neutrons with kinetic energies ranging from 0 to 15 MeV. The average neutron energy is about 2.5 MeV. The reactor core is surrounded by moderating materials (e.g. light water, heavy water, graphite and beryllium), which reduce the energies of the fission-produced neutrons. Through only a few elastic collisions with moderator nuclei, the high-energy fission neutrons rapidly lose their kinetic energy. This slowing down creates a broad distribution of neutron energies with three main components (**thermal**, **epithermal** and **fast**) as shown in Figure 4.8. The thermal component consists of neutrons in thermal equilibrium with the moderator atoms. At room temperature, the energy spectrum for thermal neutrons approximates a Maxwell–Boltzmann distribution with a most probable energy of 0.0253 eV and a most probable neutron velocity of 2200 m s^{-1}. The product of the number of neutrons per unit volume and the most probable velocity is defined as the thermal neutron flux. An upper energy limit for thermal neutrons is defined by a cadmium foil of 1 mm thickness, which effectively absorbs all neutrons below 0.5 eV but permits neutrons at higher energies to pass through. The energy threshold at 0.5 eV is frequently called **cadmium cutoff energy** for a $1/v$ radionuclide.

Immediately above 0.5 eV, the neutron spectrum consists of epithermal

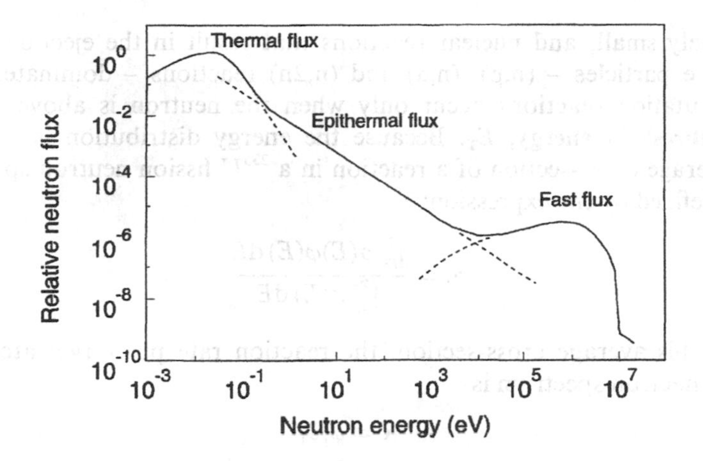

Figure 4.8 A typical neutron spectrum from a nuclear fission reactor.

neutrons (including **resonance** energy neutrons) only partially slowed down through collisions with the moderator materials. In this region, the energy distribution generally follows a $1/E$ slope beginning at the cadmium threshold energy and ranging up to about 10 keV. The epithermal neutron fluxes in most research reactors are typically on the order of 3% of the thermal neutron flux.

At higher neutron energies, the spectrum consists of the **primary fission neutrons** retaining most of the original energy acquired from fission since they are essentially unmoderated. (The primary fission neutrons are sometimes called fast neutrons, but should not to be confused with the high-energy 14 MeV neutrons from a neutron generator.) In a typical light-water-moderated reactor, the flux of primary fission neutrons is on the order of 7–10% of the thermal neutron flux.

Both thermal and epithermal neutrons induce radiative capture (n,γ) reactions. Thus, borrowing from equation (4.3), a non-rigorous but commonly used expression for the total reaction rate per target atom for (n,γ) reactions is given by

$$R = \varphi_{\mathrm{th}}\sigma_{\mathrm{th}} + \varphi_{\mathrm{epi}}I \tag{4.32}$$

where φ_{th} = thermal neutron flux (n cm^{-2} s^{-1}), φ_{epi} = epithermal neutron flux (n cm^{-2} s^{-1}), σ_{th} = average thermal neutron cross-section (b) and I = effective resonance integral (b). For convenience, the effective resonance integral, I, is defined by the expression

$$I = \int_{0.5\,\mathrm{eV}}^{\infty} \frac{\sigma(E)}{E}\,\mathrm{d}E \tag{4.33}$$

At high neutron energies, the cross-sections for (n,γ) reactions are extremely small, and nuclear reactions that result in the ejection of one or more particles – (n,p), (n,α) and (n,2n) reactions – dominate. These transmutation reactions occur only when the neutron is above a minimum **threshold energy**, E_T. Because the energy distribution is complex, the average cross-section of a reaction in a ^{235}U fission neutron spectrum, $\bar{\sigma}_f$, is defined by the expression:

$$\bar{\sigma}_f = \frac{\int_{E_T}^{\infty} \sigma(E)\varphi(E)\,dE}{\int_{E_T}^{\infty} \varphi(E)\,dE} \tag{4.34}$$

Using this average cross-section, the reaction rate per target atom in a fission neutron spectrum is

$$R = \varphi_f \bar{\sigma}_f \tag{4.35}$$

where φ_f is the average fission neutron flux.

4.5.2 Neutron generators

Neutron generators are small charged-particle accelerators designed to produce high-energy neutrons following a suitable nuclear reaction. One of the most common designs is the Cockroft–Walton accelerator, which uses the voltage difference between two terminals to accelerate deuterium ions onto a tritium-containing target. The following nuclear reaction occurs:

$$^{2}H(^{3}H,n)^{4}He$$

and yields neutrons with an average energy of 14 MeV. A typical neutron generator for research will emit about 2×10^{11} neutrons s^{-1} into 4π space. However, in a realistic irradiation geometry, the useful neutron flux densities are approximately 10^9 n cm^{-2} s^{-1}.

Although neutron generators have limited sensitivity for most research, they can be made small and portable for use in specialized work. For example, neutron generators are easily designed to fit into boreholes for oil and mineral prospecting.

4.5.3 Cyclotrons and accelerators

Cyclotrons and Van de Graaff accelerators produce energetic beams of protons, deuterons, tritons, alpha-particles, and ^{3}He particles. Secondarily, these charged particles can produce high-energy neutrons through a variety of nuclear reactions. Analytical applications with high-energy neutrons and energetic charged particles are dominated by reactions involving particle evaporation and usually produce neutron-deficient product nuclides.

Two other types of accelerators are the linear accelerator (linac) and the synchrotron, with both types used to accelerate electrons. Although electrons themselves have few analytical applications, the bremstrahlung radiation (photons) emitted when high-energy electrons interact with a tantalum or tungsten target can be used for photon activation analysis (PAA).

4.5.4 Radioisotopic sources

Two types of isotopic neutron sources are possible. One type uses an α- or γ-emitting radionuclide to induce neutron emission from a target material consisting of a low-Z nuclide with a low binding energy for neutrons (e.g. beryllium). The emitted neutrons are then employed to irradiate another target material in the same manner as Hevesy and Levi in 1936. Some common examples of these are based on the alpha emitters ^{239}Pu and ^{241}Am and the high-energy gamma emitters ^{60}Co, ^{88}Y and ^{124}Sb. The former type are capable of producing about 10^5–10^7 neutrons per second per Ci of activity and the latter type will output about 10^5 neutrons s^{-1} per Ci of activity. Both are relatively inexpensive and can easily be made portable, but their usage is mainly limited to laboratory instruction.

A second type of isotopic neutron source utilizes the spontaneous fission of transuranic elements. The primary example of this type of source is ^{252}Cf ($t_{1/2} = 2.64$ y), for which about 3.2% of the decays occur by spontaneous fission and the remainder by α decay. A 1 g source of ^{252}Cf emits 2.34×10^{12} neutrons s^{-1} into 4π space. The energy range of neutrons emitted by the spontaneous fission of ^{252}Cf is from 1 to 3 MeV. The rather short half-life for ^{252}Cf is also a disadvantage.

4.6 Detection and measurement of radiation

In order to perform activation analysis, the analyst must have a detection system capable of measuring the total number of charged particles, neutrons or gamma-rays emitted from the radioactive sample as well as the energy distribution of the radiation. Operation of any radiation detection system depends on the manner in which the incident radiation interacts with the material in the detector itself. An introduction to the fundamental mechanisms by which different radiations interact to lose their energy in matter and the properties of radiation detectors will be presented below. A more detailed discussion of the processes by which radiation interacts with matter can be found in the classic text by Evans (1955) and an excellent description of the properties of detectors is presented in Knoll (1989).

4.6.1 Interaction of heavy charged particles with matter

Heavy charged particles lose their energy in matter primarily through the Coulomb interaction with the electrons and the nuclei of the absorbing material. Collisions between charged particles and either free or bound electrons result in ionization or excitation of absorber atoms, whereas interactions with the nucleus lead only to Rutherford scattering. Because the latter are not normally significant to detector response, charged-particle detectors must rely on the interactions with electrons for their response.

Upon their entry into any absorber medium, charged particles immediately begin to interact with many electrons simultaneously. Depending on the nearness of these encounters, the interaction may either raise electrons to higher atomic excitation states within the absorber atoms or cause the electrons to be separated from the atom (i.e. ionization). Because only a small fraction of the total energy of the charged particle is transferred to each electron, the charged particle must interact with many different electrons during its passage through the absorbing matter before stopping. Except at the very end of their slowing down, the tracks of charged particles tend to be quite straight.

The rate of energy loss or the **stopping power**, $S(E)$, for a charged particle is defined as the differential energy loss divided by the corresponding differential path length

$$S(E) = -\frac{dE}{dx} \tag{4.36}$$

It can be shown that the stopping power for any incident particle is directly proportional to the incident particle's charge and inversely proportional to its energy. The range of a particle with an initial energy, E_I, is given by

$$r(E_I) = \int_0^{E_I} \frac{dE}{S(E)} \tag{4.37}$$

Particle ranges are usually expressed in units of $g\,cm^{-2}$. Stopping powers and charged-particle ranges in different materials are available in compilations by Anderson and Ziegler (1977), Ziegler (1977) and others.

4.6.2 Interaction of electrons with matter

Electrons also interact with matter through the Coulomb interaction with the electrons. In contrast to heavy charged particles, electrons lose energy at a lower rate and follow a more jagged path during their passage through absorbers. Larger deviations in their track are possible because a larger fraction of the electron energy can be lost in a single collision with atomic electrons. In addition to collisions with electrons, radiative processes such as the emission of bremstrahlung radiation can occur for higher-energy electrons.

The total stopping power for electrons is the sum of losses due to collisional and radiative processes

$$\frac{\mathrm{d}E}{\mathrm{d}x} = \left(\frac{\mathrm{d}E}{\mathrm{d}x}\right)_{\mathrm{c}} + \left(\frac{\mathrm{d}E}{\mathrm{d}x}\right)_{\mathrm{r}} \tag{4.38}$$

The ratio of energy losses is given approximately by

$$\frac{(\mathrm{d}E/\mathrm{d}x)_{\mathrm{r}}}{(\mathrm{d}E/\mathrm{d}x)_{\mathrm{c}}} \cong \frac{EZ}{700} \tag{4.39}$$

where the energy, E, is given in MeV and Z is the atomic number of the absorber. In general, radiative losses are only significant in absorber materials with high atomic numbers.

Positrons have similar energy losses and ranges as electrons; however, at the end of the positron track annihilation radiation is produced. The annihilation photons with energies of 0.511 MeV are very penetrating and typically lead to energy deposition far from the positron track.

4.6.3 Interaction of gamma radiation with matter

As illustrated in Figure 4.9, there are three significant modes by which gamma-rays interact with matter:

1. The **photoelectric effect** is an absorption process in which the incident gamma-ray interacts with an atom to eject a photoelectron and transfers all of its energy to a bound atomic electron during the process.
2. **Compton scattering** is an inelastic scattering process in which the incident gamma-ray transfers part of its energy to a bound or free electron. The scattered gamma-ray not only escapes with reduced energy but is also deflected from its original path.
3. **Pair production** is an absorption process in which the incident gamma-ray is transformed to matter in the form of an electron–positron pair.

The photoelectric effect occurs only with bound electrons, mainly from the K (80%) and L (20%) shells of the atom. When the incident gamma-ray with energy, E_γ, disappears in the interaction, it is replaced by a photoelectron with energy

$$E_{\mathrm{e}} = E_\gamma - E_{\mathrm{b}} \tag{4.40}$$

where E_{b} is the binding energy of the electron. Binding energies are small, typically a few keV to tens of keV depending on the Z of the material. The vacancy created by the ejected electron is filled by an electron from

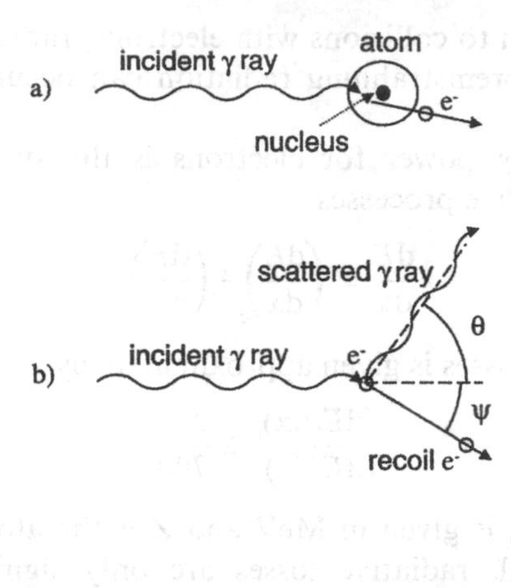

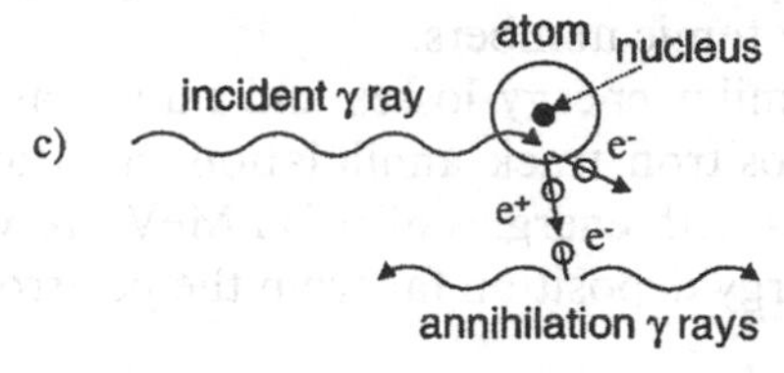

Figure 4.9 Schematic diagrams for the three principal interaction mechanisms of gamma-rays with matter: (a) photoelectric effect; (b) Compton scattering; and (c) pair production.

the next available shell and accompanied by a characteristic X-ray or Auger electron. Normally the material reabsorbs these. The photoelectric effect is the dominant interaction process for gamma-rays with $E_\gamma < 100\,\text{keV}$. The probability of photoelectric absorption decreases rapidly with increasing gamma-ray energy, but increases rapidly at higher atomic numbers. In a detector, the photoelectric effect results in a single voltage pulse directly proportional to the energy of the incident gamma-ray.

Compton scattering is the predominant interaction mechanism over the energy range most often measured in activation analysis (i.e. between 100 keV and 2 MeV). In Compton scattering, the incident gamma-ray is deflected through an angle θ with respect to its original direction. The energy lost by the scattered gamma-ray is transferred to the recoiling electron as kinetic energy. Owing to the requirement that energy and momentum be conserved, the energy of the Compton-scattered gamma-ray, $E_{\gamma'}$, can be calculated for different scattering angles from the following equation:

$$E_{\gamma'} = \frac{E_\gamma}{1 + (E_\gamma/m_0c^2)(1 - \cos\theta)} \tag{4.41}$$

where m_0c^2 is the rest-mass energy of the electron (0.511 MeV). The main difference between the photoelectric effect and Compton scattering processes in a detector is that the scattered gamma-ray has the possibility of escaping in part or entirely from the detector. The scattered gamma-ray may interact with other electrons by the photoelectric effect or a second Compton scattering or it might escape from the detector. In the former instance, all of the energy will be deposited in the detector to create a full energy peak indistinguishable from the photoelectric effect. In the latter, part of the incident energy is lost and the detector will record an energy event at less than the full energy. The Compton scattering process produces an energy continuum (i.e. background) in the detector spectrum ranging from zero up to a maximum energy called the **Compton edge**. The Compton edge occurs at an energy equal to the difference between the energy of the incident gamma-ray and the minimum possible energy for Compton scattering after backscattering at 180° from the original direction. The minimum Compton scattering energy is

$$E_{\gamma',\min}(\theta = 180^\circ) = \frac{0.511E_\gamma}{0.511 + 2E_\gamma} \tag{4.42}$$

When the incident gamma-ray energy is large ($E_\gamma \gg m_0c^2$), the escaping backscattered gamma-ray approaches a minimum value of ~0.256 keV.

Pair production is the least common absorption process, and it only occurs for high-energy gamma-rays. During pair production, an electron–positron pair is produced in the vicinity of a nucleus. Because the electron and positron have rest masses equivalent to 0.511 MeV of energy, the incident gamma-ray must have more than 1.022 MeV in order for pair production to occur. After the electron–positron pair is created, any excess energy is shared between the pair as kinetic energy. The likelihood of pair production increases with incident gamma-ray energies above 1.022 MeV, but it increases even more rapidly in absorbers of higher Z and at very high energies. The positron created by pair production annihilates with an electron, creating two photons of 0.511 MeV. The annihilation photons can escape from the detector or interact by one of the previously mentioned processes. If both annihilation photons deposit all of their energy, a full-energy event will be recorded. If one of the annihilation photons escapes, the total energy deposited in the detector will be $E_\gamma - 0.511$ MeV. The peak recorded at this energy is known as the **first-escape peak**. If both annihilation photons escape, the total energy deposited in the detector will be $E_\gamma - 1.022$ MeV and a peak known as the **double-escape peak** will be observed.

Other events that may be recorded in a detector include Compton scattering and pair production events from materials surrounding the detector and natural background radiation. A 0.511 MeV peak is likely due to annihilation photons that enter the detector after exiting a surrounding shield. If large amounts of lead shielding are used, X-rays are often observed in the region from 74 to 88 keV due to the photoelectric effect on the lead. Finally, **sum peaks** may be observed if two gamma-rays from the radioactive source enter the detector at the same time and lose all of their energy. The sum peak is equal to the sum of the energies of the individual gamma-rays. Sum peaks are most probable if the gamma-rays occur in a cascade from a single radionuclide (e.g. the 1173.2 and 1332.5 keV gamma-rays from ^{60}Co may sum to create a peak at 2505.7 keV). The probability of summing increases with count rate and close geometries.

4.6.4 Gamma-ray attenuation

Each of the gamma-ray interaction processes mentioned above serves to attenuate the gamma-rays in a beam of photons interacting with an absorber. The sum of the three processes is called the **linear attenuation coefficient** μ. For a given photon energy, the attenuation of the initial photon beam is described by a simple exponential law

$$I = I_0 \, e^{-\mu x} \tag{4.43}$$

where I is the number of photons remaining in the beam after traversing a distance x through the absorber.

The travel of photons can also be described by their average distance of travel, d, or **mean free path** through the absorber before an interaction occurs. The equation describing this is

$$d = \frac{\int_0^\infty x\, e^{-\mu x}\, dx}{\int_0^\infty e^{-\mu x}\, dx} = \frac{1}{\mu} \tag{4.44}$$

such that d is simply the reciprocal of the linear attenuation coefficient. Typical values for d range from a few millimetres to tens of centimetres in solids with gamma-rays at energies common to activation analysis. The **half-thickness**, $d_{1/2}$, of an absorber is defined as the thickness that makes $I = \frac{1}{2} I_0$, which from equation (4.43) is $d_{1/2} = 0.693/\mu$.

Linear attenuation coefficients are limited by the fact that they vary with the density of the absorbing matter (i.e. solids, liquid or gas). As a result, **mass attenuation coefficients** are more commonly used and are defined by

$$\text{mass attenuation coefficient} = \mu/\rho \tag{4.45}$$

where ρ is the density of the absorber.

4.6.5 Detectors

A typical gamma-ray spectrometer for activation analysis consists of a detector, associated electronics and multichannel analyzer (MCA). The key component of any gamma-ray spectrometer is the detector, of which there are two main types: **scintillation detector** and **solid-state ionization detector**.

A scintillation detector contains a scintillator crystal optically coupled to a photomultiplier tube (PMT). The most common type of scintillator crystal consists of NaI activated with a small amount of thallium. When gamma-rays interact with the atoms of the crystal, electrons are ejected and cause excitation in the crystal. The NaI(Tl) scintillator crystal rapidly de-excites by fluorescence creating visible light (photons) inside the crystal. When a photon of light strikes the photocathode of the PMT, an electron is released, and the electron is directed toward the first of 10 dynodes by the bias voltage. At successive dynodes, each incoming electron ejects three or four more electrons, causing a multiplication of the current. For each photon that arrives at the photocathode, about 10^6 electrons are emitted from the final dynode. Since the number of electrons is proportional to the energy of the gamma-ray, the magnitude of the output voltage pulse is also proportional to the energy of the gamma-ray. Following additional amplification and pulse shaping by a spectroscopy amplifier, the MCA sorts individual pulses to create a gamma-ray energy spectrum. Each pulse causes a single count to be added to the appropriate channel number.

A solid-state ionization detector works on the basis of ionization that occurs when gamma-rays interact with a semiconductor crystal creating free electrons and holes. The number of free electrons is proportional to the energy of the incident gamma-ray. A bias voltage across the detector crystal creates a current for collection by a charge-sensitive preamplifier. The preamplifier output is fed into a spectroscopy amplifier for amplification and pulse shaping prior to sorting by the MCA. The most popular type of solid-state detector available today is made of high-purity germanium (HPGe), which must be kept cooled by liquid nitrogen to achieve optimum response.

Detector performances are compared to one another on the basis of two main characteristics: **efficiency** and **energy resolution**. The efficiency of a detector is a measure of the percentage of gamma-rays counted in the MCA out of the number of gamma-rays entering the detector. The efficiency of an NaI(Tl) detector is greater than that of an HPGe detector due to the higher *Z* of the element iodine in the scintillation detector. However, the energy resolution of an HPGe detector is significantly superior to that of an NaI(Tl) detector. The efficiency of an HPGe detector is usually quoted relative to the efficiency of a standard NaI(Tl) detector of 7.5 cm × 7.5 cm at 1332 keV (from the decay of ^{60}Co) measured 25 cm from the source. Energy resolution is usually expressed by the full width at half-maximum (FWHM) of gamma-rays measured with the detector. The FWHM measures the narrowness of peaks such that a smaller value indicates a superior ability to separate close-lying gamma-ray peaks in a spectrum. The 1332 keV gamma-ray can also be used to compare the resolutions of both types of detectors. The FWHM of a good NaI(Tl) detector is about 75 keV at 1332 keV. A good HPGe detector for activation analysis should have an energy resolution of about 1.8 keV at 1332 keV and a relative efficiency of approximately 20–25% compared to the standard NaI(Tl) detector. In general, a larger HPGe detector is more efficient but has lower energy resolution than a small HPGe detector.

A typical gamma-ray spectrum collected by an HPGe detector at some time after a sample was irradiated is shown in Figure 4.10. The actual number of events recorded by the detector is smaller than the total number of decays due to three factors:

1. The intensities of the emitted gamma-rays are in many cases less than 100%.
2. Some of the emitted gamma-rays will not reach the detector. Owing to the isotropic nature of gamma-ray emission, only those gamma-rays that are headed in the direction of the detector can be measured.
3. A portion of the gamma-rays that arrive at the detector will pass through the detector without interacting. In other cases, some of the gamma-rays will lose part of their energies by Compton scattering and pair production.

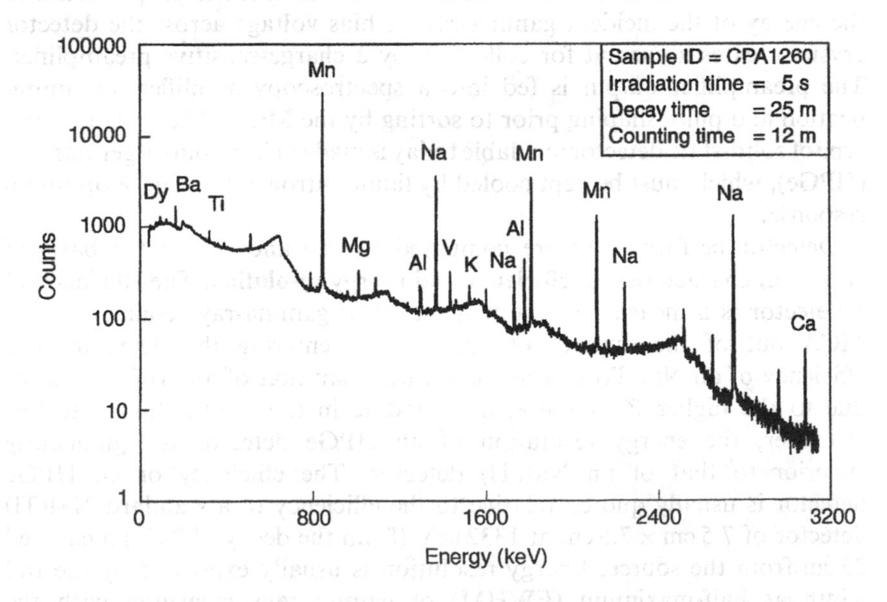

Figure 4.10 Gamma-ray spectrum for a radioactive geological specimen counted with an HPGe detector with 25% relative efficiency.

The second and third factors comprise a characteristic of the detector called the **geometric efficiency**, ε. The geometric efficiency depends on the type, shape and size of the detector crystal, the energy of the gamma-ray, the distance of the sample from the detector and the sample size. The efficiency at various energies can be experimentally determined by measuring the activities of calibrated standards for which the rates of disintegration are known to high accuracy. For a gamma-ray standard source with known activity on its date of calibration, A^0_{calib}, the number of gamma-rays measured by the detector is

$$\text{counts measured} = A^0_{\text{calib}} T_c \, e^{-\lambda T_d} P_\gamma \varepsilon \tag{4.46}$$

where P_γ = intensity of the gamma-ray, T_c = counting time and T_d = total decay time since the calibration date.

The above equation assumes that the activity of the calibration source does not change during the efficiency calibration measurement. For most HPGe detectors, a plot of the efficiency vs. energy has a maximum between 100 and 150 keV. At higher energies, the log of efficiency approximates a linear function of the log of energy

$$\log \varepsilon = b \log E_\gamma + \log a \tag{4.47}$$

where the constants a and b are dependent upon the sample-to-detector distance and the type and size of the detector crystal.

The calibration source used to determine the detector efficiency can also be used for energy calibration of the HPGe detector system. If the electronics are operating normally, the relationship between channel number in the MCA and gamma-ray energy is almost perfectly linear over the entire range of channels.

4.6.6 Gamma-ray spectroscopy

Gamma-ray spectroscopy is the most powerful technique for analysis of radioactive samples. Identification of gamma-rays provides information about the elements present in the sample. Peak areas can be used to determine the activity of nuclides and to make a quantitative determination of the amounts of different elements present in a sample.

Tables of gamma-ray energies are available for the identification of unknown peaks (Erdtmann and Soyka 1979, Glascock 1996). Several tabulations list gamma-rays according to nuclide, energy, half-life, or the type of irradiation performed. In most cases, the energy is the most important clue to its identity. However, for gamma-rays that are ambiguous on the basis of energy (perhaps two or more radionuclides have similar energies), the relative intensities (taking into account the geometric efficiency) of secondary gamma-rays and the half-life are helpful to identify the radionuclide responsible for the gamma-ray.

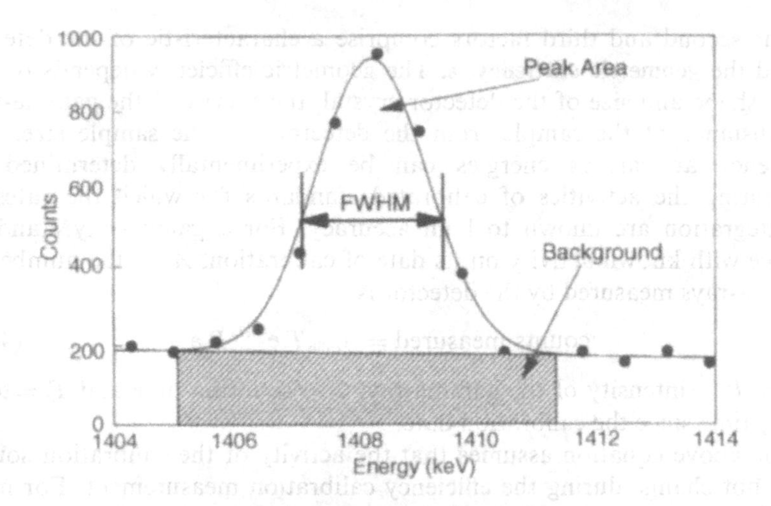

Figure 4.11 Portion of a gamma-ray spectrum showing the Gaussian fit used to determine the peak area above the background distribution. The FWHM for the 1408 keV peak is about 2.1 keV

The area under a peak in the HPGe spectrum represents the number of counts collected for that particular gamma-ray energy. Peak area divided by counting time yields the count rate. Methods for peak area evaluation vary, but all use some method to estimate the amount of background correction as illustrated in Figure 4.11. The number of net counts in a peak is sometimes called the **signal**, S, and is given by

$$S = T - B \tag{4.48}$$

where $T =$ total counts and $B =$ background counts. Finally, the standard deviation of the net counts, s.d., in the peak and the relative standard deviation, %s.d., are given by

$$\text{s.d.} = \sqrt{S + 2B} \tag{4.49}$$

$$\%\text{s.d.} = (100\sqrt{S + 2B})/S \tag{4.50}$$

The limit of detection for a peak on a background depends on how well the background is known. The usual definition of detection limit is based on the two-sigma ($\sim 95\%$) probability that a peak should be observed about the background. A publication by Currie and Parr (1988) discusses the methods for calculating detection limits now in use.

4.6.7 Interferences of gamma-rays

Although activation analysis is highly accurate and precise, there are potential interferences and other problems of which users should be aware.

Interferences include interfering nuclear reactions (primary, secondary and second-order), overlapping peaks, absorption of radiations entering or leaving the sample and miscellaneous errors.

A **primary interference reaction** occurs when the radionuclide of interest for determining a particular element can be produced from other elements in the sample matrix via different nuclear reactions. For example, in order to determine the amount of magnesium in a rock sample containing magnesium, aluminum and silicon in a neutron flux consisting of both thermal and fast neutrons, the following nuclear reactions will produce the same radionuclide: $^{26}Mg(n,\gamma)^{27}Mg$, $^{27}Al(n,p)^{27}Mg$ and $^{30}Si(n,\alpha)^{27}Mg$. The ^{27}Mg activity produced by irradiating this sample will be due to all three of the elements. The effect of the interference on the determination of magnesium will depend on the amounts of all three elements in the rock sample, the cross-section for each target nuclide, and the size of the thermal and fast neutron components of the neutron flux.

Another type of nuclear reaction that sometimes causes a primary interference is fission. When rock samples containing significant amounts of uranium are irradiated by thermal neutrons, a large number of fission-product nuclides may be produced. Some of the fission-product nuclides produced will decay to the nuclides ^{140}La, ^{141}Ce, ^{147}Nd, ^{95}Zr, and others also produced by (n,γ) reactions on these elements in the rock sample.

In some situations, simple methods to correct for primary interference reactions are possible (Glascock *et al.* 1985, 1986, Landsberger 1986). In other cases, the presence of an interfering reaction may preclude determination of certain elements.

A **secondary nuclear interference** occurs when the resulting particles from the primary reaction produce the desired radionuclide by nuclear reactions with other elements in the sample matrix. This situation occurs more often when the primary reaction involves 14 MeV neutrons, charged particles or high-energy photons than with reactor irradiations. An example of this occurs in the determination of nitrogen by 14 MeV neutrons using the $^{14}N(n,2n)^{13}N$ reaction. If the sample matrix contains a significant amount of hydrogen and carbon, the fast neutrons can induce a proton flux by scattering reactions on the hydrogen. The protons can then induce the $^{13}C(p,n)^{13}N$ on carbon and complicate the determination of nitrogen. In practice, secondary nuclear interferences are rarely serious.

A **second-order nuclear interference** occurs when the desired radionuclide is enhanced by the decay of a short-lived product of a neighboring element or by reduction of the desired radionuclide by activation if the reaction has a high cross-section and the target element becomes significantly depleted during irradiation. The latter situation may occur if the irradiation time is long such that target **burnup** occurs.

Gamma-ray spectral interferences occur when two different radionuclides emit gamma-rays with the same or nearly the same energy. For example, the radionuclides ^{51}Ti ($t_{1/2} = 5.76$ min) and ^{51}Cr ($t_{1/2} = 27.7$ d) both emit a gamma-ray at 320 keV. By knowing the history of a sample containing these radionuclides or by counting the sample twice (e.g. soon after irradiation and much later), the individual activities can be measured with good reliability. In another example, the 846.8 keV gamma-ray from ^{56}Mn ($t_{1/2} = 2.58$ h) and the 843.8 keV gamma-ray from ^{27}Mg ($t_{1/2} = 9.46$ min) often overlap. The analyst interested in determining magnesium and manganese should use the highest-resolution detector available or employ a two-measurement method similar to the one mentioned for titanium and chromium determination. Yet another possibility is to use other gamma-rays for the radionuclides of interest not affected by spectral interference.

A type of nuclear interference that can occur during the irradiation process due to absorption of the irradiating particles by certain high cross-section elements in the sample matrix is called **self-shadowing** or **self-shielding**. Sample matrices containing significant amounts of Cd and B are examples of this. The presence of these elements in the sample may result in different nuclear reaction rates in different parts of the sample. The main solution to reducing this problem is to keep the sample size small.

A type of nuclear interference that can occur during counting is due to the absorption of emitted radiation by the sample matrix. This interference is called **self-shadowing of gamma-rays** or **self-absorption**. If the measured gamma-rays are low-energy or the sample is large or contains a significant amount of high-Z elements (e.g. lead), the emitted gamma-rays may be attenuated. To overcome this problem, comparator standards of similar composition to the unknown sample matrix are often chosen, and the sizes of samples and standards are made nearly equal.

Instrument malfunctions and attempts to count samples at extremely high counting rates can lead to serious analytical problems. A careful analyst should notice these types of problems before they cause analytical difficulty.

4.6.8 Shielding and Compton suppression

Background activity is the most common cause of poor detection limits in activation analysis. The background in a gamma-ray spectrum is due to unwanted radioactivity from the surrounding environment, from the interaction of primary gamma-rays from the source with materials surrounding the detector, and from Compton scattering of gamma-rays inside the detector which escape.

Proper choice of shielding materials and their careful arrangement around the detector can reduce the contributions from the surrounding environment. When significant shielding is provided, backgrounds due to the cosmic flux and ambient sources of gamma-rays are greatly decreased, leaving the shielding materials around the detector as the most important cause of the remainder. A good-quality, low-background shield containing 10 cm or more of low-radiogenic lead surrounding a detector is typically essential.

In addition to Compton scattering, photoelectric absorption in the materials immediately surrounding the detector can lead to generation of characteristic X-rays. When the material surrounding a detector has a high atomic number, the X-ray photons are relatively energetic and can penetrate significant thicknesses of intervening material to enter the detector. In general, high-*Z* materials in the immediate vicinity of the detector should be avoided. On the other hand, the most effective shielding materials are those with high atomic numbers (i.e. lead). A graded shield is one in which the bulk of the shield is made of lead, but the inner surface is lined with successive layers of lower-*Z* materials (i.e. cadmium and copper). The successively lower-*Z* materials absorb the characteristic X-rays emitted by the bulk of the shield while emitting very low-energy and weakly penetrating X-rays of their own.

A method for reducing the Compton continuum from the spectrum involves use of a **Compton-suppression** system. By surrounding the primary detector with a second detector, events in which part of the incident gamma-ray energy is deposited in both detectors can be rejected if they both occur within a small time difference (i.e. suggesting coincidence). A modern Compton-suppression system usually involves an HPGe detector as the primary detector surrounded by an annulus of NaI(Tl). Backgrounds can be reduced by factors of 2–7, depending upon the characteristics of both detectors.

4.7 Activation analysis techniques

Activation analysis is capable of determining a large number of elements in a wide range of sample matrices. Virtually any type or size of sample (solid, liquid or gas; geological, environmental or biological; and large, medium or small) can be analyzed by activation analysis. Applications for activation analysis come from many different disciplines, including archeology, agriculture, biochemistry, environmental science, forensics, geochemistry, industry and materials science.

The general technique of activation analysis involves execution of a series of steps whose parameters are selected to optimize sensitivity and accuracy while minimizing effort and cost. A general outline of the major steps involved in performing activation analysis is as follows:

1. Select an appropriate nuclear reaction
2. Choose the irradiation facility
3. Prepare samples for analysis, including standard materials

4. Select the irradiation conditions
5. Choose the length of irradiation
6. Handle the samples after irradiation
7. Select the measurement system
8. Assess the overall accuracy and precision

Choices made during the performance of each step will determine the overall quality of the analysis and may decide whether the analysis is effective in answering the questions posed about the sample(s) under investigation. In the next section, general procedures for sample preparation, irradiation and measurement will be introduced. Several of the most common techniques of activation analysis will also be described.

4.7.1 Preparation of samples and standards

Because most activation analysis procedures are used for trace-element analysis, the process of preparing samples requires careful attention. Samples should be handled with powder-free latex gloves and Teflon tweezers to minimize contamination. When preparing samples for ultra-trace analysis, it is recommended that work be performed in a laminar flow hood or in a special clean room (filtered to remove dust particles). The samples, whose masses vary from milligrams to several grams, are weighed into irradiation containers or mounted on holders appropriate for the type of irradiation to be performed. High-purity polyethylene or polypropylene vials can be used for short irradiations (< 15 min in low to medium neutron flux densities < 10_{14} n cm^{-2} s^{-1}), but high-purity quartz vials are necessary if the irradiations are long (hours or days in medium to high flux densities > 10_{13} n cm^{-2} s^{-1}). Some large sample types (e.g. silicon wafers, slabs or crystals) intended for long irradiations at high fluxes are prepared in high-purity graphite boxes. Standards and flux monitors should be similarly encapsulated. If the samples and standards will be counted while inside their vials, blank vials should be analyzed to assess the need for background correction.

Some activation analysis procedures employ **flux monitors** irradiated with the unknowns to measure the neutron spectrum. The flux monitors are usually high-purity single-element materials (e.g. Co, Ni, Zr and/or Au) with well-known nuclear parameters. Other activation analysis procedures utilize **comparator standards** containing known amounts of the elements to be measured. The best types of **comparator** standards for activation analysis are those prepared from high-purity, stoichiometrically well-defined compounds.

Comparator standards can be made by pipeting known amount(s) of one or more elements onto a filter paper. The filter paper should be characterized before use to ensure minimal contaminants are present.

Filters need to be placed in clean vials for irradiation and, if possible, transferred to inert vials prior to counting. Some types of filters may not be transferable because of their condition following irradiation at high neutron flux. Filters can also be prepared in pellet form. The advantage of pelletization is that exact geometries can be reproduced for irradiation and counting.

If large numbers of elements are to be determined in the unknown samples, preparation of individual standards for each element becomes impracticable, because the number of standards may become excessive. As a result, carefully prepared **multi-element comparator** standards are essential for most activation analysis. **Standard reference materials** (SRMs) – certified by agencies such as the National Institute of Standards and Technology (NIST), the United States Geological Survey (USGS), the International Atomic Energy Agency (IAEA) or other sources – similar to the matrix undergoing analysis are often available. In most cases, SRMs are recommended for use as quality assurance materials to be treated as unknowns in the analytical procedure, but sometimes one SRM is used as the primary comparator standard and another SRM serves as the quality assurance material. Many different types of SRMs exist such that it is almost always possible to find an SRM material similar in composition to the unknown sample matrix. The use of comparator standards and SRMs for standardization and quality control are important as every attempt should be made to guarantee the quality of the analytical data before release for external usage (Iyengar 1989).

4.7.2 Irradiation facilities and counting techniques

Irradiations to measure short-lived radionuclides by neutron activation analysis are conducted by placing the polyethylene or polypropylene vials inside a larger polyethylene container commonly called a **rabbit**. The rabbit is made to travel from the laboratory into the reactor neutron flux through a pneumatic-tube transport system that moves the rabbit rapidly back and forth. In general, samples and standards are sent into the reactor sequentially. Some essential requirements of the system are that the irradiation geometry must be very reproducible and timing must be very precise. The pneumatic-tube transport system is often controlled by the same computer that controls the gamma-ray spectrometer used for data acquisition. Counting is normally performed in a low-background environment well away from the reactor containment.

For the analysis of longer-lived radionuclides by NAA, large numbers of samples and standards (50–100) in quartz vials are bundled together by wrapping in aluminum foil and packed into aluminum irradiation cans. Graphite boxes containing large samples can be placed into the irradiation cans directly. The irradiation cans are usually welded shut to keep the samples dry during irradiation in the reactor pool. After irradiation, the samples are allowed to cool for several hours, days or sometimes weeks to allow the unwanted short-lived radionuclides time to decay. After irradiation, the quartz vials are normally cleaned in aqua regia to remove surface contamination or the vials can be opened in order to transfer the sample material into clean non-irradiated containers. If there are large numbers of samples, counting is normally conducted with automatic sample changers that move the samples from a shielded area to the front of a detector according to commands from the data acquisition computer.

Samples irradiated in neutron or charged-particle beams are usually mounted on special holders that allow the sample beam to activate the sample and, in some instances, a reference standard or flux monitor, simultaneously. The sample holder should be made of materials that do not contribute a significant interference to the sample spectra or if post-irradiation counting is performed the sample can be removed from the holder. Reproducibility of both the irradiation and counting geometries is essential.

4.7.3 Thermal neutron activation analysis

Thermal neutrons are the dominant component of a reactor neutron spectrum and, in general, have the largest probabilities for nuclear reactions, through radiative capture (n,γ) with elements in the sample. As a result, thermal neutrons from a nuclear reactor offer the greatest analytical sensitivities for most elements such that the technique known as **thermal neutron activation analysis** (TNAA) is the most frequently employed method of activation analysis. Although some fast neutron reactions may occur in the sample due to the high-energy neutron component of the neutron spectrum, the thermal neutrons are generally responsible for the bulk of the radioactivity induced in the samples. It is generally assumed when mentioning **instrumental neutron activation analysis** (INAA) that we are referring to TNAA.

In order to calculate concentrations of elements measured in TNAA by the **absolute method**, knowledge of all nuclear and experimental parameters is required such that the following activity equation can be used:

$$A = \left(\frac{m}{M} N_A \theta\right)(\varphi_{th}\sigma_{th} + \varphi_{epi} I) P_\gamma \varepsilon SDC \tag{4.51}$$

where m = mass of sample (g), M = atomic weight (g mol^{-1}), N_A = Avogadro's number (6.02×10^{23} molecules/mole), θ = isotopic abundance, φ_{th} = thermal neutron flux (n cm^{-2} s^{-1}), σ_{th} = thermal neutron cross-section (cm^2), φ_{epi} = epithermal neutron flux (n cm^{-2} s^{-1}), I = resonance integral (cm^2), P_γ = intensity of the measured gamma-ray, ε = efficiency of detector at the energy of the measured gamma-ray, S = saturation factor $(1 - e^{-\lambda T_i})$, D = decay factor $(e^{-\lambda T_d})$, C = count factor $(1 - e^{-\lambda T_c})$ and T_i, T_d, T_c = irradiation, decay and count times, respectively, λ = decay constant (s^{-1}).

However, the absolute activation analysis procedure is rarely used because of uncertainties in the nuclear parameters. Instead, comparator methods are normally preferred. The equation used to calculate the mass of an element in the unknown relative to the comparator standard is

$$\frac{A_{sam}}{A_{std}} = \frac{m_{sam}}{m_{std}} \frac{(e^{-\lambda T_d})_{sam}}{(e^{-\lambda T_d})_{std}} \tag{4.52}$$

where A = activity of sample (sam) and standard (std), m = mass of the element and T_d = decay time.

When performing short irradiations, the irradiation, decay and counting times are normally fixed the same for all samples and standards such that the time-dependent factors will cancel. Thus, using equation (4.52), the concentration of the element in the sample becomes

$$c_{sam} = c_{std} \frac{W_{std}}{W_{sam}} \frac{A_{sam}}{A_{std}} \tag{4.53}$$

where c = concentration of element in sample and standard, and W = weight of sample and standard.

In recent years, a single-element comparator method for multi-element analysis has gained popularity. In this method, the so-called **k-zero factor**, k_0, for each radionuclide is determined experimentally by comparison with a single-element monitor. Although the most popular monitoring element is gold, other elements can also be used as the monitor. Tables of k_0 factors based on gold have been published (De Corte *et al.* 1987). The equation to calculate concentrations by the k_0 method is

$$c = \frac{A_{sp}}{A_{sp}^*} \frac{1}{k_0} \left(\frac{f + Q_0^*(\alpha)}{f + Q_0(\alpha)} \right) \frac{\varepsilon^*}{\varepsilon} \tag{4.54}$$

and c = concentration (ppm or µg/g), A_{sp} = specific activity of the comparator element (*) or sample, k_0 = k-zero factor for this isotope, f = ratio of thermal to epithermal flux, Q_0 = ratio of resonance integral to thermal neutron cross-section, α = deviation of the slope of the epithermal neutron flux from $1/E$ and ε = detector efficiency.

Thermal NAA is widely used to determine element concentrations at trace and ultra-trace levels. Many of the earliest applications of TNAA involved geochemistry and cosmochemistry because the method is one of the best for determining the rare-earth elements (REEs) in terrestrial rocks and meteorites (Haskin *et al.* 1968, Das *et al.* 1989). Meteorite data have been used to formulate theories about the process of **nucleosynthesis** that formed the elements in the abundances we observe today. One of the models of nucleosynthesis involves the so-called **Big Bang theory**.

Another of the early applications of TNAA occurred with the analysis of archeological materials (Glascock 1992, Griffin and Gordus 1967, Perlman and Asaro 1969, Sayre and Dodson 1957). Trace-element fingerprints of artifacts are often used to determine the **provenance** or origins of raw materials used to manufacture the artifacts. The information helps archeologists to reconstruct prehistoric activities such as trade and movement of peoples.

Interest in the quality of our environment has led to numerous studies of environmental samples by TNAA. Analyses of toxic metals (i.e. Hg, Cd, As, Cu and Sb) in sewage, mining runoff and air particulate samples have provided a great deal of information about the transport and monitoring of these elements in our environment (Tolgyessy and Klehr 1987).

The analysis of high-purity materials for the presence of trace contaminants is an area where TNAA excels. Ultra-trace amounts of the transition-group metals (i.e. Cr, Fe, Co, Ni, Cu, Zn, As, Sb and Au) can drastically affect the properties of semiconductors. These elements are all possible by TNAA. Other types of high-purity materials suitable for analysis by TNAA include quartz, metals, plastics and ceramics.

The sensitivity and selectivity of TNAA are excellent for use in trace-element analysis of biological materials. The spectra from irradiated human and animal tissues and plants are often dominated initially by ^{38}Cl ($t_{1/2} = 37.2$ min) and ^{24}Na ($t_{1/2} = 15$ h) and later by ^{82}Br ($t_{1/2} = 35.3$ h) and ^{32}P ($t_{1/2} = 14.3$ d). However, several trace elements with narrow tolerance limits in plants and animals are frequently measured by TNAA (e.g. Fe, Se and Zn). At the present time, there is much interest in the roles of trace elements in food composition, nutrition and health (Heydorn 1984).

Review articles have been published that describe an almost unlimited number of applications of TNAA (Ehmann and Vance 1989). Table 4.1 lists the nuclear parameters for elements suitable for measurement by TNAA. Table 4.2 lists gamma-rays frequently measured by TNAA according to increasing energy.

4.7.4 Epithermal neutron activation analysis

Most major elements follow the $1/v$ cross-section for the (n,γ) reaction throughout the energy spectrum. On the other hand, several of the less abundant elements have large resonance integrals in the epithermal energy region, and consequently are activated in relatively greater proportions by epithermal neutrons than by thermal neutrons. For example, geological samples often contain high concentrations of elements such as Na, Al, K, Sc, Cr, Mn, Fe and La, all of which are strongly activated by thermal

Table 4.1 Nuclear parameters for elements suitable for measurement by thermal neutron activation analysis*

Z	Element	Atomic weight	Nuclear reaction	Abundance of target isotope	σ_{th}(b)	I(b)	Product half-life	Energies and branching intensities for the prominent gamma-rays emitted by the product isotope
8	O	16.00	$^{18}O(n,\gamma)^{19}O$	0.00204	0.00016	0.00081	26.9 s	197.1 (95.9), 1356.8 (50.4)
9	F	19.00	$^{19}F(n,\gamma)^{20}F$	1.00	0.0095	0.039	11.0 s	1633.6 (100)
11	Na	22.99	$^{23}Na(n,\gamma)^{24}Na$	1.00	0.513	0.303	15.0 h	1368.6 (100), 2754.0 (99.9)
12	Mg	24.30	$^{26}Mg(n,\gamma)^{27}Mg$	0.1101	0.0372	0.024	9.46 m	843.8 (71.4), 1014.4 (28.6)
13	Al	26.98	$^{27}Al(n,\gamma)^{28}Al$	1.00	0.226	0.16	2.24 m	1779.0 (100)
16	S	32.07	$^{36}S(n,\gamma)^{37}S$	0.0002	0.16	0.18	5.05 m	3104.0 (94)
17	Cl	35.45	$^{37}Cl(n,\gamma)^{38}Cl$	0.2423	0.423	0.29	37.2 m	1642.7 (31), 2167.7 (42)
19	K	39.10	$^{41}K(n,\gamma)^{42}K$	0.0673	1.45	1.41	12.4 h	1524.6 (18.8)
20	Ca	40.08	$^{46}Ca(n,\gamma)^{47}Ca$	0.00004	0.62	0.81	4.54 d	1297.1 (74)
			$^{46}Ca(n,\gamma)^{47}Ca(\beta^-)^{47}Sc$				3.35 d	159.4 (67.9)
			$^{48}Ca(n,\gamma)^{49}Ca$	0.00187	1.12	0.50	8.72 m	3084.5 (92.1)
21	Sc	44.96	$^{45}Sc(n,\gamma)^{46}Sc$	1.00	26.3	11.3	83.8 d	889.3 (100), 1120.6 (100)
22	Ti	47.88	$^{50}Ti(n,\gamma)^{51}Ti$	0.054	0.171	0.115	5.76 m	320.1 (93.1)
23	V	50.94	$^{51}V(n,\gamma)^{52}V$	0.9975	4.79	2.63	3.75 m	1434.1 (100)
24	Cr	52.00	$^{50}Cr(n,\gamma)^{51}Cr$	0.0435	15.2	8.1	27.7 d	320.1 (10.1)
25	Mn	54.94	$^{55}Mn(n,\gamma)^{56}Mn$	1.00	13.2	13.9	2.58 h	846.8 (98.9), 1810.7 (27.2), 2113.0 (14.3)
26	Fe	55.85	$^{58}Fe(n,\gamma)^{59}Fe$	0.0028	1.31	1.28	44.5 d	1099.2 (56.5), 1291.6 (43.2)
27	Co	58.93	$^{59}Co(n,\gamma)^{60}Co$	1.00	37.13	74	5.27 y	1173.2 (99.9), 1332.5 (100)
			$^{59}Co(n,\gamma)^{60m}Co$	1.00	20.0	39.7	10.5 m	58.6 (2.0), 1332.5 (0.24)
28	Ni	58.69	$^{64}Ni(n,\gamma)^{65}Ni$	0.0091	1.69	1.13	2.52 h	1115.6 (14.8), 1481.8 (23.5)
29	Cu	63.55	$^{63}Cu(n,\gamma)^{64}Cu$	0.6917	4.28	4.88	12.7 h	511 (35.8), 1345.8 (0.48)
			$^{65}Cu(n,\gamma)^{66}Cu$	0.3083	2.48	2.63	5.10 m	1039.2 (7.4)
30	Zn	65.39	$^{64}Zn(n,\gamma)^{65}Zn$	0.4860	0.726	1.42	244 d	1115.6 (50.7)
			$^{68}Zn(n,\gamma)^{69m}Zn$	0.1880	0.0699	0.223	13.8 h	438.6 (94.8)
31	Ga	69.72	$^{71}Ga(n,\gamma)^{72}Ga$	0.399	4.61	30.6	14.1 h	630.0 (24.9), 834.1 (95.9)
32	Ge	72.59	$^{76}Ge(n,\gamma)^{77}Ge$	0.0780	0.05	2.0	11.3 h	211.0 (29.2), 215.5 (27.1), 264.4 (51)
33	As	74.92	$^{75}As(n,\gamma)^{76}As$	1.00	3.86	52.5	26.3 h	559.1 (44.6)
34	Se	78.96	$^{74}Sc(n,\gamma)^{75}Se$	0.0090	51.2	512	120 d	121.1 (17.3), 136.0 (59), 264.7 (59.2), 279.5 (25.2)
			$^{76}Se(n,\gamma)^{77m}Se$	0.0900	21	16	17.4 s	161.9 (52.4)

Table 4.1 (Continued)

Z	Element	Atomic weight	Nuclear reaction	Abundance of target isotope	σ_{th}(b)	I(b)	Product half-life	Energies and branching intensities for the prominent gamma-rays emitted by the product isotope
35	Br	79.90	$^{79}Br(n,\gamma)^{80}Br$	0.5069	9.85	115.9	17.7 m	616.3 (6.7)
			$^{81}Br(n,\gamma)^{82}Br$	0.4931	2.58	49.8	35.3 h	554.4 (70.8), 776.5 (83.5)
37	Rb	85.47	$^{85}Rb(n,\gamma)^{86}Rb$	0.7217	0.494	7.31	18.7 d	1076.6 (8.8)
38	Sr	87.62	$^{84}Sr(n,\gamma)^{85}Sr$	0.0056	0.690	9.14	64.8 d	514.0 (99.3)
			$^{86}Sr(n,\gamma)^{87m}Sr$	0.0986	0.770	3.17	2.81 h	388.4 (82.3)
40	Zr	91.22	$^{94}Zr(n,\gamma)^{95}Zr$	0.1738	0.053	0.268	64.0 d	724.2 (44.2), 756.7 (54.5)
			$^{94}Zr(n,\gamma)^{95}Zr(\beta^-)^{95}Nb$				35.0 d	765.8 (99.8)
			$^{96}Zr(n,\gamma)^{97}Zr$	0.0280	0.0213	5.28	16.7 h	743.3 (97.9)
42	Mo	95.94	$^{98}Mo(n,\gamma)^{99}Mo$	0.2413	0.131	6.96	65.9 h	140.5 (90.7)
44	Ru	101.07	$^{102}Ru(n,\gamma)^{103}Ru$	0.3160	1.16	4.21	39.3 d	497.1 (90.9)
45	Rh	102.90	$^{103}Rh(n,\gamma)^{104m}Rh$	1.00	11	82	4.34 m	51.4 (48.2)
46	Pd	106.42	$^{108}Pd(n,\gamma)^{109}Pd$	0.2646	8.77	253	13.7 h	88.0 (3.6)
47	Ag	107.87	$^{109}Ag(n,\gamma)^{110m}Ag$	0.4816	3.9	69	250 d	657.8 (94.6), 884.7 (72.7), 937.5 (34.4)
48	Cd	112.41	$^{114}Cd(n,\gamma)^{115}Cd$	0.2873	0.23	9.1	53.5 h	527.9 (27.5)
			$^{114}Cd(n,\gamma)^{115}Cd(\beta^-)^{115m}In$				4.49 h	336.3 (45.8)
49	In	114.82	$^{115}In(n,\gamma)^{116m}In$	0.9570	157	2638	54.2 m	416.9 (29.2), 1097.3 (56.2), 1293.5 (84.4)
50	Sn	118.71	$^{112}Sn(n,\gamma)^{113}Sn$	0.0097	0.541	26.2	115 d	391.7 (64)
51	Sb	121.75	$^{121}Sb(n,\gamma)^{122}Sb$	0.573	6.33	209	2.70 d	564.2 (69.3)
			$^{123}Sb(n,\gamma)^{124}Sb$	0.427	4.08	118	60.2 d	602.7 (97.8), 1691.0 (47.3)
53	I	126.90	$^{127}I(n,\gamma)^{128}I$	1.00	4.04	100	25.0 m	442.9 (16.9)
55	Cs	132.90	$^{133}Cs(n,\gamma)^{134}Cs$	1.00	30	390	2.06 y	604.7 (97.6), 795.8 (85.4)
56	Ba	137.33	$^{130}Ba(n,\gamma)^{131}Ba$	0.00106	9.04	184	11.5 d	123.8 (29.1), 216.0 (20), 496.3 (44)
			$^{138}Ba(n,\gamma)^{139}Ba$	0.717	0.405	0.36	84.6 m	165.8 (22)
57	La	138.90	$^{139}La(n,\gamma)^{140}La$	0.9991	9.34	11.6	40.3 h	328.8 (20.6), 487.0 (44.3), 815.8 (22.9), 1596.2 (95.4)
58	Ce	140.12	$^{140}Ce(n,\gamma)^{141}Ce$	0.8848	0.575	0.48	32.5 d	145.4 (48.2)
60	Nd	144.24	$^{146}Nd(n,\gamma)^{147}Nd$	0.1719	1.45	2.90	11.0 d	91.1 (28), 521.0 (13.1)

62	Sm	150.36	$^{152}Sm(n,\gamma)^{153}Sm$	0.267	220	3168	46.3 h	69.7 (4.85), 103.2 (28.8)
63	Eu	151.96	$^{151}Eu(n,\gamma)^{152}Eu$	0.478	5900	5564	13.3 y	121.8 (28.4), 344.3 (26.6), 1408.0 (20.8)
64	Gd	157.25	$^{152}Gd(n,\gamma)^{153}Gd$	0.0020	1100	3000	242 d	69.7 (2.3), 97.4 (27.6), 103.2 (19.6)
65	Tb	158.92	$^{159}Tb(n,\gamma)^{160}Tb$	1.00	23.8	426	72.3 d	298.6 (28.9), 879.4 (32.9), 966.2 (27.2), 1178.0 (16.2)
66	Dy	162.50	$^{164}Dy(n,\gamma)^{165}Dy$	0.282	2725	518	2.33 h	94.7 (3.6), 361.7 (0.84)
67	Ho	164.93	$^{165}Ho(n,\gamma)^{166}Ho$	1.00	58.1	636	26.8 h	80.6 (6.3)
68	Er	167.26	$^{170}Er(n,\gamma)^{171}Er$	0.149	8.85	39.1	7.52 h	111.6 (20.5), 295.9 (28.9), 308.3 (64.4)
69	Tm	168.93	$^{169}Tm(n,\gamma)^{170}Tm$	1.00	107	1552	129 d	84.3 (3.3)
70	Yb	173.04	$^{168}Yb(n,\gamma)^{169}Yb$	0.0013	3470	31000	32.0 d	63.1 (43.7), 177.2 (21.4), 198.0 (34.9)
			$^{174}Yb(n,\gamma)^{175}Yb$	0.318	128	58.9	4.19 d	113.8 (1.9), 282.5 (3.0), 396.3 (6.5)
71	Lu	174.97	$^{176}Lu(n,\gamma)^{177}Lu$	0.026	2100	1160	6.71 d	113.0 (6.4), 208.4 (11)
72	Hf	178.49	$^{180}Hf(n,\gamma)^{181}Hf$	0.351	13.5	34	42.4 d	133.0 (35.9), 136.3 (5.8), 345.9 (15.1), 482.2 (80.6)
73	Ta	180.95	$^{180}Ta(n,\gamma)^{181}Ta$	0.99988	20.4	679	114 d	67.8 (42.3), 1121.3 (35), 1221.4 (27.1)
74	W	183.95	$^{186}W(n,\gamma)^{187}W$	0.286	38.7	530	23.9 h	479.6 (21.1), 685.7 (26.4)
75	Re	186.21	$^{185}Re(n,\gamma)^{186}Re$	0.374	106	1632	90.6 h	122.4 (0.66), 137.1 (8.5)
			$^{187}Re(n,\gamma)^{188}Re$	0.626	73.2	318	17.0 h	155.1 (14.9)
76	Os	190.2	$^{184}Os(n,\gamma)^{185}Os$	0.0002	3613	1554	93.6 d	646.1 (81)
77	Ir	192.22	$^{191}Ir(n,\gamma)^{192}Ir$	0.373	924	3750	73.8 d	296.0 (28.7), 308.5 (29.8), 316.5 (83), 468.1 (47.7)
			$^{193}Ir(n,\gamma)^{194}Ir$	0.627	115	1380	19.2 h	293.5 (2.5), 328.5 (13)
78	Pt	195.08	$^{196}Pt(n,\gamma)^{197}Pt$	0.253	0.74	8.0	18.3 h	77.3 (17.1), 191.4 (3.7)
79	Au	196.97	$^{197}Au(n,\gamma)^{198}Au$	1.00	98.6	1550	2.70 d	411.8 (95.5)
80	Hg	200.59	$^{196}Hg(n,\gamma)^{197}Hg$	0.0014	3080	413	64.1 h	77.3 (18)
			$^{202}Hg(n,\gamma)^{203}Hg$	0.2980	4.35	3.8	46.6 d	279.2 (81.5)
90	Th	232.04	$^{232}Th(n,\gamma)^{233}Th$	1.00	7.26	83.7	22.3 m	86.5 (2.6), 459.3 (1.4)
			$^{232}Th(n,\gamma)^{233}Th(\beta^-)^{233}Pa$				27.0 d	300.2 (6.2), 312.0 (36)
92	U	238.03	$^{238}U(n,\gamma)^{239}U$	0.9927	2.75	284	23.5 m	74.7 (50)
			$^{238}U(n,\gamma)^{239}U(\beta^-)^{239}Np$				2.36 d	106.1 (22.9), 228.2 (10.8), 277.6 (14.2)

[*] Adapted from Glascock (1996).

Table 4.2 Gamma-rays arranged according to energy*

Energy (keV)	Branching intensity	Isotope	Half-life
63.1	43.7	^{169}Yb	32.0 d
67.8	42.3	^{181}Ta	114 d
69.7	4.85	^{153}Sm	46.3 h
88.0	3.6	^{109}Pd	13.7 h
91.1	28	^{147}Nd	11.0 d
94.7	3.6	^{165}Dy	2.33 h
103.2	28.8	^{153}Sm	46.3 h
106.1	22.9	^{239}Np	2.36 d
113.0	6.4	^{177}Lu	6.71 d
113.8	1.9	^{175}Yb	4.19 d
121.1	17.2	^{75}Se	120 d
121.8	28.4	^{152}Eu	13.3 y
123.8	29.1	^{131}Ba	11.5 d
133.0	35.9	^{181}Hf	42.4 d
136.0	59	^{75}Se	120 d
136.3	5.8	^{181}Hf	42.4 d
145.4	48.2	^{141}Ce	32.5 d
161.9	52.4	^{77m}Se	17.4 s
165.8	22.0	^{139}Ba	84.6 m
177.2	21.4	^{169}Yb	32.0 d
198.0	44.9	^{169}Yb	32.0 d
208.4	11	^{177}Lu	6.71 d
216.0	20	^{131}Ba	11.5 d
228.2	10.8	^{239}Np	2.36 d
264.7	59.2	^{75}Se	120 d
277.6	14.2	^{239}Np	2.36 d
279.2	81.5	^{203}Hg	46.6 d
279.5	25.2	^{75}Se	120 d
282.5	3.0	^{175}Yb	4.19 d
296.0	28.7	^{192}Ir	73.8 d
298.6	28.9	^{160}Tb	72.3 d
300.2	6.2	^{233}Pa	27.0 d
308.5	29.8	^{192}Ir	73.8 d
312.0	36	^{233}Pa	27.0 d
316.5	83	^{192}Ir	73.8 d
320.1	10.1	^{51}Cr	27.7 d
320.1	93.1	^{51}Ti	5.76 m
328.8	20.6	^{140}La	40.3 h
345.9	15.1	^{181}Hf	42.4 d
344.3	26.6	^{152}Eu	13.3 y
388.4	82.3	^{87m}Sr	2.81 h
391.7	64	^{113}Sn	115 d
396.3	6.5	^{175}Yb	4.19 d
411.8	95.5	^{197}Au	2.70 d
416.9	29.2	^{116m}In	54.2 m
442.9	16.9	^{128}I	25.0 m
468.1	47.7	^{192}Ir	73.8 d
479.6	21.1	^{187}W	23.9 h
482.2	80.6	^{181}Hf	42.4 d
487.0	44.3	^{140}La	40.3 h
496.3	44	^{131}Ba	11.5 d
514.0	99.3	^{85}Sr	64.8 d
521.0	13.1	^{147}Nd	11.0 d
554.5	70.8	^{82}Br	35.3 h
559.1	44.6	^{76}As	26.3 h
564.2	69.3	^{122}Sb	2.70 d
602.7	97.8	^{124}Sb	60.2 d
604.7	97.6	^{134}Cs	2.06 y
685.7	26.4	^{187}W	23.9 h
724.2	44.2	^{95}Zr	64.0 d
756.7	54.5	^{95}Zr	64.0 d
776.5	83.5	^{82}Br	35.3 h
795.8	85.4	^{134}Cs	2.06 y
815.8	22.9	^{140}La	40.3 h
843.8	71.4	^{27}Mg	9.46 m
846.8	98.9	^{56}Mn	2.58 h
879.4	32.9	^{160}Tb	72.3 d
889.3	100	^{46}Sc	83.8 d
966.2	27.2	^{160}Tb	72.3 d
1014.4	28.6	^{27}Mg	9.46 m
1039.2	7.4	^{66}Cu	5.10 m
1076.6	8.8	^{86}Rb	18.7 d
1097.3	56.2	^{116m}In	54.2 m
1099.2	56.5	^{59}Fe	44.5 d
1115.6	50.7	^{65}Zn	244 d
1120.6	100	^{46}Sc	83.8 d
1121.3	35	^{181}Ta	114 d
1173.2	99.9	^{60}Co	5.27 y
1178.0	16.2	^{160}Tb	72.3 d
1221.4	27.1	^{181}Ta	114 d
1291.6	43.2	^{59}Fe	44.5 d
1293.5	84.4	^{116m}In	54.2 m
1297.1	74	^{47}Ca	4.54 d
1332.5	100	^{60}Co	5.27 y
1368.6	100	^{24}Na	15.0 h
1408.0	20.8	^{152}Eu	13.3 y
1434.1	100	^{52}V	3.75 m
1524.6	18.8	^{42}K	12.4 h
1596.2	95.4	^{140}La	40.3 h
1633.6	100	^{20}F	11.0 s
1642.7	31	^{38}Cl	37.2 m
1691.0	47.3	^{124}Sb	60.2 d
1779.0	100	^{28}Al	2.24 m
1810.7	27.2	^{56}Mn	2.58 h
2113.0	14.3	^{56}Mn	2.58 h
2167.7	42	^{38}Cl	37.2 m
2754.0	99.9	^{24}Na	15.0 h
3084.5	92.1	^{49}Ca	8.72 m
3104.0	94	^{37}S	5.05 m

* Adapted from Glascock (1996).

neutrons. High activities of these elements in geological samples often mask the activities from low-concentration elements with high resonance integral to thermal cross-section ratios, I/σ_{th}, such as Ag, As, Au, Ga, Gd, In, Mo, Pd, Pt, Sm, Ta, Th, U and W.

The use of neutron filters made of B or Cd to shield against thermal neutrons and to selectively allow epithermal neutrons to reach the samples is called **epithermal neutron activation analysis** (ENAA). By suppressing the elements more sensitive to thermal neutrons, the relative sensitivities for the elements with high I/σ_{th} can be enhanced. In most instances, the ENAA method is accomplished by putting the samples inside irradiation containers made of B or Cd, but some reactors have permanently shielded irradiation positions for ENAA.

The enhancement in relative sensitivity by using ENAA is often expressed as an **advantage factor**. Advantage factors have been expressed in a number of ways by different authors (Bem and Ryan 1981, Parry 1984, Tian and Ehmann 1984). Different advantage factors account for the gains in relative sensitivity due to favorable ratios of I/σ_{th}, increases in signal-to-noise ratios, improvements in relative standard deviation, and taking into account the possibility of larger samples or counting samples in closer geometries by ENAA.

Epithermal NAA has been used to measure bromine and iodine in aerosols deposited on spruce needles by Wyttenbach *et al.* (1987), iodine in foods by Stroube *et al.* (1987) and several elements in geological samples by Parry (1982).

4.7.5 Fast neutron activation analysis

Fast neutron activation analysis (FNAA), with neutrons from about 0.5 MeV and up, can be helpful for determining a few selected elements when thermal neutrons cannot be used. Fast neutrons do not induce many (n,γ) reactions compared to thermal neutrons, but are instead responsible for (n,p), (n,α), (n,n′) and (n,2n) reactions. The cross-sections for fast neutron reactions are generally several orders of magnitude smaller than those for thermal neutron reactions and so the detection limits are usually much higher. Although most FNAA procedures use 14 MeV neutrons produced by a neutron generator, the fast neutron flux in a nuclear reactor is sometimes used for reactions with low threshold energies.

The main examples of use for 14 MeV neutron activation are oxygen, nitrogen and silicon using the reactions $^{16}O(n,p)^{16}N$, $^{14}N(n,2n)^{13}N$ and $^{28}Si(n,p)^{28}Al$ with cross-sections of 39, 7 and 230 mb, respectively. A review by Bild (1987) compares the use of 14 MeV neutron activation with other techniques.

4.7.6 Prompt gamma neutron activation analysis

The technique of **prompt gamma-ray neutron activation analysis** (PGNAA) utilizes the gamma-rays emitted immediately after the (n,γ) reaction with thermal neutrons. Since most nuclides have binding energies for capturing a neutron of ≈ 8 MeV, the prompt gamma-rays emitted from a compound nucleus are usually of higher energy than delayed gamma-rays. This higher initial excitation energy increases the number of nuclear levels available during de-excitation, such that prompt gamma-ray spectra are more complex. Because of the combination of a long range for neutrons in matter and the minimal absorption of high-energy gamma-rays, PGNAA is an ideal technique for large samples.

PGNAA overcomes certain limitations by (n,γ) reactions that rely on delayed gamma-rays, including:

1. Products have very short or long half-lives; for example, $^{10}B(n,\alpha)^{7}Li$ with $t_{1/2} < 10^{-14}$ s or $^{9}Be(n,\gamma)^{10}Be$ with $t_{1/2} = 1.6 \times 10^{6}$ y.
2. Product nuclides are in stable ground-state configurations; for example, $^{1}H(n,\gamma)^{2}H$, $^{113}Cd(n,\gamma)^{114}Cd$ and $^{157}Gd(n,\gamma)^{158}Gd$.
3. Product nuclides are produced with few or no gamma-rays; for example, $^{30}Si(n,\gamma)^{31}Si$, $^{31}P(n,\gamma)^{32}P$ and $^{44}Ca(n,\gamma)^{45}Ca$.

Systems for PGNAA must be designed so that irradiation and counting can be performed simultaneously. Therefore, the counting equipment needs to be placed very close to the reactor, which typically is a source of high background of gamma-rays and neutrons. Most PGNAA systems are designed around a beam of neutrons extracted from a reactor. As a consequence, PGNAA is limited to one sample at a time. The lower neutron fluxes outside a reactor necessitate longer irradiation times.

Three major inconveniences for PGNAA are background radiation, neutron scattering from the surrounding materials and neutron scattering within the sample. First, background radiation arises from interactions between the neutron beam and materials other than the sample (i.e. beam tubes, shielding materials, sample holders). High backgrounds make determination of low-intensity peaks difficult. Secondly, scattered neutrons may interact with the detector crystal to complicate the spectrum, add to background effects, distort peak shapes and degrade detector resolution. A third problem is the scattering of neutrons from hydrogenous samples. Difficulties with neutron scattering from samples containing hydrogen are least when the samples are spherically shaped (Copley and Stone 1989, Mackey *et al.* 1991). The use of **cold neutrons** and **guide tubes** to transport the neutrons to PGNAA sample counting stations further away from the reactor reduces background effects and enhances the potential sensitivity.

PGNAA has application to the same types of samples that are analyzed by delayed gamma-rays. Some of the elements reported in geological samples are present as major constituents, including Na, Al, Si, Ca, K and Fe, and others are minor constituents, including B, Cd, Co, Sm and Gd. An important industrial application of PGNAA is the analysis of low levels of boron in semiconductor materials.

Sensitivities for PGNAA are highest at reactor-base facilities but portable neutron sources enable PGNAA in different environments such as subsurface mineral exploration, well logging and bulk process streams. A number of applications for PGNAA have been reviewed by Glascock (1982).

4.7.7 Charged-particle activation analysis

Charged-particle activation analysis (CPAA) differs from NAA in several respects. Several low-Z elements with low neutron cross-sections are more easily analyzed by CPAA than NAA. Charged particles are affected by the Coulomb barrier of the nucleus and are less penetrating such that CPAA is primarily limited to thin targets or used as a surface analysis technique. Reactions with charged particles involve higher energies than most neutron reactions. Thus, a greater variety of nuclear reactions occur, and several different product radionuclides are available for measurement. Other differences generally considered disadvantages for CPAA relative to NAA are increased complexity and cost, lack of suitability for liquid samples and problems with sample heating.

In CPAA, the activating particles are charged particles generated by accelerators. The most commonly used particles are protons, deuterons, ^{3}He and alpha-particles. Because protons are the most easily accelerated, they are the favorite particle for most applications. In general, for protons with energies below 10 MeV only (p,n) and (p,α) reactions need to be considered; for protons with energies between 10 and 20 MeV, (p,2n), (p,d), (p,^{3}He) and (p,αn) reactions are more important; and for higher energies even more complex reactions will occur. Alpha-particles and other multi-charged particles have higher Coulomb barriers and must have higher initial energies to induce nuclear reactions.

If an infinitely thin sample is irradiated with charged particles, perpendicular to the surface, the rate of production is given by

$$R = ni\sigma(E)\,dx \tag{4.55}$$

where n = number of target nuclei, i = beam intensity and $\sigma(E)$ = cross-section of the reaction at energy E. Substitution of the rate of production from equation (4.55) into equation (4.26) followed by integration over an irradiation of length T_i gives the activity A:

$$A = ni\sigma(E)\,dx\,(1 - e^{-\lambda T_i}) \tag{4.56}$$

When the sample is not thin, the cross-section changes with depth, as the particle energy changes with depth. For a thick sample, the activity is given by

$$A = ni(1 - e^{-\lambda T_i}) \int_0^{r(E_I)} \sigma(E)\,dx = ni(1 - e^{-\lambda T_i}) \int_0^{r(E_I)} \frac{\sigma(E)\,dE}{S(E)} \quad (4.57)$$

In normal practice, the activation equation for CPAA is not used directly. A comparator method is used instead where the sample (x) and standard (s) are irradiated separately using charged particles of the same energy for the same length of time. Thus, the concentration in the unknown, C_x, can be determined from the following equation:

$$C_x = C_s \frac{i_s A_x r_s}{i_x A_s r_x} \quad (4.58)$$

where C_s = concentration of the element in the standard, i_s/i_x = ratio of beam intensities, A_x/A_s = ratio of measured activities and r_x/r_s = ratio of particle ranges.

The CPAA technique is used most often to determine the lighter elements B, C, N and O in metals and semiconductors for industry. Boron is usually determined with the $^{11}B(p,n)^{11}C$ reaction or less frequently by the $^{10}B(d,n)^{11}C$ or $^{10}B(p,\alpha)^{7}Be$ reactions. Carbon can be determined by using either the $^{12}C(^{3}He,\alpha)^{11}C$ or $^{12}C(d,n)^{13}N$ reactions. For the determination of nitrogen and oxygen in metals, the $^{14}N(p,\alpha)^{11}C$ and $^{16}O(^{3}He,p)^{18}F$ reactions, respectively, have been used. Vandecasteele (1988) describes a number of applications for CPAA.

4.7.8 Photon activation analysis

Photon activation analysis (PAA) is a technique that plays a minor role in activation analysis, primarily because of the small cross-sections for photonuclear reactions. PAA complements NAA by permitting light-mass and heavy-mass elements insensitive by conventional NAA methods to be analyzed. The PAA technique is also useful for samples containing elements with very large neutron absorption cross-sections (e.g. boron and cadmium) in the bulk matrix.

When a target nucleus is bombarded by photons, the nucleus can be transformed into a radionuclide by one of three possible reaction modes:

1. Photoexcitation reactions (γ,γ') that create isomeric state(s) of the target nuclide.
2. Photonuclear reactions such as (γ,n), (γ,p), $(\gamma,2n)$ and (γ,α) involving particle emission.
3. Photofission reactions (γ,f) with uranium and thorium to create fission-product radionuclides.

Nuclear reactions utilizing photons depend on the atomic number of the target atom and the energy of irradiating photons. Reaction cross-sections are typically small when compared to thermal neutron cross-sections and the photon must be above the threshold energy in order for the reaction to occur.

With the exception of a few low-sensitivity (γ,γ') reactions induced by strong ^{60}Co and ^{124}Sb radioisotopic sources, most PAA applications require photon beams with high energies and intensities. The source of photons for most PAA is bremstrahlung radiation produced by an electron linear accelerator with a beam current of several hundred microamps. For photons with energies from 15 to 20 MeV, the (γ,n) reaction is dominant; however, other reactions that can be used include (γ,p), $(\gamma,2n)$ and (γ,α). At higher photon energies, more complex reactions are induced.

The rate of production for a reaction by PAA is similar to those for other forms of activation analysis

$$R = n\int_{E_{min}}^{E_{max}} \sigma(E)\phi(E)\,dE \tag{4.59}$$

where $\phi(E)$ is the flux of photons per unit of energy interval.

Successful applications of PAA are common with the light-mass and heavy-mass elements not sensitive to NAA. Analysis of C, N, O and F is possible with detection limits below 1 ppm in large samples. Because (γ,n) reactions on these elements create radionuclides that are frequently pure positron emitters, and since pure positron emitters are only distinguishable from one another by their half-lives, radiochemical separations are frequently necessary. For heavier elements such as Pb, the PAA technique is capable of detecting concentrations below 1 ppm in small samples.

Data on the cross-sections for photonuclear reactions can be found in Forkman and Petersson (1987). Segebade *et al.* (1987) and Kushelevsky (1990) review the applications of PAA.

4.8 Special activation analysis methods

Although a majority of the elements can be determined by the routine instrumental methods described in the previous section, there are some elements or sample matrices that are unsuitable by the conventional methods of activation analysis. A few of the more specialized methods are described here.

4.8.1 Preconcentration methods with activation analysis

When part of the sample is removed prior to irradiation, the sample is considered to be **preconcentrated**. By reducing the sample volume, the sensitivity of counting is enhanced due to improved sample-to-detector geometry. For liquid samples, such as blood, urine or water, this generally means drying the sample. **Lyophilization** (i.e. freeze-drying) is normally used to remove the water, although heating to evaporate water can be used if there are no losses of the elements of interest during the process. For solid samples such as vegetation, low-temperature ashing to remove unwanted carbon, oxygen and nitrogen is possible.

The **platinum-group elements** (i.e. Ir, Os, Rh, Ru, Pd and Pt) in rock samples are not normally measured by routine methods of NAA. Their detection limits in rocks are typically several parts per million but the concentrations are often much lower.

The platinum-group elements (PGE) and gold are often dispersed heterogeneously in different mineral phases in rock samples. In order to obtain representative analyses, analysis of large specimens (e.g. 50 g) is generally recommended. Thus, a preconcentration method such as **fire assay** and collection of a nickel sulfide bead is often performed to obtain adequate sensitivity. Afterward, the bead is crushed and dissolved and the insoluble PGE and gold are collected on a filter paper. The filter paper containing them is then irradiated by TNAA, and the PGE and gold are measured with excellent sensitivities. The technique is described in more detail by Hoffman *et al.* (1978).

4.8.2 Radiochemical neutron activation analysis

The post-irradiation technique of **radiochemical neutron activation analysis** (RNAA) was quite common before the introduction of high-resolution Ge(Li) detectors in the mid-1960s, which led to the more modern and preferred instrumental NAA. However, gamma-ray spectra from some samples are dominated by a few interfering radioisotopes and the less abundant radioisotopes are difficult or impossible to measure. In biological sample matrices, the activities from ^{24}Na, ^{32}P, ^{38}Cl, ^{42}K and ^{82}Br often severely limit the analysis of other trace-element constituents. In geological samples, the activities from ^{24}Na, ^{46}Sc and ^{59}Fe are often dominating and inhibit determination of the REEs. In these cases, the full potential of NAA is not realized unless radiochemical separations after activation of the sample are used to remove the interferences or to extract the elements of interest from the sample. An advantage of post-irradiation radiochemistry over pre-irradiation chemistry is that adsorption losses and reagent contamination are not concerns.

Because the number of radioactive atoms of the element of interest in the sample is small, a non-radioactive **chemical carrier** with identical or similar properties to the element in solution is added to the solution in macroscopic amounts. The addition of a carrier provides the analyst with more manageable quantities of the sample. In addition, the amount of carrier recovered at the conclusion of the analysis versus the initial amount of carrier lets the analyst determine the efficiency or **yield** for the chemical separation procedure.

In general, RNAA requires a significant amount of labor and requires the analyst to handle large amounts of radioactivity. Examples of RNAA applied to complex biological and geological matrices are described by Fardy (1990).

4.8.3 *Cyclic instrumental neutron activation analysis*

The sensitivity of several elements with short-lived radionuclides can be enhanced by a technique known as **cyclic instrumental neutron activation analysis** (CINAA). Examples include elements such as F and Se, which create short-lived isotopes such as ^{20}F ($t_{1/2} = 11.0\,s$) and ^{77m}Se ($t_{1/2} = 17.4\,s$), respectively, from neutron capture. However, owing to their short half-lives, only a small number of radioactive atoms are produced at saturation and, as a result, counting yields a large statistical error. In order to increase the number of counts, a series of repetitive irradiations and counts followed by summing the individual gamma-ray spectra can be used to improve the counting statistics.

Use of a series of equal short irradiation times, followed by rapid transfer, measurement and re-irradiation, allows the same level of saturation to be reached for the short-lived species of interest while increasing the amounts of longer-lived species only slowly. The discrimination between short-lived and longer-lived radionuclides is enhanced most by using the shortest possible irradiation, decay and counting periods. Thus, the use of CINAA results in improved counting statistics for the short-lived radionuclide while keeping the signal-to-noise ratio low.

Cyclic activation involves four periods of time: the irradiation time, the decay time between irradiation and start of measurement, the counting time, and the transfer time required to return the sample back to the irradiation position. The total time covered by one cycle, T, is defined by

$$T = T_i + T_d + T_c + T_r \tag{4.60}$$

where T_i = irradiation time for each cycle (s), T_d = decay time for between irradiation and measurement for each cycle (s), T_c = counting time for each cycle (s) and T_r = transfer time back to the irradiation position for each cycle (s).

From equation (4.51), we know that the number of counts measured during the first count cycle, Z_1, is given by the following equation:

$$Z_1 = \left(\frac{m}{M} N_A \theta\right) \frac{(\varphi_{th}\sigma_{th} + \varphi_{epi} I) P_\gamma \varepsilon SDC}{\lambda} \tag{4.61}$$

During the second counting period, the total number of counts measured,

Z_2, will be equal to the number of new counts plus the remaining counts from the first irradiation:

$$Z_2 = Z_1 + Z_1 F = Z_1(1 + F) \tag{4.62}$$

where $F = e^{-\lambda T}$. During the nth cycle, the total number of counts measured, Z_n, is given by the geometric sum

$$Z_n = Z_1 \sum_{k=0}^{n-1} F^k = Z_1 \frac{1 - F^n}{1 - F} \tag{4.63}$$

At the conclusion of n cycles, the total number of counts in the cumulative spectrum is

$$Z_T = \sum_{i=1}^{n} Z_i \tag{4.64}$$

or when expanded

$$Z_T = \left(\frac{m}{M} N_A \theta\right)(\varphi_{th}\sigma_{th} + \varphi_{epi} I)\varepsilon P_\gamma SDC \left(\frac{n}{1 - F} - \frac{F(1 - F^n)}{(1 - F)^2}\right) \tag{4.65}$$

Optimization of the process depends on selection of appropriate times to optimize each of the time factors S, D, C and F. The optimization of CINAA has been reviewed by Spyrou (1981).

4.8.4 Gamma–gamma coincidence techniques in activation analysis

One of the more rare techniques employed in activation analysis involves the use of **gamma–gamma coincidence spectroscopy** (GGCS) in order to minimize spectral interferences. The technique relies on the condition that gamma-ray events occurring in two detectors are coincident in time such that time can be used to discriminate against those that are detected randomly. In special cases, large gains in sensitivity can be achieved by this method because the background counts can be almost entirely eliminated. One of the simplest examples is the measurement of annihilation photons emitted by positron-emitting radionuclides. Because the annihilation photons are emitted at 180° from one another, a pair of detectors on opposite sides of the sample can be used. The requirement that both gamma-rays be observed within a small time interval eliminates all but a very few accidental coincidences. Other radionuclides with gamma-rays in cascade are also possible candidates for GGCS (i.e. ^{60}Co, ^{75}Se and many others).

Assuming a decay scheme has two coincident isotropic gamma-rays in cascade, γ_a and γ_b, the equation to describe the number of coincident events measured by the system is

$$N_{ab} = \varepsilon_a \varepsilon_b A T_c \tag{4.66}$$

where N_{ab} = the total number of coincidence events counted, ε_a and ε_b = the efficiencies of detectors 1 and 2 for γ_a and γ_b, respectively, A = the activity of the radionuclide (disintegrations/s) and T_c = total counting time. Accidental events can be recorded during the coincidence resolving time, τ. The number of accidental events is approximately

$$N^*_{ab} = 2\tau\varepsilon_a\varepsilon_b A^2 T_c \quad (4.67)$$

The accidental-to-true-coincidence ratio is given by $2\tau A$. In order to minimize the number of accidental coincidences, the resolving time and detector geometries must be adjusted appropriately.

4.8.5 *Neutron depth profiling*

The distribution of certain trace elements as a function of depth can be measured in some high-purity semiconductor materials by a technique known as **neutron depth profiling** (NDP). The NDP technique employs a nuclear reaction that results in the emission of charged particles with a specific kinetic energy defined by the Q value for the nuclear reaction. If the nuclear reaction that produced the charged particle occurs inside the sample, the charged particle loses some of its kinetic energy before arriving at the detector. The difference between the initial and measured energies is related to the depth of the target nucleus inside the sample.

The NDP method was first developed by Ziegler *et al.* (1972) to determine boron depth profiles using the reaction $^{10}B(n,\alpha)^7Li$, which has a very high thermal neutron capture cross-section of 3837 b. In addition to boron, lithium, which has a thermal neutron capture cross-section of 940 b for the $^6Li(n,\alpha)^3H$ reaction, can be studied by NDP. Depth profiles of 1–10 μm have been reported by Downing *et al.* (1987).

4.8.6 In vivo *neutron activation analysis*

***In vivo* activation analysis** (IVAA) has been employed for a number of years to determine major elements in the whole body or in an individual limb. The body is usually irradiated with a reactor neutron beam, charged-particle accelerator, or a ^{252}Cf source to determine elements such as C, Ca, Cl, H, K, N, Na, O and P. Although prompt gamma-rays are usually measured during irradiation, delayed gamma-rays are sometimes detected.

The IVAA technique has been used to determine whole-body protein through the measurement of nitrogen in the body. Minor elements such as I, Fe and Mg and trace elements such as B, Cd and Hg can be measured by IVAA. Cadmium can accumulate in the livers and kidneys of certain industrial workers, and the prompt gamma-ray from the $^{113}Cd(n,\gamma)^{114}Cd$ reaction provides sensitivities of about 6.5 mg kg^{-1} in both tissues (Franklin *et al.* 1987). Other IVAA applications include the measurement of silicon in lungs of occupationally exposed workers (Kacperek *et al.* 1987) and loss of bone mass due to osteodystrophy (Krishnan *et al.* 1987).

4.9 Exercises and solutions

1. What is the decay rate of ^{40}K in 1 kg of natural potassium?

   Solution:

$$-\frac{dN}{dt} = \lambda N = \left(\frac{\ln 2}{t_{1/2}}\right)\left(\frac{mN_A\theta}{M}\right)$$

$$-\frac{dN}{dt} = \left(\frac{0.693}{1.28\times 10^9 \times 3.15\times 10^7}\right)\left(\frac{1\times 1000\times 6.02\times 10^{23}\times 1.17\times 10^{-4}}{39.1}\right)$$

$$-\frac{dN}{dt} = 3.10\times 10^4 \text{ dps} = 0.838\,\mu\text{Ci}$$

2. The radioisotope ^{47}Ca ($t_{1/2} = 4.54$ d) decays by beta emission to ^{47}Sc ($t_{1/2} = 3.35$ d), which in turn decays to stable ^{47}Ti. If a radioactive source containing 100 μCi of ^{47}Ca is initially separated from its first-daughter and second-daughter products, find the time from separation for the ^{47}Sc to reach its maximum level of radioactivity and find the total radioactivity for ^{47}Sc at that time.

   Solution:

   Differentiate the equilibrium equation (4.14) with respect to time and equate the derivative to zero:

$$\frac{dN_2}{dt} = \frac{\lambda_1}{\lambda_2 - \lambda_1} N_1^0(-\lambda_1\,e^{-\lambda_1 t} + \lambda_2\,e^{-\lambda_2 t}) = 0$$

$$t_{\max} = \frac{1}{\lambda_2 - \lambda_1}\ln\left(\frac{\lambda_2}{\lambda_1}\right) = \frac{1}{0.693/3.35 - 0.693/4.54}\ln\left(\frac{0.693/3.35}{0.693/4.54}\right)$$

$$= 5.60 \text{ days}$$

   Calculate the activity of ^{47}Sc from equation (4.15):

$$A_2 = A_1^0\frac{\lambda_2}{\lambda_2 - \lambda_1}(e^{-\lambda_1 t} - e^{-\lambda_2 t})$$

$$A_2 = 100\left(\frac{1/3.35}{1/3.35 - 1/4.54}\right)\{\exp[-(0.693/4.54)5.6] - \exp[-(0.693/3.35)5.6]\}$$

$$A_2 = 42.5\,\mu\text{Ci}$$

3. A sample containing 10 μg of chromium was irradiated in a reactor with thermal neutron flux of 10^{14} n cm^{-2} s^{-1}. The epithermal neutron flux in

this irradiation position is 2.5% of the thermal neutron flux. The abundance of the ^{50}Cr isotope is 4.35% and the thermal and epithermal neutron cross-sections are 15.2 and 8.1 b, respectively. If the sample is irradiated for 24 h, what is the activity of ^{51}Cr ($t_{1/2} = 27.7$ d) at the end of irradiation? What is the percentage of original ^{50}Cr atoms that were made radioactive during the irradiation? If the sample were irradiated continuously under these conditions, how long would the sample have to be irradiated for all of the ^{50}Cr atoms to be consumed?

Solution:
Using equation (4.28):

$$A = n(\varphi_{\text{th}}\sigma_{\text{th}} + \varphi_{\text{epi}}I)(1 - e^{-\lambda T_i})$$

$$A = \left(\frac{10^{-5} \times 6.02 \times 10^{23} \times 0.0435}{52}\right)(1 \times 15.2 + 0.025 \times 8.1)$$

$$\times 10^{-10}\{1 - \exp[-(0.693/27.7)1]\}$$

$$A = 1.92 \times 10^5 \text{ dps} = 5.18\,\mu\text{Ci}$$

The fraction of ^{50}Cr atoms made radioactive in 24 h is given by:

$$\frac{N_2}{n} = (\varphi_{\text{th}}\sigma_{\text{th}} + \varphi_{\text{epi}}I)T_i$$

$$\frac{N_2}{n} = (1 \times 15.2 + 0.025 \times 8.1) \times 10^{-10} \times (8.64 \times 10^4)$$

$$\frac{N_2}{n} = 0.00013 = 1.3 \times 10^{-2}\%$$

The time required to consume all of the ^{50}Cr atoms in the sample is:

$$T_i = \frac{1}{0.00013} = 7692 \text{ days} = 21 \text{ years}$$

4. Geological samples often contain significant amounts of aluminum, magnesium and sodium. When geological samples are irradiated in a reactor, the (n,p) and (n,α) reactions on aluminum interfere with the $^{26}Mg(n,\gamma)^{27}Mg$ and $^{23}Na(n,\gamma)^{24}Na$ reactions used to analyze Mg and Na, respectively. Calculate the interference correction factors assuming the following neutron fluxes: thermal $= 1 \times 10^{14}$ n cm^{-2} s^{-1}; epithermal $= 2.5 \times 10^{12}$ n cm^{-2} s^{-1}; and fast $= 6 \times 10^{12}$ n cm^{-2} s^{-1}. The thermal and epithermal neutron cross-sections for Mg and Na are found in Table 4.1. The fast neutron cross-sections for the $^{27}Al(n,p)^{27}Mg$ and $^{27}Al(n,\alpha)^{24}Na$ reactions are 4 and 0.725 mb, respectively.

Solution:
Assume equal amounts of Al, Mg and Na:

$$\frac{A(\text{fast})}{A(\text{thermal})} = \frac{n_1 \varphi_f \sigma_f}{n_2(\varphi_{th}\sigma_{th} + \varphi_{epi} I)}$$

$$\frac{A(^{27}\text{Al(n,p)}^{27}\text{Mg})}{A(^{26}\text{Mg(n},\gamma)^{27}\text{Mg})} = \frac{(1/26.98) \times 1.0}{(1/24.30) \times 0.1101} \times \frac{(6\times10^{12}\times4\times10^{-27})}{(1\times10^{14}\times3.72\times10^{-26}+2.5\times10^{12}\times2.4\times10^{-26})}$$

$$\frac{A(^{27}\text{Al(n,p)}^{27}\text{Mg})}{A(^{26}\text{Mg(n},\gamma)^{27}\text{Mg})} = 0.052$$

and

$$\frac{A(^{27}\text{Al(n},\alpha)^{24}\text{Na})}{A(^{23}\text{Na(n},\gamma)^{24}\text{Na})} = \frac{(1/26.98)}{(1/22.99)} \times \frac{(6\times10^{12}\times0.725\times10^{-27})}{(1\times10^{14}\times0.513\times10^{-24}+2.5\times10^{12}\times0.303\times10^{-24})}$$

$$\frac{A(^{27}\text{Al(n,p)}^{27}\text{Mg})}{A(^{26}\text{Mg(n},\gamma)^{27}\text{Mg})} = 0.000071$$

5. A standard containing 9.98 µg of vanadium and an unknown sample weighing 150 mg were irradiated simultaneously in a nuclear reactor. The standard was counted 5.0 min after the end of irradiation, and the ^{52}V counting rate was determined to be 4.44×10^3 cpm. The sample was counted 7.0 min after the end of irradiation and the counting rate for ^{52}V was 5.07×10^3 cpm. Assume the counting times were short with respect to the half-life of ^{52}V. Calculate the concentration of V in the sample, in µg/g.

Solution:
Equation (4.52) is used to solve this problem. The half-life of ^{52}V is 3.75 min:

$$\frac{4.44 \times 10^3}{5.07 \times 10^3} = \frac{9.98\,\mu\text{g}}{x\,\mu\text{g}} \times \frac{\exp[-(0.693/3.75\,\text{min})5.0\,\text{min}]}{\exp[-(0.693/3.75\,\text{min})7.0\,\text{min}]}$$

Solving for x gives 16.5 µg V in the sample. The concentration of V in the sample is:

$$16.5\,\mu\text{g Al} / 0.150\,\text{g sample} = 110\,\mu\text{g/g}$$

References

Adloff, J.-P. and Guillaumont, R. (1993) *Fundamentals of Radiochemistry*, CRC Press, Boca Raton.

Alfassi, Z.B. (ed.) (1989) *Activation Analysis*, vols I and II, CRC Press, Boca Raton.

Andersen, H.H. and Ziegler, J.F. (1977) *Hydrogen, Stopping Powers and Ranges in all Elements*, Pergamon Press, New York.

Bem, H. and Ryan, D.E. (1981) *Anal. Chim. Acta* **124**, 373–80.

Bild, R.W. (1987) *Comparison of Nuclear Analytical Methods with Competitive Methods*, IAEA-TECDOC–435, International Atomic Energy Agency, Vienna.

Brune, D. and Schmidt, J.J. (eds) (1974) *Handbook on Nuclear Activation Cross-Sections*, International Atomic Energy Agency, Vienna.

Copley, J.R.D. and Stone, C.A. (1989) *Nucl. Instrum. Meth. Phys. Res.* **A281**, 593–604.

Currie, L.A. and Parr, R.M. (1988) *Detection in Analytical Chemistry. Importance, Theory and Practice*, L.A. Currie (ed.), ACS Symposium Series 361, American Chemical Society, Washington, DC.

Das, H.A., Faanhof, A. and van der Sloot, H.A. (1989) *Radioanalysis in Geochemistry*, Elsevier Science, Amsterdam.

De Corte, F., Simonits, A., De Wispelaere, A. and Hoste, J. (1987) *J. Radioanal. Nucl. Chem.* **113**, 145–61.

De Soete, D., Gijbels, R. and Hoste, J. (1972) *Chemical Analysis, a Series of Monographs on Analytical Chemistry and Its Applications*, vol. 34, *Neutron Activation Analysis*, P.J. Elving and I.M. Kolthoff (eds), Wiley, London.

Downing, R.G., Maki, J.T. and Fleming, R.F. (1987) *J. Radioanal. Nucl. Chem.* **112**(1), 33–46.

Ehmann, W.D. and Vance, D.E. (1989) *CRC Critical Reviews in Analytical Chemistry*, vol. 20(6), CRC Press, Boca Raton.

Ehmann, W.D. and Vance, D.E. (1991) *Radiochemistry and Nuclear Methods of Analysis*, J.D. Winefordner (ed.), Wiley, New York.

Erdtmann, G. and Soyka, W. (1979) *The Gamma Rays of the Radionuclides: Tables for Applied Gamma Ray Spectroscopy*, Verlag Chemie, Weinheim.

Evans, R.D. (1955) *The Atomic Nucleus*, McGraw-Hill, New York.

Fardy, J.J. (1990) *Activation Analysis*, vol. I, Z.B. Alfassi (ed.), CRC Press, Boca Raton, pp. 61–96.

Firestone, R.B., Shirley, V.S., Baglin, C.M., Chu, S.Y. and Zipkin, J. (1996) *Table of Isotopes*, 8th edn, Wiley, New York.

Forkman, B. and Petersson, R. (1987) *Handbook on Nuclear Activation Data*, International Atomic Energy Agency, Vienna.

Franklin, D.M., Chettle, D.R. and Scott, M.C. (1987) *J. Radioanal. Nucl. Chem.* **114**(1), 155–63.

Friedlander, G., Kennedy, J.W., Macias, E.S. and Miller, J.M. (1981) *Nuclear and Radiochemistry*, 3rd edn, Wiley, New York.

Glascock, M.D. (1982) *Inst. Phys. Conf. Ser.* **62**, 641–54.

Glascock, M.D. (1992) *Chemical Characterization of Ceramic Pastes in Archaeology*, H. Neff (ed.), Prehistory Press, Madison, pp. 11–26.

Glascock, M.D. (1996) *Tables for Neutron Activation Analysis*, 4th edn, University of Missouri, Columbia.

Glascock, M.D., Tian, W.Z. and Ehmann, W.D. (1985) *J. Radioanal. Nucl. Chem.* **92**(2), 379–90.

Glascock, M.D., Nabelek, P.I., Weinrich, D.D. and Coveney, R.M., Jr (1986) *J. Radioanal. Nucl. Chem.* **99**(1), 121–31.

Griffin, J.B. and Gordus, A.A. (1967) *Science* **158**, 528.

Haskin, L.A., Haskin, M.A., Frey, F.A. and Wildeman, T.R. (1968) *Origin and Distribution of the Elements*, L.H. Ahrens (ed.), Pergamon Press, New York, pp. 889–912.

Heydorn, K. (1984) *Neutron Activation Analysis for Clinical Trace Element Research*, vols I and II, CRC Press, Boca Raton.

Hodgson, P.E. (1971) *Nuclear Reactions and Nuclear Structure*, Clarendon, Oxford.

Hoffman, E.L., Naldrett, A.J., Van Loon, J.C., Hancock, R.G.V. and Manson, A. (1978) *Anal. Chim. Acta* **102**, 157–66.

Iyengar, G.V. (1989) *Elemental Analysis of Biological Systems*, vol. I, *Biomedical, Environmental, Compositional, and Methodological Aspects of Trace Elements*, CRC Press, Boca Raton.

Kacperek, A., Evans, C.J., Dutton, J., Morgan, W.D. and Sivyer, A. (1987) *J. Radioanal. Nucl. Chem.* **114**(1), 165–72.

Knoll, G.F. (1989) *Radiation Detection and Measurement*, 2nd edn, Wiley, New York.

Krishnan, S.S., Bailey, M.T., Hitchman, A.J.W., Lin, S.C., McNeill, K.G. and Harrison, J.E. (1987) *J. Radioanal. Nucl. Chem.* **114**(1), 173–80.

Kruger, P. (1971) *Principles of Activation Analysis*, Wiley, New York.

Kushelevsky, A.P. (1990) *Activation Analysis*, vol. II, Z.B. Alfassi (ed.), CRC Press, Boca Raton, pp. 219–37.

Landsberger, S. (1986) *Chem. Geol.* **57**, 415–21.

Mackey, E.A., Gordon, G.E., Lindstrom, R.M. and Anderson, D.L. (1991) *Anal. Chem.* **63**, 288–92.

Mughabghab, S.F., Divadeenam, M. and Holden, N.E. (1984) *Neutron Cross Sections*, vol. I: *Neutron Resonance Parameters and Thermal Cross Sections*. Part A: *Z = 1–60*, and Part B; *Z = 61–100*, Academic Press, New York and Orlando.

Parry, S.J. (1982) *J. Radioanal. Nucl. Chem.* **72**(1–2), 195–207.

Parry, S.J. (1984) *J. Radioanal. Nucl. Chem.* **81**(1), 143–51.

Parry, S.J. (1991) *Activation Spectrometry in Chemical Analysis*, J.D. Winefordner (ed.), Wiley, New York.

Perlman, I. and Asaro, F. (1969) *Archaeometry* **11**, 21–52.

Satchler, G.R. (1980) *Introduction to Nuclear Reactions*, Wiley, New York.

Sayre, E.V. and Dodson, R.W. (1957) *Am. J. Archaeol.* **61**, 35–41.

Segebade, C., Weise, H.-P. and Lutz, G.J. (1987) *Photon Activation Analysis*, Walter de Gruyter, Berlin.

Spyrou, N.M. (1981) *J. Radioanal. Nucl. Chem.* **61**(1–2), 211–42.

Stroube, W.B., Jr, Cunningham, W.C. and Lutz, G.J. (1987) *J. Radioanal. Nucl. Chem.* **112**(2), 341–6.

Tian, W.Z. and Ehmann, W.D. (1984) *J. Radioanal. Nucl. Chem.* **84**(1), 89–102.

Tolgyessy, J. and Klehr, E.H. (1987) *Nuclear Environmental Chemical Analysis*, Wiley, New York.

Vandenbosch, R. and Huizenga, J.R. (1973) *Nuclear Fission*, Academic Press, New York.

Walker, F.W., Parrington, J.R. and Feiner, F. (1989) *Chart of the Nuclides*, 14th edn, General Electric, Schenectady, NY.

Vandecasteele, C. (1988) *Activation Analysis with Charged Particles*, R.A. Chalmers (ed.), Ellis Horwood, Chichester.

Wyttenbach, A., Bajo, S. and Tobler, L. (1987) *J. Radioanal. Nucl. Chem.* **114**(1), 137–45.

Ziegler, J.F. (1977) *Helium, Stopping Powers and Ranges in all Elemental Matter*, Pergamon Press, New York.

Ziegler, J.F., Cole, G.W. and Baglin, J.E.E. (1972) *J. Appl. Phys.*, **43**(9), 3809-15.

5 Inductively coupled plasma optical emission and mass spectrometry

N. DE SILVA and D.C. GREGOIRE

5.1 Inductively coupled plasma as an analytical source

5.1.1 The ICP discharge

Inductively coupled plasma (ICP) is a highly ionized gas consisting of energetic electrons, ions and atoms. In spite of the high population of ions and electrons it is macroscopically neutral. The plasma is sustained by a continuous supply of energy through electromagnetic induction, or inductive coupling, to a flowing gas [1]. Almost universally used gas for the generation of the plasma is argon, which gives the analytical ICP its unique characteristics. In principle and operation, the ICP is inherently simple, which is evident from the fact that the basic geometry of the plasma torch remains very much the same as that which appeared in the early 1960s when its potential was first demonstrated as an atom source for analytical spectrometry [2, 3].

A typical configuration of the ICP torch is shown in Figure 5.1, which consists of three concentric circular tubes (outer tube, intermediate tube and inner or injector tube) made of quartz or some other suitable material. The typical dimensions and flow rates are shown in the diagram, although their actual values can vary depending on the make of the instrument and the type of application. The plasma gas, also called the coolant gas, and the auxiliary gas are introduced tangentially to create a vortex flow pattern in the region where the plasma is generated. The auxiliary gas, which is not essential to maintain a stable plasma, can be used to make slight adjustments in the position of the base of the plasma relative to the quartz torch or the load coil. The central gas flow, which enters through the 'injector tube', is used to transport the analyte into the plasma, commonly as fine aerosol droplets. The same central gas flow typically is used for nebulization of the sample solution to generate the aerosol.

A two- or three-turn copper coil (3 mm o.d.), called the load coil, surrounds the top end of the torch, which is water-cooled and carries the

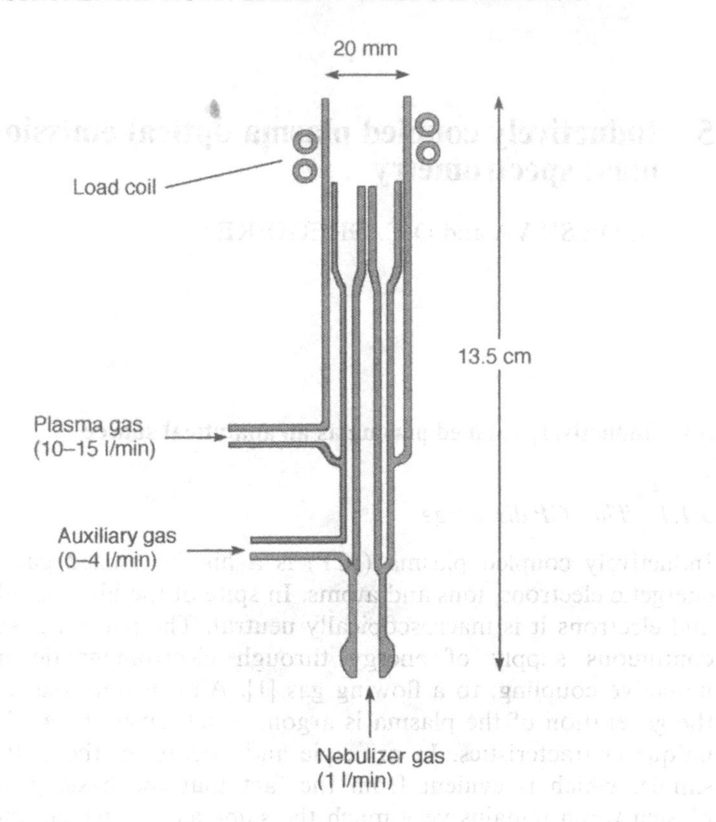

Figure 5.1 The ICP torch.

alternating current supplied by the radiofrequency (RF) power generator. The most commonly used frequencies are 27.12 or 40.68 MHz. When the RF power is applied typically at 600–1500 W to the load coil, it induces an oscillating magnetic field inside the torch, which in turn induces an oscillating intense electric field in the vicinity of the load coil. With the argon gas flowing through the torch some electrons are momentarily 'seeded' into the gas by an electrical spark using external means such as a Tesla coil. These electrons get accelerated by the alternating magnetic field and become highly energetic. The collisions between the energetic electrons and argon atoms generate argon ions and more electrons. The process continues as a chain reaction until the gas becomes highly conductive and reaches a steady state where most of the RF energy supplied to the coil is transferred to the gas to self-sustain a high-temperature plasma highly populated with electrons and argon ions. This ignition process takes only a fraction of a second. The electrically conducting ionized gas (plasma) acts

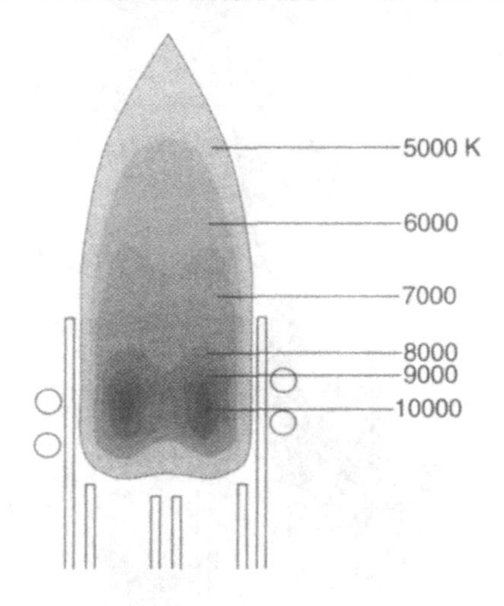

Figure 5.2 Temperature regions in the ICP.

as a shorted secondary coil of a transformer, to couple efficiently the RF energy from the load coil to heat the gas to high temperatures effectively in the vicinity of 10 000 K.

As the extremely bright hot gas exits the torch away from the induction region of the load coil, it gradually mixes with air from the atmosphere and its temperature drops rapidly within a few centimeters, giving its typical 'teardrop' shape. The nebulizer flow or the central injector gas entering at a high linear velocity literally punches a hole through the base of the discharge. Owing to cooling at the center by high-velocity gas and as a result of gradual increase in power density toward the coil, the base of the plasma is toroidal, or 'donut shaped' in form. The temperature profile within the ICP discharge is shown in Figure 5.2. The region where different processes occur along the central analytical channel is shown in Figure 5.3 [4]. The most common method of introducing analytes into the plasma is by nebulization of a sample solution into an aerosol of which a fraction of fine droplets (typically less than 10 μm) are transported into the plasma by the nebulizer gas. Liquid droplets undergo several physico-chemical processes during their passage through the hot plasma. Figure 5.4 summarizes the steps involved in the conversion of analytes in solution to gaseous excited atoms or ions.

The high temperature available in the ICP makes it one of the most efficient sources capable of breaking down virtually any type of molecule or matrix into free atoms or ions, which gives it some of its greatest analytical attributes such as simultaneous multi-element

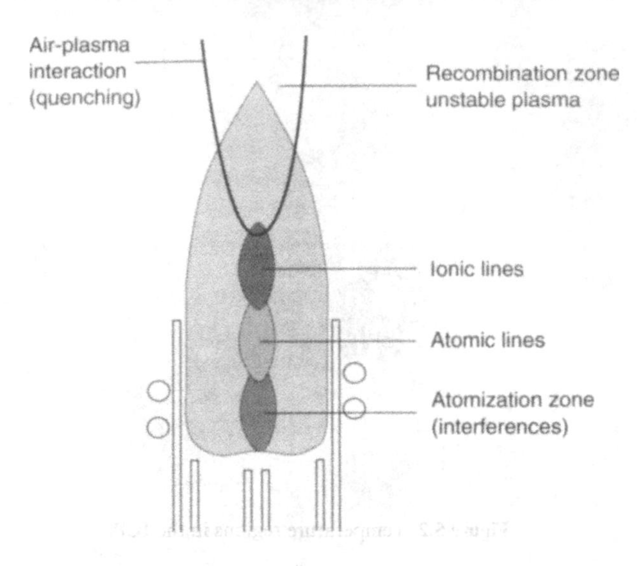

Figure 5.3 Analytical regions in the ICP.

capability, large dynamic range and relative immunity to matrix interferences. High-energy plasma provides complete desolvation and dissociation of even the most refractory compounds. It has enough energy to excite virtually every element in the Periodic Table. Some

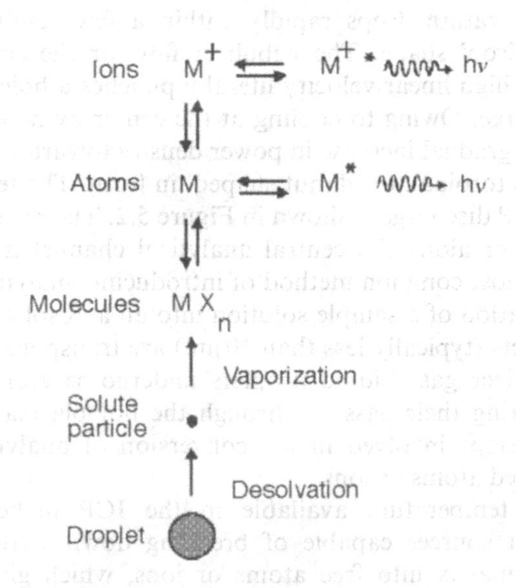

Figure 5.4 The fate of solution droplets entering the plasma.

elements that cannot be handled by other techniques such as atomic absorption (AA) (e.g. sulfur, boron, phosphorus, titanium, zirconium, etc.) can be handled by the ICP. In contrast to various combustion flames, free atoms and ions are in a chemically inert atmosphere in the ICP. As it is an electrodeless discharge, it is virtually free of contamination from the hardware such as graphite or metal electrodes and tubes.

5.1.2 Basic ICP spectrometric system

A spectrometric system can be divided into three basic components: (a) sample processor, (b) signal generator and (c) signal processor, as shown in Figure 5.5. The primary function of the **sample processor** is to convert the sample into a form that can conveniently be handled by the **signal generator**. For conventional solution analysis, the sample processor may consist of autosampler, pump, nebulizer and spray chamber. The **signal generator**, essentially the plasma, generates photons of specific wavelengths, or ions of specific masses, characteristic to each analyte element from the aerosol droplets. For photon measurement, i.e. in optical emission spectrometry (ICP-ES), the **signal processor** disperses and separates the photons according to their wavelength and quantitatively measures their intensity. In mass spectrometry (ICP-MS) the ions generated by the plasma are separated according to *m*/*e* (mass-to-charge ratio) and the intensity of ions of a desired mass is measured quantitatively.

Even though, in principle, the ICP can also be used as an atom source for atomic absorption and atomic fluorescence, because of its high excitation and ionization efficiency of producing excited atoms and ions, it is primarily used as a source for optical emission and mass spectrometry.

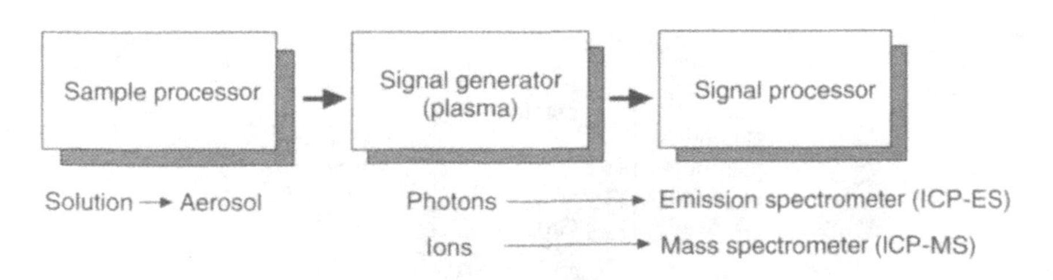

Figure 5.5 Basic ICP spectrometry system.

5.2 Inductively coupled plasma optical emission spectrometry

5.2.1 Suitability of the ICP as an emission source

In emission spectrometry the sample is subjected to high enough temperature to dissociate the matrix into atoms or ions and to cause their electronic excitation by collision. The excited atoms or ions relax from the excited state to the ground state by collision with other particles (**collisional relaxation**) or by emitting radiation (**radiative relaxation**) (Figure 5.6). The energy difference between the two energy states, and therefore the wavelength of emitted radiation, is discrete and characteristic to the element. The intensity of the light emitted at the characteristic wavelength is proportional to the concentration of the particular element.

One of the main advantages of the high excitation efficiency of the ICP is that it can populate a large number of different excited energy states for a number of elements at the same time under the same set of experimental conditions. This endows the ICP-ES its most notable feature, the capability of simultaneous multi-element determinations. However, by the same token, the emission spectra can also be quite rich in spectral lines and structure, thereby increasing the probability of spectral overlaps, which would demand an optical spectrometer with a high resolving power. The emission lines can be in the vacuum-ultraviolet (VUV, 120–190 nm), ultraviolet (UV, 190–400 nm), visible (VIS, 400–700 nm) and near-infrared (NIR, 700–850 nm). For practical reasons, the most commonly used spectral regions are UV and VIS. Spectrometers used in the VUV region need to be maintained at a vacuum to avoid absorption of VUV radiation by air, and are used for specialized applications such as sensitive determination of halogens, phosphorus and sulfur. Owing to the high efficiency of matrix decomposition and its inertness, the ICP is virtually free from sample-matrix-induced interferences, which means that calibration can be quickly and easily performed with simple multi-element aqueous standards. It is a constant emission source and its stability is unexcelled, making it capable of performing all day routine determinations with a very few straightforward calibrations. Finally, because it is an optically thin

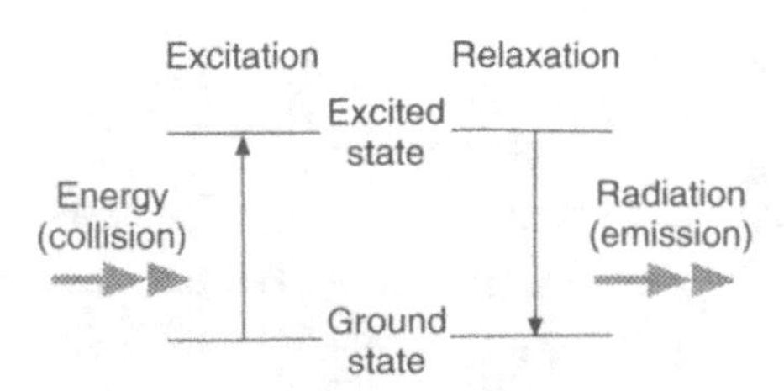

Figure 5.6 Collisional excitation and radiative relaxation.

source, ICP has exceptional linearity over a wide range of concentration without being affected by self-absorption. This large dynamic range, typically over six orders of magnitude, permits the determination of trace, minor, and major elements in the same sample simultaneously, without the need for serial dilution. In general these attributes make ICP emission spectroscopy a nearly universal technique for elemental analysis in a variety of applications ranging from geological to clinical fields.

5.2.2 Viewing options (axial and radial)

Viewing the plasma by the spectrometer can be done in two basic orientations as shown in Figure 5.7. Traditionally the center of the plasma is viewed radially in a direction perpendicular to the axis of the torch. In this orientation the observation zone in the plasma can be easily adjusted to a desired position by moving the torch along its axis. Light emitted from a narrow channel at the selected region along the central axis (Figure 5.2) enters the spectrometer through the 'fire-ball', which emits high-

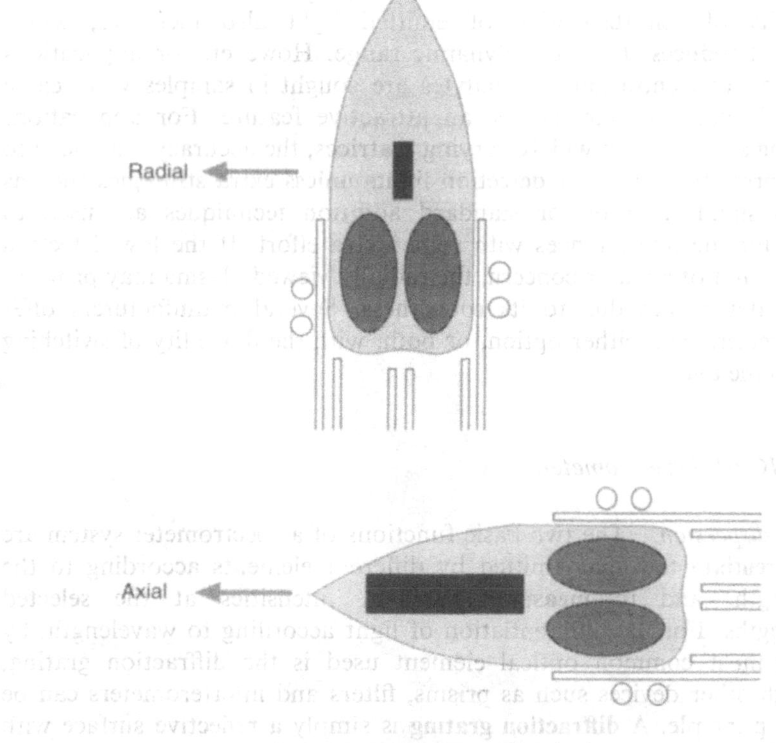

Figure 5.7 Axial versus radial viewing.

intensity background radiation. Recently, the axial, also called 'end-on', viewing orientation has also gained popularity in commercial systems, although it has been proposed for over 20 years. In this orientation, light is emitted from a deeper region along the central channel of the plasma, resulting in a bigger volume of the analytical region being sampled. Therefore, the signals produced can be much stronger than in the case of radial viewing. Secondly, the optical path lies along the central channel of the plasma, avoiding intense background radiation originating from the 'fire-ball' region surrounding the central channel. As a result, in axial viewing a more intense signal can be collected with a smaller background intensity, providing as much as an order of magnitude lower detection limits depending on the analyte element, wavelength and optical system, etc. However, there are also certain disadvantages associated with axial viewing. Since the optical path passes along the central axis, it does not allow much flexibility to the user to select the optimum viewing position. It makes the measurement more susceptible to matrix-induced signal enhancement or suppression in contrast to the radial viewing, where the viewing height in the plasma can easily be adjusted in order to obtain the optimum conditions for the least possible fluctuation of the signal due to changes in the matrix. Furthermore, as the path length increases, the possibility of self-absorption of emitted light also increases, which somewhat reduces the linear dynamic range. However, for applications where low concentrations of analytes are sought in samples with 'clean matrices', axial viewing can be an attractive feature. For applications involving samples with widely varying matrices, the accuracy may have to be compromised for lower detection limits unless extra strategies such as internal standardization or standard addition techniques are used to correct for the interferences with some extra effort. If the low detection limits are not of a major concern, the radially viewed plasma may prove to be a better choice due to its robustness. Several manufacturers offer spectrometers with either option, or both, with the flexibility of switching between the two.

5.2.3 ICP-ES spectrometers

Optical dispersion. The two basic functions of a spectrometer system are to **differentiate** the light emitted by different elements according to the wavelength, and to **measure** the light intensities at the selected wavelengths. For the differentiation of light according to wavelength, by far the most common optical element used is the diffraction grating, although other devices such as prisms, filters and interferometers can be used in principle. A **diffraction grating** is simply a reflective surface with closely spaced lines ruled by mechanical or holographic means or by

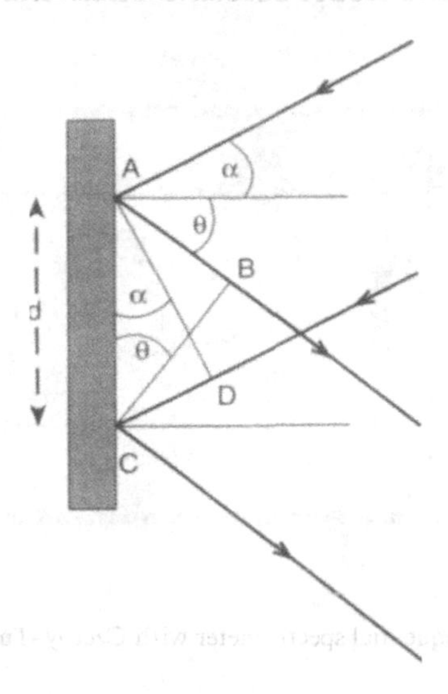

Figure 5.8 Diffraction of light from a grating.

etching. The groove density of a diffraction grating can typically be in the range 500–4500 grooves/mm depending on the type of grating, desired resolution and optical configuration. The basic formula describing the diffraction of light from the grating can be obtained by considering two parallel beams incident on two adjacent grooves as shown in Figure 5.8. The requirement for constructive interference is that the path difference of the two diffracted beams should be a simple multiple of the wavelength:

$$\text{path difference} = \text{AB} - \text{CD}$$

$$m\lambda = d(\sin\alpha - \sin\theta)$$

where d = groove spacing, λ = wavelength and m is the order (an integer). The incident angle, α, is constant for all the wavelengths. For a given wavelength constructive interference occurs at different diffraction angles, θ. Constructive interference can also occur for each integer value of m (first, second, third order, etc.) at different values of θ for the same wavelength. By physically changing the angles α or θ, or both, it is possible to select the desired wavelength and to measure its intensity using a suitable detector.

Depending on the geometrical arrangement of the dispersive element, other optical components and the detectors, commonly used spectrometers can be classified into three basic categories: (a) scanning monochromators, (b) polychromators or direct readers and (c) array spectrometers.

Scanning monochromator (sequential spectrometer). The simplest design

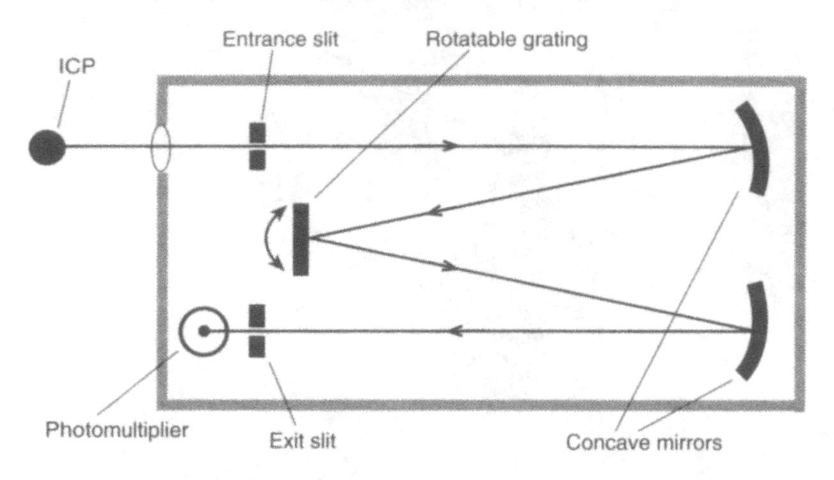

Figure 5.9 The sequential spectrometer with Czerny–Turner mounting.

among different types of spectrometer is the monochromator, which can be scanned to a desired wavelength by moving one of the optical elements, generally by rotating the grating. The most widely used type of monochromator employs a rotatable plane grating in a Czerny–Turner mounting as shown in Figure 5.9. As the grating is rotated with a computer-controlled stepper motor, the spectrum moves across the exit slit. A single photomultiplier is located behind the exit slit to measure the light emitted by the plasma at the selected wavelength. For determining several elements, the spectrometer can be scanned from one wavelength to the other and measure the emission intensity at each wavelength sequentially. It can also be scanned sequentially to a desired position to measure the background intensity. The primary advantage of the sequential spectrometer is its flexibility for accessing the wavelength of choice. Monochromators are relatively cheaper than the other types. However, multi-element measurements at different analytical lines and their backgrounds have to be performed sequentially, making it slow and consuming large volumes of samples. These systems are not very suitable for measurement of time-variable, or transient, signals.

Polychromator (simultaneous multi-element system). The basic configuration of a polychromator is shown in Figure 5.10. An image of the central channel of the plasma is focused onto the entrance slit, illuminating it with the light emitted by the plasma. The entrance slit is located on a so-called **Rowland circle**. A concave grating is tangentially mounted on the Rowland circle, which focuses the image of the entrance slit at the exit slits located at preselected positions at different angles along the Rowland circle. With this arrangement, several wavelengths, therefore several

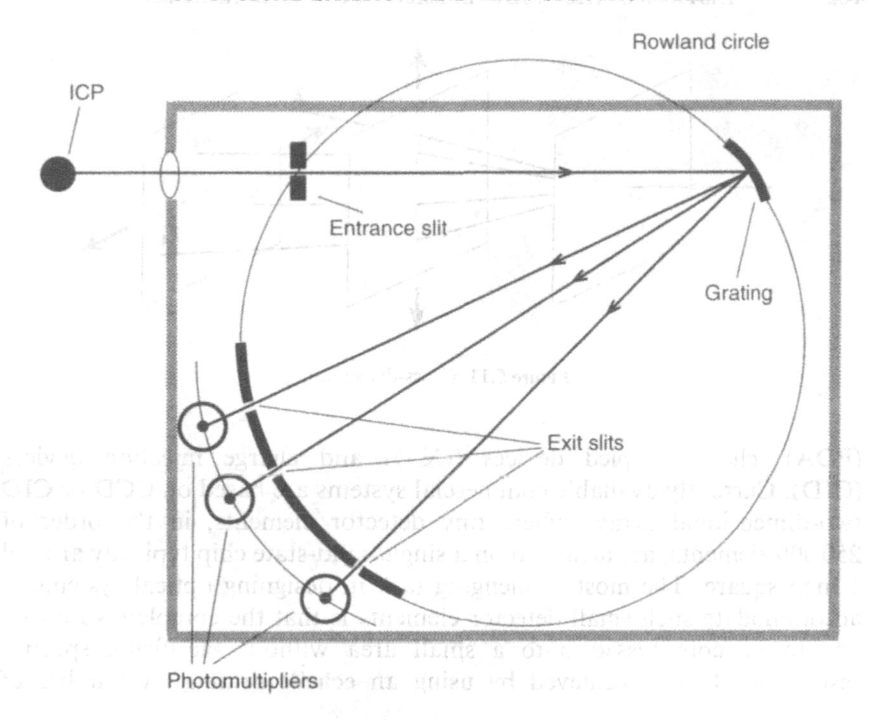

Figure 5.10 A polychromator system in the Paschen–Runge mounting.

elements, can be measured simultaneously. The exit slits are installed by the manufacturer according to customer's specifications and they are not easily changeable by the user. For the purpose of background correction and routine alignment of the spectral lines with the exit slits, the entrance slit is slightly movable with a computer-controlled stepper motor. Movement of the entrance slit allows scanning the spectrum within a narrow window around each line or 'channel'. The main advantage of the polychromator is its ability to determine a number of elements simultaneously. Commercial systems capable of determining 50 elements simultaneously are common. Thus the high sample throughput possible with polychromator systems for multi-element analysis is a significant advantage for routine operation. Its major limitation is the inflexibility of changing the analytical lines when there is a need for a measurement at an 'un-programmed' wavelength for special applications. Most commercial systems provide the option to be equipped with a combination of both sequential and multi-element systems viewing the same plasma to gain the advantages of both speed and flexibility.

Array spectrometers. A new trend in optical spectrometer design is based on using semiconductor detector arrays such as photodiode arrays

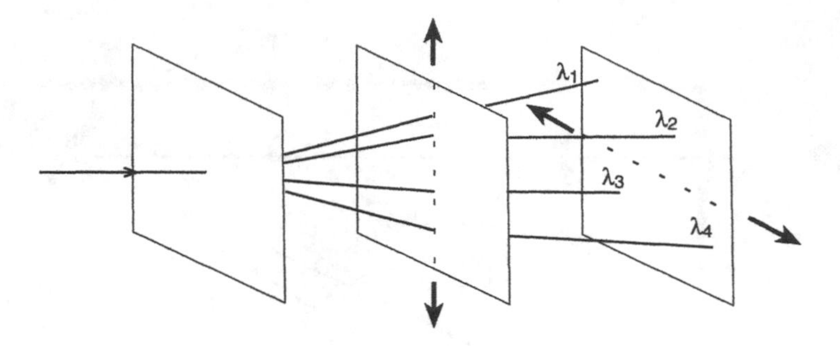

Figure 5.11 Cross-dispersion.

(PDA), charge-coupled devices (CCD) and charge injection devices (CID). Currently available commercial systems are based on CCD or CID two-dimensional arrays where tiny detector elements, in the order of 250 000 elements, are arranged on a single solid-state chip typically around 1 inch square. The most challenging task in designing optical systems to accommodate such small detector elements is that the complete spectrum has to be compressed onto a small area without sacrificing spectral resolution. This is achieved by using an **echelle grating**. A number of

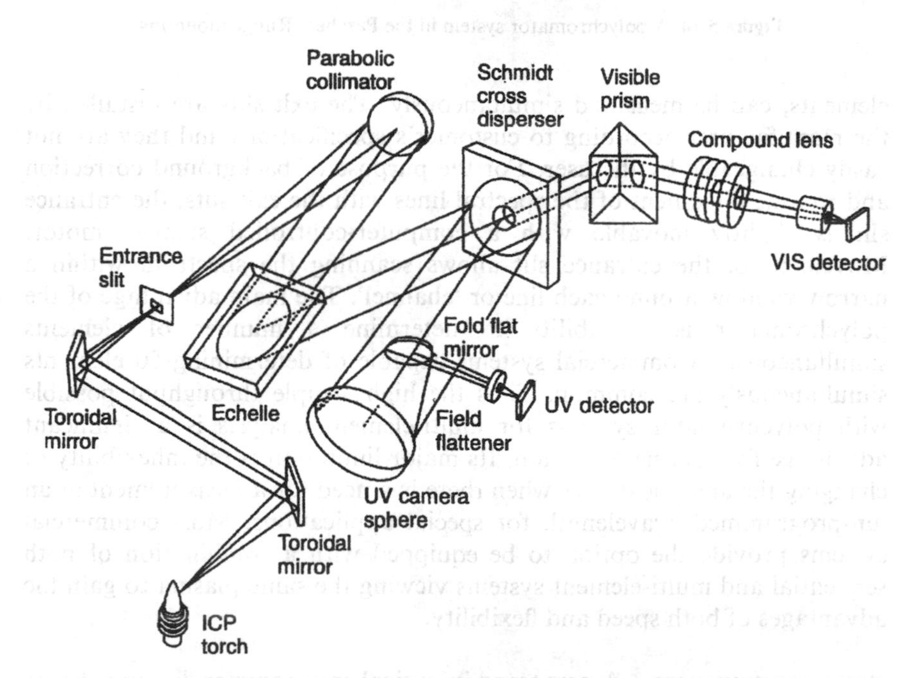

Figure 5.12 Optical diagram of Optima 300 CCD Spectrometer. (Courtesy of Perkin-Elmer Corporation.)

different spectral regions of high spectral orders can be focused onto a narrow area covered by the detector array. In order to avoid spectral order overlaps, the superimposed narrow bands of the spectrum are dispersed and separated in perpendicular direction using a second cross-disperser such as a prism to generate a two-dimensional image pattern. The principle of cross-dispersion to produce two-dimensional spectra is shown in Figure 5.11. With this arrangement, an image of virtually the whole spectrum can be focused onto the two-dimensional array for simultaneous measurement of a large number of analytical lines as well as their backgrounds. The principle of a commercial instrument is shown in Figure 5.12.

5.2.4 *Interference effects and background correction*

Although ICP-ES is relatively free from interference effects, certain precautions have to be taken to avoid possible errors, which can originate at various steps in the overall measurement process. Interference effects in ICP-ES can be classified as: (a) analyte transport interference from the sample introduction system, (b) chemical and ionization interference and (c) spectral interference.

Transport interferences. The overall transport efficiency of the sample introduction system can be affected by changes in liquid flow due to weakening of the pump tubing, gradual change in the efficiency of the nebulizer, spray chamber, desolvation system and physical conditions such as temperature. These temporal changes appear as signal drifts affecting precision and accuracy of the measurement. Transport interferences can also occur due to differences among samples and standards in their physical and chemical properties such as viscosity, surface tension, dissolved solid content and mineral acid composition. In changing from one sample to another, 'carry-over' effects due to adsorption of analyte onto the components of the sample introduction system such as pump tubing can also interfere with the measurement. Generally these problems are not serious and with proper precautions these errors can be minimized. For high-accuracy determinations, several strategies can be used to correct for such interferences.

Internal standardization. All the samples and standards can be spiked with a known concentration of an element that is not present in the sample at any significant level. If the transport efficiency remains the same for all the samples, the signal due to the internal standard should remain constant provided that the internal standard is chosen so that the analytes and the internal standard behave in a similar manner to the changes in any other parameters. Then, any changes in transport efficiency should be reflected in the internal standard signal and a correction factor can be calculated accordingly.

Standard addition. Another possible approach to minimize the errors arising from changes in transport efficiency due to changes in the matrix is to add a known concentration of analyte to the sample and take an extra measurement for each added concentration. The added known concentration and the difference in the signal can be used to calibrate the measurement. In principle, this approach can correct for transport effects as well as other physical/chemical interference due to any changes of conditions in the plasma. However, in practice, owing to the need for making several solutions and several measurements for each sample, it is not very attractive for routine analytical measurement where sample throughput can be an important consideration. For best results, it is also important to achieve a reasonable match of the spiked amount of analyte to the unknown amount already present in the sample, which requires some prior knowledge of the expected concentration of the analyte in the sample.

Matrix matching. If all the samples are prepared in the same manner, for example by using the same dissolution procedure, variation of transport behavior can be negligible as the overall matrix composition can be fairly consistent from one sample to the next. It is important to prepare the standards in the same manner to keep them in the same physical and chemical consistency as all the samples.

Chemical and ionization interferences. Owing to the high temperatures available in the ICP and its inert atmosphere, virtually any chemical bond can be broken to decompose almost any type of matrix to produce free atoms or ions. Therefore chemical interferences due to the formation of thermally stable compounds is not significant enough to cause serious difficulties, which makes the ICP-ES quite immune to chemical interferences especially at high RF powers. The presence of easily ionizable elements (EIEs) such as alkali and alkaline-earth metals at high concentrations can perturb the equilibria between species such as argon ions and electrons, altering the plasma characteristics such as effective temperature and electron number density. It can effectively change the excitation/relaxation characteristics of the analyte atoms and alter the sensitivity of the analyte depending on the amount of easily ionizable elements. Similar to chemical interferences, under normal operating conditions of the ICP these effects are insignificant, or can be minimized by matrix matching. In applications where chemical/ionization interference can be a problem because of significant variation in matrix composition, internal standardization and standard addition techniques can also be used to improve the accuracy.

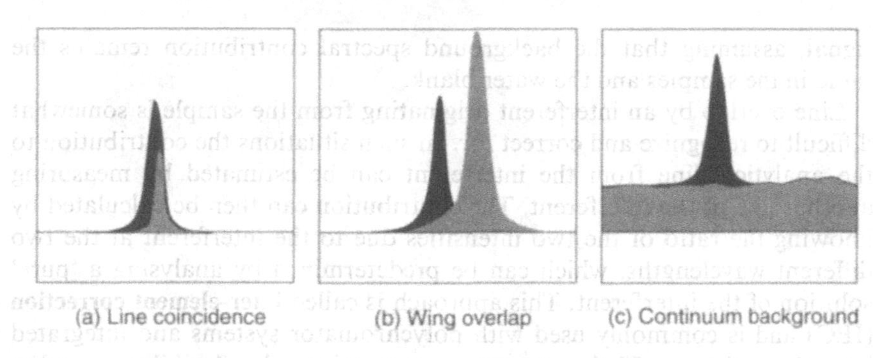

Figure 5.13 Types of spectral interferences.

Spectral interferences.

Types of spectral interference. As a result of the high atomization/excitation efficiency, ICP produces complex spectra rich in various emission lines originating from different elements. Therefore the resolving power of the spectrometer, proper selection of the analytical lines depending on the matrix and composition of the potential interferents, and proper corrections for spectral interference become important considerations. Spectral interferences can be classified into three basic categories: (a) line coincidence, (b) wing overlap and (c) continuum background. As shown in Figure 5.13, **line coincidence** occurs when the separation between the interferent line and the analyte line is small compared to the resolving power of the spectrometer. **Wing overlap** occurs when the separate lines can be easily seen but a wing of the interferent line extends into the region of the analyte wavelength. This interference can become especially significant when the interferent line intensity is much higher than the analyte line intensity, especially from matrix elements. **Continuum background** can originate from a matrix element (e.g. Al), from molecular bands such as OH bands due to the presence of oxygen and hydrogen originating from sample solution, or from the plasma itself. Such a continuum generally results in an overall shift in the intensity in the spectral region, which can be a slanted or a flat continuum.

Correction for spectral interference. Correction methods for background can be exercised to minimize the errors due to the bias in estimating the analyte signal, which becomes especially important as the signal approaches the detection limit. The method of background correction depends on the type of spectrometer and the nature and the origin of spectral interference. If the background originates from the plasma or a substance common to the blank (e.g. OH bands from water), a blank measurement can be made and simply subtracted from the analyte signal, assuming that the background spectral contribution remains the same in the samples and the water blank.

Line overlap by an interferent originating from the sample is somewhat difficult to recognize and correct for. In such situations the contribution to the analytical line from the interferent can be estimated by measuring another line of the interferent. The contribution can then be calculated by knowing the ratio of the two intensities due to the interferent at the two different wavelengths, which can be predetermined by analysing a 'pure' solution of the interferent. This approach is called **inter-element correction** (IEC) and is commonly used with polychromator systems and integrated into the software. If the spectrometer provides the flexibility, another alternative is to select a different line for the analyte of interest.

If the spectral interference is due to the extension of a wing, depending on the severity of the overlap the above approach of inter-element correction or the selection of an alternative line can be done. If there is enough separation, the overlapping portion can be treated as a slanted continuum. For spectral interference due to a flat or slanted continuum, the background at the off-peak positions suitably selected on either side of the peak can be measured and proper correction can be made. This is commonly known as **off-peak background correction**. For conventional sequential and polychromator spectrometers, this will require making two additional measurements in addition to the intensity measurement at the peak.

The most suitable approach depends on several factors, such as the type of spectrometer, the extent of overlap and the sample matrix. Whatever the correction method that is applied, as more and more measurements are used to calculate the net signal, the uncertainty of the background increases, leading to poorer detection limits. If the spectrometer provides the choice, alternative line selection can be most effective. Owing to simultaneous acquisition of the complete spectrum and the possibility of its digital storage and display of the spectral patterns, array spectrometers allow the flexibility to choose or modify the optimum spectral interference correction method as desired without doing the measurement again with modified interference correction protocols. As multiple lines for a given element are also measured simultaneously, they also provide the option of choosing alternative lines for avoiding spectral interferences. Mathematical deconvolution techniques have been used for resolving analyte signals from the interfering background with varying success.

5.2.5 Analytical performance

The ICP-ES technique is capable of making multi-element determinations at sub-ppm range with typical precisions in the order of 0.4% depending

Table 5.1 Elements and their limit of detection by ICP-ES

Element	Detection limit (μg/L)	Element	Detection limit (μg/L)	Element	Detection limit (μg/L)
Ag	3	Hf	4	Re	10
Al	2	Hg	8	Rh	5
As	12	Ho	0.5	Ru	6
Au	5	I		S	20
B	1	In	18	Sb	18
Ba	0.07	Ir	4	Sc	0.4
Be	0.08	K	10	Se	40
Bi	12	La	1	Si	5
Br		Li	0.6	Sm	7
C	65	Lu	0.1	Sn	15
Ca	0.03	Mg	0.1	Sr	0.02
Cd	1	Mn	0.3	Ta	10
Ce	8	Mo	4	Tb	5
Cl		N		Te	27
Co	5	Na	1	Th	17
Cr	4	Nb	4	Ti	0.6
Cs		Nd	2	Tl	20
Cu	1	Ni	5	Tm	1
Dy	0.5	Os	5	U	20
Er	1	P	20	V	2
Eu	0.3	Pb	15	W	17
Fe	2	Pd	7	Y	0.2
Ga	6	Pr	0.8	Yb	0.3
Gd	2	Pt	20	Zn	0.9
Ge	15	Rb	30	Zr	1.5

on the concentration and sample introduction system. Table 5.1 shows the typical detection limits obtained for conventional solution nebulization, which can vary depending on the type of spectrometer, selected analytical lines and the nature of the sample matrix. ICP-ES can be considered as a significantly robust analytical technique for varying matrices where the applicability is limited mainly by the performance of the nebulizer system.

5.2.6 *Current trends in ICP-ES*

Until recently, the types of spectrometers commonly used were photomultiplier-based sequential or simultaneous direct reader systems. In spite of the high sensitivity attainable with the PMT-based detectors, and flexibility in choosing the analytical lines, the sequential systems have traditionally been slow for scanning from one line to another or to the background positions. Although direct readers can measure a number of analytical lines simultaneously, the background measurement still has to be done by physically moving optical components such as the entrance slit.

For this reason the ICP-ES technique has not been popular for measuring transient or time-variable signals. Fast spectral scanners based on a rapidly moving refractor plate placed on the optical path have been developed and are commercially available for direct reader PMT-based systems to address the problem of measuring transient signals with a certain degree of success. With the recent introduction of array spectrometers where almost the complete spectrum including the background can be measured simultaneously and with remarkable stability, most of the limitations of selecting analytical lines and measuring transient signals have been virtually eliminated. As a result the popularity of ICP-ES as a powerful tool for coupling novel sample introduction devices and chromatographic systems is expected to grow in the future.

5.3 Inductively coupled plasma mass spectrometry

One of the more significant advances in applied atomic spectrometry that has occurred during the past 20 years has been the development of the inductively coupled plasma mass spectrometer (ICP-MS). Since 1983, over 1600 published papers and conference presentations have appeared on all aspects of ICP-MS. Why is this development of such importance?

To answer this question, one needs only to ask an analytical chemist about current trends in the demand for his/her services. Analytical chemists are increasingly asked to determine most elements of the Periodic Table at concentrations ranging from parts-per-trillion (picograms/gram) or lower to the per cent level in samples that could be as small as a few micrograms! This can only be accomplished efficiently if an analytical technique is fast, sensitive and multi-element in nature, as is ICP-MS.

Inductively coupled plasma mass spectrometry (in its most common form) takes advantage of the ionizing power of the argon plasma and the rapid scanning capabilities of the quadrupole mass spectrometer. A schematic diagram of a typical ICP mass spectrometer is shown in Figure 5.14. The instrument comprises a sample introduction system, the argon plasma, a series of ion lenses, a quadrupole mass spectrometer and a detection system.

The problem of physically sampling an argon plasma operating at 1 atm and delivering the extracted ions to a quadrupole mass spectrometer and detector system (operating at from 10^{-6} to 10^{-5} Torr) is solved by the use of high-capacity pumps and an interface that essentially links the plasma and the mass spectrometer. The interface region comprises two nickel cones each containing an aperture (orifice) of about 1 mm. The first cone, called the **sampler**, is nearly flat and is used to extract ions directly from the plasma itself. Because of the intense heat produced by the plasma, water cooling is required to keep temperatures low. The sampler

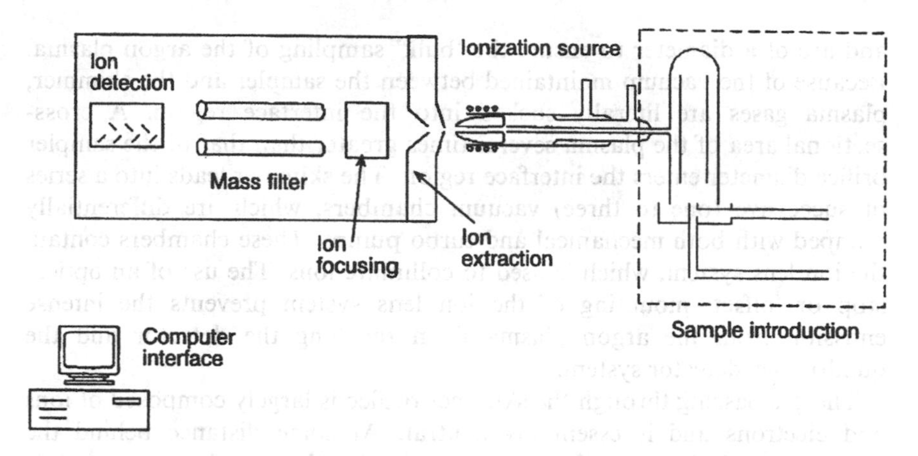

Figure 5.14 Schematic of typical ICP mass spectrometer.

is concentrically mounted relative to the next cone, called the **skimmer**, and is separated from it by 1–2 cm. The region between these cones is evacuated to a pressure of 5–10 Torr by mechanical pumping. At this point, most of the neutral (un-ionized) gas originating from the plasma sample is eliminated. A schematic of the interface region is shown in Figure 5.15. The sampler and skimmer orifice are machined to a knife-edge

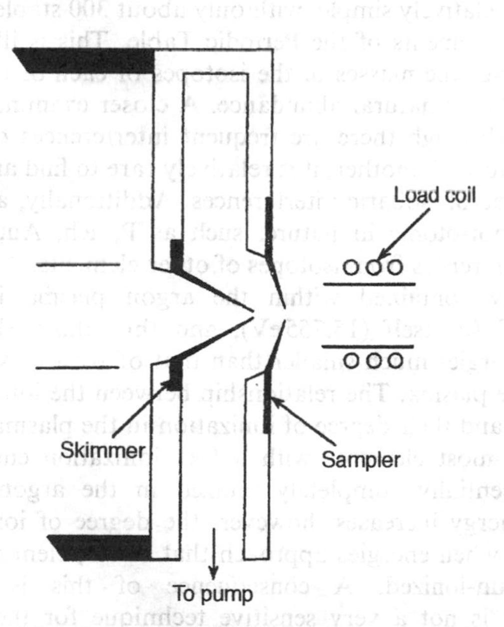

Figure 5.15 Diagram of interface region of ICP mass spectrometer showing sampler and skimmer cones. (Reproduced with permission from Douglas (1992) in *Inductively Coupled Plasmas in Analytical Atomic Spectrometry* (eds A. Montaser and D.W. Golightly), VCH Publishers, New York, p. 619.)

and are of a diameter to ensure the 'bulk' sampling of the argon plasma. Because of the vacuum maintained between the sampler and the skimmer, plasma gases are literally sucked into the interface region. A cross-sectional area of the plasma several times greater than that of the sampler orifice diameter enters the interface region. The skimmer leads into a series of successive (one to three) vacuum chambers, which are differentially pumped with both mechanical and turbo pumps. These chambers contain the ion lens system, which is used to collimate ions. The use of an optical stop or 'offset' mounting of the ion lens system prevents the intense emission from the argon plasma from reaching the detector and the quadrupole/detector system.

The gas passing through the skimmer orifice is largely composed of ions and electrons and is essentially neutral. At some distance behind the skimmer, and shortly before entering the ion lenses, the more mobile electrons are lost from the gas stream, leaving a beam of primarily positive ions. It has been calculated that approximately only 1 in 10^8 ions produced in the plasma reaches the detector.

5.3.1 Mass spectra and the argon plasma

Unlike atomic emission spectrometry, the mass spectra of chemical elements are relatively simple, with only about 300 stable isotopes possible for all of the elements of the Periodic Table. This is illustrated in Table 5.2, which gives the masses of the isotopes of each of the elements along with their relative natural abundance. A closer examination of this table shows that, although there are frequent interferences of isotopes of one element on those of another, it is relatively rare to find an element that has no isotope free of isobaric interferences. Additionally, all of the elements that are monoisotopic in nature, such as P, Rh, Au, etc., are free of isobaric interferences from isotopes of other elements.

The energy contained within the argon plasma is limited to the ionization of Ar itself (15.755 eV), and thus those elements with first ionization energies much smaller than that of Ar are expected to be fully ionized in the plasma. The relationship between the ionization energies of the elements and their degree of ionization in the plasma is given in Table 5.3. Clearly, most elements with a first ionization energy below about 8 eV are essentially completely ionized in the argon plasma. As the ionization energy increases, however, the degree of ionization decreases sharply, and when energies approach that of Ar, elements remain almost completely un-ionized. A consequence of this is that ICP mass spectrometry is not a very sensitive technique for the measurement of elements such as F and Cl.

The lowest second ionization energy for any element is 10.01 eV for Ba, a value that is close to the ionization energy of Ar itself. Elements that

Table 5.2 Relative abundances of naturally occurring isotopes*

u	*1*	*2*	*3*	*4*	*5*	*6*	*7*	*8*	*9*	*10*	*11*	*12*	*13*	*14*	*15*	*16*	*17*	*18*	*19*	*20*	*u*
H	99.99	0.01																			H
He				100.0																	He
Li						7.52	92.48														Li
Be									100.0												Be
B										18.98	81.02										B
C												98.89	1.11								C
N														99.64	0.36						N
O																99.76	0.04	0.20			O
F																			100.0		F
Ne																				90.92	Ne
u	*21*	*22*	*23*	*24*	*25*	*26*	*27*	*28*	*29*	*30*	*31*	*32*	*33*	*34*	*35*	*36*	*37*	*38*	*39*	*40*	*u*
(Ne)	0.26	8.82																			(Ne)
Na			100.0																		Na
Mg				78.60	10.11	11.29															Mg
Al							100.0														Al
Si								92.17	4.71	3.12											Si
P											100.0										P
S												95.02	0.75	4.22		0.01					S
Cl															75.40		24.60				Cl
Ar																0.34		0.06		99.60	Ar
K																			93.08	0.01	K
Ca																				96.92	Ca
u	*41*	*42*	*43*	*44*	*45*	*46*	*47*	*48*	*49*	*50*	*51*	*52*	*53*	*54*	*55*	*56*	*57*	*58*	*59*	*60*	*u*
(K)	6.91																				(K)
(Ca)		0.64	0.13	2.13		0.00		0.18													(Ca)
Sc					100.0																Sc
Ti						7.95	7.75	73.45	5.51	5.34											Ti
V										0.24	99.76										V
Cr										4.31		83.76	9.55	2.38							Cr
Mn															100.0						Mn
Fe														5.90		91.52	2.25	0.33			Fe
Co																			100.0		Co
Ni																		67.77		26.16	Ni
u	*61*	*62*	*63*	*64*	*65*	*66*	*67*	*68*	*69*	*70*	*71*	*72*	*73*	*74*	*75*	*76*	*77*	*78*	*79*	*80*	*u*
(Ni)	1.25	3.66		1.16																	(Ni)
Cu			69.09		30.91																Cu
Zn				48.90		27.81	4.11	18.56		0.62											Zn
Ga									60.20		39.80										Ga
Ge										20.52		27.43	7.76	36.53		7.76					Ge
As															100.0						As
Se														0.96		9.12	7.50	23.61		49.97	Se
Br																			50.57		Br
Kr																		0.35		2.27	Kr
u	*81*	*82*	*83*	*84*	*85*	*86*	*87*	*88*	*89*	*90*	*91*	*92*	*93*	*94*	*95*	*96*	*97*	*98*	*99*	*100*	*u*
(Se)		8.84																			(Se)
(Br)	49.43																				(Br)
(Kr)		11.56	11.56	56.90		17.37															(Kr)
Rb					72.15		27.85														Rb
Sr				0.56		9.86	7.02	82.56													Sr
Y									100.0												Y
Zr										51.46	11.23	17.11		17.4		2.80					Zr
Nb													100.0								Nb
Mo												15.84		9.04	15.72	16.53	9.46	23.78		9.63	Mo
Tc																					Tc
Ru																5.68		2.22	12.81	12.70	Ru

Table 5.2 (Continued)

u	*101*	*102*	*103*	*104*	*105*	*106*	*107*	*108*	*109*	*110*	*111*	*112*	*113*	*114*	*115*	*116*	*117*	*118*	*119*	*120*	*u*
(Ru)	16.98	31.34		18.27																	(Ru)
Rh			100.0																		Rh
Pd		0.08		9.30	22.60	27.10		26.70		13.50											Pd
Ag							51.35		48.65												Ag
Cd						1.22		0.89		12.43	12.86	23.79	12.34	28.81		7.86					Cd
In													4.16		95.84						In
Sn												0.95		0.65	0.34	14.24	7.57	24.01	8.58	32.97	Sn
Sb																					Sb
Te																				0.09	Te
u	*121*	*122*	*123*	*124*	*125*	*126*	*127*	*128*	*129*	*130*	*131*	*132*	*133*	*134*	*135*	*136*	*137*	*138*	*139*	*140*	*u*
(Sn)		4.71		5.98																	(Sn)
(Sb)	57.25		42.75																		(Sb)
(Te)		2.46	0.87	4.61	6.99	18.71		31.79		34.48											(Te)
I							100.0														I
Xe				0.10		0.09		1.92	26.44	4.08	21.18	26.88		10.44		8.87					Xe
Cs													100.0								Cs
Ba										0.10		0.10		2.42	6.59	7.81	11.32	71.66			Ba
La																		0.09	99.91		La
Ce																0.19		0.25		88.48	Ce
u	*141*	*142*	*143*	*144*	*145*	*146*	*147*	*148*	*149*	*150*	*151*	*152*	*153*	*154*	*155*	*156*	*157*	*158*	*159*	*160*	*u*
(Ce)		11.07																			(Ce)
Pr	100.0																				Pr
Nd		27.11	12.14	23.83	8.29	17.26		5.74		5.63											Nd
Pm																					Pm
Sm				3.09			14.97	11.24	13.83	7.44		26.72		22.71							Sm
Eu											47.77		52.23								Eu
Gd												0.20		2.15	14.73	20.47	15.68	24.87		21.90	Gd
Tb																			100.0		Tb
Dy																0.05		0.09		2.29	Dy
u	*161*	*162*	*163*	*164*	*165*	*166*	*167*	*168*	*169*	*170*	*171*	*172*	*173*	*174*	*175*	*176*	*177*	*178*	*179*	*180*	*u*
(Dy)	18.88	25.53	24.97	28.19																	(Dy)
Ho					100.0																Ho
Er		0.14		1.56		33.41	22.94	27.07		14.88											Er
Tm									100.0												Tm
Yb								0.14		3.03	14.31	21.82	16.13	31.84		12.73					Yb
Lu															97.40	2.60					Lu
Hf														0.18		5.20	18.50	27.14	13.75	35.23	Hf
Ta																				0.01	Ta
W																				0.13	W
u	*181*	*182*	*183*	*184*	*185*	*186*	*187*	*188*	*189*	*190*	*191*	*192*	*193*	*194*	*195*	*196*	*197*	*198*	*199*	*200*	*u*
(Ta)	99.99																				(Ta)
(W)		26.31	14.28	30.64		28.64															(W)
Re					37.07		62.93														Re
Os				0.02		1.59	1.64	13.21	16.11	26.41		41.02									Os
Ir											38.50		61.59								Ir
Pt										0.01		0.78		32.90	33.80	25.30		7.21			Pt
Au																	100.0				Au
Hg																0.15		10.02	16.84	23.13	Hg
u	*201*	*202*	*203*	*204*	*205*	*206*	*207*	*208*	*209*	*210*	*211*	*212*	*213*	*214*	*215*	*216*	*217*	*218*	*219*	*220*	*u*
(Hg)	13.22	29.79		6.85																	(Hg)
Tl			29.50		70.50																Tl
Pb				1.37		25.15	21.11	52.37													Pb
Bi									100.0												Bi
u	*221*	*222*	*223*	*224*	*225*	*226*	*227*	*228*	*229*	*230*	*231*	*232*	*233*	*234*	*235*	*236*	*237*	*238*	*239*	*240*	*u*
Th												100.0									Th
Pa																					Pa
U														0.01	0.72			99.28			U

*Note: 0.00% abundance indicates the presence of an isotope of <0.01% abundance.

Table 5.3 Degree of ionization of elements in an argon plasma

Element	Ionization energy (eV)	Degree of ionization	Element	Ionization energy (eV)	Degree of ionization
Cs	3.898	99.98	Fe	7.87	96.077
Rb	4.177	99.98	Re	7.88	94.54
K	4.341	99.97	Ta	7.89	96.04
Na	5.139	99.91	Ge	7.899	91.64
Ba	5.212	99.96	W	7.98	94.86
Ra	5.279	99.95	Si	8.151	87.9
Li	5.392	99.85	B	8.598	62.03
La	5.577	99.91	Pd	8.34	94.21
Sr	5.695	99.92	Sb	8.461	81.07
In	5.786	99.42	Os	8.7	79.96
Al	5.986	98.92	Cd	8.993	85.43
Ga	5.99	99.01	Pt	9.01	61.83
Tl	6.108	99.38	Tc	9.009	66.74
Ca	6.113	99.86	Au	9.225	48.87
Y	6.38	98.99	Be	9.322	75.36
Sc	6.54	99.71	Zn	9.394	74.5
V	6.74	99.23	Se	9.752	30.53
Cr	6.766	98.89	As	9.81	48.87
Ti	6.82	99.49	S	10.36	11.47
Zr	6.84	99.31	Hg	10.437	32.31
Nb	6.88	98.94	I	10.451	24.65
Hf	7	98.89	P	10.486	28.79
Mo	7.099	98.54	Rn	10.748	35.74
Tc	7.28	97.5	Br	11.814	3.183
Bi	7.289	94.14	C	11.26	3.451
Sn	7.344	96.72	Xe	12.13	5.039
Ru	7.37	96.99	Cl	12.967	0.4558
Pb	7.416	97.93	O	13.618	0.042545
Mn	7.435	97.11	Kr	13.999	0.2263
Rh	7.46	95.87	N	14.534	0.04186
Ag	7.576	94.45	Ar	15.759	0.01341
Ni	7.635	92.55	F	17.422	0.0001919
Mg	7.646	98.25	Ne	21.564	0.000005468
Cu	7.726	91.59	He	24.587	0.000000001
Co	7.86	94.83			

have a first ionization potential of about 10 eV are generally only about 20% ionized in the argon plasma. Most elements have second ionization potentials of about 15–20 eV, resulting in little or no electron loss in the argon plasma. This means that, except for a few elements, ICP mass spectral signals originate from singly charged ionic species and only rarely from doubly charged species. Only under circumstances when a large quantity of an element with a relatively low second ionization potential is present in the plasma does one have to be concerned with doubly charged ionic species. A typical mass spectrum for lead, which has four stable isotopes at $m/z = 204, 206, 207$ and 208, is given in Figure 5.16. Note that the resolution of the quadrupole mass spectrometer is nominally 1 u, which means that potentially serious isobaric interferences can occur with

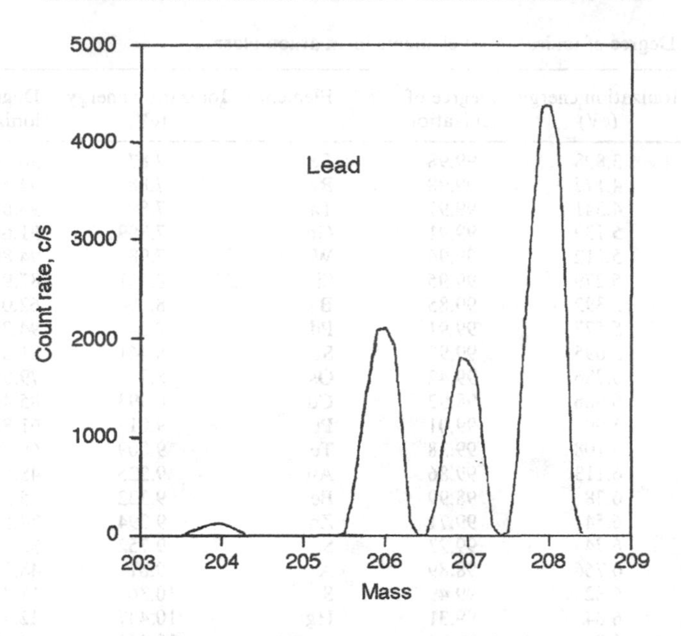

Figure 5.16 ICP mass spectrum for lead.

isotopes of other elements that have close exact masses (cf. Table 5.1). For example, ^{204}Hg and ^{204}Pb have exact masses of 203.973 467 u and 203.973 020 u, respectively. Clearly, a mass spectrometer of much greater resolving power, such as a magnetic sector instrument, would be required to resolve these signals.

In the above example, **resolution** was measured as the width of the ICP-MS signal at a given m/z measured at 10% of the maximum height of the signal. An equally important parameter describing the performance of a quadrupole mass spectrometer is the **abundance sensitivity**. A closer examination of the shape of the Pb signals in Figure 5.16 reveals that the signals themselves are asymmetrical in nature. ICP-MS signals tend to drop off sharply toward the baseline at the higher-mass side of the peak whereas the signal shows a more gradual decline at the low-mass side. The abundance sensitivity is defined as the ratio of the intensity of a given isotope signal at the central or exact mass (m/z) to the intensity 1 u below or above the exact mass. Typically, quadrupole mass spectrometers give abundance sensitivities of $>10^6$ above the exact mass and an abundance sensitivity of 10^5 below the exact mass. The practical consequence of this is that an isotope for an element present at high concentrations in solution will not generally interfere (overlap) with the measurement of a relatively much smaller signal from an element whose analyte isotope is 1 u greater. However, isobaric overlap from the 'wing' of an intense ICP-MS signal could cause an isobaric interference on an analyte isotope located 1 u

Table 5.4 Instrumental limits of detection for ICP mass spectrometry*

Detection limit (μg/L)	Elements
≤0.001	Au, Bi, Ce, Co, Cs, Dy, Eu, Er, Ga, Hf, Ho, In, Ir, La, Lu, Nb, Pb, Pr, Ra, Re, Rh, Sb, Sm, Sr, Ta, Tb, Th, Tl, Tm, U, W, Y, Yb
0.001–0.01	Ag, Al, As, Ba, Cd, Cu, Ge, Gd, Hg, I, Mg, Mn, Mo, Nd, Ni, Os, Pd, Pt, Rb, Ru, Sn, Te, Ti, V, Zn, Zr
0.1–1	Br, Fe, K, P, Si (Fe, Ca by cold plasma)
1–10	Ca
≥10	C, Cl, F, S

*Typical data for ELAN 6000.

lower. An example of this would be the interference of the intense $^{40}Ar^+$ signal on the measurement of $^{39}K^+$ (93.3% abundance). In fact, the measurement of low levels of K is not generally possible using ICP-MS unless special plasma conditions are used giving rise to a 'cold' plasma or other conditions leading to the suppression or quenching of argon ions before they reach the detector. A second example of interference of one ion signal on another would be the spectral overlap of an intense $^{12}C^+$ signal on $^{11}B^+$.

One measure of the performance of any analytical technique is the **limit of detection**, which is the smallest quantity (concentration or absolute mass) of an element that can be determined with confidence. This is usually expressed as the concentration (mass) of analyte that is equivalent to three times the standard deviation of the signal obtained for a blank. Using water as the blank, instrumental limits of detection for one of the commercially available ICP mass spectrometers are given in Table 5.4. Because most elements are fully ionized in the argon plasma, ICP-MS limits of detection for most elements vary over a relatively narrow range. This is not true for most other atomic spectroscopic techniques, because their limits of detection depend greatly on other factors such as absorption coefficients and electron transitions, that vary greatly from one element to another. For ICP-MS, variation in limits of detection for the elements can usually be linked to a combination of ionization potential and the abundance of the individual isotope used as analyte.

5.3.2 *Background spectrum and polyatomic ions*

The argon plasma is a chemically simple flame containing primarily argon and ambient atmospheric gases. The background spectrum generated by the plasma thus contains atomic ions of Ar, O, N, H and C and polyatomic ionic species made up of a combination of these elements.

Accordingly, the most intense ion signals are obtained for Ar^+, O^+ and N^+. When water is added to the plasma, usually as sample solvent, large quantities of oxygen and hydrogen are added, resulting in significant increases in the intensity of the singly charged ions of these elements. In addition to this, the presence of large quantities of oxygen and hydrogen results in the presence of such polyatomic ions as ArO^+ and ArH^+. The background spectrum of an argon plasma containing added water is shown in Figure 5.17. Note that, above $m/z = 81$, there are no polyatomic ions of consequence, and the only atomic ions present are those of Xe, which is a contaminant in the Ar supply gas. Argon itself is composed of three isotopes and, although isotopes at $m/z = 36$ and 38 have low abundance, the very large absolute number of argon ions present results in significant background signals arising from these isotopes. For example, Table 5.5 gives the interfering ions and polyatomic ions arising from argon in combination with atmospheric gases along with the isotopes of elements with which these ions interfere. Clearly, the determination of several elements such as Ca, Cr and Fe, which are nearly monoisotopic, can be made difficult by the presence of these interfering ions. Recent new developments in ICP-MS instrumentation based on using cold or shielded plasmas or hexapole mass spectrometers promise to eliminate or minimize these interferences.

When mineral acids are present, however, large quantities of elements such as Cl are added to the argon plasma, and these can now contribute to the production of polyatomic species containing chloride. For example, $^{75}ArCl^+$ and $^{77}ArCl^+$ interfere with monoisotopic $^{75}As^+$ and $^{77}Se^+$, which at an abundance of 7.6% is the only Se isotope not isobaric with an isotope of another element.

Even isotopes of elements originating from samples can be problematic if present in high enough concentrations. Concentrations of elements in the ppm (μg/g) range can result in the formation of argides. For example, high concentrations of copper in solution will produce $^{103}CuAr^+$ and $^{105}CuAr^+$, interfering with monoisotopic $^{103}Rh^+$ and $^{105}Pd^+$; the latter, at an abundance of 22.3%, is the only Pd isotope free of isobaric interferences.

The practical consequence of both of the examples given above is that chemical separations are required if one wishes to determine trace quantities of As or Se when chloride is present in the sample or trace quantities of Rh or Pd when copper is present. For both Se and Pd, alternate isotopes could be used for quantification, but corrections for isobaric interferences from other elements present in the sample (or plasma) would be required.

When aqueous samples are introduced into the plasma, relatively large quantities of oxygen are also introduced from the dissociation of water and entrained (into the plasma) air from the atmosphere surrounding the

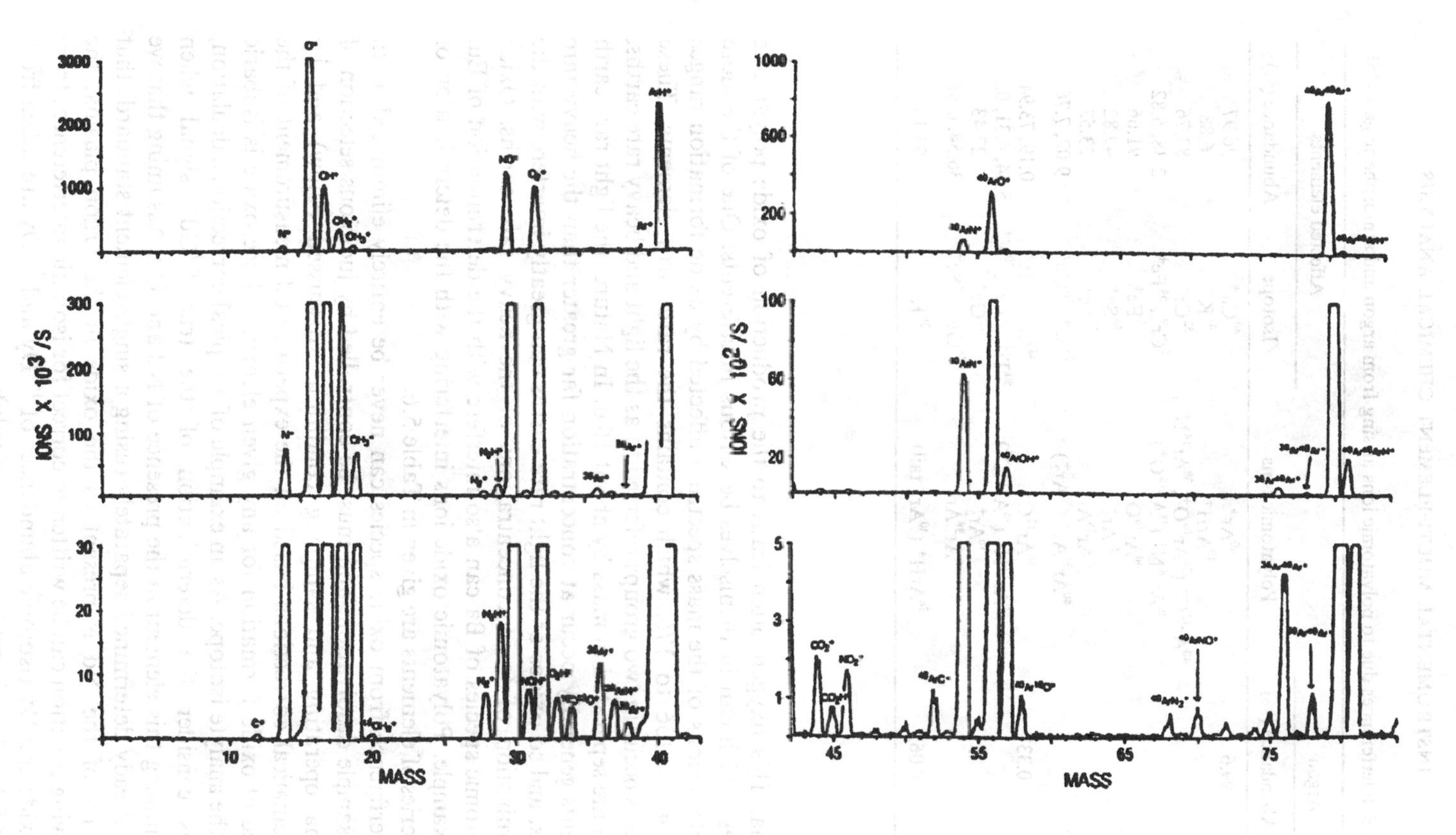

Figure 5.17 ICP-MS background mass spectra for distilled water. Spectra for each mass range are amplified (top to bottom) to show minor ions. (Reproduced from Tan and Horlick (1986) [10], with permission.)

Table 5.5 Interferences due to polyatomic ions arising from argon and atmospheric gases [5]

Argon			Affected elements	
m/z	Abundance (%)	Polyatomic ions	Isotope	Abundance (%)
40	99.6	$^{40}Ar^+$	$^{40}Ca^+$	96.97
		$^{40}ArH^+$	$^{41}K^+$	6.88
		$^{40}Ar^{12}C^+$ ($^{36}Ar^{16}O^+$, $^{38}Ar^{14}N^+$)	$^{52}Cr^+$	83.76
		$^{40}Ar^{14}N^+$ ($^{38}Ar^{16}O^+$)	$^{54}Cr^+$, $^{54}Fe^+$	2.38, 5.82
		$^{40}Ar^{16}O^+$	$^{56}Fe^+$	91.66
		$^{40}Ar_2^+$	$^{80}Se^+$	49.82
		$^{40}Ar^{38}Ar^+$	$^{78}Se^+$	23.52
		$^{40}Ar^{36}Ar^+$ ($^{38}Ar_2^+$)	$^{76}Se^+$, $^{76}Ge^+$	9.02, 7.76
36	0.337	$^{36}Ar^{12}C^+$	$^{48}Ca^+$, $^{48}Ti^+$	0.19, 73.94
		$^{36}Ar^{14}N^+$ ($^{38}Ar^{12}C^+$)	$^{50}Ti^+$, $^{50}Cr^+$, $^{50}V^+$	5.34, 4.31, 0.24
		$^{36}Ar_2^+$	$^{72}Ge^+$	27.43
		$^{36}Ar^{38}Ar^+$	$^{74}Ge^+$, $^{74}Se^+$	36.54, 0.87
38	0.063	$^{38}ArH^+$ ($^{40}Ar^+$ tail)	$^{39}K^+$	93.10

plasma. This oxygen often leads to the production of oxide polyatomic species, which can in themselves be serious interferents. One of the more notable regions of the mass spectrum affected by oxide formation ranges from $m/z = 136$ to 176, which contains the rare-earth elements. These elements occur in two groups referred to as the light and heavy rare earths, which are separated in mass by about 16 u. In Nature, the light rare-earth elements generally occur at concentration far greater than the heavy rare earths, and so oxides of the light rare earths can greatly interfere with the determination of trace concentrations of the heavy rare earths. Oxide polyatomic species of Ba can also interfere with the determination of Eu, for example. Polyatomic oxide ions interfering with the determination of this series of elements are given in Table 5.6.

Interference from oxide species can never be entirely eliminated, even with sample desolvation, and must be controlled by judicious selection of plasma operating and sampling conditions (discussed below) and by mathematical corrections based on the experimental measurement of the degree of oxide formation for any given element whose oxide is isobaric with the analyte isotope. As an example of a typical correction calculation, let us consider the determination of the true $^{160}Gd^+$ signal when determining this element in the presence of Nd and Dy. Assuming that we have already determined separately (using a single-element standard) that only 1% of the Nd is present as the oxide in the argon plasma, the following equation can be written to correct for isobaric interferences from Nd oxide and Dy (isotopic abundances of ^{160}Dy and ^{161}Dy are taken from Table 5.2 as 2.34% and 18.9%, respectively):

Table 5.6 Oxide and isobaric spectral interferences for the determination of the rare-earth elements [6]

Analyte	Mass (u)	Interfering ion
La	139	–
Ce	140	–
Pr	141	–
Nd	143	–
Sm	147	–
Eu	151	$^{135}BaO^+$
Gd	160	$^{144}NdO^+$, $^{160}Dy^+$
Tb	159	$^{143}NdO^+$
Dy	163	$^{147}SmO^+$
Ho	165	$^{149}SmO^+$
Er	166	$^{150}NdO^+$
Tm	169	–
Yb	174	$^{158}GdO^+$
Lu	175	–

$$^{160}Gd^+_{true} = {}^{160}Gd^+_{measured} - 0.01(^{144}Nd^+) - \frac{0.0234}{0.189}\,{}^{161}Dy^+$$

A summary of the different categories and types of atomic and polyatomic ion spectral interferences along with specific examples for each is given in Table 5.7.

Table 5.7 Spectral interferences in ICP mass spectrometry

Type	Mass	Interferent	Affected element
Isobaric isotopes			
analyte	138	$^{138}Ce^+$	Ba (71.7), La (0.09)
matrix (iron)	58	$^{58}Fe^+$	Ni (67.8)
Ions derived from plasma			
monatomic	40	$^{40}Ar^+$	Ca (99.94)
polyatomic	80	$^{40}Ar_2^+$	Se (49.62)
Ions derived from solvent and matrix components			
monatomic	35	$^{35}Cl^+$	Cl (75.77)
polyatomic	51	$^{35}Cl^{16}O^+$	V (99.75)
	75	$^{40}Ar^{35}Cl^+$	As (100)
Oxides	56	$^{40}Ar^{16}O^+$	Fe (91.75)
	153	$^{137}Ba^{16}O^+$	Eu (52.2)
Hydroxides	55	$^{38}Ar^{16}OH^+$	Mn (100)
	155	$^{138}Ba^{16}OH^+$	Gd (15.1)
Doubly charged ions	69	$^{138}Ba^{2+}$	Gd (60.42)
	75	$^{150}Nd^{2+}$, $^{150}Sm^{2+}$	As (100)
Argides	63	$^{40}Ar^{23}Na^+$	Cu (69.17)
	103	$^{40}Ar^{63}Cu^+$	Rh (100)

5.3.3 Optimization of instrumental operating conditions

Plasma conditions. Optimization of the instrumental parameters to obtain sensitive stable ICP-MS signals generally requires the adjustment of several instrumental parameters. The most important of these are the plasma gas flow rate, the sampling depth, plasma power, ion lens voltage(s) and liquid sample uptake rate if solution nebulization sample introduction is being used. Perhaps the most important parameter is the so-called injector or carrier argon flow rate. This is the argon flow that passes directly through the central channel of the argon plasma and carries sample aerosol or vapor. This flow is particularly important because it affects the efficiency of nebulization of the sample introduction system and sets the sampling depth. The sampling depth refers to the actual region of the central zone of the argon plasma that is physically sampled and is generally expressed as the distance from the first turn of the load coil to the sampler orifice. Earlier in this chapter, the distribution of atoms and ions within the argon plasma was discussed as well as the different temperature zones. Clearly, the best location for sampling is the plasma region of highest analyte ion density. A plot of the change in analyte (Ga) ion signal with increasing carrier (injector) argon flow rates and RF forward power is shown in Figure 5.18. Note that the curves for all of the RF power settings generally have the same shape but that their maxima are located at slightly different gas flows. Clearly, the most intense Ga ion signal is obtained only at one RF power input level, the value of which can range from about 900 W to 1300 W depending on the instrument (manufacturer) used.

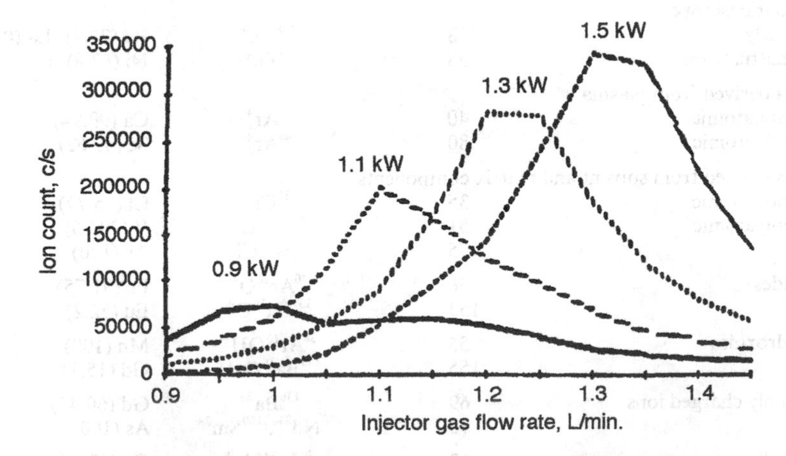

Figure 5.18 Change in Ga^+ signal as a function of injector gas flow rate and RF power. (Reproduced from Vaughan and Horlick (1987) [12], with permission.)

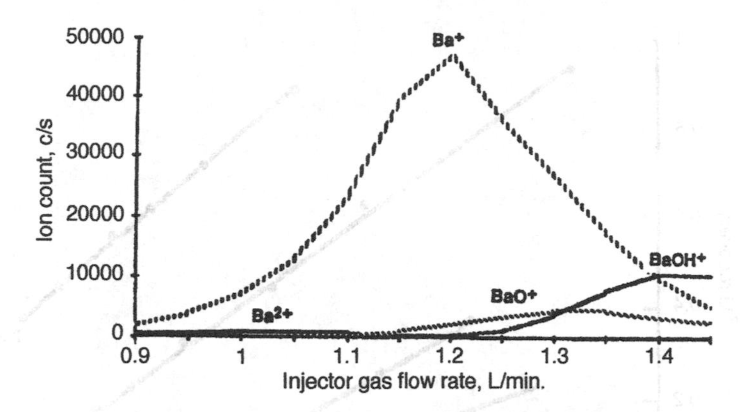

Figure 5.19 Change in signal intensity for Ba^+, Ba^{2+} and Ba oxide and hydroxide polyatomic ions as a function of injector gas flow rate. (Reproduced from Vaughan and Horlick (1986) [11], with permission.)

Signal intensities for polyatomic species are also dependent on the carrier argon flow. An interesting relationship to study is the change in signal intensity for an elemental ion (singly and doubly charged), its oxide and hydroxide, as a function of the Ar carrier gas flow (Figure 5.19). This figure shows that optimum signals for each of the species occur under different experimental conditions, and thus it is possible to minimize the effect of undesirable analyte species by judicious selection of plasma conditions.

Other parameters. The selection of ion lens voltage settings depends on the instrument (manufacturer), with individual models having from one to more than four different lenses to adjust. The process of selecting ion lens voltages and plasma operating conditions is generally accomplished by trial and error and takes only a few minutes. There is no single set of instrument operating conditions that gives optimum performance, and so it is possible to have several combinations of plasma conditions, ion lens voltages, etc., that give equivalent analytical performance.

5.3.4 *Matrix interferences*

Most samples analyzed by ICP-MS will contain so-called **matrix elements**. These elements comprise the bulk of the composition of the sample and are present in the very high ppm to per cent levels. ICP-MS, like other atomic spectrometric techniques, is not immune to matrix effects, which generally result in a loss in analyte signal intensity in the presence of concomitants occurring at high enough concentrations. The effect of adding increasing molar quantities of a series of concomitant elements on

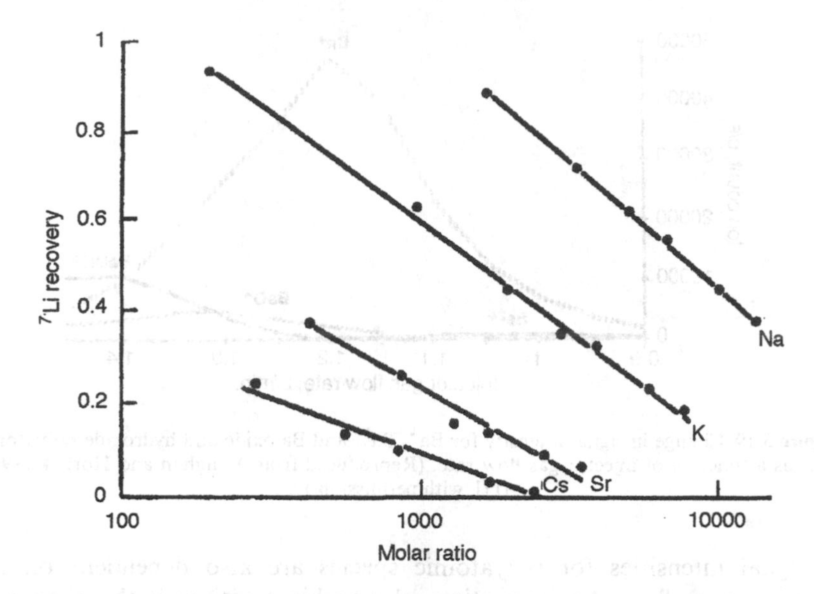

Figure 5.20 Effect of increasing molar ratio and atomic mass of added matrix components on Li^+ signal. (Reproduced from Gregoire (1987) [8], with permission.)

the ICP-MS signal for Li is shown in Figure 5.20. This figure illustrates the basic trends for matrix interferences in ICP-MS. As the atomic mass of the interfering species increases relative to the atomic mass of the analyte (from Na to Cs), the degree of signal suppression also increases. In this figure, signal suppression is measured as a 'recovery' (Y-axis), which is expressed as the ratio of the Li signal obtained in the presence of the concomitant to the Li signal obtained in distilled water. The converse is also true, which in effect means that it would be relatively easy to measure trace quantities of U in lithium metal whereas it would be difficult to determine trace quantities of Li in uranium metal. Also, as the molar ratio of concomitant to analyte ion increases, signal suppression also increases. The absolute quantity of the interfering element is also important. Keeping the molar ratio of concomitant to analyte ion constant, while decreasing the absolute quantity of both reaching the plasma, will diminish matrix interferences. This is illustrated in Figure 5.21, which shows the effect of serial dilution of a solution containing Li in the presence of larger quantities of Na, K, Sr and Cs all held at a constant molar ratio relative to Li. As the solution is diluted (moving from right to left on the graph), the Li recovery increases (interference decreases). For Na and K, interference-free conditions (recovery = 1) are reached when the Li solution is diluted by approximately a factor of 2.

The degree of ionization of a concomitant also has an effect on matrix interferences. An element (concomitant) that is not completely ionized in

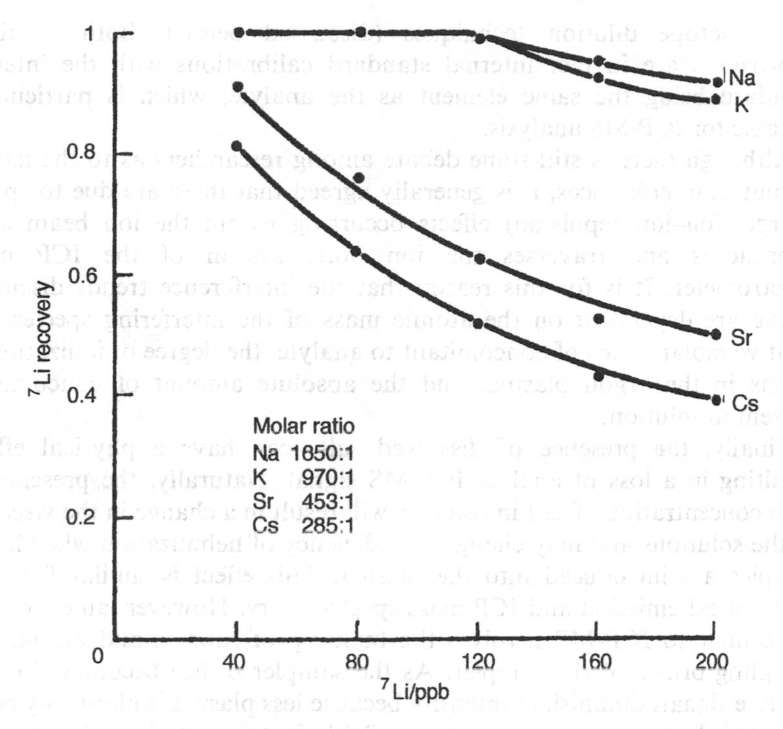

Figure 5.21 Effect of serial dilution on Li^+ signal suppression. Molar ratio of concomitants (Na, K, Sr, Cs) to analyte (Li) is kept constant. (Reproduced from Gregoire (1987) [8], with permission.)

the argon plasma will not suppress the analyte signal as much as a concomitant of the same (or nearly same) mass that is completely ionized in the argon plasma.

The control of matrix interferences is very sample-dependent. In some cases where analyte concentrations are low relative to instrument sensitivity, serial dilution of sample solutions may not be practicable. In this case, chemical separation of analyte from matrix components may be required. This is usually accomplished using ion chromatography or solvent extraction techniques. Alternatively, ICP-MS signals for an internal standard element could be used to correct for the loss in analyte sensitivity, but one must be assured that the internal standard element response is the same as that of the analyte in the presence of concomitant elements. Generally, useful internal standards are those which have a mass similar to that of the analyte and have the same degree of ionization in the argon plasma. When sufficient analyte ICP-MS signal is available even in the presence of a concomitant element, it is also possible to use the method of standard additions to correct for these effects or calibration using isotope dilution techniques (discussed below). Both of these approaches are in fact internal standard calibrations with the internal standard being the same element as the analyte, which is particularly suitable for ICP-MS analysis.

Although there is still some debate among researchers as to the nature of matrix interferences, it is generally agreed that these are due to space-charge (ion–ion repulsion) effects occurring within the ion beam as it approaches and traverses the ion optic system of the ICP mass spectrometer. It is for this reason that the interference trends discussed above are dependent on the atomic mass of the interfering species, the relative molar ratios of concomitant to analyte, the degree of ionization of atoms in the argon plasma, and the absolute amount of concomitant present in solution.

Finally, the presence of dissolved salts can have a physical effect, resulting in a loss of analyte ICP-MS signal. Naturally, the presence of high concentration of salt in solution will result in a change in the viscosity of the solutions and may change the efficiency of nebulization when liquid samples are introduced into the plasma. This effect is similar for both ICP optical emission and ICP mass spectrometry. However, an effect that is peculiar to ICP-MS involves the build-up of salts in and around the sampling orifice of the sampler. As the sampler orifice becomes clogged, analyte signals diminish in intensity because less plasma is physically being sampled. For most commercially available instrumentation, this is not a serious problem because orifice clogging occurs only at dissolved salt concentrations that are generally too high for accurate ICP-MS determinations. As a rule of thumb, it is wise to analyze only solutions that contain 0.1% or less of dissolved salts. This is especially important when running continuous analytical runs lasting 8 h or more. The use of flow injection techniques is a popular approach to reducing the total amount of salt reaching the sampler during a given time period, but flow injection cannot avoid the cumulative effects of high salt-containing sample solutions and the resulting matrix effects.

5.3.5 Quantitative analysis using ICP mass spectrometry

A number of factors must be considered when setting up for an ICP-MS analysis. These include the analytes of interest, possible spectral and matrix interferences, instrument settings and selection of internal standards and calibration strategy. Each sample type will be somewhat different from others and may require a slightly modified or even radical analytical approach. For example, for the analysis of a typical silicate geological material, powdered rock will be dissolved in a mixture of acids including hydrofluoric, perchloric and nitric acids. This is usually followed by a fusion of any residue with lithium metaborate or some other fluxing agent. The resulting solution will generally be high in concomitants such as iron, calcium, magnesium, sodium as well as other elements occurring in the high ppm to per cent concentrations.

For a completely unknown material, it is sometimes valuable to perform a semi-quantitative analysis giving information on the approximate composition of the sample. In this way, spectral interferences may be found and alternate un-interfered analyte ions can be selected when possible. Spectral interferences affecting an analyte ion can often be revealed by examining the mass spectrum of all of the isotopes of a particular element. With the exception of a small number of elements, most elements have invariant isotopic abundances in Nature. This means that, if no spectral interferences exist, then the expected natural isotope ratios for an element would be found. If the expected isotope ratio is different than expected, then there is likely a spectral interference originating from any of the sources discussed previously in the section on spectral interferences.

Aside from the known polyatomic ions present in the mass spectrum that originate from argon, water and atmospheric gases, the most serious problem facing the analyst is oxide formation. Generally, oxides of elements that have strong M–O bonds can have from 0.2 to 1% of their ions in the oxide form. The rare-earth elements are one group of elements that fits this category. In order to determine the influence of oxides on a potential analyte, one can use a solution containing only the oxide-forming element in solution. By measuring the ICP-MS signal at m/z for the elemental ion as well as the signal at +16 u, then one can calculate the MO^+/M^+ ratio and the corresponding correction factor. This correction factor is valid only for the instrumental operating conditions used during the measurement of the oxide ratio itself. These conditions should be the optimum conditions for the measurement of analyte ions.

Incorrect analytical results will be obtained if matrix suppression (or enhancement) effects go unnoticed. Although these effects are most pronounced for the lighter elements, all elements can suffer from matrix effects to varying degrees. There are several techniques that can be used to determine the presence of matrix effects. The first involves simply diluting the sample with diluent or appropriate solvent. A change in analyte signal intensity corresponding to the dilution factor indicates no matrix interference effects. If the analyte signal is proportionately too high after dilution, then a matrix suppression is occurring in the undiluted sample. However, if the analyte is present in solution at a concentration close to the limit of detection, then dilution of the sample is not practicable and the method of standard addition must be used. In this case, a small quantity, preferably equivalent to the approximate concentration of analyte in the sample itself, is added to the sample. The increase in analyte signal should be equivalent to the signal obtained when measuring the added analyte concentration in diluent solvent alone. Any change in the expected results using either of the above techniques indicates that corrective measures must be taken to avoid inaccuracies. Although it is possible to reduce somewhat matrix interference by changing the injector plasma gas flow, this approach has limited practical value, particularly when performing multi-element analysis for 40 elements or more during a single mass scan.

5.3.6 *Instrument calibration*

The use of external calibrants for quantitative analysis is the most direct means of obtaining unknown analyte concentrations. Several factors must be taken into consideration to ensure accurate results. First, the analytical response for a given concentration of analyte in the calibration standard must be the same for the equivalent concentration of analyte in the sample solution. Second, the concentration of the unknown must be within the linear dynamic range of the calibration curve. Third, the response of the ICP mass spectrometer must be constant over time or be corrected for instrumental drift by the use of an internal standard. Earlier in this chapter, we discussed the selection of internal standards and determined that these had to be close in mass and ionization potential to the analyte ions. Another equally important criterion must be that the element(s) used as **internal standards** are not present in the sample itself at measurable concentrations. When dealing with the multi-element analysis of complex materials such as rocks and minerals, only a very few elements are available to be used with confidence as internal standard elements. These elements include the platinum group elements (Pt, Pd, Ru, Ir, Os), Re and other elements of low crustal abundance such as Tl. It is advisable to determine in advance if sample solutions contain any element to be used as an internal standard. Internal standards correct for signal (instrumental) drift, plasma flicker and physical effects related to the composition of the sample. The correction is done by adding a known and constant quantity of the internal standard to all of the sample solutions as well as to blanks and standards. Rather than using the absolute signal obtained from each analyte for calibration, the ratio of the analyte signal to the internal standard signal is used.

As discussed above, the mass and the degree of ionization of the internal standard are important. This means that there is no single internal standard element that can be used for all elements of the Periodic Table. In practice, the addition of a light isotope such as ^{9}Be or the use of a plasma gas ion such as the argon dimer ($^{80}Ar_2^+$) can be used along with heavier isotopes such as ^{103}Rh or ^{187}Re.

When serious suppression effects exist and when the matrix composition of the standards is not the same as unknown solutions, the method of

Table 5.8 Isotopic abundance for natural platinum and a typical enriched platinum spike

Platinum isotope	Natural abundance (%)	Spike abundance (%)
190	0.01	0.02
192	0.79	0.04
194	32.9	95.06
195	33.8	3.76
196	25.3	0.97
198	7.2	0.15
Calculated atomic mass	195.11	194.06

standard additions can be used. Calibration using this technique involves adding known increasing quantities of analyte elements to a series of aliquots of the unknown solution. A plot of added concentration of analyte and signal intensity yields the unknown concentration at the point where the calibration line intersects the concentration axis.

Because ICP-MS is a mass spectrometric technique, the use of **isotope dilution** is possible. This technique is based on the addition to the unknown sample or sample solution of a known mass of the analyte element that has been isotopically enriched. For example, Table 5.8 gives the natural abundances for Pt metal and those for a sample isotopically enriched in ^{195}Pt. Isotope dilution is based on the measurement of the isotope ratio of the reference isotope (any analyte isotope other than the spike isotope) to that of the spike or enriched isotope in a sample, both before and after the spiking process. An equation can be readily derived which gives the relationship between the concentration of analyte in the sample and the measured ratio of the reference and spike isotopes:

$$C = \frac{M_s K(A_s - B_s R)}{W(BR - A)}$$

where C is the concentration of the analyte in ($\mu g\, g^{-1}$), M_s is the mass of the enriched isotope spike (in μg), K is the ratio of the natural to spike atomic masses, A_s is the abundance of the reference isotope in the spike, R is the experimentally measured ratio of the reference to spike isotope corrected for mass discrimination, W is the mass of the sample (in g), B is the natural abundance of the spike isotope, and A is the natural abundance of the reference isotope.

Using isotope dilution calibration, an ideal internal standard is used, which is the analyte itself, and the analysis can be completed by a single measurement of the reference to spike isotope ratio. Isotope dilution can also be used to compensate for preconcentration or analyte separation steps that are not quantitative in nature. Thus, as long as the added spike reacts in exactly the same manner as the native analyte in the sample, then isotope dilution can compensate for incomplete analyte recovery in solvent extractions, etc. The error in performing isotope dilution is minimized if the mass of the added spike is approximately no more than an order of magnitude more or less than the mass of the analyte in the sample.

5.3.7 Isotope ratio measurements

Isotope ratio measurements obtained with the quadrupole mass spectrometer have been applied to a wide variety of applications. Perhaps the most useful is in regard to isotope dilution calibration discussed above. Many applications of isotope ratios have been published in human nutrition studies, such as tracking the fate of elements such as Li, B, Cu, etc., when ingested. An important application has been the sourcing of lead in cases of lead poisoning, particularly in children. Of the four lead isotopes, only the one at $m/z = 204$ is not the product of the radioactive decay of such elements as Th and U, and each source of lead (mine, mineral, etc.) has its distinctive signature of abundances for the remaining lead isotopes. In urban areas, the source of lead in children can often be traced back to lead carbonate used as a pigment in paint or to lead used as an additive in gasolines.

Of the chemical elements, there are a relatively few whose isotopic abundances vary in Nature. Isotopic fractionation in Nature usually occurs as a result of physicochemical processes or the action of living organisms such as bacteria (light elements) or as the end-result of radiogenic processes such as the radioactive decay of nuclides.

An important point to consider when measuring isotope ratios of elements in Nature is whether or not ICP-MS is suitable to the task. The limiting factors include the concentration of the isotopes to be measured *vis-a-vis* its abundance (concentration) and the precision with which the ratio can be measured compared to the magnitude of the shift in natural materials.

The precision of isotope ratio measurement obtained using ICP-MS is limited by Poisson statistics at low concentrations and by the stability of the plasma at higher concentrations where the number of ion counts collected is not a consideration. Poisson statistics dictate that the precision of the measurement of the analyte signal intensity is proportional to the square root of the total number of counts divided by the total number of counts obtained. Assuming adequate counting times to obtain sufficient counts, the precision obtainable for isotope ratio measurements by ICP-MS is about 0.1% to 0.2%. This precision is about one to two orders of magnitude poorer than can be obtained using magnetic sector mass spectrometers such as those used in thermal ionization mass spectrometry. This means that there are a limited number of isotope systems in natural materials that can be successfully explored using ICP-MS. These isotope systems include Li, B, Os and Pb.

Table 5.9 Elements determined by various atomic spectroscopic techniques for the analysis of geological materials [7]

Technique*	Rocks, sediments, soils	Waters, brines
WD-XRF	Al, Ba, Ca, Cr, Fe, K, Mg, Mn, Na, Nb, P, Rb, Si, Sr, Ti, Zr	–
FAAS†	Ag, As, Bi, Hg, K, Li, Na, Pb, Se, Te	As, Bi, Hg, K, Na, Se, Te
GFAAS	Ag, Au, Cd, Dy, Er, Eu, Ho, Ir, Li, Nd, Pb, Pd, Pt, Rh, Ru, Sb, Sc, Sm, Tm, Y, Yb	Li
ICP-AES	Ag, Al, Ba, Be, Ca, Ce, Co, Cr, Cu, Dy, Er, Eu, Fe, Gd, Ho, K, La, Mg, Mn, Mo, Na, Nd, Ni, Pb, Sc, Si, Sm, Sr, Ti, Tm, V, W, Y, Yb, Zn, Zr	B, Ca, Fe, K, Li, Mg, Na, Si, Sr
ICP-MS	Al, Au, Ba, Bi, Cd, Ce, Co, Cr, Cs, Cu, Dy, Er, Eu, Ga, Gd, Hf, Ho, In, Ir, La, Lu, Mn, Mo, Nb, Nd, Ni, Os, Pd, Pr, Pt, Re, Rb, Rh, Ru, Sc, Sm, Sn, Sr, Ta, Tb, Th, Ti, Tl, Tm, U, Y, Yb, W, Zn, Zr $^{186}Os/^{187}Os$, $^{206-8}Pb/^{204}Pb$, $^{7}Li/^{6}Li$	Al, Ag, As, Ba, Be, Bi, Br, Cd, Co, Cr, Cs, Cu, Ga, Hg, I, In, La, Mn, Mo, Ni, Pb, Rb, Sb, Se, Sr, Th, Tl, U, V, Y, Zn, Zr $^{11}B/^{10}B$, $^{7}Li/^{6}Li$

* WD-XRF = wavelength-dispersive X-ray fluorescence; FAAS = flame atomic absorption spectrometry; GFAAS = graphite furnace atomic absorption spectrometry; ICP-AES = inductively coupled plasma atomic emission spectrometry; ICP-MS = inductively coupled plasma mass spectrometry.
† Atomic absorption spectrometry determination of As, Bi, Se and Te completed by hydride generation; Hg by cold vapor technique.

5.3.8 *Applications*

Inductively coupled plasma mass spectrometry has had a profound effect on a large number of application sciences that rely on trace-element data. User sciences range from geochemistry, archeometry and biomedicine, to the study of the environment. For example, in the geochemistry laboratory where high-quality high-volume trace multi-element analysis is required, ICP-MS has virtually replaced atomic absorption spectrometry. Elements currently determined in geological materials by different atomic spectrometric techniques are shown in Table 5.9. Note that elements such as Hg and the hydride-forming elements are still preferably determined using atomic absorption techniques, but that most elements can be determined by a combination of ICP-AES and ICP-MS. X-ray fluorescence is still the preferred technique for the determination of major elements because of the high precision (0.2%) of analytical results for elements occurring in per cent concentrations. The precision of ICP-MS determinations is about 1% or less using internal standardization.

By using techniques such as high-performance liquid chromatography and capillary zone electrophoresis, ICP-MS has been used as a detector in speciation studies. Elements such as Sn in sediments can be analyzed for their individual chemical species, as these elute from the chromatographic columns. Ion chromatography can be used to separate (on- or off-line)

Table 5.10 Specialized sample introduction techniques for ICP-MS

Flow injection techniques
- Saline waters
- Digests high in dissolved salts
- Small liquid samples
- On-line standard addition calibration

Laser ablation techniques
- Solid samples
- Analysis of micro-samples (10–100 μm)
- Zone analysis

Electrothermal vaporization techniques
- Microliter liquid sample volumes
- Organic solutions
- Solid and slurry samples
- Solutions high in dissolved solids
- Eliminate interferences due to solvent

Chromatographic techniques
- Matrix separation
- Elemental speciation
- Organometallic speciation

trace quantities of analyte from massive concentrations of dissolved salts such as those found in sea water and saturated brines.

A subject area of intense research and development is in the use of alternate sample introduction techniques. Some of the techniques developed thus far are listed in Table 5.10. These techniques are used to surmount limitations inherent to ICP mass spectrometry itself. For example, electrothermal vaporization can be used to remove from samples solvent that otherwise extinguishes the plasma or results in serious polyatomic interferences. This technique can also be used for the analysis of very small samples in either solid, slurried or liquid form. The vaporization of samples using a laser beam can be used to analyze individual mineral phases or spot areas on a sample as small as 10 μm in diameter. Information can be gained on zonation and the distribution of elements across the surface of any material. Flow injection techniques can be used to perform on-line dilutions or standard additions as well as to reduce the total quantity of dissolved solids reaching the plasma (sampler/skimmer) allowing for longer instrument running times as well as reducing the risk of sampler orifice blockage.

5.3.9 Current limitations, recent developments, future prospects

Although ICP mass spectrometry has gained popularity at an unprecedented rate, there are still limitations to the technique, which must be overcome before its full potential is realized. The existence of isobaric interferences from concomitant elements and the presence of polyatomic ions can prevent the determination of certain elements at low levels. The intense argon spectrum in conventional ICP mass spectrometry makes it virtually impossible to determine K, an important element, and elements of high ionization potential such as F are also not easily determined. Inductively coupled plasma mass spectrometric instrumentation capable of virtually eliminating or substantially attenuating the argon spectrum are currently available. These utilize 'cold' plasma conditions, which result by selecting low power and high injector gas flow rates or by the use of electrical shielding of the plasma itself. A new approach uses a small quantity of helium gas introduced into the ion beam behind the skimmer. This not only eliminates most of the argon ion signal, but also helps to break up polyatomic ions. Another approach to controlling spectral interferences is by replacing the quadrupole mass spectrometer with a magnetic sector mass spectrometer. The inherent higher resolution of these mass spectrometers makes it possible to resolve most spectral interferences. Instruments using magnetic sector mass spectrometers are available but have not been well accepted by users. This is mainly because of the cost of this equipment, which may be from two to four times the cost of quadrupole-based systems, or the relatively slow scanning (if any) capabilities for data acquisition, or both. Currently, research is under way to produce viable instruments based on other types of mass spectrometers including time-of-flight and high-resolution quadrupoles.

Signal suppression due to the presence of concomitant elements remains one of the more intractable problems associated with ICP mass spectrometry. More recent generations of instrumentation claim to have reduced these interferences, but, in practice, matrix interferences are still very serious. These interferences reduce the productivity of ICP-MS based determinations by requiring labor-intensive sample pretreatment steps or the use of specialized calibration techniques, which add error, are expensive (isotope dilution) or are wasteful of instrument time (standard additions).

As part of an on-going trend, analysts are always demanding one more order of magnitude lower limits of detection. Current ICP-MS instruments are relatively inefficient at collecting and transmitting ions to the detector. Only one in approximately 10^7 or 10^8 ions sampled by the mass spectrometer reach the detector. One can easily realize the improved limits of detection possible for an already sensitive technique should all or even most of the ions sampled be detected! Once this greater sensitivity has been achieved, the challenge will be to maintain clean-room conditions during the entire analysis and to obtain reagents and containers free of impurities.

Over the past 10 years, there have been many breakthroughs in the development and alternate sample introduction techniques. As discussed above, these techniques serve to extend the range of utility of ICP-MS while at the same time decrease or eliminate some of the polyatomic and/or matrix interferences related to specific analyses. One can only imagine the power of a new generation of ICP mass spectrometers that are super-efficient at sampling and transmitting ions to the detector, do not suffer from any interferences of any kind, and are available at a reasonable price. In our view, it is possible that such an instrument could supplant traditional fields of analysis dominated by X-ray fluorescence and ICP emission spectrometry.

5.4 Sample introduction

Although ICP-ES and ICP-MS have reached maturity in certain respects, they still remain among the most active research areas in inorganic elemental analysis. Most of the research has been focused on sample introduction aspects to address challenging applications. A variety of sample introduction techniques have been developed, which have their own advantages and disadvantages depending on the evaluation criteria such as desired analytical figures of merit, sample throughput, available sample size, convenience and nature of the sample. Although the ICP is a gaseous plasma, samples have been successfully introduced into the ICP in all three physical states, gas, liquid, or solid.

5.4.1 Liquid nebulization

In order to transport analytes that are present in solution form into the analytical region of the gaseous plasma, two basic approaches can be taken. Most commonly the sample solution is converted into an aerosol using a suitable nebulizer and gas-suspended fine droplets are transported into the ICP. The other approach is to volatilize the analyte from a small aliquot of the solution placed on a secondary support such as a graphite or metal platform using similar technology used in atomic absorption spectrometry. Almost universally samples are introduced by nebulization for routine applications. Both aqueous as well as non-aqueous (organic) solutions have been successfully introduced into the ICP as fine aerosols. Depending on the manner in which energy is supplied into the liquid for breaking it into fine droplets, the nebulizers can be classified into three basic types: (a) pneumatic, (b) ultrasonic and (c) thermospray nebulizers.

Pneumatic nebulizers. In pneumatic nebulizers a flow of gas moving through a fine orifice shears the liquid into small droplets. Three basic types of pneumatic nebulizers are used, namely concentric, cross-flow and

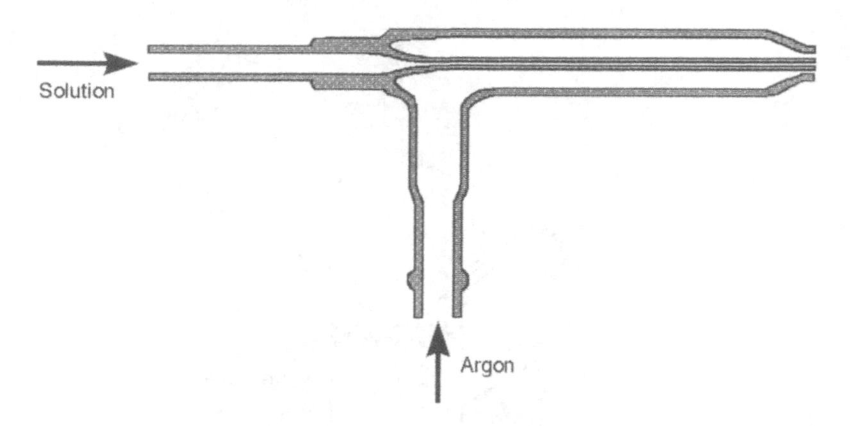

Figure 5.22 Concentric glass nebulizer.

frit nebulizers, which are shown in Figures 5.22, 5.23 and 5.24, respectively. **Concentric nebulizers**, perhaps the most common, are almost universally constructed of borosilicate glass. The inner capillary carries the solution to the tip of the nebulizer at an uptake rate of 1–2 mL/min. An argon flow of typically about 1 L/min exits the nebulizer at a high velocity through the narrow annular spacing between the internal capillary and the outer tube, creating a fine spray at the tip. In the **cross-flow nebulizer**, the solution and the nebulizer gas flow through separate capillaries, which are generally placed perpendicular to each other. Cross-flow nebulizers are considered less prone to clogging for solutions with high dissolved

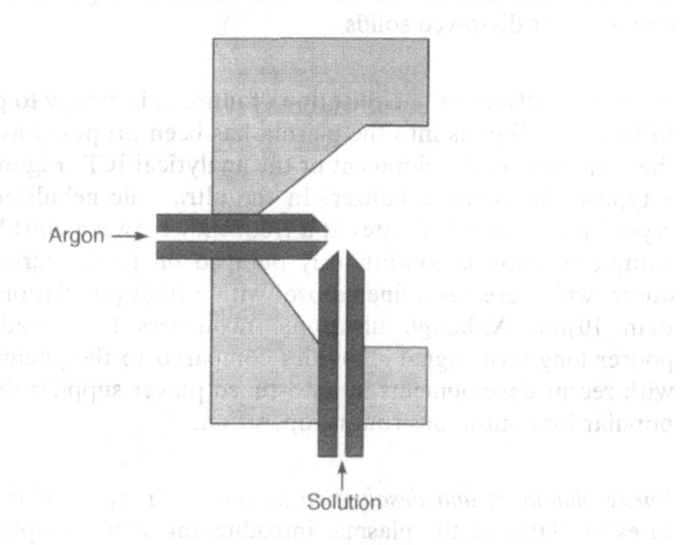

Figure 5.23 Cross-flow nebulizer.

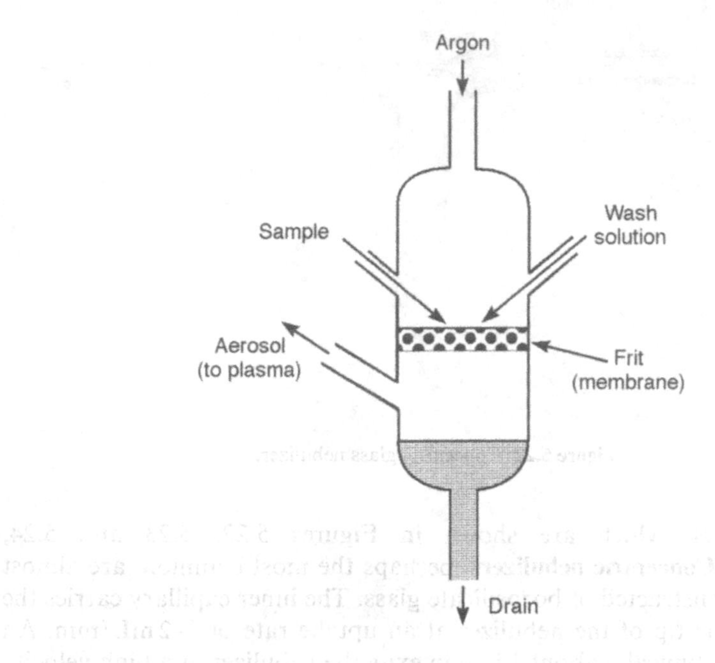

Figure 5.24 Frit nebulizer.

solids. In the **frit (grid) nebulizer** a fine mesh or a membrane of an inert material is positioned in such a way that argon gas flows through the frit while it is continually wetted by the sample solution. Certain grid nebulizers are claimed to have efficiencies as high as 90% with a high tolerance for dissolved solids.

Ultrasonic nebulizers. Utilization of ultrasonic energy to produce aerosols to introduce liquids into the plasma has been proposed as an option since the early days of development of the analytical ICP. Figure 5.25 illustrates a typical ultrasonic nebulizer. In the ultrasonic nebulizer a piezoelectric crystal (transducer) vibrates at a frequency between 100 kHz and 10 MHz. Sample solution is continuously pumped on to the surface of the transducer, which creates a fine aerosol with a high population of droplets less than 10 μm. Although ultrasonic nebulizers have traditionally shown poorer long-term signal stabilities compared to the pneumatic nebulizers, with recent developments in auto-tuned power supplies they have become popular for continuous routine operation.

Spray chambers and desolvation systems. In spite of the high temperatures available in the plasma, introduction of big droplets can make the plasma unstable, or even extinguish it. Therefore, most of the nebulizers

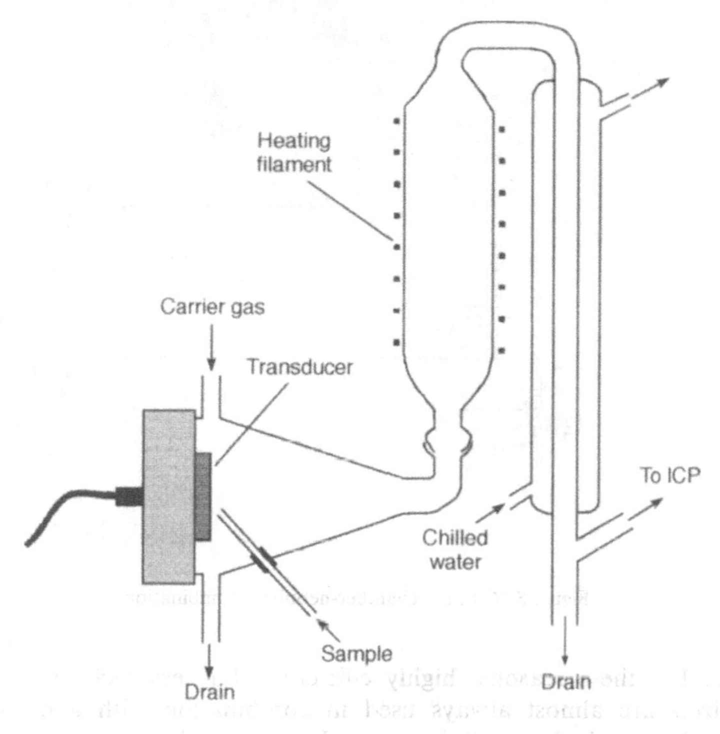

Figure 5.25 Ultrasonic nebulizer with desolvation system.

are attached to a spray chamber to filter out large droplets and transport only the fine fraction of the aerosol, typically with droplet size less than 10 μm, into the plasma. A typical spray chamber nebulizer arrangement is shown in Figure 5.26. Although this improves the stability of the analyte signal and the plasma, the spray chamber significantly reduces the transport efficiency of the overall nebulization process, typically to around 2–10%. Recently, low-flow concentric nebulizers, which can generate extremely fine aerosols without a need for a separate spray chamber, providing nearly 100% transport efficiency, have become commercially available.

Even if the solution is introduced as an extremely fine aerosol, addition of too much solvent can have deleterious effects for the analytical measurement. For example, ultrasonic nebulizers are nearly an order of magnitude more efficient in generating fine aerosol compared to conventional pneumatic nebulizers. However, in spite of the greater analyte transport efficiency, increased solvent load can also cool the plasma, reducing the overall excitation or ionization efficiency. In addition to cooling the plasma, introduction of too much water vapor can be problematic especially in ICP-MS due to the increase in population of

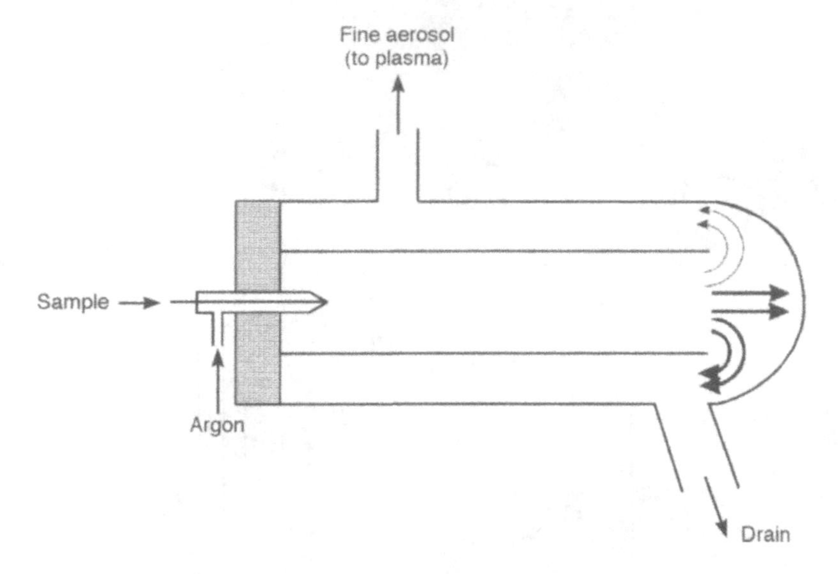

Figure 5.26 Spray chamber-nebulizer combination.

oxides. For these reasons, highly efficient nebulizers such as ultrasonic nebulizers are almost always used in combination with a desolvation system. In a basic desolvation system the generated aerosol is first passed through a heated chamber to vaporize the solvent from the fine droplets. Then the solvent vapor is preferentially removed from dry particulates (or analyte-enriched droplets) by passage through a condenser or membrane separator (Figure 5.25). These desolvation systems are claimed to be able to suppress interferences in ICP-MS caused by the formation of oxides to negligible levels.

5.4.2 Cold vapor generation

Among the approaches for gaseous sample introduction, the most common technique is **hydride generation** where analytes As, Sb, Se, Te, Ge, Pb, Sn and Bi are converted to their volatile hydrides by reacting the acidic sample solution with a reducing agent such as sodium borohydride. Mercury also can be converted to a vapor by a similar reaction with sodium borohydride or stannous chloride. The generated volatile gases are then separated from the liquid phase using a phase separator and transported as a molecular or atomic vapor into the plasma as shown in Figure 5.27. Introduction of analytes in gaseous form into the ICP can improve the detection limit mainly due to two reasons. First, beyond the generation of the volatile form of the analyte, transport efficiency can be nearly 100%, while the analytes are separated from potential interferents

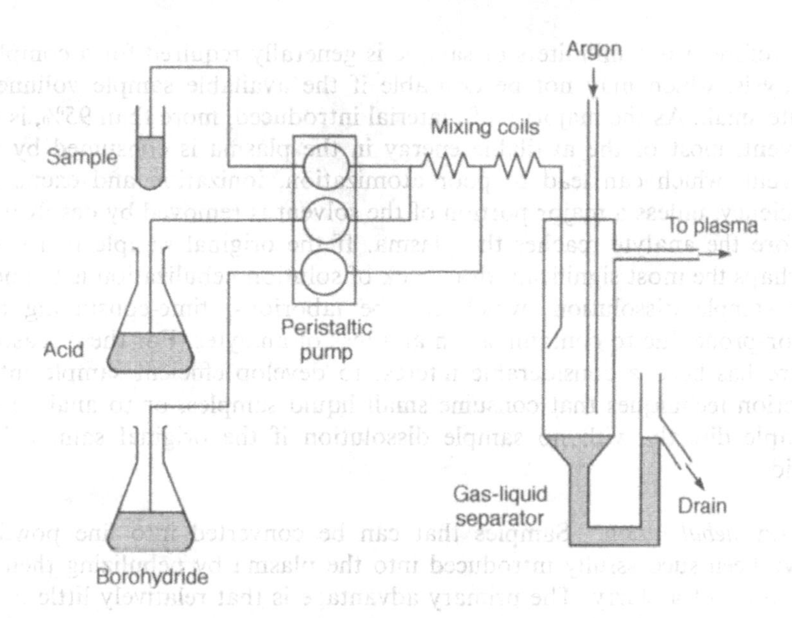

Figure 5.27 Hydride generator.

such as matrix elements. Second, the analyte enters the plasma already in molecular form, which significantly facilitates the atomization process, increasing the population of the analyte atoms in the analytical region. A practical advantage of cold vapor generation is that the automation of the system for routine operation is relatively simple. However, apart from the above advantages there can be interference and matrix effects that can affect the chemical generation of the volatile hydride, which can lead to significant errors unless proper precautions are taken. Although it was originally developed for atomic absorption spectrometry, hydride generation is widely used for both ICP-ES and ICP-MS.

In addition to volatile hydrides, some other inorganic compounds such as metal halides, organometallics and metal carbonyls can be volatile at ambient or slightly elevated temperatures. Although it is somewhat less common, such volatile forms of metals have been successfully introduced into the plasma, especially for using ICP as a multi-element detector for gas chromatographic separations.

5.4.3 Specialized approaches for sample introduction

In spite of the simplicity of operation and other benefits, sample introduction by solution nebulization has several drawbacks. Conventional nebulizer–spray chamber systems are generally poor in transport efficiency, as only a very small fraction of the sample reaches the plasma.

Therefore, a few milliliters of sample is generally required for a complete analysis, which may not be desirable if the available sample volume is quite small. As the majority of material introduced, more than 95%, is the solvent, most of the available energy in the plasma is consumed by the solvent, which can lead to poor atomization, ionization and excitation efficiency, unless a major portion of the solvent is removed by desolvation before the analyte reaches the plasma. If the original sample is a solid, perhaps the most significant drawback of solution nebulization is the need for sample dissolution, which can be laborious, time-consuming and error-prone due to contamination and loss of analytes. For these reasons, there has been a considerable interest to develop efficient sample introduction techniques that consume small liquid samples, or to analyze the sample directly with no sample dissolution if the original sample is a solid.

Slurry nebulization. Samples that can be converted into fine powders have been successfully introduced into the plasma by nebulizing them in the form of a slurry. The primary advantage is that relatively little or no hardware modification to the conventional system is necessary provided that the nebulizer can handle the amount and the size of solid particles suspended in the liquid. Aqueous standards have been used for calibration to a varying degree of success depending on the analyte element and the sample matrix. Unless the samples are ground to very fine powders, potential errors due to particle segregation and mismatch of calibration standards are possible.

Electrothermal vaporization. Electrothermal vaporization (ETV) combines some of the important advantages of flameless atomic absorption spectrometry and ICP emission or mass spectrometry. Sample volumes as small as a few microliters can be efficiently introduced into the ICP, providing very low absolute detection limits. Electrothermal vaporization is especially attractive for ICP-MS as the solvent, which can lead to interferences due to formation of molecular ions such as oxides, can be selectively removed by a separate drying cycle before the analytes are vaporized into the plasma. Solid samples can also be introduced directly in the form of slurries by electrothermal vaporization, although the sample size can be extremely small, which may lead to poor reproducibility due to sample heterogeneity.

Direct sample insertion. Similar to electrothermal vaporization, small aliquots of liquids or slurries are placed on an external graphite or metal probe, which is directly inserted into the high-temperature region of the plasma. Different levels of heating can be achieved by computer-controlled positioning of the probe inside the plasma. The center tube of the torch has to be modified to accommodate the sample insertion probe. Although the volatilization characteristics can depend on the nature of the probe and type of sample, 100% analyte transport efficiency can be easily achieved.

Laser ablation. A solid sample is placed in an enclosed chamber with a gas inlet and the outlet connected to the ICP torch with a plastic tubing. A laser is focused on to the surface of the solid to sputter the material as fine particles or vapor from the surface. The target area can be selected using a microscope. Owing to the nature of the laser beam, laser ablation can be advantageously used to analyze localized spots on solid surfaces such as fine grains on geological materials. A common problem in laser ablation is the poor reproducibility and difficulty of calibration due to lack of direct information on the amount of analytes reaching the plasma. Internal standardization using matrix elements can be used to improve precision and to estimate indirectly the amount of material entering the plasma. The relatively high cost of hardware has also been a limiting factor for the widespread popularity of laser ablation.

Spark ablation. Similar to laser ablation, electrical sparks can be used to ablate material from solid surfaces. A dry aerosol of fine particles is generated, which is swept away into the plasma by the carrier gas. As the sample needs to be an electrically conducting material, the spark technique has been mainly popular for direct analysis of metals. Non-conducting materials can be mixed in powder form with copper or graphite powder to make conducting pellets, which can be used as the sample electrode.

Direct powder introduction. Solid samples available as powders have been introduced to the ICP as a dry aerosol. Known quantities of materials, individual particles to milligram quantities, can be introduced at a preselected computer-controlled delivery rate. Signal normalization is done directly using the amount of the powder delivered into the plasma. Calibration has to be done using standard powders of similar material.

References

1. Reed, T.B. (1961), *J. Appl. Phys.* **32**, 821–825.
2. Greenfield, S., Jones, I.L.I. and Berry, C.T. (1964), *Analyst*, **89**, 713–720.
3. Wendt, R.H. and Fassel, V.A. (1965), *Anal. Chem.* **37**, 920–922.
4. Mermet, J.M. and Poussel, E. (1995), *Appl. Spectrosc.*, **49**, No. 10, 12A–18A.
5. Beauchemin, D. (1992), *Spectroscopy*, **9**, 12–16.
6. Doherty, W. (1989), *Spectrochim. Acta*, **44B**, 263–280.
7. Gregoire, D.C. (1994), *Spectroscopy*, **9**, 12–17.

8. Gregoire, D.C. (1987), *Spectrochim. Acta*, **42B**, 895–907.
9. Montaser, A. and Golightly, D.W. (eds) (1992), *Inductively Coupled Plasmas in Analytical Atomic Spectrometry*, VCH Publishers, New York.
10. Tan, S.H. and Horlick, G, (1986), *Appl. Spectrosc.*, **40**, 445–460.
11. Vaughan, M.A. and Horlick, G. (1986), *Appl. Spectrosc.*, **40**, 434–445.
12. Vaughan, M.S. and Horlick, G. (1987), *J. Anal. At. Spectrom.*, **2**, 765–772.

6 Electroanalytical methods

P.C. HAUSER

6.1 Introduction

Electroanalytical techniques encompass a wide range of quite diverse methods. Some of these methods are selective (such as potentiometry) while others are almost completely non-specific (e.g. conductometry). Similar variations in other performance characteristics such as dynamic range or detection limit are found. For some methods a signal that is linear with the logarithm of the concentrations, or the activity, is obtained while others show the more usual direct relationship with analyte concentration. In many instances the electrochemical methods provide a unique means for analysis, while in others they are an alternative to spectroscopic or chromatographic methods. One important feature is common, however, to all electroanalytical methods. In all instrumental methods of analysis a chemical signal (concentration or amount of analyte) is transformed into a physical signal (voltage, current, conductivity, radiation, heat, pressure, weight, etc.), which is then transformed into an appropriate electrical signal and processed by the electronic instrument. In many cases an excitation signal is needed to obtain the information from the sample or a separation has to be carried out (e.g. chromatography). This is illustrated in Figure 6.1. The processed signal is displayed or printed (plotted) by the instrument in a manner that can be understood by the operator. The detector that transforms the physical signal into an electrical signal is the input transducer of the electronic instrument.

The considerable advantage of electrochemical instruments is the fact that the physical signal is obtained directly in electronic form amenable for further processing by the instrument. No detector other than electrodes are required. If excitation of the sample is required this usually takes the form of the simple application of a voltage to an electrode. This keeps the instrumentation generally fairly simple and therefore less expensive, than for example spectroscopic instruments, and it is often possible to construct portable battery-powered instruments which may be taken into the field for on-site analysis. Examples of such compact portable instruments can be found for even the more sophisticated methods such as adsorptive stripping voltammetry, which is capable of delivering among the best

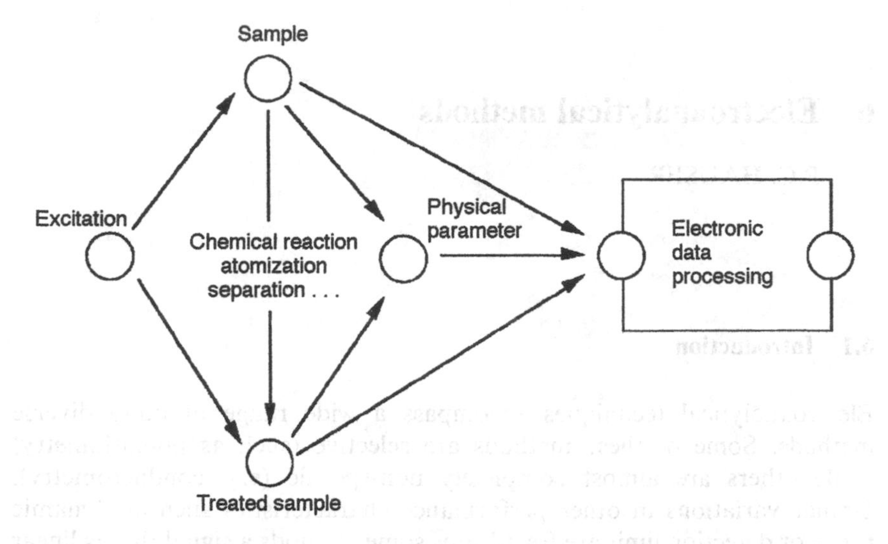

Figure 6.1 Signal conversion in instrumental analysis.

detection limits achievable with any analysis method. The relative simplicity of the instrumentation often also allows easy automation and the construction of unattended data-logging devices. In many cases electroanalytical means are furthermore used for detection in methods such as chromatography, capillary electrophoresis and flow-injection analysis.

Electrochemical instrumentation often represents a chemical sensor, which is usually defined as a probe that delivers an electrical output directly and reversibly related to the concentration of a species of interest. The selective sensor approach represents one possible ideal solution to an analytical problem. In fact the oldest, best established and most widely used chemical sensors are mostly electrochemical devices. This is not surprising given the direct nature of the signal transduction.

A further advantage of electrochemical methods is the generally low sample consumption. In many cases it is possible to carry out a measurement without altering the sample so that, for example, on-line monitoring of a product stream is feasible. It is also often possible to distinguish directly between different oxidation or complexation states so that speciation analysis may be performed. This is often not possible with spectroscopic methods and may require, for example, a chromatographic separation. Another fact of electrochemical measurements is often the response to the activity rather than to the concentration of the analyte, a feature that may or may not be desired.

Spectroscopic methods possess an inherently higher selectivity as quantized transition processes are exploited there. However, this advantage can often not be realized in practice because of spectral overlaps. Electrochemical methods of analysis may be carried out mainly on aqueous solutions but some methods extend to the gas phase as well. On the other hand many spectroscopic methods can be applied to the direct analysis of solid samples. Electrochemical methods rely in almost all cases on reactions on electrode surfaces. The instrumental signal does therefore not necessarily represent the bulk of the solution, a potential source of error. The electrode reaction itself may alter the composition in the vicinity. Another disadvantage of electrochemical methods is the fact that often a reference electrode is required, which is a potential source of measurement error.

The techniques can be divided largely into two groups, methods without current flow and methods with current flow. In potentiometry, perhaps the most simple method of analysis of all instrumental methods, electrode potentials are measured in the absence of any electron-exchange reactions. In conductometry, a current is measured, but it is the mobility and number of charge carriers in a matrix that are of relevance and not an electrode reaction. In electrogravimetry and coulometry the analyte is reduced or oxidized in an electrode reaction, a so-called Faraday reaction, and the total amount of analyte is determined by either the weight of the product or the charge required for total consumption. The most complex relationship between the measured quantity and the concentration is realized in the last two of the methods, amperometry and voltammetry. Here a concentration-dependent current is measured; the difference between the two methods is that in amperometry the applied voltage is held constant, while in voltammetry this parameter is varied during the analysis. The individual methods will be introduced in the order of increasing complexity. However, some fundamental aspects are dealt with first.

6.2 Fundamentals

6.2.1 *Activity*

Owing to interaction of ions in solution, some of their electrochemical properties cannot take full effect. In such cases the use of the activity rather than the concentration of the ion is more appropriate.

The **activity** a of an ion is related to its concentration through the activity coefficient:

$$a = \gamma c \tag{6.1}$$

where $c =$ concentration and $\gamma =$ activity coefficient. However, it is naturally not possible to make up solutions of single ions and activity coefficients for single ions can only be an approximation.

The **activity coefficient**, γ_i, for an ion i, depends on the total ionic strength I of the solution and is given by the **Debye–Hückel equation**:

$$\log \gamma_i = -\frac{An_i^2\sqrt{I}}{1 + Ba_i\sqrt{I}} \tag{6.2}$$

A and B are medium- and temperature-dependent parameters, corresponding to 0.509 and 0.328, respectively, in water at 25°C. The parameter a_i is specific for the ion size and is a measure of the diameter of the hydrated ion; n_i is the charge on the ion. This equation is applicable to dilute aqueous solutions up to $I \approx 0.1\,\text{mol}\,\text{L}^{-1}$.

I is defined by:

$$I = 0.5 \sum_i c_i n_i^2 \tag{6.3}$$

The summation is taken over all the ions present in the solution. The activity coefficient is thus dependent on the total concentration of all ions in the solution. For very dilute solutions (less than about $1 \times 10^{-4}\,\text{mol}\,\text{L}^{-1}$), the activity coefficient approaches 1 and therefore a can then be taken to be equivalent to c. At concentrations of $0.1\,\text{mol}\text{L}^{-1}$ typical activity coefficients are approximately 0.7 for monovalent ions and about half of that value for divalent ions.

6.2.2 *Nernst equation*

Many, but not all, electroanalytical methods involve redox (electron transfer) reactions at working electrodes:

$$\text{Ox} + ne^- \rightleftharpoons \text{Red}$$

Ox denotes the oxidizing (or oxidized) species and Red the reducing (or reduced) species. These reactions can be characterized by their redox potential, which is given by the well-known **Nernst equation**. This relates an electrode potential and the activities of the two forms of the redox-active species:

$$E = E^\circ + \frac{RT}{nF} \ln\left(\frac{a_{\text{Ox}}}{a_{\text{Red}}}\right) \tag{6.4}$$

where E = redox potential, E° = standard potential for the reaction, R = molar gas constant ($8.314\,\text{J}\,\text{K}^{-1}\,\text{mol}^{-1}$), F = Faraday constant ($96\,487\,\text{C}\,\text{mol}^{-1}$), T = absolute temperature (K), n = number of electrons transferred in the redox reaction, a_{Ox} = activity of the oxidized species and a_{Red} = activity of the reduced species.

Many redox reactions involve protons or hydroxide ions, so that the redox potential is dependent on pH. An example is the reduction of permanganate according to the following reaction:

$$MnO_4^- + 8H^+ + 5e^- \rightleftharpoons Mn^{2+} + 4H_2O$$

The resulting Nernst equation for this reaction also has to include the pH value:

$$E = E^\circ + \frac{RT}{5F}\, 2.303\left[\log\left(\frac{a_{(MnO_4^-)}}{a_{(Mn^{2+})}}\right) - 8 \times \text{pH}\right] \qquad (6.5)$$

The redox potential is of relevance to all redox reactions, many of which do not involve an electrode, and the Nernst equation applies equally to all of these. However, electroanalytical methods always involve an electrode as these are required for signal transduction into the electronic domain.

6.2.3 *Electrochemical cells*

It is never possible to apply or measure an electrode potential by itself. Instead one has to work with a potential difference (or voltage) between two electrodes. Fundamentally two types of electrochemical cells can be distinguished: galvanic cells and electrolytic cells. The Nernst equation applies equally to both cases. In galvanic cells the two electrodes are used to produce energy in the form of batteries and fuel cells, and these have a limited role in electroanalytical chemistry. In electrolytic cells an external voltage is applied to a cell in order to force a redox reaction on the electrodes. This is illustrated in Figure 6.2. Note that for galvanic cells the electrode designations and processes are different.

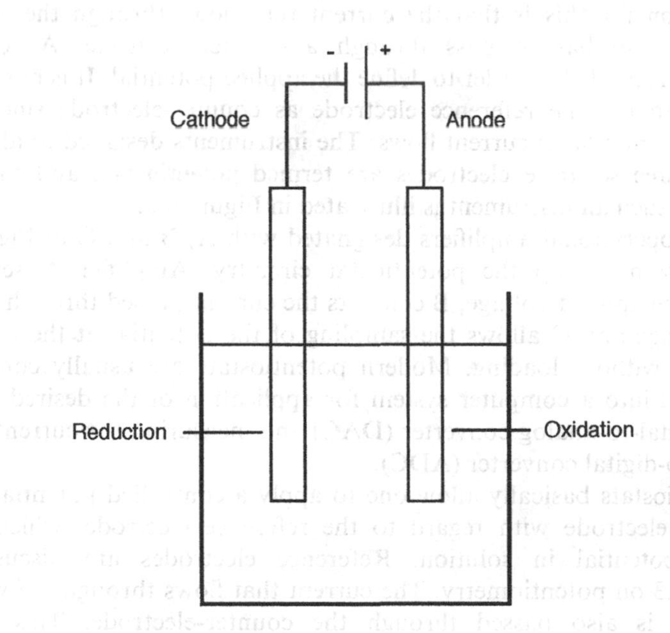

Figure 6.2 Electrolytic cell.

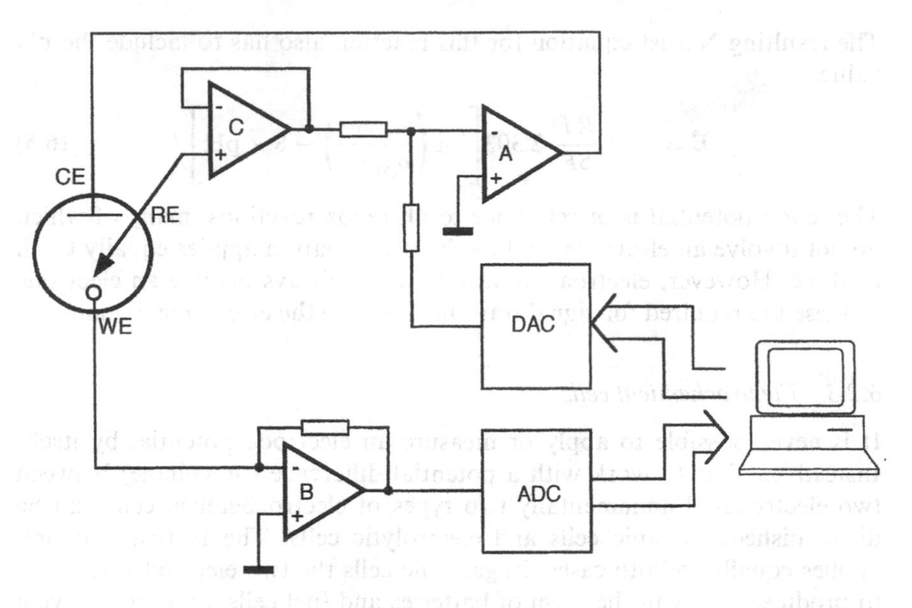

Figure 6.3 Schematic of a potentiostat.

Three-electrode cells. Cells comprising three electrodes are required when a precise potential is to be applied and the resulting current is measured. The reason for this is that the current that flows through the working electrode also has to pass through a counter-electrode. A reference electrode is needed in order to define the applied potential. It is not usually possible to use the reference electrode as counter-electrode since these cannot tolerate large current flows. The instruments designed to allow the use of three separate electrodes are termed **potentiostats** and the basic design of such an instrument is illustrated in Figure 6.3.

Three operational amplifiers designated with A, B and C in Figure 6.3 essentially make up the potentiostat circuitry. Amplifier A serves to control the applied voltage, B converts the current passed through the cell to a voltage and C allows the sampling of the potential at the reference electrode without loading. Modern potentiostats are usually completely integrated into a computer system for application of the desired voltage via a digital-to-analog converter (DAC) and measuring the current via an analog-to-digital converter (ADC).

Potentiostats basically allow one to apply a controlled potential to the working electrode with regard to the reference electrode, which has a known potential in solution. Reference electrodes are discussed in section 6.3 on potentiometry. The current that flows through the working electrode is also passed through the counter-electrode. This usually consists of a relatively large-area inert metal electrode and its potential is allowed to reach the value required that will force some reduction or oxidation to set in to carry the current. The nature of this redox reaction is not normally defined and in the absence of other redox-active species will be the decomposition of the solvent (hydrogen or oxygen evolution in the case of aqueous solutions).

6.2.4 *Faraday's law*

The total charge passed between two electrodes for the electrolytic reaction of an analyte is given by the following form of the **Faraday equation:**

$$Q = it = zFcV \tag{6.6}$$

where Q = charge in coulombs, i = current in amperes, t = time in seconds, z = number of electrons exchanged in the redox reaction, F = Faraday constant ($96\,487\,\mathrm{C\,mol^{-1}}$), c = molar concentration and V = volume. This form of the equation is appropriate for methods where all of the analyte present in the sample is consumed (electrogravimetry, coulometry). As will be seen below, a different form is more appropriate for reactions where only part of the analyte is consumed in the analytical step.

6.2.5 *Current limitation*

The current passed through an electrode in an electrochemical reaction may be limited either by the rate of transport (mass flux) of the analyte species to the electrode surface or by the kinetics of the electrode reaction itself. The former case is much more prevalent and deserves a more detailed discussion.

In principle, charged species can be brought to the electrode surface by three different processes: diffusion, migration and convection. The **extended Nernst–Planck equation** describes this situation:

$$J_i = \underbrace{-u_i c_i \frac{\mathrm{d}\mu_i}{\mathrm{d}x}}_{\text{diffusion}} \underbrace{- z_i u_i c_i F \frac{\mathrm{d}\varphi}{\mathrm{d}x}}_{\text{migration}} + \underbrace{z_i v}_{\text{convection}} \tag{6.7}$$

Diffusion is caused by the concentration difference between the bulk of the solution and the electrode surface, migration is induced by electrical potential gradients (electrical field), and convection refers to the transport of a species by a flow (stirring) of the solution. In equation (6.7), J_i = mass flux of ion i, u_i = mobility of ion i, c_i = concentration of ion i, μ_i = chemical potential of ion i, x = distance perpendicular to the electrode surface, z_i = charge on ion i, F = Faraday constant, φ = absolute potential and v = flow velocity perpendicular to the electrode surface.

Faraday's law may be used again to relate the measured electrode current to the mass flux of the electroactive species toward the electrode:

$$i = Aj = AF \sum n_i J_i \tag{6.8}$$

where i = current, A = electrode area, j = current density (A cm^{-2}) and n_i = number of electrons exchanged in the redox reaction of species i.

In analytical electrochemistry the contributions by migration and convection are often deliberately minimized by adding an inert electrolyte (reduces the electrical field) and by not stirring the solution. The current is therefore only determined by diffusion to the electrode surface.

After applying the redox potential to the working electrode (**potential step experiment**) the concentration of analyte in the vicinity of the electrode surface becomes depleted. This situation is illustrated in Figure 6.4. The concentration c is plotted against distance x from the electrode surface. Initially the concentration is constant throughout the solution (c_{bulk}). After starting the reaction a depletion sets in depicted by the curves at t_l, t_2 and t_3. After a short while a steady-state situation is reached and a layer depleted in concentration of about 0.1–0.3 mm thickness (or less in stirred solutions or in flowing streams) is obtained. For fast electrode kinetics, the concentration directly at the electrode surface decays to zero. In the so-called Nernst approximation (see Figure 6.5) a final linear concentration profile extending to the thickness δ_N, the Nernst layer thickness, is assumed.

A mass flux (J) is given by the concentration gradient (dc/dx) and the diffusion coefficient (D) using **Fick's first law** of diffusion:

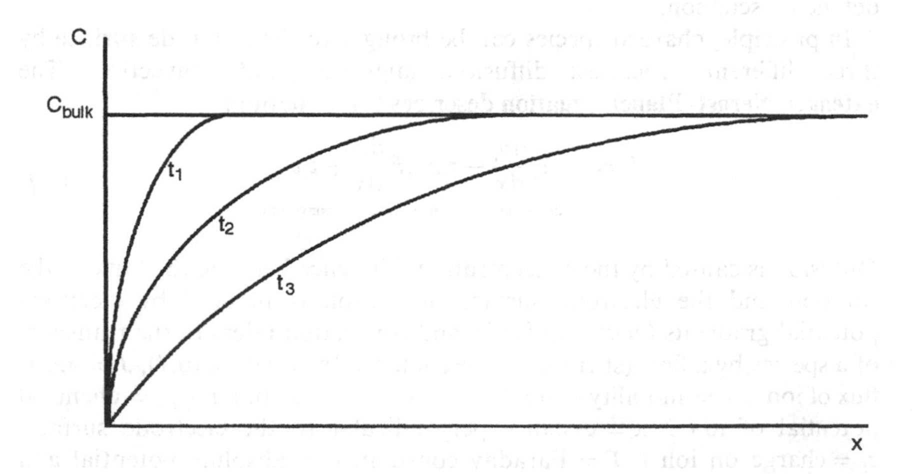

Figure 6.4 Potential step experiment.

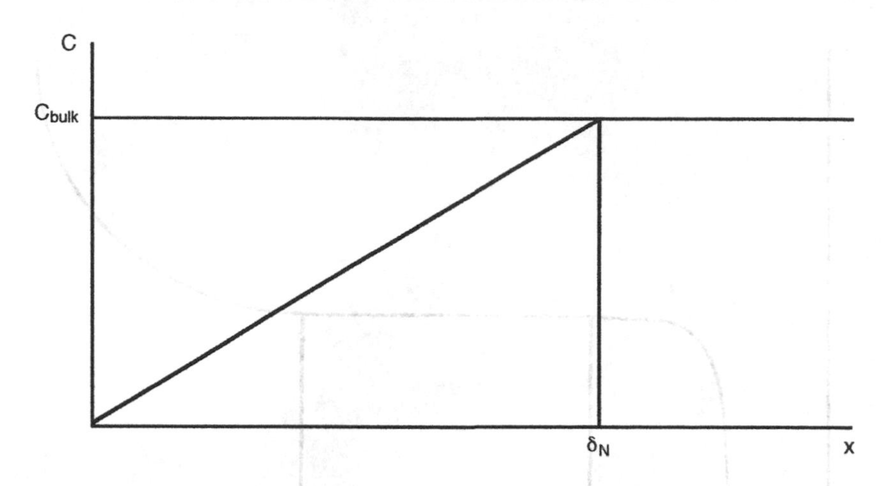

Figure 6.5 Nernst approximation of the diffusion layer.

$$J = -D\frac{dc}{dx} \tag{6.9}$$

For the Nernst approximation this can be rewritten in the following form:

$$J = -D\frac{C_{\text{bulk}}}{\delta_N} \tag{6.10}$$

The current is obtained by substituting J in Faraday's law:

$$i = AnFD\frac{C_{\text{bulk}}}{\delta_N} \tag{6.11}$$

A linear relationship between the redox current and the concentration is therefore obtained for constant applied potentials. This fact is employed in the electroanalytical method of **amperometry**.

For the diffusion-limited case, the behavior illustrated in Figure 6.6 is obtained when altering the applied voltage. For low applied reducing or oxidizing voltages no redox reaction occurs. On increasing the voltage, a sharp rise in current results when the redox potential of the sensed species is reached. A current plateau is then obtained since the transport process is independent of the applied voltage. If the voltage is increased further, a different reaction such as the reduction or oxidation of a secondary species in the sample or eventually the decomposition of the solvent will set in. It is important that the plateau for the analyte is relatively wide so that the method is not affected by small changes in the applied voltage. Discrimination against interferences can be obtained by choosing the appropriate potential. This, however, does not allow one to suppress species that are reduced or oxidized at lower

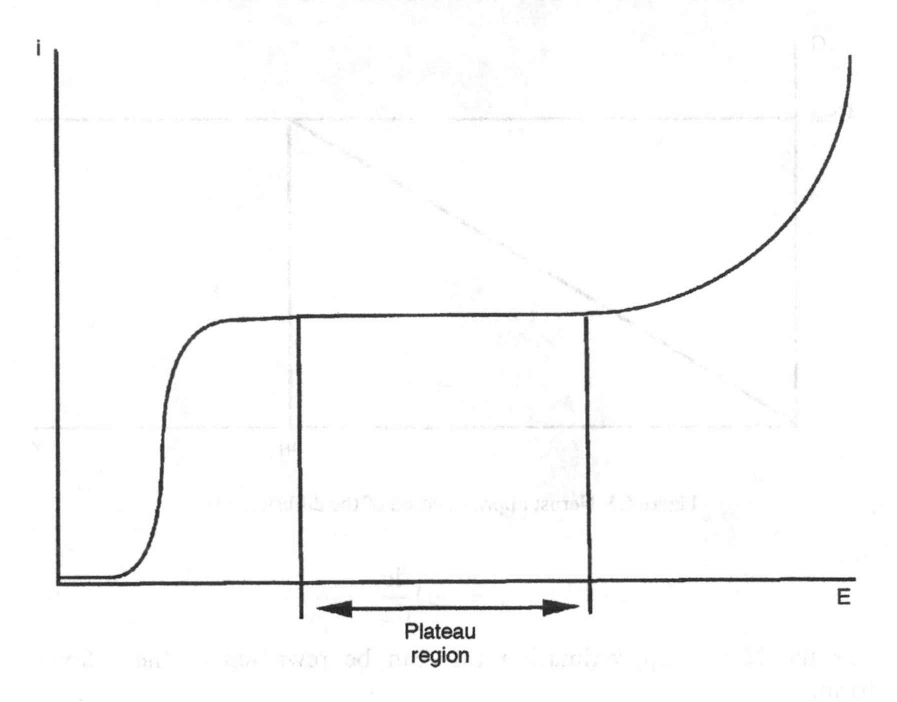

Figure 6.6 Current limitation due to diffusion.

potentials than the analyte, but it is often possible to obtain the desired value by subtracting a background current.

Other methods (**voltammetry**) involve a change in applied potential. The current as a function of time after the start of application of the electrode potential in a potential step experiment can be described for flat electrodes by the **Cottrell equation**. Here again the Nernst approximation is made use of, but the diffusion layer thickness is taken to be increasing with time ($\delta_N = f(t)$). So

$$J = c_{\text{bulk}}\sqrt{\frac{D}{\pi}}\frac{1}{\sqrt{t}} \tag{6.12}$$

$$i = c_{\text{bulk}}AnF\sqrt{\frac{D}{\pi}}\frac{1}{\sqrt{t}} \tag{6.13}$$

The measured current is therefore again directly proportional to concentration, which is desired for many electronanalytical methods. Current limitation by the kinetics of the electrode reactions is only obtained when the rate of diffusion is exceptionally fast (such as in the gas phase). This situation can in principle also be employed for analytical purposes as the reaction kinetics are usually concentration-dependent, but such applications are rare.

6.2.6 *Electrical double layer*

To the surface of an electrode that has excessive positive or negative charges, (hydrated) ions of the appropriate opposite charge will be attracted. Thus a so-called electrical double layer is formed, which is depicted in Figure 6.7. This double layer represents an electrical capacitance as known from electronic circuitry. The charge, q, stored in a capacitor is given by the capacitance, c, and the voltage:

$$q = cV \tag{6.14}$$

Therefore, when the potential applied to an electrode is altered, a current has to flow until the capacitance for the new applied voltage is satisfied. The kinetic behavior on a step change of applied potential (a **chronopotentiometric experiment**) can be described by viewing the system as equivalent to an RC circuit as known from electronics. Such an arrangement shows the following temporal behavior:

$$i_c = \frac{\Delta E}{R} e^{-t/RC_{DL}} \tag{6.15}$$

This so-called **charging current** is often superimposed on currents obtained

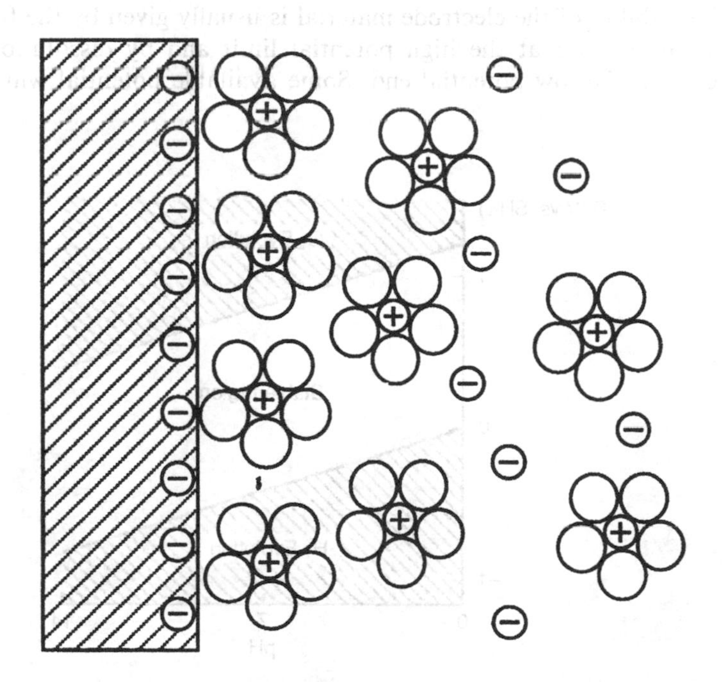

Figure 6.7 Electrical double layer at an electrode surface.

from redox reactions (**faradaic currents**) and such currents are also generally referred to as non-faradaic currents. It may therefore interfere with the current of interest. Some schemes to minimize this influence based on the different kinetic behavior are employed as will be seen later.

6.2.7 *Working ranges of electrodes and solvents, and pH dependence*

The range of potentials that can be applied to a working electrode is limited by the electrochemical stability of the solvent and the electrode material used (i.e. their oxidation and reduction potentials). For the usual aqueous solvent this relates to the evolution of oxygen and hydrogen. These processes are given by the following two equations:

$$6H_2O \rightarrow O_2 + 4H_3O^+ + 4e^-$$

$$2H_2O + 2e^- \rightarrow H_2 + 2OH^-$$

Both of these processes are pH-dependent and the stability plot (**Pourbaix diagram**) given in Figure 6.8 is obtained. A potential window, 1.23 V wide, is therefore obtained regardless of pH, but its range is shifted with pH. The use of organic solvents allows this range to be extend and some solvents (such as acetonitrile, dimethylformamide or dichloromethane) allow extremely large potential windows that are about 4 V wide.

The stability of the electrode material is usually given by the formation of an oxide layer at the high potential limit and by dissolution of the material at the low potential end. Some available potential windows for

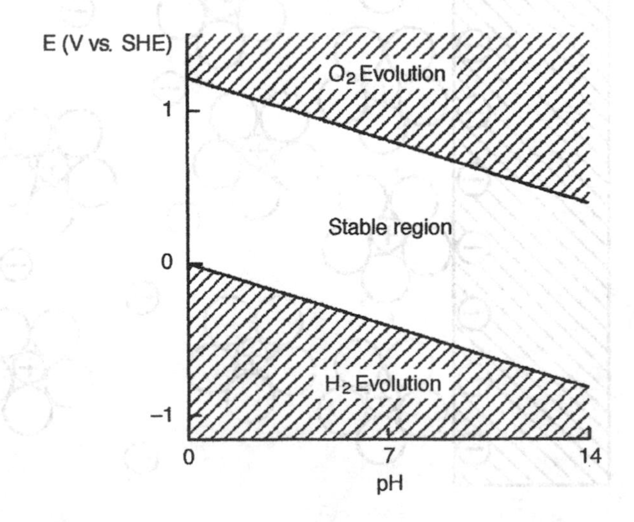

Figure 6.8 Electrochemical stability boundaries of water as a function of pH.

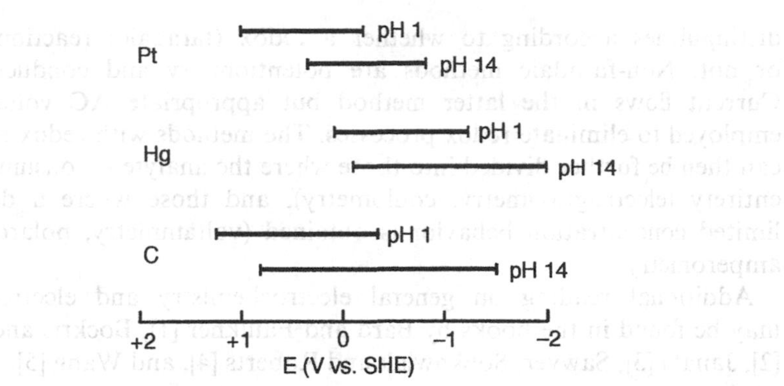

Figure 6.9 Available potential range for three electrode materials.

typical electrode materials are given in Figure 6.9. Note that mercury is a very useful electrode material. On mercury, hydrogen evolution in aqueous solution is hindered (**overpotential**), which means that the useful range at reductive potentials is extended. Also, the fact that mercury is a liquid allows the construction of electrodes with easily renewable surface. Dissolved oxygen often acts as an interference due to its reduction and then has to be removed by passing nitrogen or another inert gas through the solution.

6.2.8 *Classification of electroanalytical methods*

Electroanalytical methods may be classified along different lines. One possible such scheme, which is followed here (see Figure 6.10),

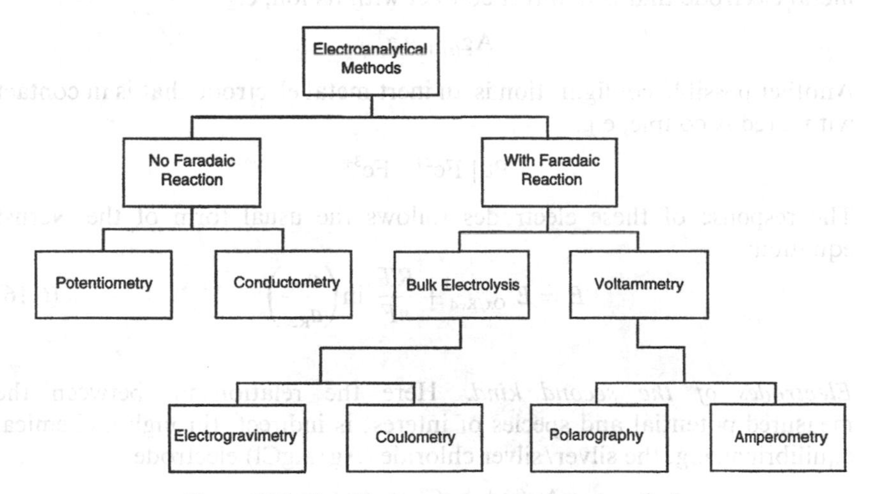

Figure 6.10 Classification of electroanalytical methods.

distinguishes according to whether a redox (faradaic) reaction occurs or not. Non-faradaic methods are potentiometry and conductometry. Current flows in the latter method but appropriate AC voltages are employed to eliminate redox processes. The methods with redox reactions can then be further divided into those where the analyte is consumed in its entirety (electrogravimetry, coulometry), and those where a diffusion-limited concentration behavior is obtained (voltammetry, polarography, amperometry).

Additional reading on general electrochemistry and electroanalysis may be found in the books by Bard and Faulkner [1], Bockris and Reddy [2], Janata [3], Sawyer, Sobkowiak and Roberts [4], and Wang [5].

6.3 Potentiometry

6.3.1 *Fundamentals*

In potentiometry, as the name implies, potentials are measured. In most cases this is carried out under conditions at which no current flows, although there are exceptions (for example the potentiometric stripping analysis briefly introduced in section 6.6 on voltammetry). In this chapter only zero-current potentiometry is discussed. As no voltages are applied to the cell, these are in principle galvanic cells; however, any current flow is suppressed by the external measuring circuitry. A systematic introduction to different classes of electrodes follows.

Electrodes of the first kind. Electrodes of the first kind may consist of a metal electrode that is in direct contact with its ion, e.g.

$$Ag_{(s)} \mid Ag^{+}_{(aq)}$$

Another possible configuration is an inert metal electrode that is in contact with a redox couple, e.g.

$$Pt \mid Fe^{2+},\ Fe^{3+}$$

The response of these electrodes follows the usual form of the Nernst equation:

$$E = E^{\circ}{}_{\mathrm{Ox/Red}} + \frac{RT}{nF}\ \ln\left(\frac{a_{\mathrm{Ox}}}{a_{\mathrm{Red}}}\right) \tag{6.16}$$

Electrodes of the second kind. Here the relationship between the measured potential and species of interest is indirect, through a chemical equilibrium, e.g. the silver/silver chloride (Ag/AgCl) electrode

$$Ag(s) \mid AgCl(s),\ Cl^{-}(aq)$$

Here again the potential is defined by the activity of Ag^+ at the Ag electrode. However, this is given by the solubility product K_s for AgCl:

$$AgCl \rightleftharpoons Ag^+ + Cl^-$$

$$K_s = \frac{a_{Cl^-} a_{Ag^+}}{a_{AgCl}} \tag{6.17}$$

The activity of Ag^+ can then be obtained through rearrangement and substituted into the Nernst equation:

$$a_{Ag^+} = K_s \frac{a_{AgCl}}{a_{Cl^-}} \tag{6.18}$$

$$E = E^\circ - \frac{RT}{nF} \ln\left(\frac{a_{Ag}}{a_{Ag^+}}\right) \tag{6.19}$$

$$E = E^\circ - \frac{RT}{nF} \ln\left(\frac{a_{Ag} a_{Cl^-}}{K_s a_{AgCl}}\right) \tag{6.20}$$

Because the activity for solids is unity this simplifies to:

$$E = E^{\circ\prime} - \frac{RT}{nF} \ln a_{Cl^-} \tag{6.21}$$

Membrane electrodes. Some membranes are permeable to certain ions, and this causes the development of potentials. Owing to their selective response, membrane electrodes have found widespread use in the form of ion-selective electrodes, which will be discussed in more detail below (section 6.3.3). The inside of the membrane cannot be contacted with a metallic wire directly, but must also be in contact with a solution (the internal filling solution). The solution itself is then contacted with an electrode of the second kind as internal reference.

The response of electrodes based on such ion-permeable membranes follows the Nernst equation in the following form:

$$E = E^\circ \pm \frac{RT}{nF} \ln a_i \tag{6.22}$$

It is more convenient to use the decadic logarithm and this can be achieved by rewriting the equation as follows:

$$E = E^\circ \pm \frac{2.303RT}{nF} \log a_i \tag{6.23}$$

The term before the log function is then the **response gradient**, often termed s, to the logarithm of the activity of the ion. The sign is positive for cations and negative for anions; n is the charge on the ion.

For electrodes with an ideal response to monovalent ions ($n = 1$), the gradient s is given by:

$$s = \frac{2.303RT}{F} = 59.16\,\text{mV}\ (25°\text{C}) \quad \text{or} \quad 58.16\,\text{mV}\ (20°\text{C}) \tag{6.24}$$

Such membranes are never perfectly selective. An extended form of the Nernst equation is used to describe the behavior in the presence of interfering ions. This is called the **Nicolsky–Eisenman equation**:

$$E = E^\circ + \frac{RT}{nF} \ln\left(a_i + \sum_j K_{ij}^{\text{pot}} a_j^{n/p} \right) \tag{6.25}$$

where a_i = activity of the primary ion of charge n (mol L^{-1}), a_j = activity of an interfering ion of charge p (mol L^{-1}), K_{ij}^{pot} = potentiometric selectivity coefficient and summation is over all interferents present.

The selectivity coefficient can be determined by comparing the potential obtained in a solution of the primary ion alone with that obtained in a solution of an interfering ion alone (SSM, separate solution method) or in mixed solutions of primary and interfering ions (MSM, mixed solution method). The value is very important to predict whether a given concentration of interferent(s) will cause an appreciable error on the measurement. The sum of the products of the activities of the interferents with their respective selectivity coefficients should generally be less than 1% of the lowest expected concentration of the analyte in the sample.

Cell voltage. In potentiometry, the cell voltage or so-called **electromotive force** (EMF) is determined, which is the difference between two electrode potentials. One of the electrodes therefore has to assume the function of a reference electrode. Such a potentiometric cell is illustrated in Figure 6.11. The measured cell EMF is the difference between the potentials, E, of the measuring electrode and the reference electrode:

$$\text{cell EMF} = E_{\text{Meas}} - E_{\text{Ref}} \tag{6.26}$$

To allow comparisons between different electrodes, the standard hydrogen electrode (SHE) was chosen as the standard electrode of 0 V against which all other potentials are referenced. This electrode involves the following reaction, which is sensed with a platinum electrode, and this is therefore an electrode of the first kind:

$$H_2 \rightleftharpoons 2H^+ + 2e^-$$

The hydrogen-ion concentration is maintained at 1 mol L^{-1} and hydrogen is bubbled over the electrode at a pressure of 1 atm.

Reference electrodes. The standard hydrogen electrode is not very practicable and therefore other reference electrodes have been introduced. Reference electrodes are widely used and are required for all electrochemical methods (not only for potentiometry) where potentials need to be known with accuracy.

Practical reference electrodes consist of electrodes of the second kind,

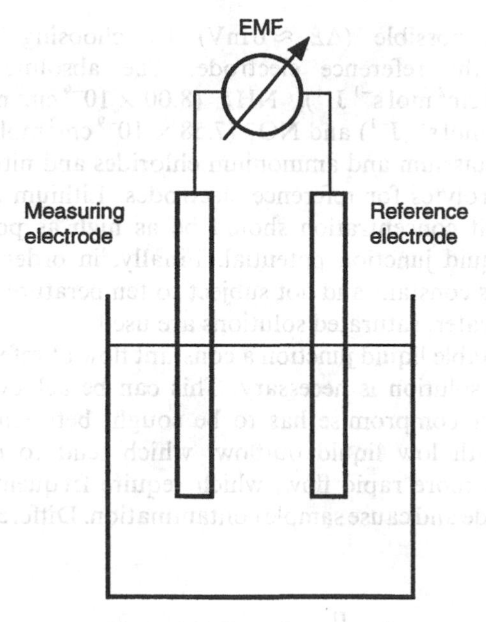

Figure 6.11 Potentiometric cell.

such as the Ag/AgCl or calomel electrode, in a body that contains a filling solution that is in contact with the sample through a so-called liquid junction. The calomel electrode (Hg_2 | Hg_2Cl_2, Cl^-) employs mercurous chloride instead of silver chloride. The filling solutions usually consist of saturated chloride solutions and the widely used saturated calomel electrode is often denoted as SCE.

Liquid junction potentials. Reference electrodes make contact with the sample solution via a liquid junction. A small but not negligible potential can arise when two different solutions are in contact, the **liquid junction potential**. This is caused by the diffusion rates (or mobilities) of ions being different (diffusion potential).

Liquid junction potentials can be calculated from the mobilities of the ions involved (which can be obtained from conductivity measurements) using the Henderson or Planck equations. The **Henderson equation** is:

$$\Delta E_J = \frac{\sum_i n_i u_i (a_i - a_i^*)}{\sum_i n_i^2 u_i (a_i - a_i^*)} \frac{RT}{F} \ln\left(\frac{\sum_i n_i^2 u_i a_i^*}{\sum_i n_i^2 u_i a_i}\right) \tag{6.27}$$

where n_i = charge of ion i, u_i = absolute mobility of ion i (cm^2 mol s^{-1} J^{-1}), a_i = activity of ion i in sample solution (mol L^{-1}) and a_i^* = activity of ion i in the salt bridge (mol L^{-1}).

In practice, the liquid junction potential should be kept as small and as constant as possible ($\Delta E_J \approx 0\,\text{mV}$) by choosing an appropriate electrolyte for the reference electrode. The absolute mobilities for K^+ ($8.00 \times 10^{-9}\,\text{cm}^2\,\text{mol}\,\text{s}^{-1}\,\text{J}^{-1}$), NH_4^+ ($8.00 \times 10^{-9}\,\text{cm}^2\,\text{mol}\,\text{s}^{-1}\,\text{J}^{-1}$), Cl^- ($8.11 \times 10^{-9}\,\text{cm}^2\,\text{mol}\,\text{s}^{-1}\,\text{J}^{-1}$) and NO_3^- ($7.58 \times 10^{-9}\,\text{cm}^2\,\text{mol}\,\text{s}^{-1}\,\text{J}^{-1}$) are very close, so that potassium and ammonium chlorides and nitrates are recommended as electrolytes for reference electrodes. Lithium acetate can also be used. The salt concentration should be as high as possible, again to minimize the liquid junction potential. Finally, in order to have a concentration that is constant and not subject to temperature fluctuations and evaporation of water, saturated solutions are used.

To achieve a stable liquid junction a constant flow of reference electrolyte into the sample solution is necessary. This can be achieved with various configurations. A compromise has to be sought between liquid junction arrangements with low liquid outflow, which tend to clog easily, and junctions with a more rapid flow, which require frequent refilling of the reference electrode and cause sample contamination. Different arrangements

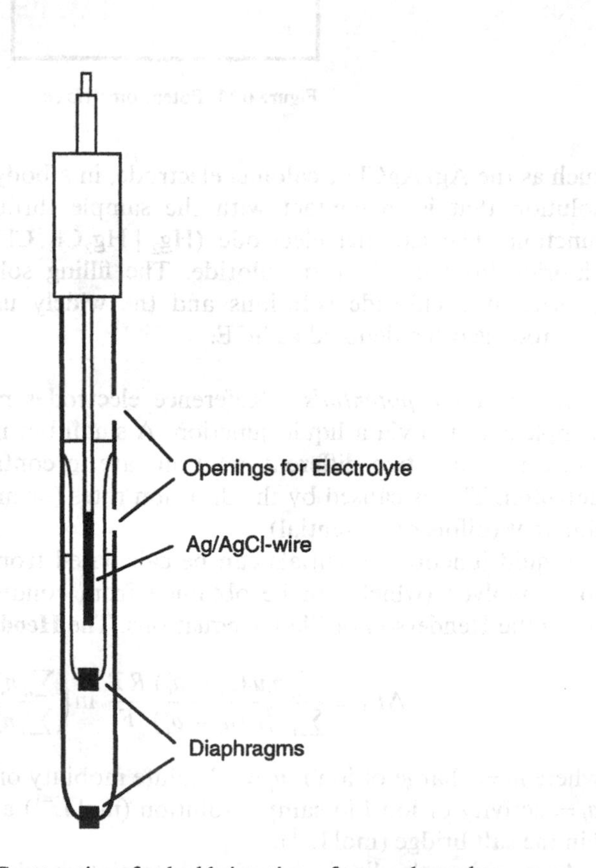

Figure 6.12 Construction of a double junction reference electrode.

are used, and different constructions are also found depending on the electrochemical method for which the reference electrode is used.

In particular in potentiometry, careful attention has to be paid to the possibility of sample contamination by the reference electrolyte. Therefore liquid junctions with slow flow are preferred. So-called free-flowing liquid junctions, which consist of a single hole of a small bore, appear to give the best performance. However, if an ion that is used as reference electrolyte is to be measured with a potentiometric cell then this solution is not normally adequate. Double junction reference electrodes are available that allow the use of two solutions, an inner filling solution forming the electrode of the second kind and the outer solution in contact with the sample. One is then quite free in the choice of the outer filling solution. An additional liquid junction potential exists between the two filling solutions, but this is constant because the solutions do not vary. The practical construction of a double junction reference electrode is illustrated in Figure 6.12.

Potentiometric cells with membrane electrodes. The measuring chain consists of a number of different phases and potentials arise at all the interfaces. Such a cell is illustrated in Figure 6.13. For this cell the measuring chain can be written as follows:

$$Ag \mid AgCl, Cl^-, M^+ \mid membrane \mid M^+ \parallel Cl^-, AgCl \mid Ag$$

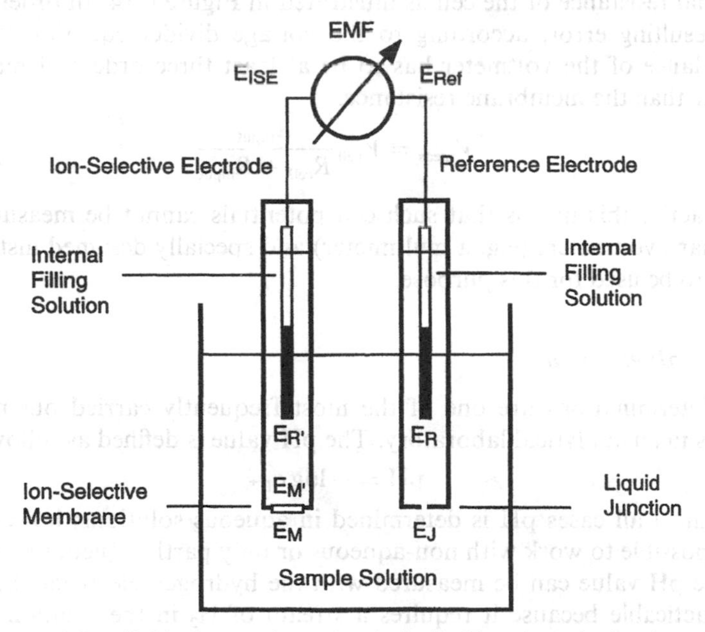

Figure 6.13 Detailed look at an electrochemical cell with an ion-selective electrode.

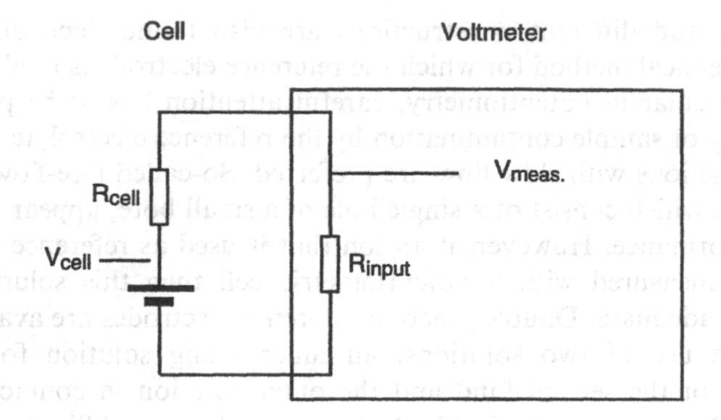

Figure 6.14 Impedance consideration for potentiometric measurements.

Each interface between different phases where a potential arises is denoted with a vertical bar. Liquid junctions are commonly indicated with a double vertical bar. M^+ denotes here the analyte cation. Since all concentrations other than the analyte concentration are constant, the response of the system is to the analyte alone.

Most ion-selective electrodes possess a fairly high membrane impedance (10^6 to $10^{10}\Omega$, depending on the membrane material). These can be viewed as voltage sources (galvanic cells) with an associated internal resistance. The input impedance of a voltmeter forms a voltage divider with the internal resistance of the cell as illustrated in Figure 6.14. In order to limit the resulting error, according to the voltage divider equation the input impedance of the voltmeter has to be at least three orders of magnitude higher than the membrane resistance:

$$V_{\text{meas}} = V_{\text{cell}} \frac{R_{\text{input}}}{R_{\text{cell}} + R_{\text{input}}} \tag{6.28}$$

In practice this means that such cell potentials cannot be measured with standard voltmeters (e.g. a multimeter) and specially designed instruments have to be used for this purpose.

6.3.2 *pH electrodes*

pH determinations are one of the most frequently carried out measurements in an analytical laboratory. The pH value is defined as follows:

$$\text{pH} = -\log a_{\text{H}^+} \tag{6.29}$$

In almost all cases pH is determined in aqueous solutions; however, it is also possible to work with non-aqueous or only partly aqueous solvents.

The pH value can be measured with the hydrogen electrode, but this is impracticable because it requires a stream of H_2 in the solution and has interference by other redox couples that may be present.

The most commonly used electrode for pH measurements is the **glass membrane electrode**. These electrodes were developed early this century and are perhaps the oldest form of chemical sensor in use. The glass membrane electrode for measuring pH really is a type of ion-selective electrode, but because of its importance it is discussed here separately. The drawing of a pH electrode is given in Figure 6.15. The membrane consists of a special glass that is responsive to a_{H^+} and is usually blown in the form of a bulb. Most commonly used are combination electrodes where the pH electrode and reference electrode are combined into one body. Different shapes are available such as micro- and semi-microelectrodes (to allow the measurement of small solution volumes), electrodes with long skinny barrel (to measure the pH of solutions contained in e.g. volumetric flasks) and electrodes with flat membrane (to measure the pH of moist surfaces).

The glass membrane pH electrodes work satisfactorily between about pH 1 and 10. At lower values the liquid junction potential becomes significant due to the high H^+ concentration (the so-called **acid error**); at higher pH the interference from other cations (such as Na^+, K^+) according

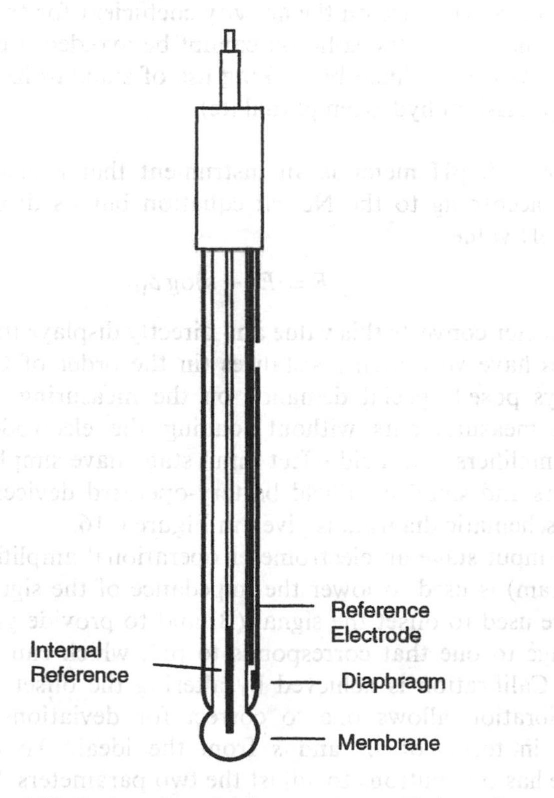

Figure 6.15 Construction of a glass membrane pH electrode.

to the Nicolsky–Eisenman equation becomes significant. When the electrode is in solution, the glass develops a gel layer (hydration layer) about 10^{-4} mm thick which is necessary for its functioning. The electrode is usually stored in the reference electrolyte to avoid drying out of the gel layer. However, when the gel layer becomes too thick, after a few months, the response becomes sluggish (remedy: etching in 5% aqueous HF for about 2 min). Continued storage in aqueous solutions increases the rate of aging. A glass electrode has a typical lifetime of approximately 1–2 years depending on its usage.

Other pH electrodes such as the quinhydrone electrode (based on the pH dependence of the quinone/hydroquinone redox couple), metal oxide or hydroxide electrodes (based on Sb, Bi, Pd or Ir) or those based on ionophores (such a tertiary amine in a PVC matrix) have been reported but are not widely used.

Calibration. In practice pH electrodes (or rather complete measuring cells) need to be calibrated against solutions of known hydrogen-ion activity. However, because of the fact that a single ion solution is not possible, an interference on the activity coefficient for the hydrogen ion by the anion making up the solution cannot be avoided. For this reason, the pH scale has been defined by making use of standardized buffer solutions (such as potassium hydrogen phthalate).

pH meters. A pH meter is an instrument that measures the electrode potential according to the Nernst equation but its display is labelled in terms of pH value:

$$E = E^\circ + s \log a_{H^+} \qquad (6.30)$$

A pH meter converts this value and directly displays the pH value. Glass electrodes have very high resistances (in the order of GΩ), which in the early days posed special demands on the measuring circuitry to allow potential measurements without loading the electrode. Modern operational amplifiers with field-effect input stage have simplified the design of pH meters and small handheld battery-operated devices are available. A possible schematic diagram is given in Figure 6.16.

At the input stage an electrometer operational amplifier (amplifier A in the diagram) is used to lower the impedance of the signal. Two amplifier stages are used to offset the signal (B) and to provide gain (C) to convert the voltage to one that corresponds to pH, which can then be displayed directly. Calibration is achieved by altering the offset and gain settings. This calibration allows one to correct for deviations of the electrode response in terms of E° and s from the ideal. An analog pH meter therefore has two buttons to adjust the two parameters. Sometimes a third button is found, labelled 'Temperature', to allow one to correct for the

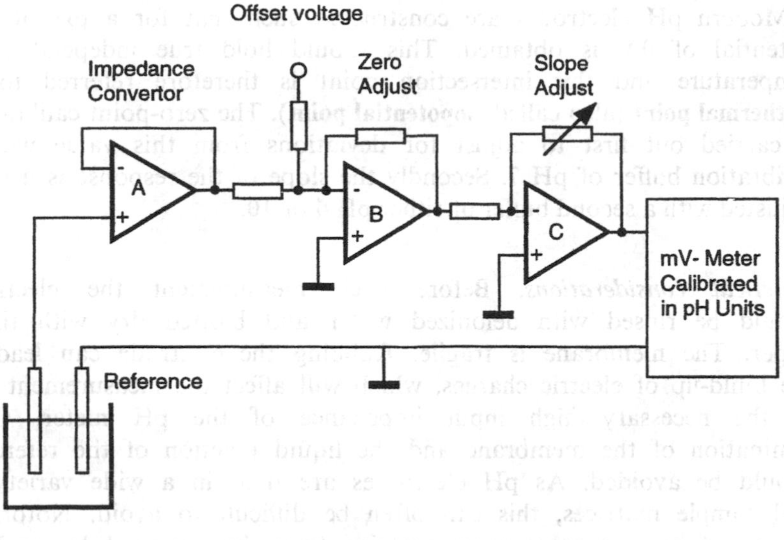

Figure 6.16 Possible circuitry of a pH meter.

influence of T on s, but the slope adjust can also be used for the same purpose. Modern microprocessor-controlled pH meters carry out these adjustments automatically. When calibrating, at least two buffer solutions of different pH are needed to allow the corrections for deviations of the two variables in the equation. However, it is not possible to arrive at correct settings unless the proper procedure is adhered to. This is illustrated by Figure 6.17.

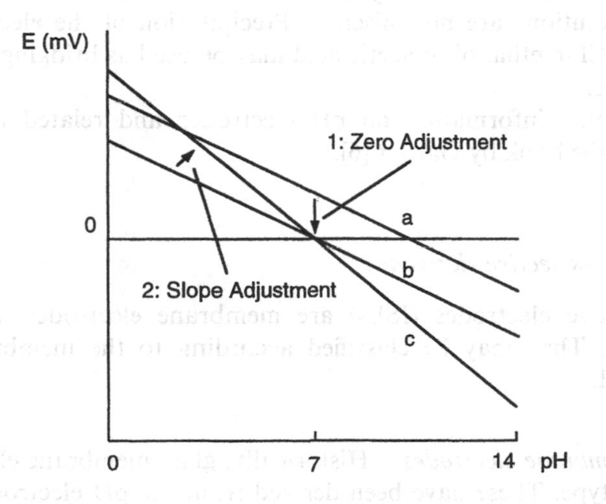

Figure 6.17 Calibration of a pH meter.

Modern pH electrodes are constructed such that for a pH of 7 a potential of 0 V is obtained. This should hold true independent of temperature and this intersection point is therefore referred to as **isothermal point** (also called **isopotential point**). The zero-point calibration is carried out first to adjust for deviations from this value with a calibration buffer of pH 7. Secondly the slope of the response is usually adjusted with a second buffer of either pH 4 or 10.

Practical considerations. Before each measurement the electrode should be rinsed with deionized water and blotted dry with tissue paper. The membrane is fragile. Rubbing the electrode can lead to the build-up of electric charges, which will affect the measurement due to the necessary high input impedance of the pH meter. Contamination of the membrane and the liquid junction of the reference should be avoided. As pH electrodes are used in a wide variety of real sample matrices, this can often be difficult to avoid. Notorious are the following substances: proteins (may be removed by soaking the electrode in pepsin/HCl for several hours), fats and other organics (may be removed by a brief rinse with acetone or ethanol), sulfides (may cause precipitation of Ag_2S in the liquid junction as the filling solution always contains some Ag^+), solutions containing HF, and solutions of low electrolyte content (because of the low conductivity the measurement is prone to the build-up of static charges, in particular when stirred).

When organic or partly organic solutions are used, the measured pH value is relative only since the pH scale is defined for aqueous solutions. Problems arise at the liquid junction when the internal electrolyte and the sample solutions are not miscible. Precipitation of the electrolyte might occur. LiCl in ethanol or acetic acid may be used as bridging electrolyte in such cases.

Additional information on pH electrodes and related topics can be found in the book by Galster [6].

6.3.3 *Ion-selective electrodes*

Ion-selective electrodes (ISEs) are membrane electrodes, as is the pH electrode. They may be classified according to the membrane material employed.

Glass membrane electrodes. Historically, glass membrane electrodes were the first type. These have been derived from the pH electrode by varying the composition of the glass membrane to obtain selectivities for ions

Table 6.1 Crystalline membrane electrodes

Membrane composition	Analyte ion
$AgCl, Ag_2S$	Cl^-, Ag^+
$AgBr, Ag_2S$	Br^-
AgI, Ag_2S	I^-
$AgCN, Ag_2S$	CN^-
Ag_2S	Ag^+, S^{2-}
CuS, Ag_2S	Cu^{2+}
CdS, Ag_2S	Cd^{2+}
PbS, Ag_2S	Pb^{2+}

other than protons. Electrodes for the following ions have been obtained in this way: Li^+, Na^+, K^+, Tl^+ and Ag^+.

Crystalline membrane electrodes. Pressed pellets based on Ag_2S are used to construct electrodes for Ag^+ and S^{2-}. Their response is governed by the low solubility product of Ag_2S similar to an electrode of the second kind. However, the advantage of a membrane electrode over an electrode of the second kind is the fact that a membrane electrode is not interfered with by redox-active species. Any bare metal of an electrode of the second kind accessible to redox-active species will act as an electrode of the first kind to these species and mixed potentials would be the result. The mobility of Ag^+ in the membrane provides conductivity to the membrane. Mixed salt membranes with the salts shown in Table 6.1 are used to obtain electrodes for other anions and cations.

A further important crystal-membrane-based ion-selective electrode is the one for the determination of fluoride ions. This ISE has been very successful, not only because the electrode functions very well with regard to selectivity and stability, but also because there are few good methods for the determination of fluoride ions. Spectrophotometric methods usually require separation of fluoride from the matrix by distillation as hydrofluoric acid and the only alternative instrumental method is ion chromatography. The fluoride ion-selective electrode is based on a single crystal of LaF_3 that has been doped with EuF_2. This renders the membrane conductive for fluoride ions. The response mechanism is clearly based on a crystal lattice effect and not on solubility as is the case for the silver-salt-based membranes as discussed above. This is evidenced by the fact that the LaF_3 electrode does not respond to La^{3+} ions.

Liquid membrane electrodes. These types of membranes are probably the most widely used ones. Here an organic matrix in which a selective recognition molecule is dissolved makes up the membrane. The species responsible for the response is termed a **carrier molecule** or an **ionophore**.

The membrane is a lipophilic material, which used to be relatively viscous organic solvents immiscible with water held in a porous material. Modern membranes are actually solid, consisting mostly of poly(vinyl chloride) (PVC) with about 70% of plasticizer, but the name liquid membrane is still used mainly for historical reasons. Usually about 1–3% of ionophore is contained in these membranes. These may be either charged or neutral species.

The selectivity pattern of these electrodes is generally determined by two parameters: the lipophilicity of the ions and any specific interaction of the ionophore with the target ions. The lipophilicity is governed by the hydration energy of the ions, which is a function of their charge and size. The smaller and more highly charged the ion, the lower its lipophilicity. It is possible to construct electrodes that derive their selectivity entirely from this feature. Membranes that contain positively charged ion exchange sites such as quaternary amine ions respond to anions with the following selectivity sequence, the so-called **Hofmeister series**: $ClO_4^- > SCN^- > I^- > NO_3^- > Br^- > NO_2^- > Cl^- > HCO_3^- > SO_4^{2-} > HPO_4^{2-} > F^-$. Such electrodes are often sold as nitrate ISEs, nitrate being an important analyte, and the other more lipophilic, potentially interfering ions are not likely to be present at significant levels in real sample matrices.

In order to create electrodes which show a selectivity pattern that differs from the lipophilicity sequence, a strong specific interaction with a ligand in the membrane has to be present. This has been achieved for many cations with neutral carrier molecules. The first of these was the use of the naturally occurring substance valinomycin in the construction of electrodes for potassium ions. This substance regulates the passage of potassium through cell walls. In ISEs it imparts a very high selectivity for this ion. Similarly, a natural species (nonactin) was found that responds selectively to ammonium ions. Following these successes, a large effort was undertaken to synthesize ligands that would provide selectivity for many other cations, and liquid-membrane-based electrodes for cations such as Na^+, Ca^{2+}, Li^+, Mg^{2+}, H^+ and others are now available. Some of these have found widespread use in clinical analysis for the determination of blood electrolytes (K^+, Na^+, Ca^{2+}, etc.) where they have often replaced the more cumbersome flame photometric method. Membranes based on these neutral carriers include a small amount of a lipophilic anion in the membrane, which improves the response, suppresses the response to anions and lowers the membrane resistance.

The search for highly selective anion sensors has not been as successful. Some organometallic complexes show selectivity that deviates from the Hofmeister pattern, but the quaternary-amine-based liquid membrane remains the only really widely used representative for negatively charged species. The structures of some ionophores used for ISEs and the corresponding ions are given in Figure 6.18.

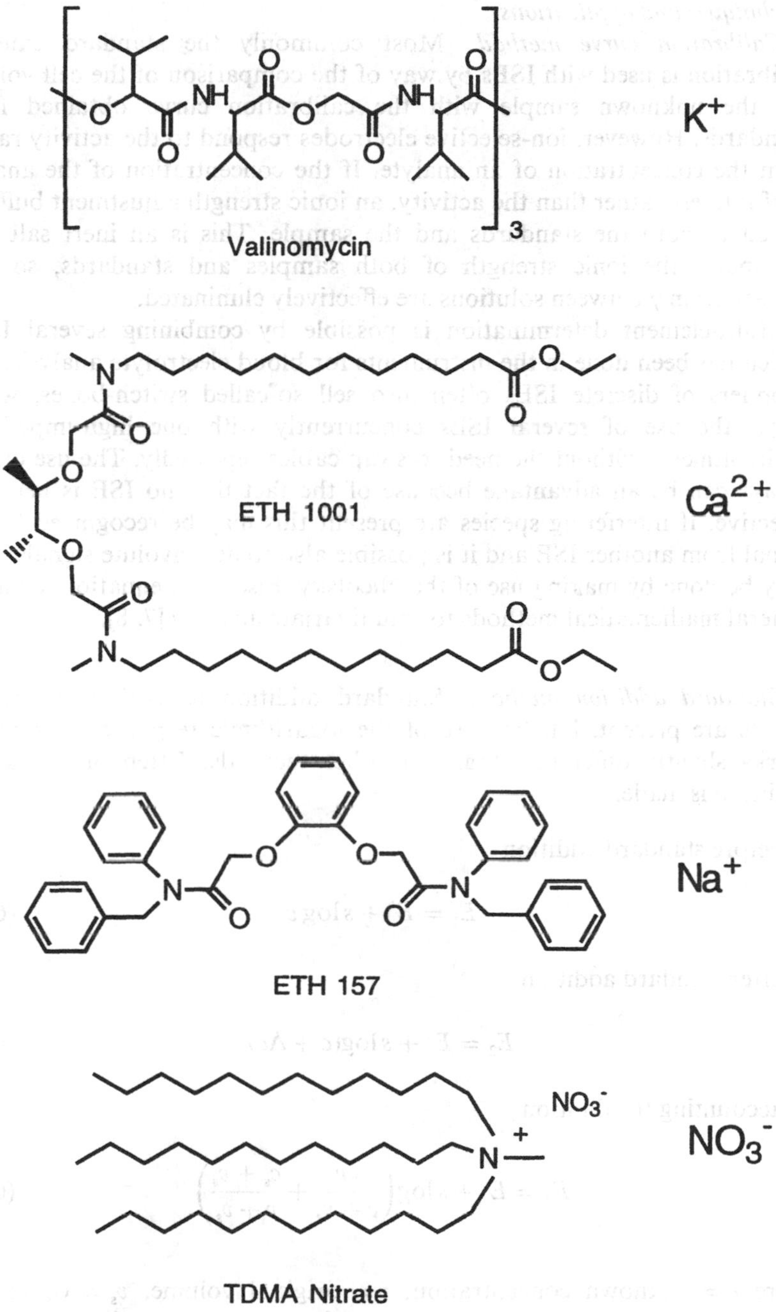

Valinomycin K^+

ETH 1001 Ca^{2+}

ETH 157 Na^+

TDMA Nitrate NO_3^-

Figure 6.18 Some typical ionophores used with PVC membrane electrodes.

Techniques and applications.

Calibration curve method. Most commonly the standard external calibration is used with ISEs by way of the comparison of the cell voltage for the unknown sample with the calibration curve obtained from standards. However, ion-selective electrodes respond to the activity rather than the concentration of an analyte. If the concentration of the analyte is of interest rather than the activity, an ionic strength adjustment buffer is added to both the standards and the sample. This is an inert salt that dominates the ionic strength of both samples and standards, so that variations in γ between solutions are effectively eliminated.

Multi-element determination is possible by combining several ISEs, which has been done in the instruments for blood electrolyte analysis. The suppliers of discrete ISEs often also sell so-called switch-boxes, which allow the use of several ISEs concurrently with one high-impedance millivoltmeter without the need to swap cables repeatedly. The use of ISE arrays can be an advantage because of the fact that no ISE is perfectly selective. If interfering species are present this may be recognized by the signal from another ISE and it is possible also to deconvolute signals. This may be done by making use of the Nicolsky–Eisenman equation, or using general mathematical methods for multivariate analysis [7, 8].

Standard addition method. Standard addition is used when matrix effects are present, but because of the logarithmic response behavior, it works slightly differently than for other methods. Often only a single addition is made:

- before standard addition

$$E_1 = E^\circ + s \log c \tag{6.31}$$

- after standard addition

$$E_2 = E^\circ + s \log(c + \Delta c) \tag{6.32}$$

- accounting for dilution

$$E_2 = E^\circ + s \log\left(\frac{cv}{v + v_s} + \frac{c_s + v_s}{v + v_s}\right) \tag{6.33}$$

Here c = unknown concentration, v = original volume, v_s = volume of standard added and c_s = concentration of standard. The value of c can be obtained from the difference in the potential before and after the addition:

$$\Delta E = s \log\left(\frac{cv/(v+v_s) + c_s v_s/(v+v_s)}{c}\right) \tag{6.34}$$

$$c = \frac{c_s v_s}{(v+v_s) \times 10^{\Delta E/s} - v} \tag{6.35}$$

For better precision, several standard additions can be made and the result is obtained by linear regression from a special plotting procedure (**Gran plot**). In both cases, the activity coefficient needs to stay constant, so that in the absence of an ionic strength buffer only small additions can be made.

Titrations. For high precision (i.e. $< 1\%$) titrations can be used with the ISE serving as an end-point detector. Here the selectivity is determined by the titration reaction (often precipitation or complexation) rather than the ISE. Typical applications include: acid–base titrations using a pH electrode for end-point detection; the determination of F^- with $La(NO_3)_3$ by precipitation as LaF_3 using a LaF_3-based electrode; the determination of Ca^{2+} or water hardness with EDTA using a Ca^{2+} electrode; and the determination of halides with a Ag_2S-based membrane electrode by precipitation with Ag^+. In the latter case it is possible to carry out a fractionated titration as separate end-points for the different halides are obtained because of the different solubilities.

Pre-treatment. Pre-treatment is often carried out in the form of the addition of an ionic strength adjustment buffer to eliminate differences in the activity coefficient between sample and standard solutions as discussed above. Occasionally other forms of pre-treatment are necessary to overcome potential interferences. An example is the TISAB solution (total ionic strength adjustment buffer) used with the fluoride electrode. This solution consists usually of $1\,\text{mol}\,L^{-1}$ of NaCl, $1\,\text{mol}\,L^{-1}$ of an acetic acid/sodium acetate buffer at pH 5, and $4\,\text{g}\,L^{-1}$ of CDTA (cyclohexylene dinitrilotetraacetic acid). Ionic strength buffering is achieved by the NaCl. The pH buffer is necessary as the electrode is interfered with by OH^- at high pH and at low pH the fluoride is converted to HF (pK_a 3.2). CDTA serves to complex cations such as Al^{3+} and Fe^{3+} which otherwise will bind free fluoride. In the determination of nitrate with a Hofmeister electrode it is sometimes necessary to remove chloride (by precipitation with AgF or Ag_2SO_4), which is a potential interferent because of its ubiquity even though the electrode is less selective for chloride.

Ion-selective electrodes are also often used as detectors in flow-through methods. These include detection in flow-injection analysis, a general means of automation of analysis methods by injecting the sample into a flowing stream. When using an ISE, this arrangement may be termed flow-injection potentiometry and the flow-injection system serves as a more convenient sampling method than physically moving the ISE into different sample containers with associated rinsing and blotting dry of the electrodes. It is possible to extend this to an array of several ISEs for simultaneous determinations. Clinical blood electrolyte analyzers, which use several ISEs and are widely employed, often use a flow-through arrangement. Furthermore ion-selective electrodes have been used as detectors for multi-species analysis in ion chromatography [9] and capillary electrophoresis [10, 11].

Additional information on ion-selective electrodes and related topics may be found in the books by Freiser [12], Morf [13], and Koryta and Stulik [14].

6.3.4 *Other potentiometric sensors*

Ion-selective electrodes as introduced above represent the most widely used application of potentiometry. However, some different special devices have also found use and deserve attention.

Ion-selective field-effect transistors. Ion-selective field-effect transistors (ISFET) (see for example Janata [3]) are the outcome of the desire to combine potentiometric sensing membranes directly with the measuring electronics in order to obtain compact and simple devices. Field-effect transistors are semiconductor components that show a dependence of the current passed on the potential applied to a so-called gate. By directly placing an ion-selective membrane onto such a gate it is possible to eliminate the electrolyte solution, internal reference electrode and electrical connections. A fairly large research effort has been carried out in this regard and presently pH ISFETs are commercially available incorporated into pen-shaped devices for pH measurements, which also include the entire readout electronics including display. The pH sensitivity is obtained by using insulators such as Al_2O_3, which adsorbs protons on the surface to an extent that is dependent on pH.

The surface reactions can be summarized as follows:

$$SO^- + H^+ \rightleftharpoons SOH$$
$$SOH + H^+ \rightleftharpoons SOH_2^+$$

where S stands for the surface.

Severinghaus-type gas probes. These types of sensor (Figure 6.19) are used to measure concentrations of dissolved gases. A gas-permeable membrane is exposed to the solution. Behind the membrane a small volume of electrolyte is contained that reacts with the gas that diffuses through. The ionic product of this reaction is determined with an ion-selective electrode.

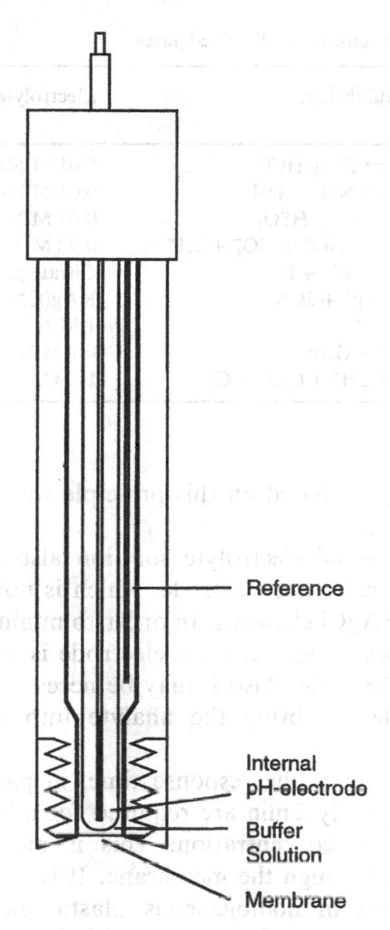

Figure 6.19 Severinghaus electrode for the determination of dissolved CO_2.

The original application of this type of sensor was for the determination of carbon dioxide according to the following equilibrium:

$$CO_{2(aq)} + H_2O \rightleftharpoons HCO_3^- + H^+$$

The activity of the protons in the internal electrolyte solution will depend on the CO_2 concentration as follows:

$$a_{H^+} = \frac{K_a a_{CO_2}}{a_{HCO_3^-}} \tag{6.36}$$

If the internal filling solution contains a high concentration of hydrogencarbonate (0.1 mol L^{-1} $NaHCO_3$) then the increase in hydrogencarbonate due to the diffused carbon dioxide can be neglected and

$$a_{H^+} = K_a a_{CO_2} \tag{6.37}$$

Table 6.2 Potentiometric sensors for dissolved gases

Analyte species	Chemical equilibrium	Electrolyte	Sensing electrode
CO_2	$CO_2 + H_2O \rightleftharpoons H^+ + HCO_3^-$	0.01 M $NaHCO_3$	H^+
NH_3	$NH_3 + H_2O \rightleftharpoons NH_4^+ + OH^-$	0.01 M NH_4Cl	H^+
SO_2	$SO_2 + H_2O \rightleftharpoons H^+ + HSO_3^-$	0.01 M $NaHSO_4$	H^+
NO_2	$2NO_2 + H_2O \rightleftharpoons NO_3^- + NO_2^- + 2H^+$	0.02 M $NaNO_2$	H^+, NO_3^-
H_2S	$H_2S + H_2O \rightleftharpoons HS^- + H^+$	Citrate pH 5	S_2^-
HCN	$Ag(CN)_2^- \rightleftharpoons Ag^+ + 2CN^-$	$KAg(CN)_2$	Ag^+
HF	$HF \rightleftharpoons H^+ + F^-$	1 M H^+	F^-
HOAc	$HOAc \rightleftharpoons H^+ + OAc^-$	0.1 M NaOAc	H^+
Cl_2	$Cl_2 + H_2O \rightleftharpoons 2H^+ + ClO^- + Cl^-$	H_2SO_4	H^+, Cl^-

Sensors for other gases based on this principle were developed later, and are listed in Table 6.2.

Note that the internal electrolyte solution also needs to provide the counter-ion for the reference electrode, which is not of the junction type (e.g. Cl^- for an Ag/AgCl element). In order to minimize interference from other diffusible gases an ion-selective electrode is used wherever possible rather than a pH electrode. Also it may be necessary to adjust the pH of the sample in order to bring the analyte into its neutral molecular (diffusible) form.

A critical parameter is the response time, in particular to decreasing concentrations. Typically 2 min are required for a 99% response time for a 10-fold increase in concentration. This is mainly governed by the diffusion of the gas through the membrane. It is found that the diffusion coefficients for gases in homogeneous plastic membranes are on the order of 10^4 times smaller than in air. This led to the development of so-called air-gap electrodes. However, similar performance can be achieved with a simpler design by employing a microporous PTFE membrane. This material is strongly hydrophobic and therefore not wetted by the aqueous solutions.

The lambda probe. In order to optimize the fuel efficiency of cars and to minimize pollution it is necessary to control the stoichiometry (air/fuel ratio) very closely to keep it at its theoretical value (14.7:1). Fuel-rich combustion leads to poor efficiency, hydrocarbon emission and the production of CO, while lean combustion will cause the formation of NO_x. Furthermore, the conversion efficiency of catalytic converters is also highest at this ratio. To ensure this optimum value, an oxygen sensor in the exhaust system is needed that provides the feedback to a control loop that incorporates the carburettor.

The lambda (λ) value is used to define the actual stoichiometry:

$$\lambda = \frac{\text{actual air/fuel}}{\text{air/fuel at stoichiometry}} \tag{6.38}$$

The λ value for a stoichiometric air/fuel ratio is therefore 1.

The lambda probe [15] makes use of the following redox reaction:

$$O_2 + 4e^- \rightleftharpoons 2O^{2-}$$

This reaction occurs at high temperature on a platinum electrode (which acts as a catalyst) and is dependent on the partial pressure (pO_2) of oxygen. An electrolyte that takes up the O^{2-} ions is needed. The usual electrolyte is yttrium-stabilized zirconia (YSZ, $ZrO_2;Y_2O_3$). This ceramic material conducts O^{2-} at temperatures above about 100°C. Usually the exhaust itself is used to heat the sensor to its working temperature. Ambient air in contact with an identical oxygen electrode serves as the reference so that the following electrochemical cell is obtained:

$$O_{2(\text{sample})}, \text{Pt} \mid ZrO_2;Y_2O_3 \mid \text{Pt}, O_{2(\text{reference})}$$

The cell potential is then given by the following form of the Nernst equation:

$$E_{\text{cell}} = \frac{RT}{F} \ln\left(\frac{pO_{2(\text{sample})}}{pO_{2(\text{ref})}}\right) \tag{6.39}$$

6.4 Conductometry

Conductance G (or L) is the inverse of resistance and has the unit S (siemens; also ohm^{-1}, Ω^{-1}, or mho). The measured conductance G of a cylinder of material with length l and cross-section A is determined by the specific conductance (or conductivity) κ of the material. This is illustrated in Figure 6.20. The specific conductance κ is thus given by:

$$\kappa = \frac{Gl}{A} \tag{6.40}$$

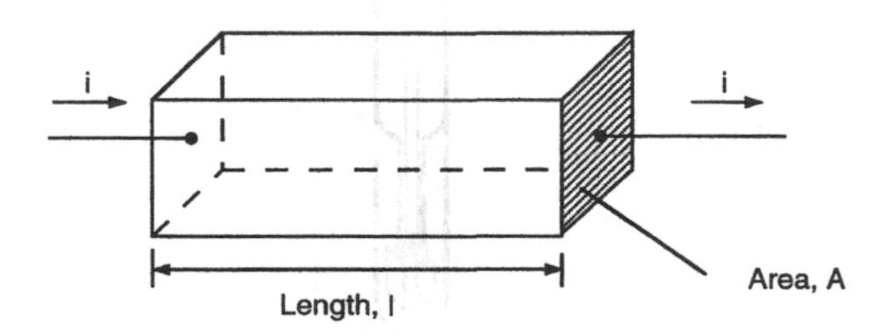

Figure 6.20 Measurement of conductivity.

The dimensions are traditionally expressed in centimetres, so that the common units of specific conductance are $S\,cm^{-1}$ or $\Omega^{-1}\,cm^{-1}$. Related to this value is the specific resistance or resistivity. Conductometric cells for the measurement of the conductance of solutions usually employ two equal flat parallel electrodes. A tip-type electrode is shown in Figure 6.21. Cells that are filled with solutions or flow-through variants are in usage as well.

The ratio l/A for a given cell is called the **cell constant** K. The specific conductance of a solution can therefore be obtained from the measured conductance by multiplication with the cell constant:

$$\kappa = GK \tag{6.41}$$

KCl solutions of known specific conductances are normally used to determine the cell constant of measuring electrodes. Most instruments can then be calibrated with the cell constant in order to obtain directly the more useful specific conductance values of solutions.

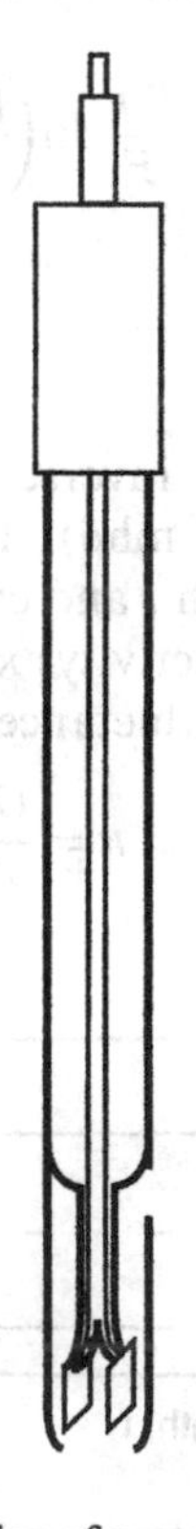

Figure 6.21 Construction of a conductometric probe.

Table 6.3 Limiting ionic specific conductances, λ (S cm^2 mol^{-1}), for some ions

Cations	λ	Anions	λ
H^+	350	OH^-	198
Li^+	39	F^-	55
Na^+	50	Cl^-	76
K^+	74	Br^-	78
Rb^+	78	I^-	77
Cs^+	77	CN^-	82
NH_4^+	73	NO_3^-	71
Mg^{2+}	106	SO_4^{2-}	160
Ca^{2+}	120	Acetate	41

For known solutions κ can be derived from the limiting ionic conductivities, λ, of the solutes:

$$\kappa = \sum \lambda_i c_i \tag{6.42}$$

Table 6.3 gives the limiting ionic conductivities for a number of ions. As can be seen, the values are not very different and this indicates that the method is not specific.

To avoid electrode reactions (faradaic reactions) AC voltages are employed rather than DC voltages. It is found that frequencies around 1000 Hz work best. For lower frequencies electrode reactions are not completely eliminated and for higher frequencies spurious impedance effects interfere with the measurement. It is however also possible to operate with frequencies in the radiofrequency range. The advantage of these methods is that the electrodes do not need to be in direct contact with the solution. The cell is made a part of an oscillator circuit (hence the name **oscillometry** is also used) whose frequency varies with the conductance of the solution.

Owing to the non-specific nature of the method its applications often lie in the determination of purity. In the preparation of ultra-pure water (such as for use as a solvent in an analytical laboratory) the proper functioning of the purification unit (ion exchanger, distillation) is often monitored by measuring the specific conductance in a flow-through cell. Pure water has a specific conductance of about 5×10^{-8} S cm^{-1} (or a specific resistance of 18 MΩ cm^{-1}). Similar applications are found in the monitoring of the concentration of pure solutions (e.g. acids, bases) or the salinity (e.g. of sea water).

Conductance measurements are also used for the end-point detection in acid–base titrations. Titration curves in the shape of the letter V are obtained, the end-point being indicated by the minimum in conductance.

This arises because of the large conductivity of protons and hydroxide ions compared to other ions, which leads to a dip at neutral pH.

One important application of conductance measurements is in the detection of peaks in the multi-species analytical method of ion chromatography. The usual optical methods of detection known from high-performance liquid chromatography (HPLC) do not work well for inorganic ions, while conductance measurement (due to its non-specificity) presents a good universal method. The background conductance of the eluent is removed by electronic subtraction or by physically removing the eluent ion prior to detection in a so-called suppressor.

One special application of conductance measurements is found in the form of conductometric gas sensors. Here the variation of conductance of a solid material on exposure to different gases is employed. The most common and commercially available representative is the semiconducting oxide sensor based on SnO_2 or ZnO_2. The surface conductance of this material changes when exposed to electron-donating (reducing) or electron-withdrawing (oxidizing) gases. The nature of the response mechanism makes this not very selective, but this pattern may be influenced to some extent by doping the substrate. The semiconducting material needs to be heated to about 200°C for functioning, which is achieved by including an electrical heating coil in the sensor body. This causes a fairly high power consumption of the system. Nevertheless, these types of sensors are fairly commonly used for industrial monitoring applications and are readily available (e.g. electronic component suppliers). A common application is the detection of CO build-up.

6.5 Electrogravimetry and coulometry

In both of these methods the analyte is exhaustively reduced or oxidized at a working electrode. In electrogravimetry the product is weighed, and in coulometry the total electrical charge required is determined. Both of these measurements can be carried out to an accuracy of 0.1% or better. The methods are thus very useful when high accuracy is required, an advantage shared with ordinary gravimetric analysis. Also common is the fact that no calibration curves are required; rather the measurements rely on the calibration of balances or coulometers and exact knowledge of atomic weights and the Faraday constant. On the other hand, the selectivity is limited in that any species present that can be reduced (or oxidized) more easily than the analyte will interfere. It is sometimes possible, however, to remove such interfering species electrolytically prior to the analytical step.

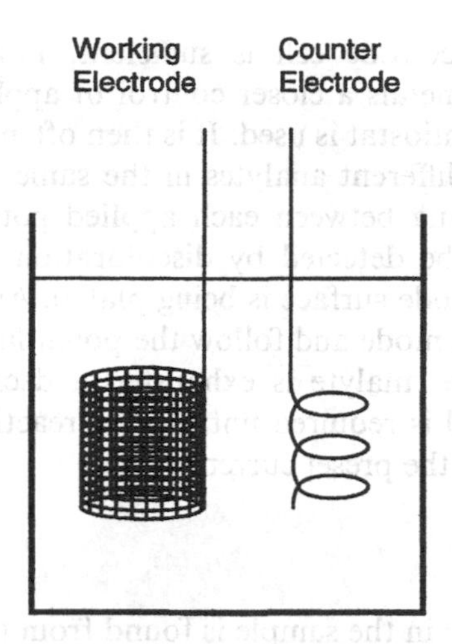

Figure 6.22 Electrogravimetric cell.

6.5.1 Electrogravimetry

A potential is applied across an electrode pair to cause deposition of the analyte. The working electrode is usually a tubular platinum mesh electrode (see Figure 6.22). The electrode is weighed before and after deposition in order to obtain the mass of the analyte. For this reason it is essential that the reduced or oxidized analyte adheres very well to the electrode. Therefore the method is mainly limited to the determination of metals by their reduction. Exceptions are the determination of Cl^- or Br^- by the formation of AgCl or AgBr on a silver anode or the deposition of lead as PbO_2 on a platinum anode. In order to ensure the deposition of a smooth layer of the analyte it is often necessary to control the pH and to include complexing agents in the solution (as in electroplating methods). A number of typical methods and conditions are listed in Table 6.4.

If the sample matrix is simple, i.e. only one reducible metal is present,

Table 6.4 Applications of electrogravimetry

Analyte	Deposit
Ag^+	Ag
Cu^{2+}	Cu
Ni^{2+}	Ni
Zn^{2+}	Zn
Mn^{2+}	MnO_2 (anodic)
Pb^{2+}	PbO_2 (anodic)
Br^-	AgBr (anodic)

then a simple two-electrode cell is sufficient. For samples comprising mixtures of different metals a closer control of applied working potential is required and a potentiostat is used. It is then often possible subsequently to determine several different analytes in the same sample. The electrode is removed for weighing between each applied potential. Completion of deposition can often be detected by discoloration of the solution or by testing if a fresh electrode surface is being plated. Another possibility is to use a constant-current mode and follow the potential required to maintain the current. When the analyte is exhausted a decrease (or increase for oxidation) in potential is required until a side reaction (such as hydrogen evolution) can sustain the preset current.

6.5.2 Coulometry

The amount of analyte in the sample is found from the total charge, which can be obtained by integrating the current passed over time with an electronic integrator, or if the current is kept constant by measuring the time until completion of the reaction. The completeness may be indicated again by a change in the required potential to maintain the current.

The advantage of coulometry compared to electrogravimetry is that the former also works when no solid, weighable, deposits are obtained. It has in common with electrogravimetry (and gravimetry in general) its high accuracy and precision because electrical charges can be measured easily and accurately. Calibration is generally not required.

Coulometric titrations are indirect methods where the titrant is produced electrolytically. This is usually achieved by using a constant-current source for generation of reagent and the measurement of time until the end-point is reached. The product of the two values then directly gives the total charge passed, which can be related to the amount of analyte through the stoichiometry of the titration reaction.

Advantages compared to classical titration are that it is not necessary to standardize a reagent solution (as long as care is taken to ascertain a 100% current efficiency) and that unstable reagents can be used as they are produced *in situ*. It is also possible to extend coulometric titrations to smaller amounts of analytes than possible with conventional titration. Accuracies of 0.1% are possible as with normal coulometry. End-point detection can in principle be achieved by any method used for classical titration. For coulometry, potentiometric detection is often employed. The apparatus may be completely automated, and this is actually achieved more easily than for conventional titrations including burettes.

The counter-electrode reaction may interfere and in this case this electrode has to be physically separated from the sample solution by using a frit as liquid junction or a flow-through arrangement that only delivers the product from the desired electrode to the sample.

All of the four types of titrations have been implemented coulometrically (i.e. acid–base, precipitation, complexometric and redox titrations). Acid–base titrations are achieved by generating protons or hydroxide ions from the solvent water by electrolysis (the hydrogen and oxygen evolution reactions). A recent development has been the integration of such titrators on a silicon substrate using manufacturing techniques known from microelectronics [16].

A further application of coulometry is the detection of reducible or oxidizable species in HPLC or ion chromatography. Here it is simply the magnitude of the current that is measured as the electroactive species is fluted through a flow cell. In order to ensure complete reaction, the working electrode is made large with respect to the cell volume. Coulometry is then part of a separation method, which by its nature is a multi-element analysis method.

6.6 Voltammetry and amperometry

In both of these methods a potential is applied to an electrochemical cell in order to cause a faradaic reaction and the resulting (usually) diffusion-limited current is measured. The relationship between the analytical concentration and the measured signal is therefore established indirectly via the dependence of the rate of diffusion on the concentration as discussed earlier. In voltammetry the applied potential is varied and a current–voltage plot is acquired (a so-called **voltammogram**), which yields information on both the identity of the species and the concentration. Multi-species analysis is readily possible. In amperometry the applied potential is fixed, and this method may therefore be regarded as a subgroup of the range of voltammetric methods.

6.6.1 Voltammetry

Voltammetry encompasses a group of methods that are widely used in electrochemical studies because they yield information not only on the concentration but also on the identity of substances. The term is derived from *volt-am*pere-*ometry*. For analytical applications the mercury electrode is widely used and methods employing a dropping mercury electrode are usually classified as polarography. Polarography may be considered a subclass of voltammetry. This method is discussed first.

Polarography. Mercury as electrode metal is employed for two main reasons. First, mercury shows a large overpotential for hydrogen evolution, which allows the reduction, and determination, of metals at potentials not accessible with other electrode materials. Second, owing to

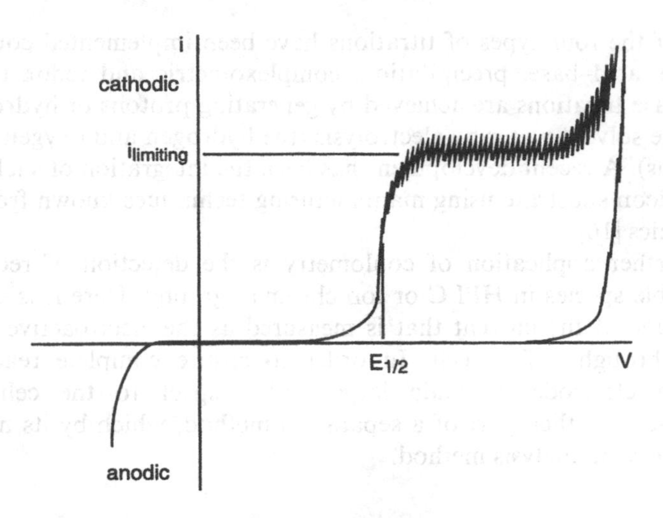

Figure 6.23 Polarogram.

its liquid nature it is possible to construct electrodes with readily renewable surface so that any contamination or poisoning effects are effectively eliminated. Such electrodes are based on a capillary at the end of which a mercury drop is formed. If a slow flow is maintained this leads to the dropping action of mercury. These electrodes are then called **dropping mercury electrodes** (DME).

If the electrode potential is scanned to more negative potentials in the presence of reducible species, **polarograms** as in Figure 6.23 are obtained. The polarogram is characterized by the polarographic wave obtained in the presence of a reducible species. A sudden increase in current is obtained when the redox potential of the analyte is reached. The potential at which half of the maximum current is reached is termed the **half-wave potential** and is characteristic of the species found. This value is related to the standard redox potential for the species but is not completely identical to it because the reduced form is not obtained as the pure metal, but dissolved in mercury as an amalgam. Consider, for example, the case of cadmium reduction:

- standard electrode reaction

$$Cd^{2+} + 2e^- \rightleftharpoons Cd_{(s)} \qquad E^\circ = -0.403\,\text{V}$$

- on the mercury electrode

$$Cd^{2+} + 2e^- \rightleftharpoons Cd_{(Hg)} \qquad E_{1/2} = -0.6\,\text{V}$$

The value of the half-wave potential may also show a dependence on the solution composition, in particular if complexing agents are present.

The current reaches a plateau because of diffusion limitation and in polarography the terms 'diffusion current' or 'limiting current' are used to refer to the analytical signal. The fluctuations on the plateau are due to the dropping action of the mercury. With every new droplet, there is initially a minimum in current because of the smaller electrode area, which then increases as the area increases. This behavior can be described by the **Ilkovic equation**, which is an extension of the Cottrell equation to allow for the time dependence of the size of the spherical mercury electrode:

$$i = knD^{1/2}m^{2/3}t^{1/6}c \qquad (6.43)$$

where i = current (μA), k = combination of several constants (e.g. F and d_{Hg}, and has the value of 706), n = number of e^- involved in the reaction, D = diffusion coefficient of ion ($cm^2\,s^{-1}$), m = mass flux of mercury ($mg\,s^{-1}$), t = lifetime of a mercury droplet (s) and c = concentration ($mmol\,L^{-1}$).

In order to ensure that the mass flux to the electrode is only due to diffusion, migration is eliminated by adding an inert background electrolyte, and convective transport is avoided by not stirring the solution. It is also essential to keep the electrode size small compared to the total volume of the solution so that only a negligible fraction of the analyte present is consumed in the reaction and the solution composition is effectively not altered by the measuring process.

The available potential range is from about 0 V to −1.5 V vs. SCE and is limited by hydrogen evolution on the negative side:

$$2H_2O + 2e^- \rightleftharpoons H_{2(g)} + 2OH^- \qquad (-1 \text{ to } -1.5\text{ V vs. SCE})$$

and by anodic dissolution of mercury at the positive side:

$$Hg \rightleftharpoons Hg^+ + e^- \qquad (0 \text{ to } +0.1\text{ V vs. SCE})$$

Dissolved oxygen interferes because it is reduced under the conditions in polarography and therefore has to be removed by passing nitrogen or argon through the container for a few minutes before the analysis.

Differential pulse polarography. In normal polarography as described above so-called **residual currents** are obtained. This is mainly due to double-layer charging of the mercury droplet electrodes. This background current limits the detection limits to approximately $10^{-5}\,mol\,L^{-1}$. It is, however, possible to achieve a better discrimination between faradaic current and charging current with pulse voltammetric methods, differential pulse voltammetry being the most widely utilized variant. As the electrode area is not constant in polarography, the kinetic behaviors of both the faradaic and charging currents are not trivial. However, for each individual droplet life the faradaic current shows an increase ($i_{diff} \propto t^{1/6}$)

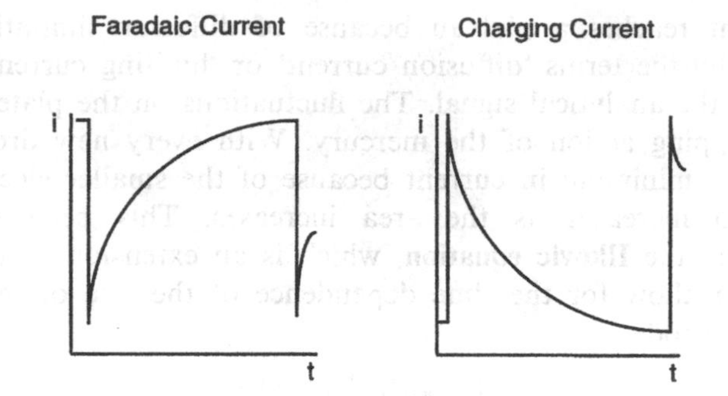

Figure 6.24 Current behavior during the lifetime of a mercury droplet.

while the charging current shows a decrease ($i_{cap} \propto t^{-1/3}$). This behavior is illustrated in Figure 6.24.

Some discrimination can therefore be achieved by current sampling just before the end of each droplet life. An even better distinction is achieved by superimposing potential pulses of typically 50 ms duration and 5–100 mV amplitude every 0.5–5 s on a linear potential scan. This pulsing scheme is illustrated in Figure 6.25.

The current is sampled before application of each pulse (S_1) and again during the last 10 ms of each pulse (S_2) when the charging current is lowest. The difference in current for each pulse is then plotted against applied potential. This results in the polarographic waves appearing as peaks on a smooth background similar to the first derivative of normal polarography, as shown in Figure 6.26.

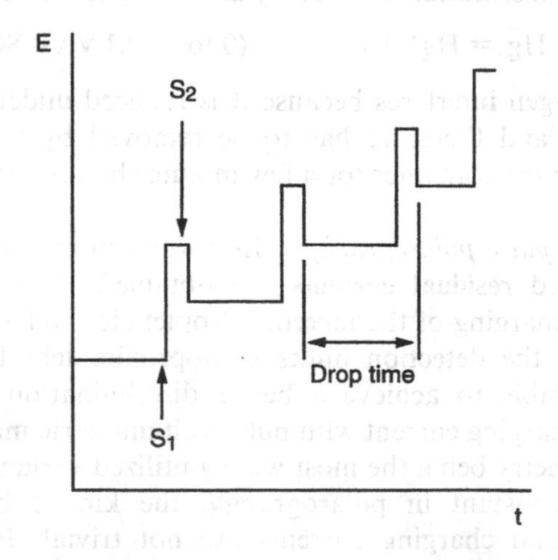

Figure 6.25 Pulsing scheme in differential pulse polarography.

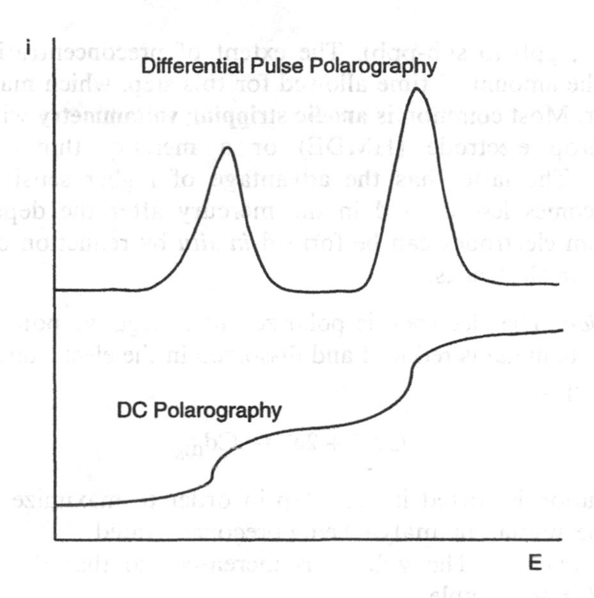

Figure 6.26 Comparison of DC and DPP polarographs.

Impressive detection limits of 10^{-8} mol L^{-1} are obtained in this way. A further improvement is the better resolution of the half-waves. It is possible to distinguish species with half-wave potentials separated by as little as 50 mV (compared to 200 mV required in normal polarography). Multi-species analysis is also readily available.

Applications. Applications are primarily in the determination of metal ions. The determination of traces of such species as lead, cadmium, copper, zinc, etc., in industrial effluents, foodstuffs, soil and biological materials is an important application of the technique. However, it is also possible to determine electroactive anions and many organic substances that contain functional groups. The method also allows the speciation analysis between different oxidation states of metal ions (e.g. between Fe^{2+} and Fe^{3+}), complexation states or even geometrical isomers (e.g. maleic and fumaric acids). Oxygen is usually an interferent, but a polarograph might also be used for the determination of dissolved oxygen.

Stripping voltammetry (inverse voltammetry). This term encompasses a range of related procedures where the analyte is first deposited electrochemically on an electrode and then the charge is measured in a second dissolvation step. The detection limit is lower with these methods than with any of the polarographic techniques because the deposition step causes a preconcentration of the analyte (detection limit: 10^{-8} to

10^{-9} mol L^{-1}, ppb to sub-ppb). The extent of preconcentration is determined by the amount of time allowed for this step, which may be as long as one hour. Most common is **anodic stripping voltammetry** with a hanging mercury drop electrode (HMDE) or a mercury thin-film electrode (~ 100 μm). The latter has the advantage of higher sensitivity, as the analyte becomes less diluted in the mercury after the deposition step. Mercury film electrodes can be formed *in situ* by reduction of a mercury salt on carbon electrodes.

1. *Deposition.* The electrode is polarized to a negative potential, so that the analyte metal is reduced and dissolved in the electrode as amalgam. For example,

$$Cd^{2+} + 2e^- \rightleftharpoons Cd_{(Hg)}$$

The solution is stirred in this step in order to maximize the flux and hence the amount of analyte being preconcentrated.

2. *Anodic stripping.* The voltage is increased, so that the metal is re-oxidized. For example,

$$Cd_{(Hg)} \rightleftharpoons Cd^{2+} + 2e^-$$

The current in this second step is measured by any of the polarographic methods (but note that the process and voltage are reversed). Here the solution is kept quiescent in order to obtain a diffusion limited current related to concentration.

Variations of this procedure are possible by the inclusion of reagents to form precipitates or complexes, and the extension of such schemes to the determination of certain anions has also been described. The main application of these techniques is in the trace analysis of heavy metals in environmental samples. The excellent detection limits achieved by this method are otherwise only obtained with very much more expensive spectroscopic means. However, extremely clean working conditions and sampling techniques are required to avoid contamination of samples. Multi-element determination is readily possible, similarly to differential pulse polarography.

A relatively recently introduced variation to the stripping techniques is the method of **potentiometric stripping** [17]. Here the stripping step is carried out by application of a constant current to the working electrode or the use of an oxidizing agent such as ambient oxygen. The potential required to maintain the constant oxidation rate is monitored. When plotting the stripping potential against time this leads to steps for each analyte, the lengths being proportional to the analyte concentrations.

Further reading on voltammetric methods can be found in Bond [18], Smyth and Vos [19], and Wang [20].

6.6.2 *Amperometry*

Amperometry is not normally used as a direct method for solution analysis, because the voltammetric methods give more information on the sample with little extra effort. Nevertheless, amperometric methods have found widespread use for certain applications, notably amperometric titrations, amperometric detectors in chromatography and amperometric sensors.

Amperometric titrations. Here a working electrode serves as an instrumental means for the detection of the end-point in a titration, similarly to the use of an ion-selective electrode for potentiometric end-point detection. The reason for performing a titration rather than direct amperometry or voltammetry is the better precision that can be achieved. Furthermore, it is often possible to determine species that are not electroactive via an amperometric titration by monitoring the concentration of a reagent that is electroactive.

Either the titrant, the added standard or a product of the titration reaction is reduced or oxidized at the working electrode. The possible titration curves resulting when the current is monitored are illustrated in Figure 6.27. A typical example is the titration of chloride with silver ions. AgCl is precipitated as long as there is chloride present. After the end-point the excess concentration of Ag^+ increases linearly and is detected amperometrically by its reduction to Ag.

Amperometric detectors in HPLC and IC. For application in high-performance liquid chromatography and ion chromatography, special flow-through cells are used that employ a glassy carbon or precious-metal working electrode. The cell volumes are kept as small as possible to avoid extra-column mixing and loss of sensitivity as well as to achieve chromatographic resolution. A wide variety of species can be detected in this way by reduction or oxidation.

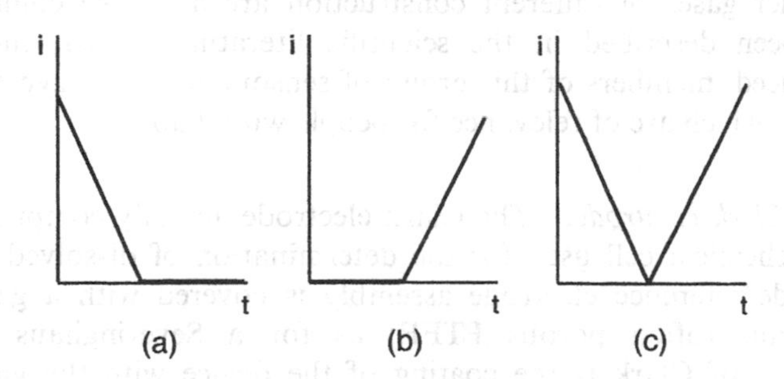

Figure 6.27 Amperometric titrations: (a) the analyte is electroactive; (b) the reagent is electroactive; and (c) both analyte and reagent are electroactive.

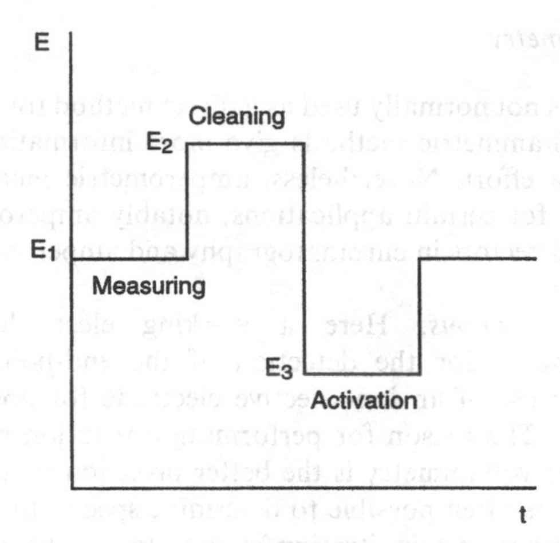

Figure 6.28 Amperometric pulsing scheme employed for amperometric detectors.

Some reaction products have been found to foul the surface of the electrode. In such cases it may be possible to clean the electrode electrochemically by applying a pulsing scheme as shown in Figure 6.28. After measuring the analytical current, the electrode potential is ramped to a high potential where oxidation of the electrode surface occurs. In a third step the potential is lowered so that the metal oxide is reduced, yielding a clean metal surface, ready for measuring at a suitable intermediate potential. The difference between amperometric detection in chromatography and the coulometric detection discussed above is that here the usual diffusion limitation is present while this is not the case in coulometric detection.

Amperometric sensors. The most prominent representative is the widely used Clark electrode for the determination of dissolved oxygen. Sensors for other gases of different construction are available commercially or have been described in the scientific literature. Other, more recently introduced, members of this group of sensors are the enzyme sensors for glucose, which are of relevance for people with diabetes.

The Clark electrode. The Clark electrode actually comprises an entire electrochemical cell used for the determination of dissolved oxygen. An electrode/reference electrode assembly is covered with a gas-permeable membrane (often porous PTFE) as for a Severinghaus probe. The invention of Clark is the coating of the device with the gas-permeable membrane. Prior to its introduction, open cells were used, which were susceptible to interfering electrode reactions and electrode poisoning. Such cells are incidentally still used for certain applications where membrane contamination is a problem such as in sewage treatment. The construction of a Clark electrode is illustrated in Figure 6.29.

The cell consists of two electrodes only, a working electrode and a counter-electrode. The electrode reactions are the following:

- cathode, working electrode

$$O_2 + 2H_2O + 4e^- \rightleftharpoons 4OH^-$$

- anode, counter-electrode

$$Ag + Cl^- \rightleftharpoons AgCl + e^-$$

It is possible to use a two-electrode system rather than a three-electrode

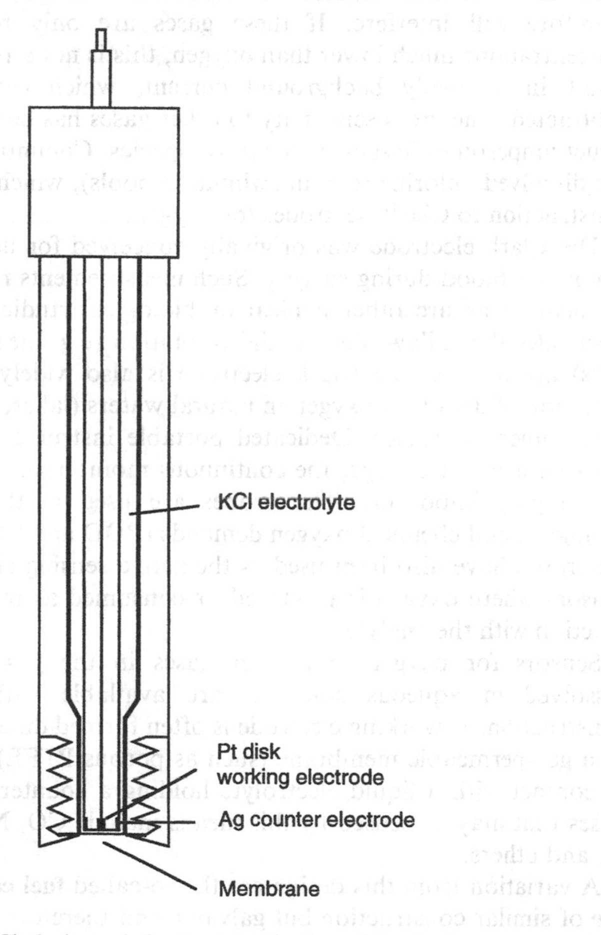

Figure 6.29 Clark electrode for the determination of dissolved oxygen.

cell because, due to the lack of possible interferents, the exact potential applied to the working electrode is not critical. The applied voltage is in the range from −600 to −900 mV vs. Ag/AgCl.

Commonly, porous PTFE (poly(tetrafluoroethylene)) is used as the membrane but other materials, such as polyethylene, silicon rubber and PVC, can be used. The sensitivity is higher and the response time of the sensor is shorter when the membrane is thinner and has higher permeability for oxygen. For the same reason it is also important to keep the gap between membrane and working electrode small. The permeability is a function of both solubility and diffusion rate of the gas in the polymer, but is usually dominated by the latter. The membranes can usually be replaced by the user and are commonly held in place with an O-ring.

Gases that have a reduction potential lower than that of oxygen on the precious-metal cathode, such as Cl_2, NO and CH_2O, will be reduced and therefore will interfere. If these gases are only present at constant concentrations much lower than oxygen, this is not a problem, since it will result in a steady background current, which can be electronically subtracted. The cross-sensitivity to other gases has been employed to construct amperometric sensors for those species. Commonly used are sensors for dissolved chlorine (e.g. in swimming pools), which are very similar in construction to Clark electrodes for oxygen.

The Clark electrode was originally conceived for use in on-line *in vivo* studies on blood during surgery. Such measurements remain an important application as are other clinical or biological studies on tissue. Microelectrodes that allow high spatial resolution (e.g. measurements in single cells) are in use. The Clark electrode is also widely used in the determination of dissolved oxygen in natural waters (lakes, rivers, sea water) in environmental studies. Dedicated portable instruments are available. In the treatment of sewage, the continuous monitoring of oxygen content is also highly important. The probes are used in the determination of biological and chemical oxygen demands (BOD and COD) of water. Clark electrodes have also been used as the active sensing element in enzymatic sensors where oxygen is produced or consumed as part of the enzymatic reaction with the analyte.

Sensors for oxygen and other gases in the gas phase rather than dissolved in aqueous solutions are available with slightly different construction. A working electrode is often formed directly on the back side of a gas-permeable membrane (such as porous PTFE), which is then also in contact with a liquid electrolyte holding a counter-electrode. Reactive gases that may be sensed by this means include CO, NO, NO_2, H_2S, SO_2, H_2 and others.

A variation from this design are the so-called **fuel cell sensors**. The cells are of similar construction but galvanic and therefore resemble a fuel cell, which may be used for electricity generation from e.g. hydrogen. The counter-electrode is a lead anode, which is oxidized in the process. These are fairly common for oxygen sensing. Another application is the sensing of ethanol vapors. Some of the handheld devices available for breath alcohol determination operate by this means. The electrode reactions are as follows:

- anode

$$CH_3CH_2OH \rightarrow CH_3CHO + 2H^+ + 2e^-$$

The cathode reaction does not involve an electrode that is consumed, but the reduction of atmospheric oxygen from the surrounding ambient air.

- cathode

$$O_2 + 4H^+ + 4e^- \rightarrow 2H_2O$$

Amperometric enzyme sensors. Amperometric enzyme sensors are an example of so-called 'modified electrodes'. This term is used to describe electrodes that have been coated with some material that induces selectivity to analytes by catalyzation of an electrochemical process. An important example are sensors for glucose. Glucose can be catalytically oxidized to glucolactone by glucose oxidase immobilized on an electrode surface:

$$\text{glucose} + \text{glucose oxidase}_{(Ox)} \rightarrow \text{glucolactone} + \text{glucose oxidase}_{(Red)}$$

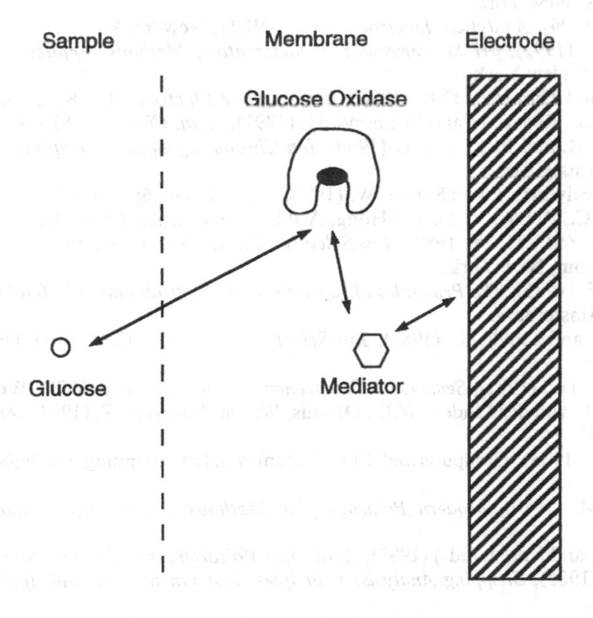

Figure 6.30 Enzymatic glucose sensor.

However, the current produced in this reaction cannot be sensed if glucose oxidase alone is attached to the electrode surface. A mediator is needed in the form of a small and highly mobile redox-active molecule that shuttles the electrons from the active center of the enzyme to the metal electrode. Redox couples consisting of small molecules such as ferrocene/ferrocenium are employed for this purpose. This is shown schematically in Figure 6.30.

The glucose sensor is of particular interest because of the clinical importance of this parameter. Some of the portable analyzers currently used by people with diabetes to monitor their blood glucose level are based on amperometric sensors, others on optical principles. The long-term aim is to develop an implantable sensor that controls an insulin pump.

Other enzymatic sensors have a huge potential field of application in the simplification of clinical analysis for other species (e.g. urea, creatinine) and in the control of industrial processes.

References

1. Bard, A.J. and Faulkner, L.R. (1980), *Electrochemical Methods, Fundamentals and Applications*, John Wiley, New York.
2. Bockris, J.O'M. and Reddy, A.K.N. (1970), *Modern Electrochemistry*, Plenum, New York.
3. Janata, J. (1989), *Principles of Chemical Sensors*, Plenum, New York.
4. Sawyer, D.T., Sobkowiak, A. and Roberts, J.L. Jr (1995), *Electrochemistry for Chemists*, John Wiley, New York.
5. Wang, J. (1994), *Analytical Electrochemistry*, VCH, New York.
6. Galster, H. (1991), *pH Measurement, Fundamentals, Methods, Applications, Instrumentation*, VCH, New York.
7. Otto, M. and Thomas, J.D.R. (1986), *Ion-Selective Electrode Rev.* **8**, 55–84.
8. Forster, R.J., Regan, F. and Diamond, D. (1991), *Anal. Chem.* **63**, 876–82.
9. Haddad, P.R. and Jackson, P.E. (1990), *Ion Chromatography, Principles and Applications*, Elsevier, Amsterdam.
10. Nann, A., Silvestri, I. and Simon, W. (1993), *Anal. Chem.* **65**, 1662–7.
11. Hauser, P.C., Renner, N.D. and Hong, A.P.C. (1994), *Anal. Chim. Acta* **295**, 181–6.
12. Freiser, H. (ed.) (1978, 1980), *Ion-Selective Electrodes in Analytical Chemistry*, vols 1 and 2, Plenum, New York.
13. Morf, W.E. (1981), *The Principles of Ion-Selective Electrodes and of Membrane Transport*, Elsevier, Amsterdam.
14. Koryta, J. and Stulik, K. (1983), *Ion-Selective Electrodes*, Cambridge University Press, Cambridge.
15. Göpel, W. *et al.* (1991), *Sensors, A Comprehensive Survey*, vol. 2, VCH, Weinheim.
16. Kolev, S.D., van der Linden, W.E., Olthuis, W. and Bergveld, P. (1994), *Anal. Chim. Acta* **285**, 247–63.
17. Jagner, D. (1983), Computerised Flow Potentiometric Stripping Analysis, *TRAC* **2** (3), 53–5.
18. Bond, A.M. (1980), *Modern Polarographic Methods in Analytical Chemistry*, Dekker, New York.
19. Smyth, M. and Vos, J. (eds) (1992), *Analytical Voltammetry*, Elsevier, Amsterdam.
20. Wang, J. (1985), *Stripping Analysis: Principles, Instrumentation, and Applications*, VCH, New York.

7 Atomic absorption spectrometry

I.Z. PELLY

7.1 Introduction

This chapter is aimed at the reader who has only a basic knowledge of the technique (some basic undergraduate course of instrumental analysis) and no real experience in using the method. Theory, descriptions of the instruments and methods of analysis will be described, mainly paying attention to the implications for experimental work, so that the reader who wants to use atomic absorption spectrometry for analytical work may reduce the necessary trial and error. If one is interested in enlarging his knowledge of the physics and theory relevant to the various sections, one should turn to appropriate textbooks or journal articles. In several sections a few experiments are suggested in order to illustrate some points. The resulting figures are given so that the reader will know what to expect generally, as the exact results depend on the spectrometer used (and the operator). The experiments should be carried out only after the reader has read all the sections, gained some experience in operating the spectrometer and has knowledge of the safety precautions (and is willing to carry them out).

Atomic absorption spectrometry (AAS) is a method for the analysis of elements in solution (mainly). It is very sensitive, can detect different elements (67) and can detect elements in the range of a few ppm or less (flame atomization) or in the range of a few ppb or less (electrothermal atomization). In most cases it is not important in what molecular form the metal exists, as in this analysis the total concentration of the metal is measured in all its molecular forms in the sample. The analysis of the element can be carried out in the presence of many other elements without, usually, having to separate the analyzed element from other elements in the sample, an advantage that makes the process simpler and saves a lot of time and errors.

The analysis is carried out on an atomic absorption spectrometer but detection limits and the sensitivities of the instrument are not the only factors determining the quality of the results (precision and accuracy). This is determined equally, or even to a higher extent, by the treatment the sample received before it was analyzed by the instrument, i.e. by sampling, homogenization or splitting, weighing, dissolution or melting, etc. Extreme care should be taken in executing these operations in order to utilize fully the capabilities of the instrument.

The sample introduction process (liquid aerosol generation, electrothermal vaporization, flow injection, laser ablation, etc.), which conditions the sample to be passed to the flame or furnace (plasma in an inductively coupled plasma, ICP), may be considered the Achilles heel of the analysis and so largely determines the accuracy of the analysis (Browner and Boom 1984).

AAS is not a magic instrument that solves all elemental analysis problems. It is not that all one has to do is to reduce the sample to the form needed for introduction into the instrument, press the button and collect the results. With experience gained, problems have been identified, the limitations of the method have been realized and its usefulness, compared with other major methods, has been determined.

The atomic absorption spectrometer was one of the first modern commercial instruments for trace-element analysis. Since its appearance it has become the 'work horse' of analytical laboratories. Later, other methods, namely X-ray fluorescence (XRF) and inductively coupled plasma (ICP) spectrometry, have threatened its position and become relatively widespread. Still, it seems that to this day AAS in all its forms (flame, electrothermal, hydride generation and Zeeman) remains the most commonly used method and a few thousands of spectrometers are still sold each year (Grosser and Schneider 1996).

When one wants to buy a major instrument for elemental analysis, one has to consider (in addition to the price of the instrument): what one mainly needs the instrument for, i.e. analysis of major elements or analysis of trace elements (and at what level – ppm or ppb); whether samples have a generally constant matrix (for instance, water or alloys) or not; the number of samples per unit time; operational expenses, etc. (Slavin 1986).

7.1.1 *Literature*

Atomic absorption spectrometry is a well established analytical method. Consequently, one can find a vast amount of literature dealing with various aspects of AAS. There are hundreds of specific applications of atomic absorption analysis and several thousands of references to articles in the literature concerning these applications. When one wants to carry out an analysis (or to develop a new method), unless one has specific instructions for the analysis, the best advice is to start with a thorough scan of the literature (or compiled references). This should be done either to find an existing method or to find similar cases and based on this to decide whether to use the method or to develop a new method (knowing now what the specific problems are and how they were solved by others) without wasting time to invent the wheel again.

One also should remember that many analytical methods are subject to strict analytical procedures resulting from legislation (implemented for instance by governments or the European Community). If an analytical method is required to comply with regulations, then the relevant literature of guidelines and procedures should be obtained and followed.

The theory of the method and a description of the various parts of the spectrometer can be found in textbooks dealing with instrumental analysis methods (for instance Skoog 1985, Mann *et al.* 1974, Strobel 1973, Ewing 1969, Pelly 1994), or books wholly dedicated to AAS (for instance Slavin 1968, Kirkbright and Sargent 1974, Alkemade and Herrmann 1979).

A very short description of the theory, in addition to a description of important experimental aspects, can be found in the manufacturers' manuals (for instance Perkin-Elmer, Varian-Techtron a, Varian-Techtron b) obtained with the instrument. These books, usually referred to as 'cookbooks', describe the detailed experimental conditions for the analysis of each element that can be determined by AAS in aqueous solutions or in other different materials, for instance geological materials, biological materials, blood and serum, urine, beer, cement, alloys, food products, plastic materials, etc.

Varian Co. publishes articles dealing with new AAS instrumental techniques and new analytical methods for analysis of elements in different environments or in different matrices, in a series called *Varian Instruments at Work – Atomic Absorption*.

Perkin-Elmer Corp. publishes a bibliography of articles dealing with AAS that were published in various scientific journals and updates it from time to time. In recent years the updating of the bibliography has been published every six months in the journal *Atomic Spectroscopy* (published up to 1980 under the name *Atomic Absorption Newsletter*) together with new original articles concerning AAS (and ICP) methods and many reviews on different subjects. More than 20 000 numbered references (AAS and ICP) were compiled and also indexed under several titles such as instrumentation, elements, geochemistry, etc. Sometimes, bibliographies concerned with a special subject are published in this journal, for instance, compiled references for the years 1973–1989 on the subject of 'Chemical modification in electrothermal atomization, Atomic absorption spectrometry', published in 1991.

The journal *Analytical Chemistry* has a section called *Analytical Chemistry Reviews* or R pages (the letter R follows the page number), which contains a chapter in which progress in AAS in the preceding two years is reviewed (there are also chapters with reviews of other analytical methods). Every alternate year these application reviews are published according to instrumental methods (i.e. a chapter of theories, instrumentation and applications of AAS, ICP, thermal analysis, etc.). In the years in between they are reviewed according to research fields, describing the different methods used in the discussed field (i.e. different instrumental methods used in the analysis of polymers, or of geological samples, or of explosives, etc.) reported in the preceding two years.

Articles dealing with AAS (both theoretical and experimental aspects) can also be found in several other journals, for instance, *Applied Spectroscopy, Analytica Chimica Acta, Talanta, Fresenius' Zeitschrift für Analytische Chemie* (also includes articles in English), *Journal of Analytical Atomic Spectrometry, Spectrochimica Acta* R (formerly called *Progress in Analytical Spectrometry*) and *Analytical Chemistry*. In addition, paragraphs dealing with AAS methods can be found in the experimental sections of articles in other journals, dedicated to other branches of science.

7.1.2 *Quality of results*

One should realize that there are many factors that determine the quality of results. Some factors, such as the intensity of radiation from the radiation source, the fuel–oxidant composition of the flame, the stability of the instrument reading, etc., determine the sensitivity, the detection limit and the repeatability of the instrument readings. Good sensitivity or detection limit do not mean that the result of the analysis is correct, as this depends also on other factors such as how representative the samples are, the matrix of the sample, the treatment the samples received before the determination in the instrument, etc. The reliability of the results depends on many factors and all should be taken care of in the analysis.

The **sensitivity** for a certain element, for the specific instrument one uses, is a term that shows the change in absorption due to a change in concentration or, in other words, the slope of the calibration curve (absorbance against concentration) and is given in the analytical methods book ('cookbook') of the instrument. It is defined as the concentration of an element in aqueous solution (in μg/mL) that absorbs 1% (absorption) of the incident radiation (in absorbance units this is 0.0044). The sensitivity is useful for checking if the operating conditions are optimal and result in the same sensitivity described in the instrument manual. One can measure the sensitivity using a solution that will give a higher absorbance. For instance if for an element a concentration of 1 mg/mL is needed for 0.440 absorbance, the sensitivity is 0.01 mg/mL (or 10 μg/mL). One can use the sensitivity to determine the working range needed. If the sensitivity is 10 μg/mL then the absorbance of 1 mg/mL will be 0.440 and that of 2 mg/mL 0.880 (provided all this is in the linear range).

Usually one does not measure the sensitivity as defined. The analytical methods books or the operation programs of computerized instruments tell the user what concentration is needed to give an absorbance of 0.200

(characteristic concentration) and one can (with an aqueous solution having the specified concentration of the element) check if this absorbance is really obtained. In addition, using the sensitivity, one can calculate the expected absorbance for the concentration range of the standards for the calibration curve and of the unknowns (if one can estimate their concentration range) and check if it is really obtained.

Some 'cookbooks' do not state the sensitivity but rather the optimum working range (the concentration range that will yield 0.200–1.000 absorbance). The relevant terms for graphite furnace work are listed in section 7.4.2.

The **detection limit** depends on the signal-to-noise ratio: the higher the noise, the higher should be the signal in order to be distinguished as something above the noise – an analytical signal. The detection of an element, for the specific instrument one uses, is a term that shows the lowest concentration that can be measured, under favorable conditions, with a certain degree of reliability and is given in the analytical methods book ('cookbook') of the instrument. One can find different definitions in the literature. Here we refer to it as the concentration in aqueous solution of an element (in μg/mL) that can be detected with a 95% certainty, i.e. a reading equal to twice the standard deviation of a series of at least 10 determinations near or at blank level. This is important as one can measure the signal of a solution concentrated 10-fold and divide the result by 10. In this case the experiment was carried out under much more favorable conditions. Had it been carried out with the right concentration, maybe nothing above noise level would have been detected.

Detection limits and sensitivities are the results of the instrument design, the quality of its parts and the operational parameters. All these do not ensure a 'true' result. This also depends on other factors such as how representative the sample is, grinding, splitting, weighing, dissolution, transfer by pipets, etc. The important thing to remember is that what counts is not only the instrument's reading but the whole analytical method. It is recommended to read Chapter 3 in this book on precision and accuracy.

In atomic absorption analysis the precision may (and should) be checked in several ways. A method to check the precision of the instrument's reading (checking the change in instrument readings with time) is reading one of the solutions every few samples. A method to estimate the precision of the whole analytical procedure is analyzing one sample (after grinding and mixing) several times including weighing and dissolution or melting and calculating the standard deviation. Another method is preparing duplicates of some unknowns and analyzing them while they are mixed among other samples (with samples named and their order arranged in such a way that the operator will not know that he or she is running the duplicate – to avoid bias).

One can and should always check the results of a new analytical procedure using atomic absorption spectrometry, by analyzing certified standards exactly the way one intends to analyze one's samples and comparing the results with those stated for the standards (for the types of relevant standards see section 7.8.2).

The matrix of a complex material may influence the spectrometer's reading, so when the absorbance expected from the sensitivity is not obtained, one should prepare pure aqueous solutions of the analyzed element and check if the spectrometer operates under optimal conditions and the desired sensitivity and detection limit are obtained. If not, something is wrong with the instrument or the operation mode. If the results are what they should be, then the reason for the bad results of the analysis is probably interferences due to the complex matrix (which are quite common in atomic absorption spectrometry) and one should look for ways to lower the material's interferences.

7.2 Theory

Only the bare minimum of theory will be given in this section. Readers interested in a more thorough theoretical treatment of the theory and of some subjects discussed in the following sections should refer to the proper textbooks (for examples see section 7.1.1).

Under normal conditions atoms are in their lowest possible energy state, the ground state. If valence electrons of ground-state atoms absorb energy, the electrons are raised to a higher energy level (excited state). There are allowed transitions and forbidden transitions. Transitions that start or end in ground-state levels are called resonant transitions. There are also transitions between excited states (non-resonant) but these are not used in atomic absorption. Different atoms need to absorb different energies for the transitions among the different levels, i.e. absorb energies at different wavelengths (called absorption lines) and these energies are used as fingerprints to identify the different atoms. The total amount of radiation absorbed depends (among other factors) on how many atoms there are to absorb it.

In atomic absorption spectrometry the energy supplied for the transitions is radiation from a specially built light source, emitting radiation with wavelengths that the analyzed atom can absorb. The analyzed solution is aspirated into the spectrometer, turned into an aerosol and passed into a flame where the sample is dissociated into ground-state atoms. Radiation from the light source, at a wavelength that can excite the analyzed atom, is passed through the flame where it is absorbed by the analyte atoms. The radiation is measured before absorption (incident radiation) and after absorption (transmitted radiation) and the logarithm of the ratio is proportional to the concentration of the analyte.

The **Beer–Lambert law** mathematically describes the absorbance of light passing through a sample solution (in atomic absorption – the population of atoms in the flame) as a function of the length of the optical path l through the sample (length of the flame) and the concentration c of the absorbing species (ground-state atoms):

$$I_t = I_0 e^{-\varepsilon lc}$$

$$A = \log(I_0/I_t) = \varepsilon lc$$

where A is the absorbance (also called optical density), I_0 the incident radiation power, I_t the transmitted radiation power and ε the absorptivity of the analyzed atom (absorption coefficient at the wavelength used for the analysis). Actually atomic absorption spectrometers read the amount of light transmitted through a blank solution (containing all the components of the unknown solution except for the analyzed atom) and subtract it from the amount of light transmitted through the sample (a known standard or an unknown sample) and the result (I_t, the net amount absorbed by the analyte atoms) is shown in absorbance units.

This equation means that when analyzing the same type of atom (for instance Cu) in samples of unknown concentrations and in standard solutions of known concentrations, where ε and l remain the same, the absorbance will be a linear function of the analyzed atom's concentration. It is not always linear throughout all the concentrations range, and analytical work is essentially limited to the range where the relationship does not differ much from linearity.

There is no need to know the values of ε and l as atomic absorption is a **comparative technique**. One measures a set of standards of known concentrations of the analyzed atoms, prepares an absorbance vs. concentration line – a calibration graph (in modern spectrometers this is done by the instrument's computer) – and for each absorbance reading of samples with unknown concentrations finds the respective concentration (again, done by the computer).

Contrary to emission spectrographs or ICP spectrometers, which are simultaneous, i.e. can measure the signals of many elements at the same time, atomic absorption spectrometers are sequential, measuring one element at a time. One has to measure for instance copper in all the samples then nickel in all the samples, etc., otherwise one has to change the wavelength and recalibrate separately for each element in each sample. This sequential system makes the analyses much longer (operator time).

7.3 Major components and instrument types

All atomic absorption spectrometers include some major basic components (whose shape, mode of operation, efficiency, etc., may change

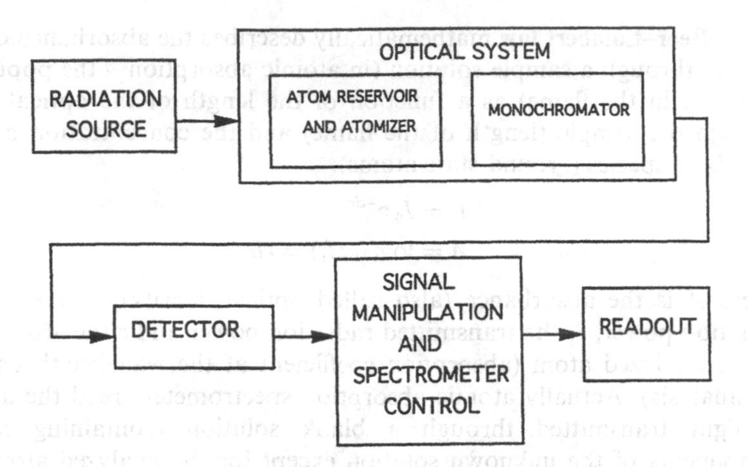

Figure 7.1 A schematic diagram of the main components of an atomic absorption spectrometer.

from one manufacturer to another). The main components are shown in Figure 7.1.

1. *Radiation source.* Usually a hollow cathode lamp, which emits light of the specific wavelengths needed to excite the analyzed atoms.
2. *Optical system.* To direct the light from the radiation source through the ground-state atoms in the atomizer to the monochromator and then to the detector.
3. *Ground-state atom reservoir and atomizer.* There are two types: (a) a system for flame atomization which includes a nebulizer that aspirates the analyzed solution and passes it to the spray chamber where it is turned into a fine aerosol and passed on to the burner where the solvent is evaporated and the analyzed compound is dissociated into ground-state atoms; and (b) a system for electrothermal atomization (known also as graphite furnace, graphite tube atomizer, carbon rod atomizer or flameless atomization) which includes a specially coated graphite tube (or cup) that acts as a sample holder and is heated by an electric current passing through two carbon electrodes, thus evaporating the solvent and dissociating the analyzed compound into ground-state atoms.
4. *Monochromator.* A grating that selects the desired wavelength out of all the wavelengths coming from the radiation source and passes it on to the detector.
5. *Detector.* A photomultiplier that translates the intensity of the light from the hollow cathode lamp, before and after absorption by the analyzed atoms, into an electric current and amplifies the current.
6. *Signal manipulation device.* In modern instruments this is a computer

that translates the response of the detector into analytical results, and calculates concentrations (compared to a calibration curve) and standard deviations, etc. (in addition to acting as a system control – selecting the desired wavelength, controlling fuel-oxidant ratios and flow rates, etc.; see section 7.9.2).

7. *Readout device.* Older instruments used meters, recorders or digital readout devices. In modern instruments the computer's screen and a printer are used.

Generally atomic absorption spectrometers are divided into two types, i.e. single-beam and double-beam instruments (each containing all the major components mentioned above).

In a **single-beam instrument** there is only one beam of light, which passes from the source into the optical system and flame and then to the photomultiplier. Any change in the light source intensity or in flame characteristics will influence the final signal.

In a **double-beam instrument**, the light from the source passes through a beam splitter and is split into two beams of about equal intensity. One passes through the flame etc. (I_t) as in a single-beam instrument, and the other passes along the same optical system etc. but not through the flame (i.e. it is not influenced by the analyte) and serves as a reference beam (I_0 in the Beer–Lambert equation). The intensity of each beam is read intermittently by the photomultiplier and their ratio is amplified and displayed. Thus, any change in the light source intensity, the detector sensitivity or the electronic amplification will affect both beams and the effect will be canceled.

The measurements of a double-beam instrument are much more stable – less noise (a real benefit to the operator's nerves) – the result of which is that higher amplifications are possible so that lower concentrations can be measured. On the other hand, this is not a real double-beam compensation as the reference beam does not pass through the flame and changes in the flame (for instance absorption by molecules formed from the solvent at high temperatures) or momentary changes in the nebulizer uptake will not be compensated for. In addition, the double-beam system means more parts (a beam splitter, mirrors, half-silvered mirrors, etc.) and a bigger, complicated instrument, hence its high price compared to that of a single-beam instrument (but if one can afford it, it is more than worth the difference).

One system of a beam splitter is a rotating disk with alternate quadrants removed. The spaces (missing quadrants) pass the radiation into the sample path. The remaining quadrants, which have a mirrored surface, reflect the light into the reference path, where through an array of mirrors the beam avoids the flame and then a half-silvered mirror recombines it into the optical path of the sample beam. The action of the beam splitter also modulates the beams into intermittent beams, so that only they will be read and not the continuous emission from the flame (see section 7.4.1). In another type of beam splitter system the splitter consists of a partially aluminized quartz plate. A part of the light is reflected to the flame and the other part passes through the splitter, through an array of mirrors, directly to the monochromator. A rotating reflecting chopper is located in front of the monochromator entrance slit, where it alternately interrupts the sample beam to direct the reference beam into the monochromator. In this way a two-beam system with modulation is achieved.

In addition to differences in the components, instruments of different manufacturers differ in software – the amount of automation, special 'tricks', data handling and report managing.

7.3.1 Radiation sources

Atomic absorption lines are very narrow. If a relatively wide bandwidth reaches the reading device (photomultiplier), the narrow absorption line will only be a small fraction of it, resulting in a reduced sensitivity and Beer's law will not be obeyed. This will result in a reduced correlation with concentration, i.e. a non-linear calibration curve. In order to measure the absorption under conditions that will guarantee a high sensitivity, radiation with a very narrow wavelength band, at the absorption peak, is needed.

The best source for a single sharp wavelength would be a laser source for all wavelengths, a source that does not exist, so no commercial instrument is built with a laser.

The simplest way would seem to be the isolation of one wavelength from a source emitting light over a wide wavelength range (a continuum source) by using a monochromator. The problem is that the width of the band separated by the monochromator is wider than the width of the atomic absorption peaks, so the light source must be a line source, but usually it emits more than one line. A monochromator and a slit are therefore used, in addition to the line source, in order to select the needed wavelength region from all the close wavelengths that the radiation line source and the flame emit (Figure 7.2).

The most commonly used source is the hollow-cathode discharge tube better known as the **hollow-cathode lamp** (HCL). The emission from this line source has a bandwidth narrower than the matching absorption peak. The HCL is specific to the metal to be analyzed (i.e. a Fe lamp for analyzing iron, a Ni lamp for analyzing nickel, etc.). For convenience and in order to save money, some cathodes are built of several elements (usually two or three), a possibility only where the lines used for analysis do not overlap. These multi-element lamps may result in the need for a better monochromator (more expensive) and may shorten the lamp's life.

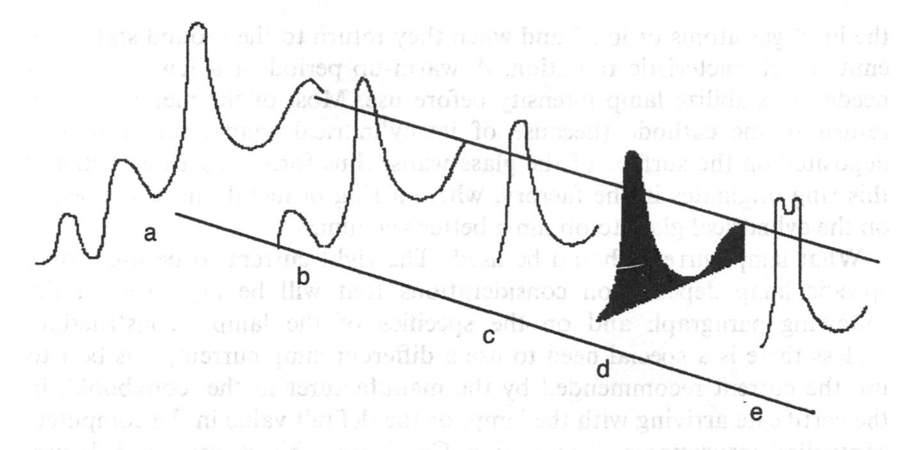

Figure 7.2 A schematic diagram showing wavelength selection: (a) wavelengths emitted by a hollow-cathode lamp; (b) wavelengths passed by a monochromator; (c) a narrow emission line; (d) an absorption line at the same wavelength as (c) is wider; and (e) self-absorption (self-reversal).

Nevertheless, some multi-element lamps are very common (such as Na–K, Ca–Mg, etc.).

The HCL is built of a sealed glass cylinder, one end of which is a quartz window and to the other end an anode and a cathode are sealed (Figure 7.3). The tube is filled with an inert gas (argon or neon) at a pressure of a few torr. The anode is made of tungsten. The cathode is a narrow hollow cylinder made of, or lined with, the metal (or alloy) of the element(s) to be analyzed and whose spectrum the lamp is to emit.

A potential difference is applied between the electrodes causing a current between them. Electrons leave the cathode, collide with atoms of the inert gas and ionize them. These positive ions hit the cathode and as they have enough kinetic energy they eject atoms from the surface (sputtering). Some of these metallic atoms are in excited states (probably by collisions with

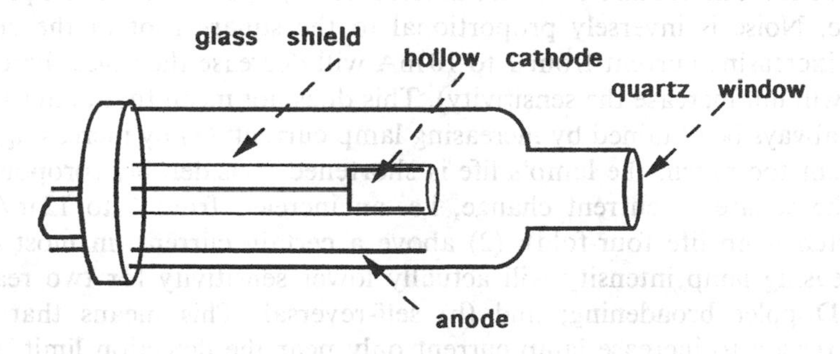

Figure 7.3 A schematic diagram of a hollow-cathode lamp.

the inert gas atoms or ions) and when they return to the ground state they emit the characteristic radiation. A warm-up period of a few minutes is needed to stabilize lamp intensity before use. Most of the metallic atoms return to the cathode (because of its cylindrical shape) but a part is deposited on the surface of the glass walls, thus forming a ring. Actually this ring originates in the factory, where a ring of metal film is deposited on the cylindrical glass to obtain a better vacuum.

What lamp current should be used? The right current to be used for a specific lamp depends on considerations that will be explained in the following paragraph and on the specifics of the lamp's construction. Unless there is a special need to use a different lamp current, it is best to use the current recommended by the manufacturer in the 'cookbook', in the certificate arriving with the lamp, or the default value in the computer-controlled instruments. For Varian Co. lamps this is usually 5–10 mA and for Perkin-Elmer Corp. lamps 10–20 mA.

Experiment 7.1

The change of Cu^{2+} absorbance as a function of the hollow-cathode lamp's current. An example of the expected results is shown in Figure 7.4.

Prepare a solution containing 5 ppm Cu^{2+}. Measure the absorbance at wavelength $\lambda = 324.8$ nm in an air–acetylene flame. Measure the absorbance with the following hollow-cathode lamp currents: 2, 3, 4, 6 and 9 mA. Make sure you do not use a current higher than that allowed for your lamp. At each current wait for stabilization of the lamp's emission, zero the instrument's reading, and read the absorbance of a blank solution and then that of the Cu^{2+} solution. Plot the absorbance as a function of the current.

There are several factors to be considered when deciding what lamp current to use. If the lamp current is too small, then the analytical signal will be low and require excessive amplification, which also will amplify the noise. Noise is inversely proportional to the square root of the current (i.e. increasing current from 1 to 10 mA will decrease the noise three-fold, but will not increase the sensitivity). This does not mean that better results will always be obtained by increasing lamp current: (1) by increasing lamp current too much, the lamp's life is shortened considerably (proportional to the square of current change, i.e. an increase from 6 to 12 mA will shorten lamp life four-fold); (2) above a certain current, in most cases, increasing lamp intensity will actually lower sensitivity for two reasons: (a) Doppler broadening; and (b) self-reversal. This means that there is a reason to increase lamp current only near the detection limit, where noise may be too high.

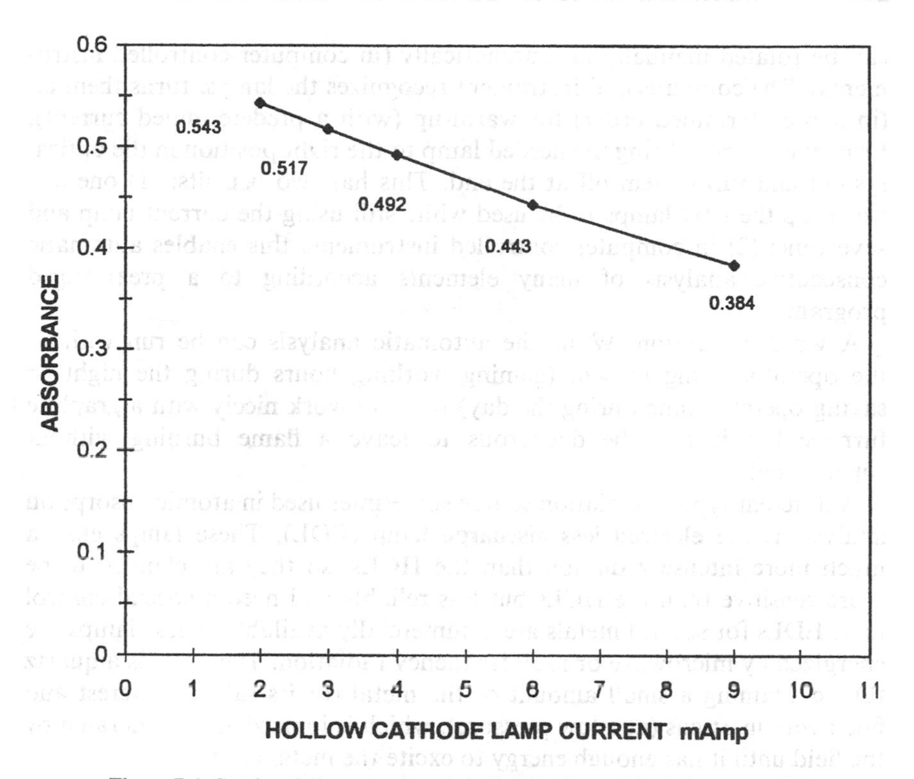

Figure 7.4 Cu absorbance as a function of the hollow-cathode lamp current.

Doppler broadening of emission line width with increasing currents happens because atoms emitting the radiation in the lamp move randomly relative to the observer (in this case the monochromator). The apparent wavelength emitted by the atoms will be higher or lower according to the direction (Doppler effect) and, instead of a narrow line, a wider band will be obtained.

Self-reversal (or self-absorption) happens because higher current will form more ground-state atoms and these will absorb a part of the radiation from the emitting atoms. The absorption is greater at the wavelengths in the center of the broadened emission line (which, as a result, will look like a saddle – see Figure 7.2) and hence a lowering of the intensity occurs. The amount of this effect depends on the element and the lamp's construction.

Modern instruments can use coded lamps, i.e. the computer recognizes the lamp (by an array of pins near the electrodes at the bottom of the lamp) and applies the right current, without the operator having to supply the data.

Modern instruments have a turret for four or eight lamps. The turret can be rotated manually or automatically (in computer-controlled instruments). The computerized instrument recognizes the lamps, turns them on (in a predetermined order) for warm-up (with a predetermined current), turns the turret to bring the needed lamp to the right position in the optical system and turns them off at the end. This has two benefits: (1) one can warm up the next lamps to be used while still using the current lamp and save time; (2) in computer-controlled instruments this enables automatic consecutive analysis of many elements according to a prearranged program.

A word of caution. While the automatic analysis can be run without the operator being present (gaining working hours during the night or saving operator time during the day) this can work nicely with a graphite furnace but it may be dangerous to leave a flame burning without supervision!

A different type of radiation source sometimes used in atomic absorption analysis is the **electrodeless discharge lamp** (EDL). These lamps emit a much more intense radiation than the HCLs, so they are claimed to be more sensitive than the HCLs but less reliable and need a special control unit. EDLs for several metals are commercially available. These lamps are energized by microwave or radiofrequency radiation. The lamp is a quartz tube containing a small amount of the metal (or its salt) of interest and filled with inert gas (at a low pressure), which is ionized and accelerated by the field until it has enough energy to excite the metal atoms.

A different type of lamp used in atomic absorption analysis is the **deuterium lamp**, which emits continuous radiation in the ultraviolet region. This lamp is used for correction of background caused by species other than the analyzed element (see sections 7.6 and 7.7). As the output of radiation in the visible range is low, this lamp is used for elements whose analytical lines are below 350 nm.

The construction of a deuterium lamp is different from that of the HCL. It is filled with deuterium, has a metal anode, the cathode is an electron-emitting thermionic cathode and there is a restrictive aperture between them, causing an area of high excitation and high light emission. This lamp is used simultaneously with the element's HCL. In former instruments a very careful tedious alignment of the two was needed, otherwise an erroneous correction resulted as the gaseous system (flame or arc) is very inhomogeneous. In modern instruments this alignment is very simple and in computer-controlled instruments the current for the deuterium lamp and the ratio of intensities of the two lamps are automatically controlled.

7.3.2 Wavelength selection system

As described above, the first step of isolating a specific wavelength is taken by using a line source, i.e. a hollow-cathode lamp emitting only a small number of wavelengths.

An additional wavelength selection system (a **monochromator**) is needed to isolate a very narrow emission line from all the emitted wavelengths and produce a beam of radiation of high spectral purity, the wavelength of which can be varied at will, over a wide range. The quality of a monochromator is measured by the purity of the radiation it passes, in other words, by the ability to resolve adjacent wavelengths and by the intensity of the radiation passed.

The main components of such a system are: an entrance slit, a collimating mirror (causing the beam to travel as parallel rays), a dispersive device (separating the polychromatic light into its monochromatic components), a focusing mirror and an exit slit.

There are several types of wavelength selection system arrangements (referred to in the literature as mounts): the Littrow, Ebert and Czerny–Turner. They differ from one another in compactness, number of mirrors used, the use of prisms and focus arrangements, etc.

A slit consists of two exactly parallel metal jaws, carefully machined to have very sharp edges. In some instruments there are several slits with a fixed aperture and the operator has to use the one that is best for the analyzed element. In most instruments the distance between the jaws can be varied as needed. In the common type of slits both jaws move relative to each other and the center remains fixed. The actual distance between the jaws is called the mechanical slit width. Another term used is the spectral slit width (or spectral bandwidth), which is the range of wavelengths (polychromatic light) that passes through the slit for a given monochromator setting.

The entrance slit acts virtually as a light source to the system (passing the light arriving from the hollow-cathode lamp). A lens or a mirror (which is usually used in commercial instruments) is placed after the slit in such a way that its distance from the entrance slit equals its focal length, so that it will collimate the light arriving from the slit, i.e. pass it on as parallel rays. A second lens or mirror is placed before the exit slit in such a way that its distance from the slit equals its focal length, so that the parallel rays arriving from the first mirror will be focused at the focal plane of the second mirror, i.e. at the surface containing the exit slit. The net result is that the narrow rectangular image of the first slit – a bright line, hence the term emission line – will be formed on the surface containing the second slit. Usually the exit slit is adjusted to the dimensions of the entrance slit and the system is arranged in such a way that the image will be formed on the exit slit and the light passed from the slit to the detector. The dispersive unit is located between these two slit–mirror systems. The dispersive unit resolves the light into separate wavelengths so that several rectangular images (spectral lines) of the entrance slit – one for each wavelength – will be formed on the surface that contains the exit slit. By rotating the dispersive unit, the image formed by the desired wavelength is brought to overlap the exit slit and will be passed on to the detector. In computerized instruments the rotation is done automatically so that the prescribed wavelength reaches the slit. After that, a careful scan (called peaking), in small increments, is carried out to locate the exact point of the emission line's peak.

The dispersive unit was a prism in former instruments and in modern instruments it is a **grating**. One of the reasons a grating is preferred to a prism is that the spectrum produced by a grating is linear (it has an equal length for the same range of wavelengths, i.e. an equal length for a 100 nm range throughout the whole UV and visible spectrum) while that of a prism is not. As a result, in order to obtain a linear change of wavelengths reaching the detector, a grating has to be rotated in a linear way while a prism should be rotated in a non-linear way, which is much more complicated and expensive.

If the monochromator passes the desired wavelength λ the image should exactly fill the exit slit. The optimal width of the slit (for the desired wavelength of each element) will be determined by the existence of close lines and their distance from the desired line, and the choice of the slit width is a compromise. If the slit is wide, part of the adjacent images will also pass the slit, but a wide slit is needed to increase the light throughput to the monochromator, to increase signal-to-noise ratio. On the other hand, good separation of wavelengths will be obtained with narrow slits. If the separation is not good the calibration line will be curved (see explanation in section 7.3.1). In the 'cookbooks' the width for each useful wavelength is recommended, usually between 0.1 and 1.0 nm. In computerized instruments a default value of a recommended slit width is automatically used (it can be changed by the operator if necessary and should be changed if a good reason exists).

If monochromatic light passes through a very narrow slit, the slit acts as a light source and the light, after passing through the slit, instead of forming the image of the slit, will form a diffraction pattern, which consists of a central bright band bordered by alternating dark and bright bands of decreasing intensity and quite diffuse borders. If many close, very narrow slits are used, interference occurs and the central and subsidiary bright bands are split into several sharp bright lines with most of the intensity in the central lines. The greater the number of slits, the sharper, narrower and brighter are the illuminated lines.

A grating is essentially an array of a great number of slits. The first gratings consisted of fine wires stretched across a frame, acting as a series of slits. Today a grating is a series of parallel equidistant straight lines (grooves) cut into a plane surface (coated with highly reflective aluminum), each of which acts as a slit – a light source. For the UV and visible range there are usually 1200–2000 lines/mm, but the number may be lower or higher. Each groove in the grating gives rise to a diffracted beam and these diffracted beams then interfere with one another to produce the final pattern. There are two types of gratings: transmission grating (where light falls on one side of the grating and is emitted as separate wavelengths on the other side) and reflection grating (where the lines are actually ruled on a mirror and incidence and dispersion occur at the same side of the grating). In modern instruments reflection gratings are used.

Master gratings are prepared by ruling the lines on a hard polished surface with the aid of a properly shaped diamond tool (having elaborate controls to ensure the necessary precision). From this master, replica gratings are prepared, by applying a layer of parting agent to the master, a layer on which a film of aluminum is deposited. To this film a glass or quartz base is attached with a suitable cement and the replica is separated from the master.

In modern instruments the so-called holographic grating is used. The grating is obtained by using laser technology to form the grooved surface. Much better line shapes and line dimensions, a much higher groove density and bigger gratings can be formed, thus causing purer spectral lines, at a relatively low cost.

The dependence of grating dispersion on interference means that there are angles in which the diffracted rays are destroyed (dark lines) and angles in which the rays are reinforced. The condition for reinforcement is described by the **grating equation**

$$n\lambda = d(\sin i \pm \sin \theta)$$

where λ is the wavelength of the radiation, d the distance between grooves, which is the reciprocal of the number of grooves per unit distance, i the angle between the normal to the surface and the line of incidence of the light beam, θ the angle between the normal to the surface and the line of dispersion of wavelength λ, and n an integer number called the order. Incident light may be diffracted on either side of the normal. When i and θ are on the same side of the normal the + sign applies and when they are on opposite sides the − sign applies.

Zero order, $n = 0$, corresponds to the direct reflection ray (reflection from a mirror) which will occur at an angle of $-i$, i.e. an angle equal to i on the other side of the normal to the surface. It does not depend on the wavelength and all wavelengths in the incident beam will be included in the zero-order ray.

For different wavelengths, the angle of dispersion θ is different, i.e. each wavelength will be dispersed at a different angle. This means that a beam of polychromatic light will be dispersed into monochromatic components. A beam of polychromatic light falling upon the grating will be dispersed into a series of spectra located symmetrically on each side of the zero-order position; the closest spectrum is the first order, the next the second order, etc. Each order consists of a whole spectrum, in which the shorter wavelengths are at the smaller angles.

Resolution is the smallest wavelength interval that a grating is capable of isolating. **Resolving power** is a measure of the ability to separate one wavelength from another, nearly identical, wavelength, or the ability to distinguish between two adjacent slit images having a small wavelength difference. Many books do not distinguish between these two terms. The resolving power R of a grating can be determined in two ways:

1. By the equation $R = nN$, where n is the order and N is the total number of grooves. It is not the density of lines that determines R but their total number, i.e. a 5 cm, 1200 lines/cm grating will have in the third order ($R = 18\,000$) the same resolving power as a 10 cm, 1800 lines/cm grating in the first order. One should pay attention that it is not the total numbers of grooves on the grating that counts but the total number of grooves *upon which light falls* by the optic system, a difference that may give room to false claims of high R. The dependence upon N enables working with bigger, coarse gratings having a lower groove density (easier to manufacture and cheaper).
2. By the equation $R = \bar{\lambda}/\Delta\lambda$, where $\bar{\lambda}$ is the mean wavelength of the two lines to be resolved and $\Delta\lambda$ is the difference in wavelength of the two lines that can just be distinguished as two lines.

These two equations enable calculation of the resolving power and hence the combination of order and number of grooves needed to separate two lines, or to calculate if a given n and R can resolve two given wavelengths. For instance, to just separate the sodium doublet lines 5890 and 5896 Å, the resolving power needed is $5893/6 = 982$.

Dispersion is the separation of polychromatic light into monochromatic components. The effectiveness can be described in different terms: (1) angular dispersion $d\theta/d\lambda$ (radians/Å), the change in angle of dispersion by the grating θ per unit wavelength – the greater it is, the better the separation; (2) linear dispersion $dl/d\lambda$ (mm/Å), the lateral separation of the slit images per unit wavelength – the greater it is, the better the separation; (3) reciprocal linear dispersion $d\lambda/dl$ (Å/mm) – the smaller it is (less images/mm), the better the separation.

The grating equation shows that there will be overlapping lines; for instance, wavelengths 1000 Å of the third order, 1500 Å of the second order and 3000 Å of the first order will be dispersed at the same angle, as they have the same value of $n\lambda$. If for resolving power considerations one has to use third-order 1000 Å, one should take care to eliminate first-order 3000 Å, second-order 1500 Å, fourth-order 750 Å, etc. This can be accomplished by putting in front of the grating a filter that will pass on to the grating only wavelengths close to the desired one, or a prism (called an order sorter) that will pass on only the desired (or a part of it) order.

One of the measures of a monochromator's quality is the energy throughput or the amount of radiation passed on to the detector. In the same grating, the higher the order, the better the dispersion. On the other hand, the higher the order, the lower the intensity of the radiation (with the greatest intensity being in the first order). This can be changed by changing the shape of the grooves in such a way that the intensity is concentrated mainly around a predetermined angle. Instead of the grooves being symmetric, they are grooved (the common term is blazed) in such a way that they have the form of steps having broad faces from which reflection occurs and narrow faces that are unused. There is an angle β between the broad faces and the surface of the grating, or, in other words, between the normal to the surface and the normal to the broad face. This type of grating is called an echelette grating and β is the **blaze angle**. This angle will determine where the intensity will be concentrated. The diffracted energy will be concentrated around the diffraction angle for which the ordinary law of reflection from the individual faces is most nearly obeyed. By optimization of the blaze angle as much as 90% of the available energy of a particular wavelength can be made to fall in a single order. The monochromator can be blazed for two different wavelengths to provide high energy at both the UV and visible ranges.

7.4 Atomization

7.4.1 *Flame atomization*

The purpose of the atomization system is to generate uncombined analyte atoms, which in turn will be excited by radiation having the characteristic wavelength λ. The atomization system is very important in determining the sensitivity, detection limit, precision and accuracy of the determinations. There are momentary variations in the excitation system so that conditions are not exactly the same during the analysis, and the stability of the excitation system is one of the major parameters determining the quality of results. At the beginning of the analysis several adjustments should be done (see the paragraphs just before Experiment 7.2) to raise sensitivity, but the aim is not to obtain maximum signal or maximum sensitivity as such, but to obtain the highest possible sensitivity while remaining under stable conditions (and acceptable signal-to-noise ratio).

One of the methods of atomization used in atomic absorption analysis is flame atomization (the other being electrothermal atomization). The system (Figure 7.5) includes a nebulizer, a spray chamber (with a liquid trap), a burner and a flame.

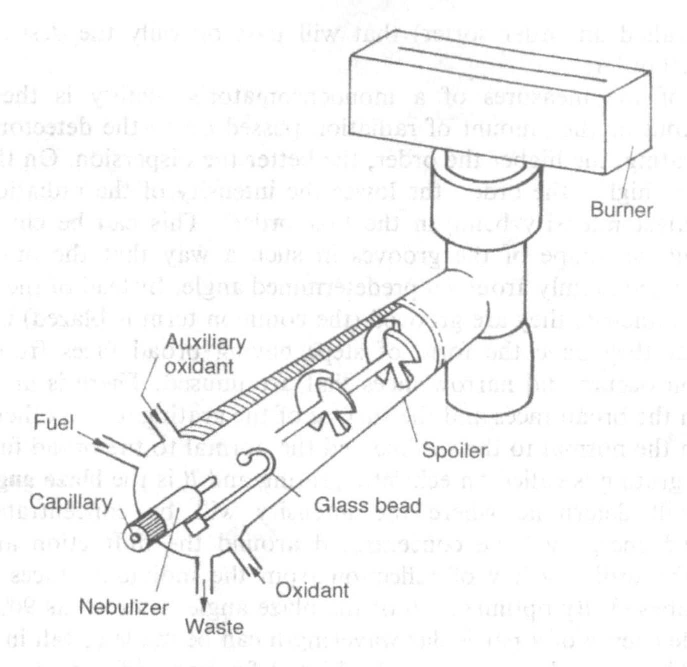

Figure 7.5 A schematic diagram of a nebulizer–spray chamber–burner system (with a flow spoiler and a glass bead).

The nebulizer sucks up the sample solution into the spray chamber where it is broken into a very fine aerosol, which is passed on to the flame. There it is heated to a temperature high enough to evaporate the solvent and then to dissociate the compound into its constituent atoms. The latter are then excited by the hollow-cathode lamp radiation and the absorbance is measured. A delay time (aspirating the solution to the flame without reading the absorbance) of a few seconds before measuring each sample is necessary to clean the system from the remains of the former solution and to stabilize the reading. The instrument takes a great number of readings per second. Measurements a few (usually 3–5) seconds long are usually taken to obtain the mean and the relative standard deviation of the sample's concentration.

Ordinary **nebulizers** are fixed nebulizers (fixed solution uptake rate) but there are also adjustable nebulizers (adjustable uptake rate). They are recommended for work with organic solvents, which act both as solvent and as fuel. In a fixed nebulizer, there will be excess of fuel during aspiration of an organic solvent and the acetylene flow rate has to be lowered. The result is that when aspiration stops there will be too little fuel and the flame may go out. The adjustable nebulizer solves this problem.

The oxidant flows through the nebulizer and draws solution (from the sample vessel, through a plastic capillary tube) as a mist of fine droplets (using the venturi effect, in a similar way to perfume sprays), which is then mixed with the incoming fuel and passed on to the burner. There is an optimum size for an analyte droplet which will allow it to be evaporated completely and atomized during the short time the droplet is in the flame. This is considered to be in the range of 2–10 μm. Small droplets will be carried to the flame while the big drops will be drained away to a liquid trap. The larger the part of the solution used, the higher will be the sensitivity and the detection limit will be better. The amount utilized will depend on the distribution of drop size, i.e. the percentage of droplets having the right size. The aim is to increase the percentage of droplets having the right size, and this is done by breaking the bigger droplets into smaller ones. In some instruments the spray from the nebulizer hits flow spoilers (also called baffles) in the form of fixed propellers in the **spray chamber**, and is broken into smaller droplets. A better way seems to be the glass bead (or impact bead) used in other instruments. This is a bent thin glass rod a few centimeters long, with a bead at its end (with a special coating), in the spray chamber opposite the nebulizer. The spray hits the bead and is shattered into fine droplets. The absorbance depends critically on the distance of the bead from the nebulizer venturi (reaches a maximum and then drops) and the optimum distance depends upon the physical properties of the solution. The distance of the glass bead from the nebulizer has to be adjusted at the beginning of the analysis, while aspirating a standard solution of the analyzed element, until maximum stable signal is obtained. Nowadays there are instruments that use both types of spoilers. Though the amount of droplets of the right size in the spray can almost be doubled, only about 10% of the solution is utilized while the rest is drained. There used to be total consumption burners in which most of the sample was passed to the flame. However, the energy in the flame was not high enough to evaporate and dissociate this amount of solution during the short residence time in the flame, and this resulted in low sensitivity and many interferences. In the modern pneumatic nebulizers only about 10% of the solution reaches the flame, but this is done under conditions that favor stability

Burners are made of special metals that can resist corrosion by acidic or basic solutions reaching the burner at the high operating temperatures. Some instruments used to have a plastic material film coating the inner part of the burner, to avoid corrosion, but no longer, as it was found to cause a memory effect. Solution droplets adhered to the film and later renebulized and passed on to the flame while the next solution was aspirated.

There are special burners for air–acetylene and for nitrous oxide–acetylene flames, and they should not be mixed as explosion may occur. The burner's design depends on parameters such as gas flow rates, burning rate, fuel–oxidant type, efficient heat dissipation so that the burner will not become red hot (some burners have cooling fins), etc. The burner's slot dimensions depend on the fuel–oxidant for which it is intended. The burning velocity is the speed of propagation of the flame front in the gas mixture. A certain correlation must be maintained between burning velocity and the combustible mixture flow velocity. Higher flow velocities can cause the flame front to blow away, while lower ones can cause a flashback (the flame propagates within the burner), which can lead to explosion.

Several gas mixtures have been used as oxidants and fuels. Because of safety, cost, usefulness and convenience, the main gas combinations used in modern commercial instruments are as follows:

1. *Air–acetylene.* Compressed air is used (from a laboratory central supply, a commercial cylinder or a small laboratory compressor). This is a relatively hot flame (about 2300°C) that causes partial ionization of alkali and alkaline-earth elements. Used for most elements.
2. *Nitrous oxide–acetylene.* The hottest flame used (about 3000°C). Used for refractory elements and to reduce chemical interferences. May cause severe ionization, and an ionization suppressor may have to be added.

In addition, the following gas combinations have been used in the past but are rarely used today.

3. *Air–propane.* A relatively low temperature. Was used mainly for alkali metals, which can be ionized in hotter flames. Severe chemical interferences may occur and noise level is relatively high. Used today only for flame emission analysis (flame photometer) of the alkali metals.
4. *Air–hydrogen.* A relatively low temperature. A reducing medium fit for arsenic, selenium and tin, for which this flame is mainly used. Severe chemical interference may occur.

The quality of the gases to be used depends upon the purity of the gases available locally. In some places commercial acetylene is good enough and in some places extra-pure acetylene should be used. The compressed air may contain water and oil and the system should contain a trap to remove them from the gas. A compressor that does not emit oil vapors should be used.

There are three types of **flames** whose characteristics are determined by fuel–oxidant ratios: (1) fuel-lean or oxidizing flame (the hottest); (2) fuel-rich or reducing flame (the coolest); and (3) chemically balanced or stoichiometric flame (about midway between the first two). These flames can be identified by the height and color of the inner part of the flame (not a cone as in ordinary laboratory burners but a long inner part parallel to the length of the burner).

The properties of the analyzed atom, its molecular association and the final products it may form in the flame will determine what type of flame has to be used (recommended in the 'cookbook' or as default parameters in computerized instruments). There are several types of atoms according to their behavior in a flame:

1. Easily atomized atoms (such as potassium, sodium, lead, etc.) that can be analyzed in air–acetylene flame (provided the ionization problem is taken care of).
2. Atoms that can be determined in both air–acetylene and nitrous oxide–acetylene flames, but not as effectively. For instance, if a solution of calcium nitrate is analysed, it will decompose in both flames into CaO, but a much higher proportion of this oxide will dissociate into ground-state atoms in the nitrous oxide flame and a higher signal will be obtained. The analysis can be carried out also in air–acetylene flame but many interferences may occur and the fuel–oxidant ratio is important.
3. Refractory elements that almost will not be atomized by a cool flame and have to be atomized in a nitrous oxide flame.
4. Elements for which not only temperature but also the chemical nature of the flame (oxidizing or reducing) determines the dissociation, so that a specific fuel–oxidant ratio is needed. For instance, Mo in a stoichiometric air–acetylene flame will remain as an oxide but in a fuel-rich flame, in spite of it being cooler, it will be reduced to the free atom. Similarly, a strongly reducing nitrous oxide–acetylene flame is needed to effectively atomize silicon.
5. Atoms that react with other species in the sample. For instance, calcium in the presence of phosphate will form a compound that is difficult to break down. Calcium chloride is volatilized and dissociated easily. Two solutions with equal concentrations of calcium will absorb different amounts of radiation, depending on the anion. Changing the flame characteristics or adding a releasing agent may solve the problem. For instance lanthanum, when added in high concentrations, will bind the phosphate, thus releasing the calcium (see section 7.6).

For atomic absorption, ground-state atoms that can be excited by radiation are needed. A high enough temperature is needed for vaporization and dissociation but, if the flame temperature is too high, the atoms may receive enough energy for ionization to occur (which, depending upon the atom, temperature and concentration, may reach the range of 80%). This may be a serious problem for elements with low ionization energies. The electron will leave the atom completely and will not take part in excitation by the radiation process. The usual remedy is the addition (to samples and standards) of high concentrations of an ionization suppressor, a salt of an easily ionized element (such as K), which will saturate the flame with electrons and inhibit ionization (see section 7.6).

Each element has an optimal absorption as a function of location in the flame (for instance, absorption increasing with height, decreasing with height or increasing to a maximum and then decreasing with height). This is caused by two opposing effects. Raising the temperature should (due to increased evaporation and dissociation) increase the total number of atoms in the flame and hence the signal. On the other hand, the intensity may be lowered because of increased excitation (by flame and not by the hollow-cathode lamp), ionization and Doppler broadening. Maximum signal will be obtained if the light beam from the HCL passes into the flame through the maximum population zone, which is not identical for all elements. For any given combination of elements, matrix, sample aspiration rate, flame type and burner, the population density of absorbing atoms will vary throughout the flame in a manner characteristic of that combination.

Changing fuel–oxidant ratios will also change the flame temperature profile (hence the location of maximum population) and characteristics (oxidizing, reducing or stoichiometric). In computerized instruments there are (for each element) default values for flow rates of air, acetylene and N_2O, so as to obtain a high sensitivity, and the right amounts are delivered by a control unit. One can change these values when needed, for example to compensate for matrix effects, simply by typing in the new values to be used.

Changes in sample flow rate change the number of atoms in the flame and can change the sensitivity. If more solution reaches the flame there will be more dissociated atoms, but the additional solution is not effectively evaporated and atomized because the solution cools the flame, thus lowering the signal. The system tends to be saturated and the curve describing absorption as a function of intake rate rises and then flattens at rates higher than about 6 mL/min.

There are several **adjustments** that have to be made at the beginning of each analysis, while aspirating a standard solution of the analyzed element. At the beginning of the analysis of each element the height of the burner relative to the light beam has to be adjusted (care should be taken not to raise the burner into the light beam) until maximum stable signal is obtained. The burner position has to be adjusted horizontally (backward and forward) to align the center of the burner slot with respect to the light beam, until maximum stable signal is obtained. The burner position has to be adjusted rotationally, to align the length of the burner with the light beam (so that the whole length of the flame will be utilized) until maximum stable signal is obtained. At the beginning of the analysis of each element the nebulizer has to be adjusted (while aspirating a standard solution of the analyzed element) to an optimum flow rate until maximum stable signal is obtained. At the beginning of the analysis of each element the fuel–oxidant ratio has to be adjusted (while aspirating a standard solution of the analyzed element) so as to obtain maximum stable signal.

For each element there are several optimal working ranges (a linear calibration curve) at different wavelengths. For instance, for Cu: 2–8 µg/mL at 324.7 nm, 6–24 µg/mL at 327.4 nm, 70–280 µg/mL at 222.6 nm, etc. One may have to dilute the element's solutions to obtain concentrations in the working range to be used.

Experiment 7.2
The curvature of a calibration curve. An example of the expected results is shown in Figure 7.6.
Prepare solutions containing 2, 5, 8, 12 and 15 µg/mL Cu^{2+}. Measure the absorbance at wavelength $\lambda = 324.8$ nm in an air–acetylene flame. Plot the absorbance as a function of Cu^{2+} concentration and note the deviation from a straight line. From the plot find the deviation from the characteristic concentration given in the 'cookbook' of your spectrometer. From the linear part of the curve calculate the sensitivity (absorbance per unit concentration) and from that calculate the characteristic concentration.

In many cases, when the concentration is higher than the optimum range, it is worth while (instead of diluting the samples) to rotate the burner at a certain angle. The distance that the light beam travels in the

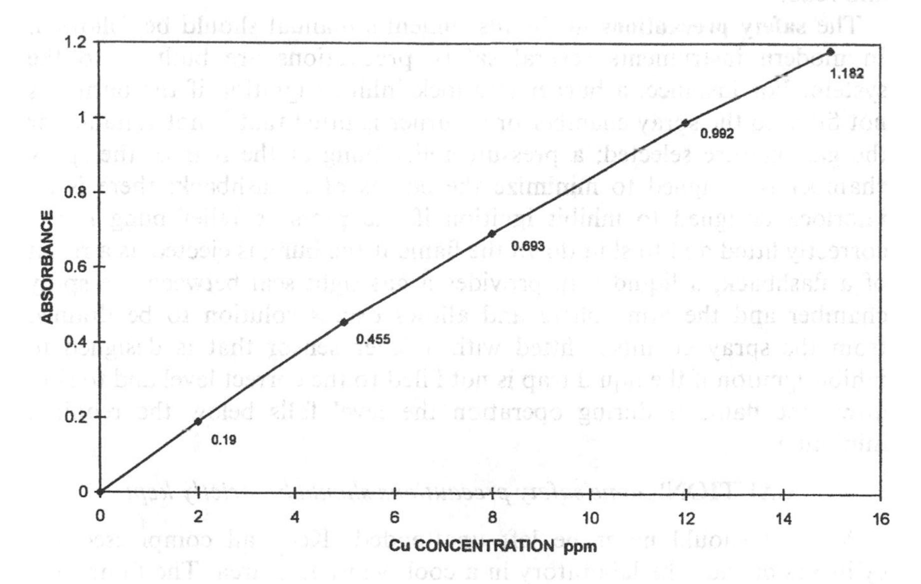

Figure 7.6 Cu absorbance as a function of concentration.

flame is shortened and the beam meets fewer atoms, resulting in the same absorbance as if the concentration were lower. The angle of rotation can be adjusted to reach absorbance values in the range of one of the optimum working ranges for lower concentrations and the analysis is carried out at that angle. In order to be able to do this, one has to know the true range of concentrations (for instance, by diluting and measuring the lowest and highest concentrations) and to prepare calibration standards that will fit that range when they are measured at the same rotation angle.

A difference in physical properties (such as surface tension or viscosity) between unknown solutions and calibration standards will result in different flow rates (for the same instrument setting) and hence in different signals for equal concentrations. Usually these types of interferences should be corrected by physical means (such as matching temperatures) or careful matching of sample and standards (for details see section 7.6).

The analyte atoms in the flame, in addition to the absorption of energy from the HCL, also emit their characteristic radiation at the same wavelength (due to the formation of excited atoms by the flame). The problem is solved by modulation of the system. One way is to modulate the output of the source (the power to the source is modulated) so that its intensity fluctuates at a constant frequency. Another way is by use of a chopper (between the HCL and the sample) alternate quadrants of which are removed. Rotation of the disk causes the beam to be chopped at the desired frequency. In both cases the HCL radiation is modulated but the emission at this wavelength is not. Only the modulated light is amplified and read.

The **safety precautions** in the instrument's manual should be followed. In modern instruments several safety precautions are built in to the system. For instance: a burner interlock inhibits ignition if the burner is not fitted to the spray chamber or a burner is fitted that is not suitable for the gas mixture selected; a pressure relief bung at the rear of the spray chamber is designed to minimize the effects of a flashback; there is an interlock designed to inhibit ignition if the pressure relief bung is not correctly fitted and to shut down the flame if the bung is ejected as a result of a flashback; a liquid trap provides a gas-tight seal between the spray chamber and the atmosphere and allows excess solution to be drained from the spray chamber, fitted with a lever sensor that is designed to inhibit ignition if the liquid trap is not filled to the correct level and to shut down the flame if during operation the level falls below the required minimum.

CAUTION! *Some safety precautions should be strictly kept.*

A flame should never be left unattended. Keep all compressed gas cylinders outside the laboratory in a cool, ventilated area. The fumes and vapors produced in the analysis can be toxic and should be extracted from the instrument by an efficient exhaust system. Always use a liquid trap filled to the correct level to avoid explosions. A special type of burner is needed for each type of flame. An air–acetylene burner should never be used for a nitrous oxide–acetylene flame. Acetylene can react with copper, or copper alloys, to give acetylides, which may explode. Hence, no copper (or brass) pipes should be used with this fuel and special pressure regulators, not containing copper, should be used. Do not use oxygen–acetylene flames. Acetylene cylinders should not be used completely, as the acetylene is dissolved in acetone, and when the acetylene pressure in the cylinder is low, acetone will also be evaporated, entering the burner and may explode. Always stop using the cylinder when some pressure still remains (for the exact amount check the manufacturer's instructions). Nitrous oxide can cause spontaneous combustion of oil and care must be taken to clean connections and the pressure regulator from oil. When N_2O emerges from the cylinder the cooling effect can freeze the gas and a regulator with a built-in heater should be used.

Operators should familiarize themselves with these and many other safety precautions specified in the instrument's manual *and follow them.*

7.4.2 Electrothermal atomization

This technique is known by many names such as electrothermal atomization, graphite furnace, carbon rod, graphite tube atomization, etc. The function of electrothermal (electrically heated graphite furnace) atomization (as well as that of flame atomization) is the generation of free analyte atoms ready to be excited by the right wavelength radiation. This is a technique that is complementary to flame atomization, rather than a technique that replaces it. Flame atomization is simple, reproducible and good results are obtained, but the method has a sensitivity limitation. For a large number of elements, furnace techniques are more sensitive than flame methods and can determine concentrations typically a hundred times smaller than those possible by flame. The analysis is much longer (may reach 2–3 min per sample) and the necessary hardware costs much more than that for flame excitation. Sample volumes in the microliter range are used by the spectrometer. As combustible gases are not used, a furnace system can safely be left unattended overnight (provided no other trouble develops). Sampling efficiency of the flame is low as almost 90% of the sample is drained, residence time in the flame is very short (a very small fraction of a second) and dissociation depends both on temperature and on the chemical nature of the flame (reducing, oxidizing or stoichiometric). In electrothermal atomization, despite the very small size of the sample, absorption is high as all of it is utilized and the residence time of the atoms in the optical path is long (about a second) so that the concentration of analyte atoms is high. This means that the proper absorption working range (about 0.1–0.8 absorbance) can be reached with lower nominal concentrations. The small sample volume (usually 10–40 μL) requires very precise measurement and introduction into the furnace, and this may become the limiting factor for reproducibility. At first manual injection with micropipets with interchangeable plastic tips (to prevent sample cross-contamination) was used, but this proved to be not good enough. For reproducible results an automatic injector is a must!

A modern **autosampler** (automatic injector) can inject into the furnace a programmed volume of a blank solution, a chemical modifier and a sample solution and can prepare (and inject) the programmed amounts of standard solutions by injecting into the furnace the right amounts of diluent and of the analyzed element stock solution (up to a constant volume). It can also rinse the injector plastic tip between samples. Standard additions analysis is often used for the furnace analysis of samples. This involves the preparation of several solutions for each sample. Automatic preparation of these solutions by the autosampler saves time and nerves (and errors).

A word of caution. When one checks tables of electrothermal atomization detection limits, the numbers are very impressive, but it has to be remembered that what is really important is the actual working range, which is much higher.

In the 'cookbook', relevant data describing the instrument's performance for a specific element may be given in different forms:

1. *Typical characteristic concentration*, which is the weight in grams of the element that would typically yield 0.0044 absorbance (1% absorption) in the peak height mode (under specified conditions), which is in the few picograms range, based on a 20 μL aliquot (which is in the range of a few ng/mL – ppb).
2. *Typical response*, for instance, 10 μL of 5.0 ng/mL is expected to yield 0.3 absorbance.
3. *Analytical working range*, for instance (for a 20 μL aliquot) 0.5–12 ng/mL (which are expected to yield 0.3–0.8 absorbance).

In modern instruments the furnace is interchangeable with the flame nebulizer–burner unit and only small adjustments (for correlation with the optical path) are needed.

The **sample holder** (Figure 7.7) is a replaceable (expendable) graphite tube which is contained in a water-cooled cell fitted at each end with quartz windows. A stream of inert gas (argon or nitrogen) flows through the cell protecting the graphite from oxidation. The graphite tube is a cylindrical tube, open on both sides (the optical path) with a hole in the center of the upper part of the tube for sample injection and through which the inert gas emerges with the products of drying and ashing. The tube fits between a pair of graphite electrodes (the form of which differs according

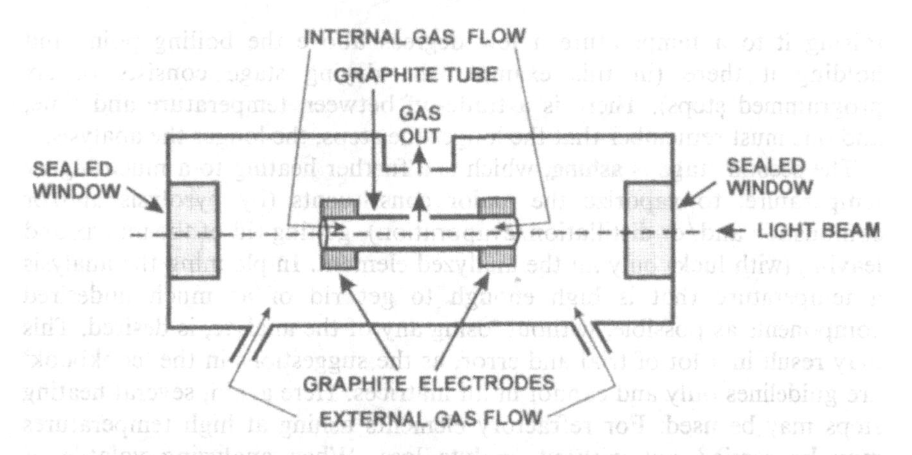

Figure 7.7 A schematic diagram of a graphite furnace. Toggle mechanism, cooling water, springs, connections, etc., are not shown.

to the manufacturer). A toggle mechanism locks the graphite tube in place, to eliminate gaps between the tube and the electrodes (due to graphite corrosion) so that the same voltage will result in the same temperature. The tube is electrically heated in programmed steps by a high current (low voltage) passing along its length. The temperature is measured by an external sensor (a pyrometer) or by automatic voltage–current measurements.

The graphite tubes are coated with pyrolytic carbon (obtained by passing through the cell an inert gas and a hydrocarbon, at a high temperature), which seals pores in the tube. This hard surface prevents soaking of the sample into the graphite (which would affect reproducibility and will lower sensitivity), and also prolongs tube life and minimizes carbide formation, which can prevent analysis of refractory carbide-forming elements such as Mo, Ti, U, V, etc.

The analytical process consists of three stages: drying, ashing and atomization. In the first stage (**drying**) the solvent is evaporated. The sample is gently dried by programmed heating of the graphite tube. If the temperature is raised too fast to temperatures near or above the boiling point, the sample will sputter, losing material (which also may stain the quartz window). Because of this, if by mistake a wrong sample (or amount) was introduced, one should be careful not to use the 'clean' option as this raises the temperature immediately to very high temperatures (close to those of atomization) and sputtering will occur. This mode should be used for cleaning only before analysis or after atomization, after most of the sample has been evaporated. The best results in a drying stage may be obtained, for example, by raising the temperature slowly to a temperature below the boiling point and holding it there for a time, then slowly raising it to the boiling point and holding it there, and then slowly raising it to a temperature a few degrees above the boiling point and holding it there (in this example the drying stage consists of six programmed steps). There is a trade-off between temperature and time, and one must remember that the longer the steps, the longer the analysis.

The second stage is **ashing**, which is a further heating to a much higher temperature, to vaporize the major constituents (by pyrolysis and/or combustion and/or distillation/evaporation), getting rid of the matrix and leaving (with luck) only all the analyzed element. In planning the analysis a temperature that is high enough to get rid of as much undesired components as possible, without losing any of the analyte, is desired. This may result in a lot of trial and error, as the suggestions in the 'cookbook' are guidelines only and cannot fit all matrices. Here again, several heating steps may be used. For refractory elements ashing at high temperatures may be carried out without analyte loss. When analyzing volatile or relatively volatile elements, there is a serious danger that at high enough ashing temperatures part of the analyzed element will be lost. Sometimes quite low ashing temperatures may solve the problem, causing volatilization of the interfering component and not of the analyte. But then, maybe not all the matrix will be destroyed, and at the atomization stage this may result in non-specific background absorption, which needs correction (see sections 7.6 and 7.7). Sometimes the use of a modifier can help, as it may form with the analyte a compound stable at higher temperatures, facilitating ashing at higher temperatures without analyte loss (see section 7.8.1). In the case of an organic matrix, sometimes the use of oxygen as an auxiliary gas in the ash stage will help in getting rid of the matrix more easily.

Finally, in the third **(atomization)** stage the tube is very quickly (at rates of up to 2000°C/s) heated to between 2000 and 3000°C (depending on the element) to atomize the remaining material and produce (in the optical path) a transient cloud of atoms to be analyzed by atomic absorption. During atomization, the inert gas flow is stopped so that all the atom population will remain in the tube and not be swept out. A small amount of inert gas still remains in the tube and it takes time for oxygen to enter and oxidize the graphite.

During the atomization stage, absorbance rises very quickly and then falls, so that the peak exists very briefly and may exist only for a fraction of a second. Nowadays, with computerized instruments, the measurement can be done easily and accurately. For the analysis, peak area or peak height is measured and compared to those of standards.

Repeatedly injecting and drying small aliquots of a dilute sample increases the amount of the analyte in the tube prior to atomization, thus increasing the analytical signal. This means a better detection limit. With many injections the sample may be partly smeared on the walls of the tube and not concentrated in one place at the bottom of the tube, causing a wider and lower signal. In such cases (which can be checked on the computer screen) one must use the peak area measurement mode of the computer.

In computerized instruments programming of the different steps becomes a lot easier by using the graphics screen. The background and atomic signals are graphically displayed together with peaks obtained in the drying and ashing stages (as the volatiles and solid particles block a part of the radiation) and with the furnace temperature program (T vs. t). The operator can see the changes during the various steps. The need for higher temperatures or longer heating times may immediately be obvious.

Complex samples or samples with concentrations close to the detection limit may require an investment of a day or two to prepare the right analysis method, by trial and error, finding the right time durations, the right temperatures, the right rates, to decide whether a modifier is needed and what modifier, etc.

For some elements better results are obtained by using a **L'vov platform**. This is a pyrolytic graphite platform with a central depression to contain the sample. The platform is placed on the tube's inner bottom, under the injection opening, touching the tube by lobes at the ends of the platform. The sample is injected and deposited on the platform's depression instead of the furnace wall. During atomization the platform temperature lags behind that of the furnace wall and reaches the designated temperature only after the tube has reached a stable higher temperature. The sample is vaporized into a stable higher-temperature environment. Recombination of the vapor components is avoided, thus reducing interferences. For instance, in the determination of lead in a salt matrix, the lead vapors may reach a lower-temperature zone, recombine with the salt anion (even redeposit) and not take part in the atomization and excitation. With the platform, the evaporated lead atoms will find a stable high temperature and recombination will not take place. The use of the platform may also cause a great reduction of background absorption due to the higher temperatures experienced by the sample after vaporization. A better reproducibility is obtained. The L'vov platform is useful for the determination of volatile elements. Refractive elements that require high atomization temperature are better determined without the platform.

The platform's atomization temperature should be that recommended by the 'cookbook'. As the measured temperature is that of the tube wall (which is higher), sometimes a higher tube temperature should be used so that the platform should reach the recommended one.

In many instances, samples that are near the detection limit of flame atomization and have a matrix that causes interferences can still be analyzed by using furnace atomization. One can dilute the sample, say 100 times, get rid of the matrix effect completely and still a high signal may be obtained by electrothermal atomization. Even if the signal is small, it is still worth diluting, as the matrix effect is relatively lowered much more.

There may be a light-scattering problem, as small solid particles (smoke) may be in the tube during atomization and scatter the light beam, thus increasing absorbance. At short wavelengths this may be a problem and background correction has to be applied.

If one has no idea of the sample's concentration and the related signal size, the way to find it is to inject a small sample and check, increase the sample and check, and so on, till a peak is seen. If one starts with a big sample, the tube may become totally contaminated and many cleaning cycles or a change of tube may be needed.

The sample has a matrix and this may cause a problem – what to use as a blank solution. If, as usual, a solution containing only the solvents is used, this may not compensate for the matrix remains and this may be critical. One can prepare a standard additions calibration line of the sample, dilute it by two and prepare another standard additions calibration line. If the lines are parallel then the blank can be used; if not, it means that this blank cannot be used for this analysis. One can try adding the proper concentrations of the major constituents of the matrix to the blank solution and determine what causes the problem.

There are situations of chemical interferences where the standard additions method has to be used. This has to be done in the linear part of the calibration curve otherwise one does not know how to do the extrapolation. One can check in the 'cookbook' to find the sensitivity for the pure analyte, for instance 0.0044 absorbance for 5 ppm analyte. This means 0.44 absorbance for 500 ppm (provided it is still in the linear range of the calibration curve). If for example the sample yielded 0.200 absorbance, then one can use additions of 100, 200 and 300 ppm so that the readings will be in the range of 0.3–0.5 absorbance and three points for the standard additions line will be obtained.

7.5 Hydride generation

Hydride generation is used for the analysis (especially traces) of arsenic, antimony, tin, selenium, bismuth, tellurium and lead (and vapor generation is used for mercury). The method is used to separate and thus preconcentrate analytes from sample matrices by a reaction that turns them into their hydride vapors (AsH_3, SbH_3, SnH_4, SeH_2, BiH_3, TeH_2 and PbH_4), which are passed into the atomizer.

These elements were formerly determined by atomic absorption spectrometry, the sample being introduced by direct solution nebulization. There were problems such as absorption lines of some elements in the far-ultraviolet region, a very high background absorption in air–acetylene flame, chemical reactions in the cooler argon–hydrogen flame, etc.

At first the hydrides were generated by a metal (usually Zn)–acid reaction to liberate hydrogen atoms, which combined with the analyte to form the hydride. This was a very slow reaction that resulted in broad absorption peaks, not all of the elements formed the hydrides and some elements had to be prereduced.

The common method today is $NaBH_4$–acid reduction to liberate hydrogen atoms, which combine with the analyte to form the hydride (from all the above-mentioned elements). After a short period (the borohydride reduction is quite fast – several seconds) the volatile hydride is swept (by an inert gas) into the atomizer where the hydride is dissociated. A precondensation (liquid nitrogen) is used to concentrate the analyte before the atomization, thus sharpening the response.

Pentavalent arsenic has to be reduced (by KI in the borohydride solution) to trivalent arsenic, preferably separately, before the reaction, as it takes time. Copper, nickel, cobalt, oxidizing agents, etc., interfere with hydride formation and have to be dealt with.

The atomizer can be a flame (usually argon–hydrogen), a graphite furnace or a heated quartz device, which is a quartz tube heated to several hundred degrees. The radiation passes from the source through the tube to the monochromator.

Automated hydride generator systems (continuous flow with a peristaltic pump or a pressurized reagent pumping system), which are attached as an accessory to the spectrometer, are manufactured by some spectrometer manufacturers.

Hydride generation improves the detection limits by a factor of 10 to 100 (for lead there seems to be a small change only) and reduces chemical and spectral interferences. Determinations of traces of these elements in air particulates, sewage, alloys, food, biological and geological materials, etc., have been reported.

The quartz vapor generator can be used for mercury analysis, without forming the hydride. Mercury compounds in acid are reduced to the free element with stannous chloride and the mercury vapor is swept by an inert gas into the quartz tube (no need of heating for dissociation). There are instruments in which the mercury is preconcentrated by collecting it on a gold gauze and then all the collected amount is heated and swept into the quartz cell, thus improving the detection limit to the ppt level (good for ultra-trace analysis of mercury in drinking water).

7.6 Interferences

A few of the interferences were dealt with in previous sections and will be mentioned here briefly again, to see the whole picture.

One type of interference is **instability** of the hollow-cathode lamp output with time. This interference can be corrected by the use of double-beam instruments. The light from the radiation source is split into two beams, one passing through the flame and sample, the other serving as a reference to which the first beam is compared. Changes in lamp output will affect both beams (see section 7.3).

Some of the factors that may cause the absorbance of the samples to be different from those of the standard solutions with the same concentrations are **physical interferences** such as viscosity. The latter may be caused for instance by temperature or composition difference between sample and standards (i.e. samples taken out of a refrigerator and standards at room temperature, acids or organic materials in the sample and not in the standards, etc.). Their effect may be to change the rate of solution uptake or the aerosol droplet size, etc. Copper in an aqueous solution will have a calibration curve with a slope very different from that of Cu in the organic solvent methyl isobutyl ketone (MIBK). In graphite furnace analysis (micro-quantities) small variations in physical properties may have a large influence on the result. For instance, difference in viscosity can cause a difference in spreading on the inside surface of the graphite tube, and this may change the absorbance, because the residence time of atoms in the optical path will depend on how far from the center atomization takes place. (Because of this spreading, organic solvents should be injected deep into the tube, very close to the bottom, and for some, preheating is preferred.) One has to match (as much as possible) these properties of the samples to those of the standards. If because of the dissolving procedure the solutions have a relatively high acid concentration, the calibration standards should contain the same amount of the same acid. If the aqueous sample solutions contain a certain amount of inorganic salts or of an organic material, the same amount should be added to the standards. For instance, if for the analysis of trace amounts of Cu in blood serum, the Cu is leached into 12.5% trichloroacetic acid (TCA), the standards should contain 12.5% TCA (Pelly 1992). If samples of skin cancer cells containing Pt are kept in a refrigerator and one uses freshly prepared Pt standards at room temperature, then, before the analysis, the samples should be allowed to reach the temperature of the standard solutions. In some cases, using the standard additions method may solve the problem. In many cases dilution of the sample solves the problem. The analyte concentration is lowered but the interference may be lowered much more, or even canceled.

Experiment 7.3
The importance of matching the temperature of the calibrating solutions with that of the sample. An example of the expected results is shown in Figure 7.8.

In 100 mL volumetric flasks prepare solutions containing 2, 4, 5, 6,

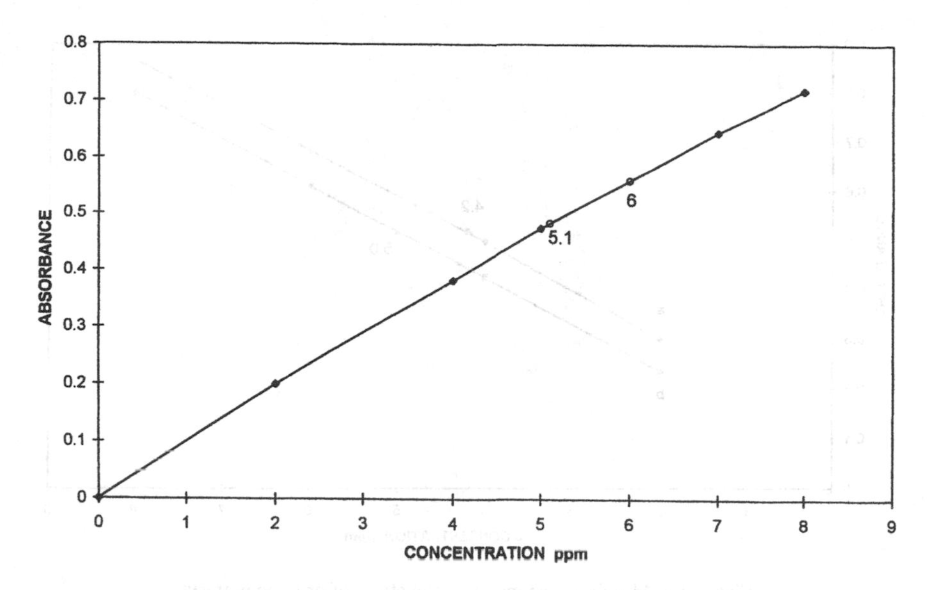

Figure 7.8 Matching the matrix – the influence of temperature.

7 and 8 ppm Cu^{2+}. Transfer about half of the 6 ppm solution to another clean dry flask and keep it for a few hours in a refrigerator (make sure it is cold when you measure it) while keeping the other half and the rest of the solutions at room temperature. Measure the absorbance at wavelength $\lambda = 324.8$ nm in an air–acetylene flame. Prepare a calibration curve with the 2, 4, 5 and 8 ppm solutions and measure the two 6 ppm solutions. Note the different concentrations obtained for what should be the same concentration.

Experiment 7.4

The importance of matching the matrix of the calibrating solutions with that of the sample solution. An example of the expected results is shown in Figure 7.9.

In 100 mL volumetric flasks prepare two sets of Cu^{2+} solutions containing: (a) 2, 4, 6 and 8 ppm Cu^{2+}; and (b) 2, 4, 5, 6 and 8 ppm Cu^{2+}, each containing 5 g of sugar. Measure the absorbance of the solutions at wavelength $\lambda = 324.8$ nm in an air–acetylene flame. Prepare two calibration curves from the 2, 4, 6 and 8 ppm solutions, and measure the absorbance of the 5 ppm solution with each curve. Note the difference between the concentrations obtained for the same solution.

In atomic absorption spectrometry it is assumed that the absorbance is

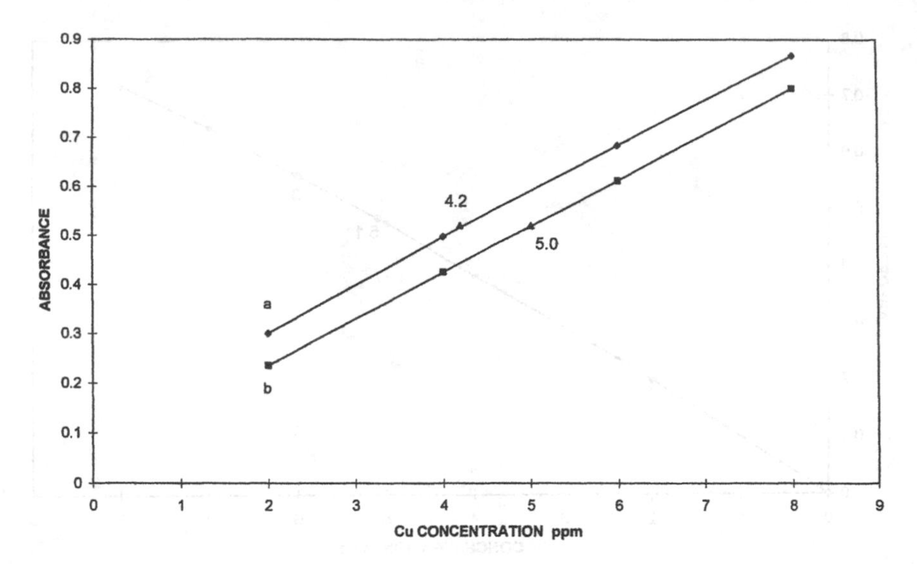

Figure 7.9 Matching the matrix – the effect of organic material.

caused by the analyzed element only and is proportional to its concentration. An interference occurs when the proportionality between the absorbance and the concentration varies as the analytical signal is also changed by factors other than concentration. The measured absorbance may be higher than expected, and will consist of the atomic absorbance and absorbance caused by other sources – **background absorption** (mostly in the ultraviolet region and little in the visible). A background correction will be needed to obtain the true absorbance due to the analyte only. Sometimes the total absorbance will be lower than expected, due to chemical reactions of the analyte, and other types of corrections will be needed.

In flame atomization, background absorption is usually important when working near the detection limit or at short wavelengths. Sometimes the problem can be avoided by change of operational parameters like flame temperature or fuel–oxidant ratio (to decompose oxides), by measuring at a longer wavelength where molecular absorption is not high, etc., or by the use of an instrumental background corrector. The most common commercial instrumental methods are the continuous source correction (deuterium background correction) and correction based on the Zeeman effect (see section 7.7).

The major causes for background absorption (i.e. absorbance caused not by the analyte) are spectral interferences and chemical interferences.

Spectral interferences that cause background absorption occur when, during atomization, there is another species whose absorption lines are very close to the analyte absorption (for instance, Cu 3247.54 Å and Eu 3247.53 Å), so that resolution by the monochromator is impossible and the result is as if there is a higher analyte concentration. This type of interference depends very much on the interfering element/analyte concentration ratio. Emission lines of hollow-cathode lamps are very narrow so that overlapping because of absorption of close lines of other elements is not common (close emission lines of the elements in the sample, caused by flame temperature, are not modulated and are not read). In such a case usually absorption at another wavelength of the analyte should be used (for the former example, Cu 3273.96 Å, where Eu 3272.77 Å will not interfere). Sometimes, working with a narrower slit may solve the problem, or standard solutions with a matching matrix may be used (if the amount of the interfering element is known), or the standard additions method may be used, or instrumental background correction may solve the problem. In furnace analysis there is a real problem in the analysis of Se in blood (1960.90 Å) where iron is present and has about 15 peaks (structured background). In the analysis of, for instance, 0.5% Cu in steel, iron causes a problem. The iron has to be taken out (extraction), or, if the amount of iron is known, standard solutions with a matching matrix are prepared, or the standard additions method may be used, or sometimes an instrumental background correction may be applied (depends also on the method of correction).

The spectral interference in flames is usually caused not by another element but by a broad absorption band (non-specific absorbance) caused by other molecules, combustion products, which originate in the flame or in the matrix. This may occur in flames when the flame is not hot enough to decompose all the molecular compounds that were formed by the sample in the flame. These molecules will absorb light, and if the spectrum overlaps the analytical wavelength this will be measured, increasing the signal. The interfering absorbance may also be caused by molecules that are reaction products of the matrix, for instance the broad absorption band of CaOH (OH may come from the flame), which overlaps the absorption line of Ba. The dissociation of CaOH by a nitrous oxide–acetylene flame will eliminate the interfering band.

Interference absorbance can also be caused by radiation-scattering particles in the flame (for instance by the formation of metal oxides of refractory elements), thus lowering the beam intensity. Working with a fuel-rich flame may eliminate the formation of the scattering oxide.

The flame itself (without a sample) produces absorption, especially at low wavelengths. To correct for absorbance of absorbing products originating in the flame, i.e. fuel–oxidant–solvent system (for instance the formation of OH by water dissociation), measurements with a blank solution should be made and subtracted (in modern instruments by the computer) from each analyte reading.

Spectral interferences by matrix products are a much greater problem in a graphite furnace than in flame atomization. In the ashing stage one has sometimes to use a low temperature, not high enough to decompose all the matrix, to avoid loss of a relatively volatile analyte. Leftover molecular species will be present during atomization, and an incomplete breakdown during the brief atomization in the graphite furnace will result in broad molecular absorption bands. In samples containing halogens, stable halides with wide absorption bands can be formed. Incompletely decomposed organic material can form absorbing molecules or particles which can scatter light (including carbon particles from a much used tube), etc. In these cases, the use of a background corrector may be necessary. For some cases of furnace analysis the use of a chemical modifier can solve the problem (see section 7.8.1).

Chemical interferences are caused by various chemical processes, occurring during atomization, that alter the absorption of the *analyte*. Chemical interferences are more common than spectral interferences. The effect usually can be minimized by use of proper operating conditions.

One type of chemical interference is ionization of analyte atoms (mostly alkali and alkaline-earth elements) due to high flame temperatures. While the effect is small in an air–acetylene flame it is very high in an acetylene–nitrous oxide flame (may reach ranges of 80%, depending on the element), thus removing them from the ground-state atom population (the ions will absorb at other wavelengths). One solution, though not a common one, is to work with a colder flame, but this may cause other problems. The method to overcome this problem is to add to the analyzed solution another element, such as Na, K or Cs, which is ionized more easily (or about the same) at a very high concentration compared with that of the analyzed element. The **ionization suppressor** has to be added to the unknown solutions as well as to the calibration standard solutions. For instance, for the analysis of trace amounts of Ca (in the range 1–4 μg/mL) in the nitrous oxide–acetylene flame, the solution is made to contain 2000 μg/mL potassium, which will cause signal enhancement due to suppression of ionization. In the analysis of traces of strontium, the sensitivity may be greatly increased by the addition of an ionization suppressor – potassium ions (with the same amount of potassium, the sensitivity may be greatly increased by raising the temperature, using a nitrous oxide–acetylene instead of air–acetylene flame). The effectiveness of the suppressor may change with its concentration and the right amount has to be determined experimentally (absorption vs. concentration, till a plateau is found) or, preferably, taken from a 'cookbook'. One has to be careful not to cause harm by this addition. The added potassium chloride may contain, for instance, some Na (depends on the purity of the material used, say 0.01%). If one analyzes a very low concentration of Na, this may cause an error. In this case a very pure potassium chloride powder or solution has to be used.

Experiment 7.5
The effect of ionization suppressor. An example of the expected results is shown in Figure 7.10.

In 100 mL volumetric flasks prepare three sets of Sr^{2+} solutions containing: (a) 3, 5, 7 and 9 ppm Sr^{2+}; (b) the same Sr^{2+} concentrations as in set (a), but each solution also containing 1000 ppm K^+ (191 mg KCl in each flask); and (c) the same Sr^{2+} concentrations as in set (a), but each solution also containing 5000 ppm K^+ (953 mg KCl in each flask). Measure the absorbance of all solutions at wavelength $\lambda = 460.7$ nm in (i) air–acetylene flame and (ii) nitrous oxide flame. Plot a graph showing the absorbance as a function of Sr^{2+} concentration for each of the six sets of measurements.

Another type of chemical interference is caused by the fact that, in the hot environment during atomization, many dissociations and associations

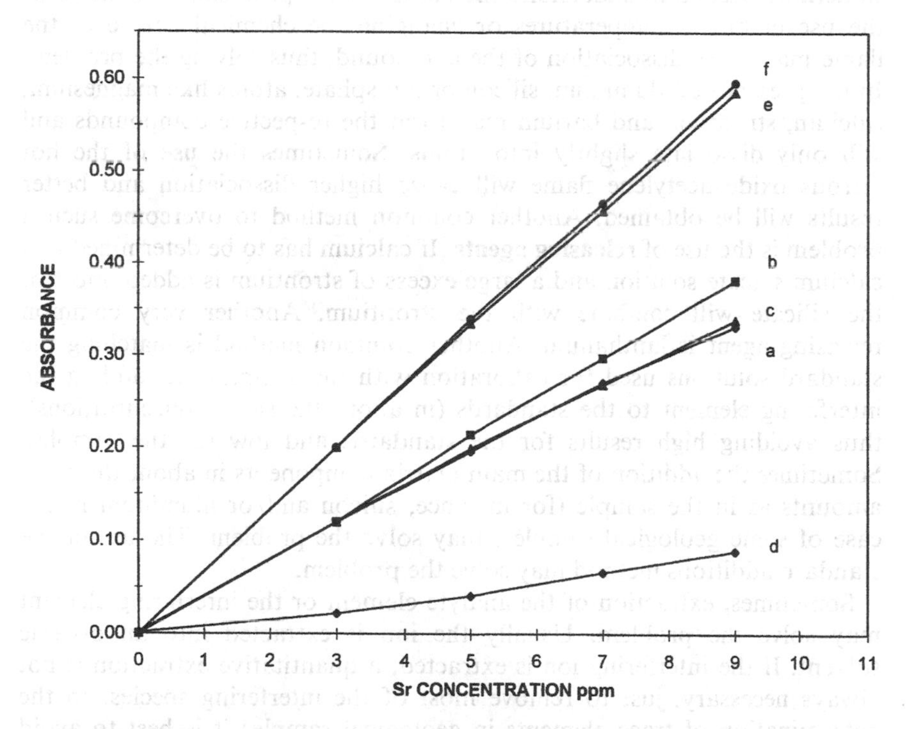

Figure 7.10 The effect of ionization suppressor (on Sr^{2+} solution): (a) air–acetylene, no K^+; (b) air–acetylene, 1000 ppm K^+; (c) air–acetylene, 5000 ppm K^+; (d) nitrous oxide–acetylene, no K^+; (e) nitrous oxide–acetylene, 1000 ppm K^+; (f) nitrous oxide–acetylene, 5000 ppm K^+.

take place (see section 7.4.1), among them reactions of the analyte leading to the formation of molecular species, for instance metal oxides and hydroxides. Some of these are stable and their molecular bands are wide, bright and prominent in the spectra of the sample's components (for instance alkaline-earth oxides). Changing the analytical conditions, for instance working with a fuel-rich flame, may cause the dissociation of the oxide. Na in the presence of HCl can form NaCl, thus decreasing the sodium atom concentration. The absorption of V may be enhanced by the presence of Al or Ti in a fuel-rich flame. The concentration of oxygen and OH is relatively small, and as the other metals use a part of it, a part of the vanadium oxide is decomposed, increasing its atomic absorbance. In fuel-lean flames there is enough oxygen so that the addition of Al and Ti will not make much of a difference.

In another type of chemical interference, low results will be obtained because of reactions (in the flame) of the analyte with anions forming low-volatility compounds. A solution containing calcium nitrate together with aluminum nitrate may form in the flame a new compound, $CaAlO_4$, which does not dissociate even at 3000°C. A good knowledge of chemistry is sometimes needed to understand the nature of the problem. In some cases the use of higher temperatures or changing the chemical nature of the flame may cause dissociation of the compound, thus solving the problem. In the presence of aluminum, silicon or phosphate, atoms like magnesium, calcium, strontium and barium may form the respective compounds and will only dissociate slightly into atoms. Sometimes the use of the hot nitrous oxide–acetylene flame will cause higher dissociation and better results will be obtained. Another common method to overcome such a problem is the use of **releasing agents**. If calcium has to be determined in a calcium silicate solution and a large excess of strontium is added, most of the silicate will combine with the strontium. Another very common releasing agent is lanthanum. Another common method is matching the standard solutions used for calibration with the samples, i.e. adding the interfering element to the standards (in about the same concentrations), thus avoiding high results for the standards and low for the samples. Sometimes the addition of the main matrix components in about the same amounts as in the sample (for instance, silicon and/or aluminum in the case of some geological samples) may solve the problem. The use of the standard additions method may solve the problem.

Sometimes, extraction of the analyte element or the interfering element may solve the problem. Usually the ion is extracted into an organic solvent. If the interfering ion is extracted, a quantitative extraction is not always necessary, just to remove most of the interfering species. In the determination of trace elements in geological samples it is best to avoid interferences by getting rid of the silicon (evaporating with HF) but unless a specific compound of an analyzed element with silicon can be formed, the separation does not have to be quantitative, just removing the bulk of the silicon. The same may be applied to the extraction of iron into isobutyl acetate as the chloride complex, in the determination of trace metals in iron ores. If the analyte itself has to be extracted, care must be taken to ensure complete removal (for instance an exact pH, a specific organic extractor, etc.) and the method should be checked with solutions of known analyte concentrations.

7.7 Instrumental background corrections

For a good background correction one must measure the background at the same wavelength as that at which the analyte is measured and at the same time. All commercial instruments measure the analyte and the background one after the other (a very short time difference) and not simultaneously. The time difference between the two measurements is very important, especially in furnace analysis, and should be as small as possible (ideally none) as the background can change very fast (a fraction of a second).

All background corrections are carried out by making one measurement of the total absorbance (analyte and background) and one measurement of background only and then subtracting it from the total (one can actually measure the total, the background and the analyte absorptions). As the background may change with time, it seems that one could measure the background before the absorbance peak and after the peak and use the average. This may work only if the change is linear, but no one can guarantee that this is the case.

Four background correction methods have been used in commercial instruments, of which one is not used any more, one is rare and only two are widespread, the continuum (deuterium) correction and a correction based on the Zeeman effect.

The oldest, cheapest (and not used any more) is the **two-line correction method**. At one wavelength the total absorption (analyte and background) is measured and at another wavelength, as close as possible to the analyte line (but not absorbed by the analyte), the background absorption is measured, assuming that the background is the same for both wavelengths. The second line may be another line from the same element of the hollow-cathode lamp (e.g. for Al HCL, first line at 3093 Å and second line at 3070 Å), from the filler gas of the same HCL (the first line Ba at 5536 Å and the second line Ne at 5409 Å) or from another HCL (the first line Na at 5890 Å and the second line by a Mo HCL at 5883 Å). There is a big time difference between the two measurements, depending on how fast one can change wavelengths (may reach a few seconds). This means that the method should be used for flame atomization and not for furnace analysis, as two electrothermal atomization heatings will be needed and no one can guarantee the same background in both. One can use two monochromators simultaneously but this will complicate the instrument structure and will be expensive (and still will have a big time difference compared to modern methods).

Another method, the newest (**Smith–Hieftje**), rare in commercial instruments, is based on **self-reversal** (Maugh 1983). An HCL needs a certain, small, amount of current to emit radiation at precise, narrow wavelengths. At higher currents the emission lines are broadened and at still higher currents there is a high concentration of unexcited atoms of the lamp's element, which absorb a great part of the HCL's emitted desired wavelengths (see section 7.3.1). The emission peak will look like a saddle and the result is that the source emits light at wavelengths other than those theoretically absorbed by the analyte, close to and on both sides of the theoretical wavelengths. This liability is used for background correction. The hollow-cathode lamp is cycled through periods (each precisely timed) of low and high currents. At low currents the total absorbance (background and analyte) is determined, while at high currents the background only is determined and then subtracted from the total. The method requires special HCLs (to prevent arcing at high currents) and a special electronic system.

The most widespread method is the **deuterium** (or **hydrogen** in other instruments) **background correction.** This needs a deuterium lamp to emit continuous radiation (see section 7.3.1). The output of the deuterium lamp is quite low at wavelengths above about 350 nm and this type of correction is not used above the ultraviolet region.

The principle of this method is that the slit width is kept sufficiently wide. An absorption line of the analyte is typically very narrow in relation to the total spectral bandpass of the monochromator, so that the background absorption of the sample may be measured using a suitable continuum source, but the fraction of the continuous source that is absorbed by the atoms of the sample (the analyte absorption line) is negligible. The radiation from the continuous source and the HCL is passed alternately through the analyte vapor (in the flame or the graphite tube). Measurements of the total absorption are made with the analyte HCL, while measurements with the deuterium lamp give the background absorption only (which is subtracted by the instrument's computer).

The HCL and the deuterium lamp must be aimed at exactly the same line along the flame, as the sample vapor may be very inhomogeneous. This alignment used to be a hard task but in modern instruments the lamp is installed at the factory and only small adjustments are needed before use.

When the background is not a continuum but consists of a fine-structure molecular spectrum (structural background), errors may occur by corrections with a continuum source, as a narrow line may overlap the analyte resonance line, which may then contribute a big part of the measurement. There are only a few such cases. If this happens one should make sequential measurements (and subtraction) of the absorption of the sample and of a blank solution whose composition with respect to the interfering species is very carefully matched to that of the sample (but does not contain the analyte).

Another type of background correction (Maugh 1977, Brown 1977, Fernandez *et al.* 1980) is based on the **Zeeman effect** (the splitting of spectral lines in a strong magnetic field). The advantages are: correction for spectral interferences, correction over the complete wavelength range, correction for structural background, correction for high background absorbances (in the range of 2 compared to 0.7 absorbance with deuterium correction) and the use of a single line source with no possibility of misalignment. Less sample preparation (to reduce the concentration of materials that can cause interferences) is needed. This type of correction is particularly used with graphite furnaces.

This correction requires a strong magnet, a polarization device, etc., and these cannot be bought as accessories and attached on an existing commercial instrument (unless one wants to build one's own instrument). One has to buy an electrothermal atomic absorption spectrometer manufactured for this type of correction. This instrument is a very expensive instrument, which is justified only in special cases, like the determination of ultra-trace metals in drinking water (water standards), or the determination of trace metals in a difficult, problem-causing matrix like sea water (high background because of reflecting salt particles) or blood, urine, milk, etc. (where the decomposition of the organic matrix may lead to the formation of absorbing molecules and very high background). For most analyses, deuterium background correction (alone or combined with the use of a chemical modifier or with extraction) is enough.

There are two types of Zeeman effect: (1) a *normal* Zeeman effect (where both terms involved in the transition are singlets, or the values for both Landé g values are identical – e.g. some transitions of Ca, Cd, Ba, Be, Mg, Sr and Zn); and (2) an *anomalous* Zeeman effect (where the terms are not singlets and the g values are not equal – e.g. some other Zn transitions), where more splittings occur, either relatively simple (Ge, Pb, Sn) or very complicated (Cr, Mo) splittings. We shall deal with the simplest case of a singlet transition, a case of a normal effect.

In this case the analyte spectral line is split into three: one central line (π) at the original wavelength, with half the intensity of the original line, which is polarized parallel to the direction of the magnetic field; and two lines, one on each side of the first line ($+\sigma$ and $-\sigma$, not at the analyte wavelength), each with a quarter of the intensity of the original line, which are polarized perpendicular to the magnetic field. The background is insensitive to the magnetic field.

There are several methods to use this splitting for background correction but in all commercial instruments the magnetic field is perpendicular to the optical path (transverse).

In some instruments the magnet surrounds the hollow-cathode lamp and the correction is at a wavelength slightly different from that of the analyte. The light emitted from the HCL (and not from the analyte atoms) is split into π and σ components. A rotating disk transmits alternately the π and σ components of the radiation. With the π component (the theoretical wavelength) the total absorbance and with the σ component the background absorbance are measured. In this method the σ component does not read the background at the same place as the π component reads the analyte (but at the edges of the analyte peak, depending on the amount of separation caused by the magnetic field).

In most instruments the magnet surrounds the graphite tube. In this method, correction is performed at the resonance line, not at a slightly different wavelength, which is an important advantage. In one method the graphite atomizer is surrounded by the permanent magnet that splits the analyte line (again, the simple case of splitting into one π and two σ components). Unpolarized light from the HCL is passed through a rotating polarizer, which separates the beam into two plane-polarized waves, perpendicular to each other. These waves pass into the graphite tube. During that part of the cycle in which the radiation is polarized parallel to the field (π), the total absorbance is measured. When the radiation is polarized at 90° to the field (σ), absorbance of the background (caused by molecular absorption and by scattering) is measured and subtracted from the total. In another method where the magnet surrounds the graphite atomizer, a polarizer is inserted in the optical path to remove the π component. The measurements are very fast (100 measurements per second). The total absorbance is measured with the magnet off (no splitting). The background absorbance is measured with the magnet on (σ components only while the π component is removed). A number of background measurements before and after total absorbance measurement and a polynomial interpolation are used to find the background at the time the correction should take place.

In Zeeman atomic absorption correction there may be a curve rollover, where the curve of absorption vs. concentration rises, reaches a peak and falls again, so that one absorbance reading will fit two concentrations. To avoid the problem the concentration working range is limited, depending on field strength (the maximum absorbance is predetermined by the manufacturer and can be changed by the user after investigation of high-concentration standards) and the computer will indicate when the maximum permissible absorbance is exceeded.

7.8 Modifiers, standards and chemicals

7.8.1 Modifiers

If the analyte and the matrix are volatilized at similar temperatures, it is difficult to find an ashing temperature far enough from the atomization temperature. In this case the use of chemical modification should be considered as it may allow ashing (getting rid of the matrix) without losing a relatively volatile analyte.

The aim is to find a reaction in which the analyte forms a less volatile compound and to know its decomposition temperature and intermediate compounds. Reducing the volatility of the analyte will allow a higher ashing temperature to remove all matrix materials that produce background absorption and form a single compound of the analyte, to ensure a single large atomic peak.

The modifier can be premixed with the sample or, if it will form a precipitate, added to the sample in the furnace.

A few examples will illustrate the point. In the analysis of trace metals in sea water, NaCl causes a high background absorption. Ammonium nitrate will react with sodium chloride and get rid of the chloride as the volatile ammonium chloride. Addition of a large excess of phosphoric acid to Pb, Cd and Zn will raise the atomization temperature. The addition of EDTA to lead will result in a complex with atomization temperature lower than that of a nitrate or chloride matrix. The peak of atomization temperature for lead in 1% NaCl is 910°C, with 0.5% EDTA 560°C and with 5% phosphoric acid 1180°C. Nickel is typically used to stabilize arsenic and selenium (by forming metal selenides) to higher temperatures.

A relatively new matrix modifier is palladium, introduced into the furnace as Pd solution, 50–1000 mg/L, preferably as nitrate. The development of a successful analytical method requires the analyst to optimize the palladium concentration and temperature program carefully. Increasing the palladium concentration shifts the atomization signal to higher temperatures, up to a limit (depends on the matrix) where broad irregular signals with reduced sensitivity will be obtained. For aqueous samples 50–200 mg/L may be sufficient, while for more complex samples 200–1000 mg/L may be necessary.

If the solution contains oxidizers such as nitric acid, sodium sulfate, sulfuric acid, etc., Pd becomes a poor matrix modifier, so the addition of a reducing agent is required. The reducing agent guarantees that the Pd will be reduced to the metal form early in the temperature program, before the analyte is volatilized. The metal retains the analyte element (probably by the formation of an intermetallic species) until a higher gas-phase temperature is achieved. Scanning electron microscopy showed that the size and distribution of the Pd on the graphite surface (hence the effectiveness as a modifier) depend on the reduction method used.

With reducing agents such as hydroxylamine hydrochloride or, better, ascorbic acid, Pd can raise ash temperatures for several semimetallic elements by 400–800°C and by 200–500°C for some transition metals (Voth-Beach and Shrader 1986, 1987). Ascorbic acid precipitates Pd so rapidly that palladium and ascorbic acid must be introduced separately into the furnace, so the use of 5% hydrogen in 95% argon (a standard mixture supplied commercially) was recommended to be a better reducing agent (Beach 1987).

7.8.2 Standards and chemicals

Two types of standards are used in atomic absorption spectrometry: one type is chemicals (solids or solutions) used for calibration curves; the other type is used for checking the analytical procedures and the accuracy of the results. Atomic absorption is a comparative technique, so the accuracy of quantitative measurements depends on the accuracy and the condition of the standards employed (i.e. matching standard solutions and samples in respect of physical and chemical properties), otherwise the basis for comparison is destroyed. If matching cannot be obtained, the standard additions method should be used.

The calibration standards can be very pure solid chemical compounds (as an example, $CaCO_3$, KCl, etc.) or metals (such as Ni, Cu, etc., as wires or powder), which are weighed, dissolved and diluted to obtain (master) standard solutions of known concentrations of the elements in question. The various 'cookbooks' describe, for each element, a method to prepare a standard solution. The master standard solutions, which are prepared for use over a long time, have to contain a high concentration of the element (usually 1000 mg/L metal) and these master solutions are diluted to prepare the low-concentration solutions needed for the calibration curves. These dilute solutions can be used the same day or for a maximum of a few days.

When many of the metals are exposed to air their surface is oxidized quite easily. The next time they are to be used, if they can still be used at all, they have to be etched or 'purified' in another way before use. In addition, with solid standards there are 'technician errors' of weighing and dilutions. These may cause severe problems as they may be discovered (if at all) after a research program or a set of analyses was finished. Using commercial certified standard solutions is very easy and avoids most of these problems.

Nowadays prepared elemental master standard solutions (usually 1000 mg/L) spectroscopically pure for atomic absorption analysis can be bought from various commercial sources. There are standards for aqueous matrices or oil-dissolved standards to match samples in an oil matrix, etc.

The second type of standards includes materials of known composition, for checking analytical procedures and the accuracy of the results. There are several kinds of this type of standard, to fit the various fields using AAS. For example, there are standards of serum containing Cu ions: Norm (in the range of normal serum Cu concentrations) and Path (in the range of pathological concentrations). There are also standards containing desired concentrations of elements (organometallic compounds) in an organic oil matrix, for the analysis of metal wear in engine lubrication oils. Geochemical standards are rocks (such as granites, basalts, syanites, etc.) of which very large amounts (tens or hundreds of kilograms) are taken, ground very finely, mixed very thoroughly and analyzed.

Geochemical standards of rocks and minerals are available from many sources, in different countries, from both institutional and commercial sources. Two of the first compilations of values for geological standards are Flanagan (1973) and Abbey (1977), together with discussions of subjects such as the determination of usable values, verification of values and statistical problems. *Geostandards Newsletter* is a journal (since 1977) dedicated to analytical methods and values of geochemical reference samples. Every few years a compilation of working values for reference rocks and minerals is published.

These certified standards, so-called 'international' standards, are quite expensive, so that 'home' standards are frequently used. A rock sample is finely ground, thoroughly mixed and analyzed (the method is checked with a certified standard). This 'home' standard is then used whenever a standard is needed when a new method is developed (but for the last time, when the certified standard is used).

The purity of reagents is one of the things that determine the accuracy of the analysis, especially in the case of ultra-trace element analysis, for which all reagents (and distilled water!) should be of the highest purity available. These include acids for dissolution, fluxes for melting, chemicals used as ionization suppressors, releasing agents or chemical modifiers, etc. Special care must be attached to the quality of the water used, for dilutions and preparation of reagents, especially when analyzing trace levels. In most cases distilled water or ordinary deionized water can be used, but sometimes the element to be analyzed has to be removed from the water by a better procedure.

Special attention must be paid to the blank solution, which is a most (if not the most) important standard. This solution should contain the appropriate amounts of all the components of the sample solution, except for the sample itself, as they may contain trace amounts of the analyzed element, which should be subtracted from each measurement (for instance, the same water used for dilutions, acids used for dissolution, flux used for melting, organic material added to the sample, ionization suppressor, etc.).

One should not forget the level of cleaning needed for laboratory-ware and the laboratory itself (for instance – no dust).

7.9 Sample preparation and automation

7.9.1 Sample preparation

Sample preparation is one of the critical factors determining the quality of the analysis results. Any error in preparation will be expressed in the analysis results. Most samples cannot be analyzed as obtained and need some treatment before analysis.

The first step may be the collection of **representative samples** (fresh uneroded geological, biological, medical or industrial samples). If the sample does not represent the population, no matter how accurate the analysis, the results are useless.

The next step may be grinding the samples. There are several grinding techniques and the grinders are made of different materials, some of which may contaminate the samples, depending on the material the grinder is made of and the hardness of the samples. In the analysis of trace elements this can be a problem.

The next step (for most samples) may be the reduction of the samples to solution in order to be presented to the spectrometer and aspirated into the flame. In a graphite furnace sometimes analysis of solid samples is possible. The problem in such cases (and that is why solid-state analysis is almost never used) is that the part of the sample actually used for analysis cannot be considered to be representative of the whole sample as only a few milligrams of a large sample can be used in the furnace).

The following is a brief listing of the more common sample preparation techniques. The reader should turn to Chapter 1 on sample preparation for further details.

The sample may be reduced to solution in several ways (depending on the sample) such as **dissolution** (in organic or inorganic solvents, in an acid or in a combination of acids – including HF to decompose silicates and volatilize silicon as a fluoride, etc.) or **melting** (with sodium carbonate, sodium hydroxide or lithium metaborate, etc., as a flux, and then acidifying). Dissolution can be carried out in a beaker (glass or teflon, etc., using a flame), or in a teflon bomb (inside a metallic jacket, using an electric oven) where dissolution is assisted also by the evolved pressure, or using a microwave oven with an open teflon vessel for organic samples (to prevent dangerous accumulation of high pressure resulting from degradation of organic materials), or in the case of inorganic materials (usually rocks and minerals) with a closed teflon bomb (without a metallic jacket) where dissolution is assisted by the evolved pressure. Melting is usually carried out in Pt crucibles, Pt–5%Au crucibles (not wetted by silicate melts) or Zr crucibles (with NaOH flux silicates can be melted at relatively low temperatures without losing relatively volatile trace elements), etc.

A relatively new method is **laser ablation**, where a laser beam is used to evaporate a selected part of the sample (which can be very small, such as a selected mineral, etc.).

Even if the sample is originally a solution, pretreatment may be needed, for instance dilution, preconcentration, the use of chelating resins (ion exchange), extraction to remove interfering elements or to separate the wanted elements, addition of ionization suppressor or a releasing agent, etc.

7.9.2 Automation

The different atomic absorption spectrometers on the market differ from each other in the amount and sophistication of automation (relating to analytical procedures, instrumental parameters, data processing and reporting) though several properties are common nowadays (in one form or another) in all modern instruments. Some operations need the operator's decisions and input to the computer and some are controlled by default values, which can be overridden by the operator.

Several examples of automation were described in previous sections. The following is a partial list of examples of automation that can be found in atomic absorption spectrometers (by no means in all of them).

Automation may include provision for permanent storage and retrieval of complete analytical procedures. Automatic analyses can be carried out where the computer commands the instrument to retrieve an analytical program, set up the proper physical (optical and atomizer) parameters, perform calibrations (ordinary or standard additions method, with automatic blank subtraction), draw calibration graphs, analyze unknowns, print the results and carry out statistical data processing.

Automation of spectrometer physical parameters may include wavelength selection and peaking, spectral bandwidth selection, controlling lamp turret position to select the desired lamp, lamp current, selection of the continuum background source and its intensity, correction of background absorption, selection of oxidant, determining and controlling gas flow rates, ignition (for flame atomizers), selection of heating cycle programs for the graphite furnace (i.e. programming the dry, ash and atomization stages) and operation of an autosampler (together with options for automatic analysis with handheld samples or non-automatic analysis with automatic sampler).

There may be many options from which the operator may choose, such as measuring absorbance, flame emission, ordinary calibration curve, a standard additions calibration, scale expansion (scale expansion does not make the measurement more accurate, as the noise is also amplified, just easier to read very small absorptions, which may be important if one wants to dilute the sample to overcome interferences), reading peak height, peak area, integration (reading during a specified duration – for instance 10 s, at 0.1 s increments and giving the average), precision-optimized measurement time (where the operator asks for a specified precision, for instance 0.5% relative standard deviation and the computer reads only long enough to reach the specified precision), sequence control to determine the order in which the elements will be analyzed automatically (each with calibration curve), determine the last sample after which one can program an automatic turning off of the flame, printing of results after all measurements (sequential or multi-element) or after each reading (during run), automatic labeling (naming of samples), the possibility to edit analytical results and calculate final concentrations in the sample considering sample weight or dilutions.

Automation makes life easier, makes the analysis simpler to carry out (also reducing operator errors), helps in sustaining precision and (maybe the most important) allows repeating analysis of samples at different times under almost identical conditions. A very important feature is automatic control of many safety precautions.

References

Abbey, S. (1977), *1977 Edition of 'Usable' Values*, Geological Survey of Canada, Paper 77–34.

Alkemade, C.Th.J. and Herrmann, R. (1979), *Fundamentals of Analytical Flame Spectroscopy* (translated from German), Adam Hilger, Bristol.

Beach, L.M. (1987), *Spectroscopy*, **2**, 21.

Brown, S.D. (1977), *Anal. Chem.*, **49**, 1269A–1281A.

Browner R.F. and Boorn, A.W. (1984), *Anal. Chem.*, **56**(7), 787A–798A.

Ewing, G.W. (1969), *Instrumental Methods of Chemical Analysis*, McGraw-Hill Kogakusha, Tokyo.

Fernandez, F.J., Myers, S.A. and Slavin, W. (1980), *Anal. Chem.*, **52**, 741–746.

Flanagan, F.J. (1973), *Geochim. Cosmochim. Acta*, **37**, 1189–1200.

Grosser, Z.A. and Schreiber, C.A. (1996), *At. Spectrosc.*, **17**(6), 209–210.

Kirkbright, G.F. and Sargent, M. (1974), *Atomic Absorption and Fluorescence Spectroscopy*, Academic Press, London.

Mann, C.K., Vickers, T.J. and Gulick, W.M. (1974), *Instrumental Analysis*, Harper and Row, New York.

Maugh, T.H., II (1977), *Science*, **198**, 39–40.

Maugh, T.H., II (1983), *Science*, **220**, 183.

Pelly, I. (1992), *Clin. Chim. Acta*, **213**, 51–59 and unpublished results.

Pelly, I. (1994), Determination of trace elements by atomic absorption spectrometry, in *Determination of Trace Elements*, Alfassi, Z.B. (ed.), pp. 145–190, VCH, Weinheim.

Perkin-Elmer Corp., *Analytical Methods for Atomic Absorption Spectrophotometry*.

Skoog, D.A. (1985), *Principles of Instrumental Analysis*, 3rd edn, Saunders, Philadelphia.

Slavin, W. (1968), *Atomic Absorption Spectroscopy*, Interscience, New York.

Slavin, W. (1986), *Anal. Chem.*, **58**(4), 589A–598A.

Strobel, H.A. (1973), *Chemical Instrumentation*, Addison-Wesley, Reading, Mass.
Varian-Techtron Pty Ltd, *Analytical Methods for Flame Spectroscopy*.
Varian-Techtron Pty Ltd, *Analytical Methods for Graphite Tube Atomizers*.
Voth-Beach, L.M. and Shrader, D.E. (1986), *Spectroscopy*, **1**, 49.
Voth-Beach, L.M. and Shrader, D.E. (1987), *J. Anal. At. Spectrom.*, **2**, 45.

8 X-ray fluorescence analysis

P. WOBRAUSCHEK, P. KREGSAMER and M. MANTLER

8.1 Introduction

8.1.1 Background

X-ray fluorescence analysis (XRF) is a powerful analytical technique for the qualitative and quantitative determination of chemical elements in a sample. The operating principle of XRF is based on the irradiation of the sample with ionizing radiation, thus inducing excited states of the atoms in the analyte. The response of the atoms during the de-excitation process is the emission of X-rays having well defined energies or wavelengths characteristic for an element. (The characteristic energies of many elements of the Periodic Table can be found in Table 8A.1 in the Appendix.) Complementary to the emission of characteristic X-rays is the emission of Auger electrons, which is the dominant process for light elements. The term 'fluorescence' expresses the analogy with the excitation of atoms during irradiation with ultraviolet (UV) radiation or by thermal excitation of the optical states by a flame and emission of fluorescence light in the visible regime. The identification by the color of the emitted radiation is accomplished by an optical spectroscope. As humans are not able to 'see' X-rays, sophisticated detectors are necessary to detect the characteristic lines of the elements.

The excitation source for XRF can be any type of ionizing radiation, either electrically charged particles or high-energy photons. Accelerators or radioactive isotopes are used to deliver electrons, protons, α-particles, etc.; X-ray tubes, radioactive isotopes or synchrotron storage rings act as photon emitters. First steps are currently being taken to establish X-ray lasers and plasma sources for the excitation of light elements.

If a sample is excited by an appropriate source, a number of characteristic lines are emitted for each element, forming a spectrum in terms of counts versus either wavelength or energy, depending on the detection system. Wavelength and energy are equivalent and can be correlated by

$$E = \frac{1.2398}{\lambda} \tag{8.1}$$

where the energy E is in kiloelectronvolts (keV) and the wavelength λ in nanometers (nm). The spectrum contains the analytical information to determine the elements qualitatively and quantitatively. The wavelength or energy of a characteristic line corresponds according to Moseley (1913, 1914) to the atomic number Z of the emitter and thus enables the determination of the element investigated. The count rate of the characteristic line is related to the concentration of the element.

Often the non-destructiveness of XRF is praised, but this property is certainly applicable only in a few cases. The reason is that the fluorescent X-rays are effectively induced and emitted only in near-surface layers. For many specimens the relevant information depth is in the range of some 100 μm. Obviously homogenization processes are important in many cases. Typical sample preparation techniques are grinding and consecutive pressing of pellets, glass fusion and acidic dissolution in pressure vessels.

Simultaneous or sequential determination of almost all chemical elements of the Periodic Table, from Na ($Z = 11$) to U (92), with standard spectrometers is possible. Lighter elements can be analyzed with special instrumentation. The usual dynamic range of XRF extends from 100% to μg/g, depending upon the atomic number of the element and the matrix, where sometimes in the case of line interferences this lower limit cannot be reached. Major components can be identified within a few seconds counting time (energy-dispersive XRF), but 1000 s or more may be necessary for low concentrations or traces. In routine applications with wavelength-dispersive spectrometers, quantitative analysis of major constituents is often possible within one minute per element or less. Dedicated spectrometers, which allow a fixed set of elements to be analyzed simultaneously, e.g. in steel or non-ferrous alloy factories, reduce complete preparatory and analytical turn-around cycles to a very few minutes from taking a sample from the smelt.

Interest in the analysis of environmental samples like the trace-element contents in the atmosphere, soil, sediment, vegetation, river, drinking, lake or sea water, etc., has given a strong impact to reduce the detection limits down to ng/g. With technically optimized equipment (special excitation sources and geometries) for medium-Z elements, the range of femtograms (fg, 10^{-15} g) has been reached.

Elemental distribution studies have been performed, analyzing point by point across an area of interest with finely focused beams. The exciting radiation has dimensions of micrometer diameter and is produced by X-ray optical devices with an X-ray tube or synchrotron storage ring as source. Total reflection XRF has been applied successfully to the surface analysis of Si wafers for the detection of metallic contaminations, where detection limits in the range of 10^8–10^{10} atoms/cm^2 are demanded. One spectacular application of (energy-dispersive) XRF was the completely remotely controlled rock analysis of geological formations on Mars.

Generally it can be said that XRF is an analytical tool for the determination of major, minor and trace elements in a great variety of samples.

8.1.2 Excitation of characteristic radiation and absorption of X-rays

Photoabsorption of X-rays with energy $h\nu$ is usually associated with an inner-shell ionization of the atom. A photoelectron is thereby emitted from the atom and carries the energy $h\nu - E_B$. Here E_B is the binding energy (ionization energy) of the electron. Resonant absorption, where the electron is lifted to a higher energy level within the atom (which is the common effect in optical spectrometry), is extremely unlikely because photons of exactly matching energies are quite rare and all neighbor shells of inner electrons (for which transition probabilities would be reasonably high) are generally already occupied.

Excitation is followed by relaxation. If the K shell was ionized, the most probable electron transition into the K hole is $K \leftarrow L_{III}$, which is associated with emission of a $K\alpha_1$ photon, followed in probability ranking by $K \leftarrow L_{II}$ ($K\alpha_2$), $K \leftarrow M_{III}$ ($K\beta_1$), and so forth. The new holes created by these transitions are successively filled from outer shells in a cascade (accompanied by a corresponding series of emitted fluorescent lines of rapidly increasing wavelengths) until finally a free electron from the continuum is captured. The sum of energies of all emitted lines is of course the ionization energy of the original photoelectron.

Instead of the emission of a photon, an electron is sometimes emitted with a kinetic energy $\Delta E - E_B$, where ΔE is the difference of energies of the relaxing electron. This is called **Auger effect**. Its probability is around 50% or less for elements with medium and higher atomic numbers, but becomes dominating for light elements (99.9% in the case of carbon). Because of the resulting low X-ray intensities, the Auger effect is the main limiting factor in light-element analysis.

Absorption of photons by a thin layer of material is described by $dn = -n\bar{\mu}\,dt$ where n is the number of photons per unit area and unit time (photon flux), of which dn are absorbed while penetrating a thin layer of thickness dt of a material characterized by the (total, linear) **absorption coefficient** $\bar{\mu}$. This is equivalent to

$$n = n_0 \exp(-\bar{\mu}t) \qquad (8.2)$$

for absorption by a thick piece of material (n and n_0 are the photon fluxes behind and in front of the absorber, respectively, and t is the thickness). Note that this law applies only to monochromatic photons, because $\bar{\mu} = \bar{\mu}(E)$ is a function not only of the absorbing material but also of the photon energy.

The coefficient $\bar{\mu}$ summarizes all possible photon interactions, $\bar{\mu} = \bar{\tau} + \bar{\sigma}$, where $\bar{\tau}$ is photoabsorption (i.e. excitation of atoms and subsequent emission of fluorescent radiation), and $\bar{\sigma} = \bar{\sigma}_{coh} + \bar{\sigma}_{inc}$ is the contribution by coherent scattering and incoherent (Compton) scattering (Figure 8.1). Most tables in the literature show mass absorption coefficients μ that are obtained by dividing the linear coefficients by the density ρ, e.g. $\mu = \bar{\mu}/\rho$ (the same applies to τ and σ). Note that the mass absorption coefficient of a material that is composed of several elements with weight fractions w_i is

$$\mu = \sum_i w_i \mu_i$$

Employment of mass absorption coefficients suggests replacing the thickness by the area-related mass $\hat{m} = m/A$ (mass m per unit area A) and rewriting the absorption law as

$$n = n_0 \exp(-\mu\hat{m}) \tag{8.3}$$

respectively (note that $t = m/\rho A$).

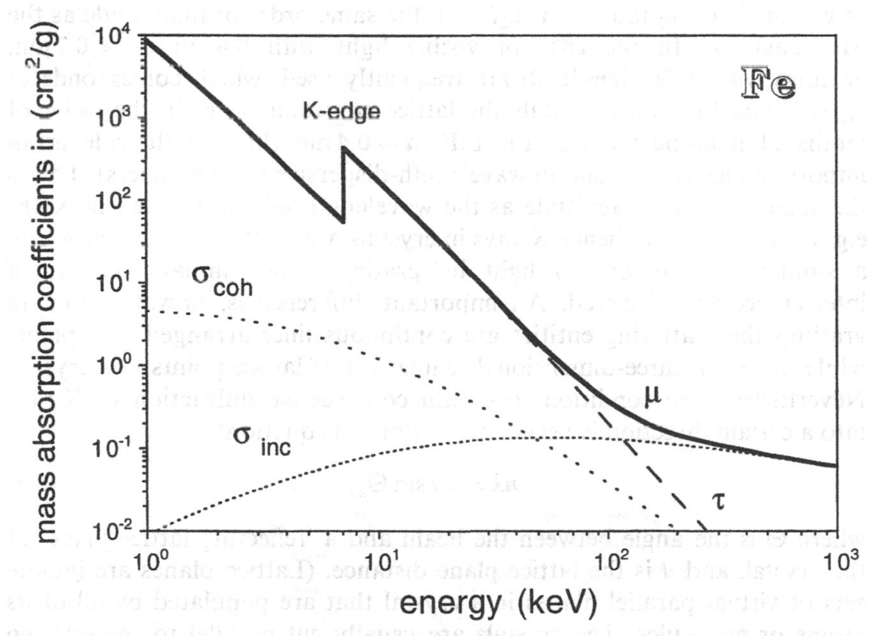

Figure 8.1 The total mass absorption coefficient (μ) is composed of the dominant photoabsorption (τ), and coherent and incoherent scattering (σ). For this example of iron as absorber the K-shell absorption edge is at 7.11 keV.

8.2 Wavelength- and energy-dispersive XRF

8.2.1 Wavelength-dispersive XRF

The basic principle of a wavelength-dispersive (WD) spectrometer is very similar to optical spectrometers with a grating. A grating with a line distance a_G diffracts monochromatic radiation of wavelength λ into a series of discrete angles, for which the condition

$$n\lambda = a_G(\sin\alpha_n + \sin\beta_n)$$

is satisfied. The integral number n is called **order of diffraction**. The explanation for this effect is that the partial waves scattered by each line of the grating interfere constructively or destructively depending upon their phase difference; their superposition is constructive when the above condition is met. Transmission as well as reflection gratings are commonly in use (note that the angles of the incident or observed beam, α and β, are always measured with respect to the surface vector of the grating.) For symmetrical reflection ($\alpha = \beta$) it follows that

$$n\lambda = 2a_G \sin\alpha_n$$

In order to obtain observable patterns in the X-ray range, the distance of grating lines a_G must be roughly in the same order of magnitude as the wavelength λ. In the case of visible light with $0.4\,\mu m < \lambda < 0.7\,\mu m$, gratings with 2400 lines/inch are frequently used, which corresponds to $a_G \approx 10\,\mu m$. For most crystals the lattice constants a are in the order of tenths of nanometers, e.g. for LiF, $a \approx 0.4\,nm$ (lithium fluoride is an important analyzer crystal in wavelength-dispersive spectrometers). This is the same order of magnitude as the wavelength of X-rays used in XRF, e.g. $\lambda_{Cu\,K\alpha} \approx 0.15\,nm$, hence X-rays in crystals are scattered by the atoms in a similar way as optical light by gratings, and similar patterns of interference are observed. An important difference is, however, that in gratings the scattering entities are continuous lines arranged in a plane, while there is a three-dimensional regular set of lattice points in a crystal. Nevertheless, the condition to obtain constructive diffraction of X-rays into a certain direction is very similar (Bragg's equation):

$$n\lambda = 2d\sin\Theta_n \tag{8.4}$$

where Θ is the angle between the beam and a 'reflecting lattice plane' of the crystal, and d is the lattice plane distance. (Lattice planes are infinite sets of virtual parallel planes in a crystal that are populated by all of its atoms or molecules. The crystals are usually cut parallel to the selected reflecting lattice plane, in which case Θ is also the angle of the beam with the surface.)

Experimental set-up. The specimen is illuminated by X-rays from an X-ray tube and emits the excited fluorescent radiation into all directions. A small fraction passes a set of parallel metal plates (collimator) and forms a quite parallel beam with a well defined angle Θ to the surface of the dispersing crystal (Figure 8.2). The crystal 'reflects' (in fact, diffracts) into a detector only those wavelengths which satisfy Bragg's equation

$$\lambda_n = \frac{2d}{n} \sin \Theta$$

Most crystal spectrometers are designed as goniometers, where the crystal can be rotated by Θ while the detector moves by 2Θ (mirror geometry). The fluorescent spectrum is sequentially analyzed by continuously scanning (or step scanning) through a range of Θ–2Θ (**sequential spectrometer**). Tables (or computers) are then used to relate the reflection angles (or wavelengths, as obtained from Bragg's equation) to emission lines and chemical elements (qualitative analysis). Note that in most instrument settings and tables the values of 2Θ and $2d$ are indicated rather than Θ and d.

For quantitative analysis the angles are set sequentially to peak and background positions. In industrial applications, where always the same set of elements is analyzed and speed is important, **simultaneous spectrometers** are sometimes used. In such instruments several analyzer units (up to 30) are built into the spectrometer, each consisting of a crystal and a detector at fixed angles Θ and 2Θ, respectively, and collimators. Thereby several lines (and background positions) can be measured at the same time.

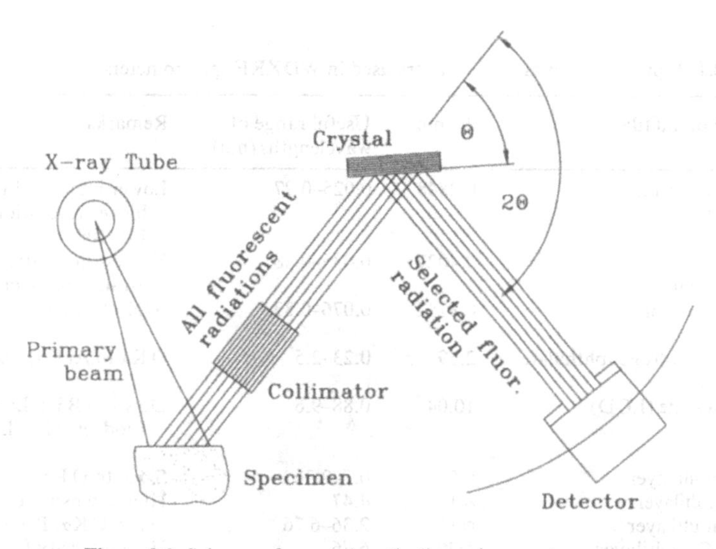

Figure 8.2 Scheme of a wavelength-dispersive spectrometer.

Vacuum of around 0.2 mbar is generally necessary for longer wavelengths, i.e. above ≈ 0.2–0.3 nm. Note that measurement of fine powders and liquids in an evacuated chamber is generally not recommended.

Crystals and multilayer structures. About 70 crystals and types of multilayer structures have been reported in the literature as being useful for one application or another (Table 8.1). They are primarily selected according to their *d*-spacing (in the case of multilayers, *d* denotes the distance between two adjacent layers), but 'reflectivity' (intensity ratio of diffracted and incident beams) and resolution are important secondary properties as well as price and availability, chemical stability, sensitivity to temperature fluctuations, hygroscopy, toxicity, or sufficient mechanical elasticity for being bent to focusing monochromators.

In typical spectrometers the reflection angle 2Θ can be varied within about 10° < 2Θ < 150° (depending on construction details). This defines a certain range of wavelengths λ which can be measured with a crystal of a given *d*-spacing. For crystals with large *d*-spacings this range is situated at long wavelengths and vice versa. The largest practical *d*-spacings are around 2*d* ≈ 2.5 nm (TAP) in natural crystals and 2*d* ≈ 20 nm (even wider for special applications) in multilayers; the shortest is 2*d* ≈ 0.16 nm (α-quartz (502) reflection).

It is generally more difficult to find suitable crystals for the long than for the short wavelengths. For wavelengths above 2.5 nm multilayer structures (Love and Scott 1987, Nicolosi *et al.* 1987, van Eenbergen and Volbert 1987, Huang *et al.* 1989) are employed as dispersive devices. One

Table 8.1 Typical crystals and multilayers used in WDXRF spectrometers

Crystal or multilayer	2*d* (nm)	Useful range of wavelengths (nm)	Remarks
Lithium fluoride (LiF (110))	0.2848	0.025–0.27	Lower intensity but higher resolution than LiF 100
Lithium fluoride (LiF (100))	0.4027	0.035–0.38	Highest intensity, widest λ-range of all crystals
Pentaerythritol (PET)	0.8742	0.076–0.83	Al Kα to Cl Kα
Thallium hydrogenphthalate (TAP)	2.59	0.23–2.5	O Kα to Al Kα, toxic
Lead stearate (LOD)	10.04	0.88–9.6	Down to B Kα Langmuir–Blodgett film (LBF)
W/Si multilayer	≈5.5	0.7–2.36	Si Kα to O Kα
V/C multilayer	≈10	4.47	High intensity for C Kα
Ni/C multilayer	≈15	2.36–6.76	O Kα, C Kα, B Kα
Mo/B_4C multilayer	≈20	6.76	High intensity for B Kα

type, the Langmuir–Blodgett films (LBF), are made from monomolecular layers of surfactant molecules that are formed on a water surface. By a special technique they are transferred and stacked, monolayer by monolayer, onto a solid substrate. Lead stearate is an example that is often found in electron microprobes. The second type consists of a set of around 50 layer pairs of alternating light elements and elements with high atomic numbers,[1] and is manufactured by sputtering or evaporating techniques. Silicon single crystals are generally used as substrates.

Detectors for wavelength-dispersive XRF. In most cases a **scintillation counter** (SC) is used for the harder radiation and **gas-flow counters** (FC) for wavelengths above 0.2–0.3 nm, but there is no sharp limit. Note that the range of FCs corresponds also to the vacuum range. In some instruments both counters can be used simultaneously in a tandem configuration for a widely overlapping wavelength range. In this case the harder radiation that is not absorbed by the gas of the FC enters the SC, which is positioned behind the FC.

The resulting pulse height distribution is roughly Gaussian with a full width at half-maximum (FWHM) of ≈15–50% of the energy (the better values are for FCs). This is just sufficient to distinguish (i.e. accept or reject) photons of the various orders of reflection, λ_n, which enter the detector simultaneously at a given reflection angle Θ, by electronic circuitry (single-channel pulse height analyzer, PHA).

8.2.2 Energy-dispersive XRF

In energy-dispersive X-ray fluorescence spectrometry (EDXRF) the identification of characteristic lines is performed using detectors that directly measure the energy of the photons. In the simplest case an electron is ejected from an atom of the detector material by photoabsorption (Figure 8.3). The loss of energy of this just created primary electron results in a shower of electron–ion pairs in the case of a proportional counter, optical excitations in the case of a scintillation counter, or showers of electron–hole pairs in a semiconductor detector. The resulting detector signal is proportional to the energy of the incident photon, in contrast to wavelength dispersion in which the Bragg reflecting properties of a crystal are used to disperse X-rays at different reflection angles according to their wavelengths. Although energy-dispersive detectors generally exhibit

[1] The intensity of coherently scattered radiation is proportional to the number of electrons in the atoms (i.e. the atomic number), hence only the contribution of the heavy layers is significant. Because they form regularly arranged, discontinuous scattering entities, the partial waves from the individual layers superpose constructively or destructively as a function of reflection angle.

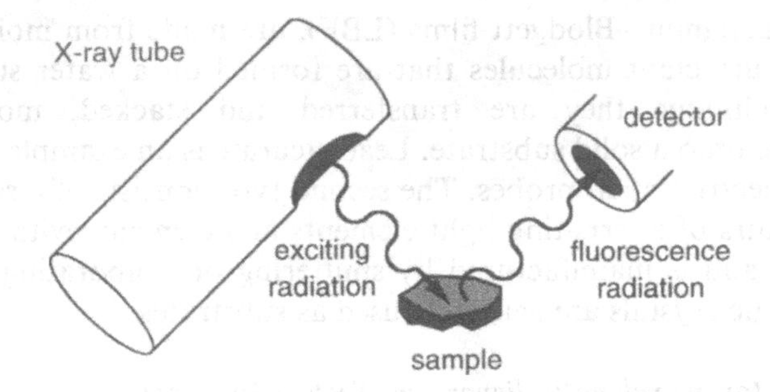

Figure 8.3 Principle components of an energy-dispersive XRF system: source (X-ray tube), sample, detector.

poorer energy resolution than wavelength-dispersive analyzers, they are capable of detecting simultaneously a wide range of energies.

The most frequently used detector in EDXRF is the **silicon semiconductor detector**, which nowadays can have excellent energy resolution. The two other types of detectors, mentioned above, with their poorer energy resolution are limited to special cases where certain features of semiconductor detectors are not acceptable. Also the germanium semiconductor detector with its comparable characteristics has a major drawback for conventional XRF: inherently the escape peaks of intense lines can obscure other lines of interest.

For light elements with neighboring atomic numbers the separation of characteristic lines can only be maintained with very good detectors, but the interference between the Kβ line of an element and the Kα line of the next higher-Z element can usually only be resolved by computer aid. In many cases wavelength-dispersive spectrometers do not suffer from these interferences, and thus it seems that there would be little use for energy-dispersive systems. However, the wavelength-dispersive spectrometers have the disadvantage of the need for precise collimation, which results in a greatly reduced geometrical efficiency for the detection. Also the simultaneous multi-element capability of EDXRF therefore gives it an important role in X-ray analysis.

A lithium-drifted silicon crystal (**Si(Li) detector**) consists of a p-i-n structure (Figure 8.4), referring to the p-type contact on the entry side, the intrinsic active volume, and the lithium-diffused n contact, typically less than 10 mm in diameter and about 3–5 mm thick. The donor lithium has been drifted under the influence of an electric field in order to compensate p-type impurities in the original silicon. When a reverse bias (in the range of 500–1000 V) is applied to the device, the drifted region acts as an insulator with an electric field throughout its volume. A photon reaching the active volume generates photoelectrons, as well as Compton and Auger electrons, which lose energy by producing ionization in the form of

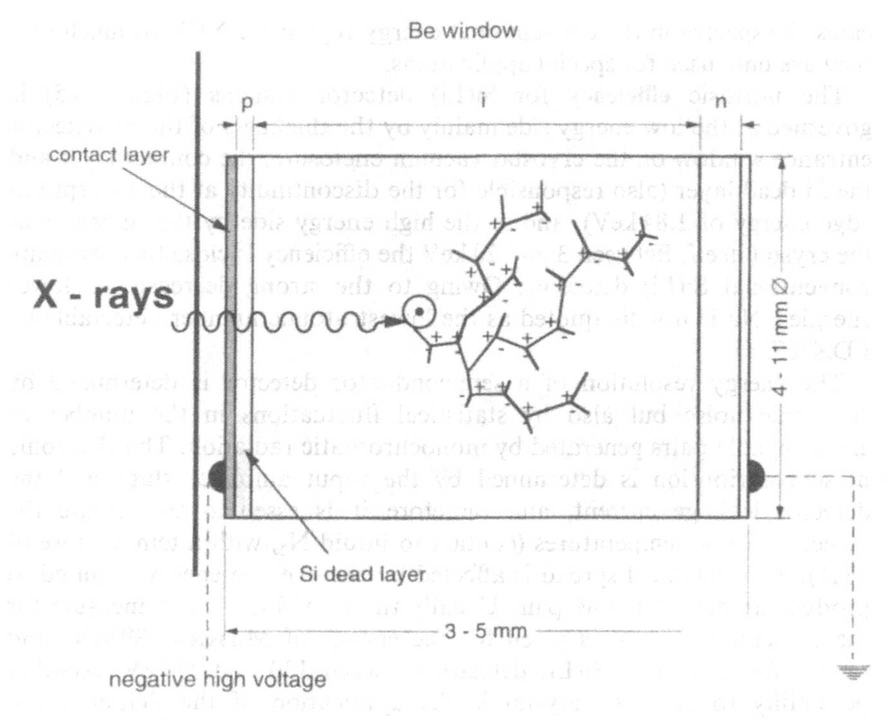

Figure 8.4 Cross-section of an Si(Li) detector crystal with its p-i-n structure and the production of electron–hole pairs.

electron–hole pairs. The free charges are swept away by the applied bias and collected within a time of typically 25–100 ns. Since the average energy to generate an electron–hole pair is well defined (e.g. 3.7 eV for an Si(Li) detector), the *total number of charges is directly proportional to the energy* of the incident photon. The frequency of such events is proportional to the number of photons (photon count rate).

In an ideal detector the charge is collected completely and the response of the system would be a single peak, containing only counts of a certain energy and no other counts elsewhere in the spectrum. In practice some artifacts are observed: incomplete charge collection, resulting in a tail to the lower-energy side of a principal peak (low-energy tailing), can be strongly reduced by the design of the crystal. The escape of photons (or secondary particles) from the detector surface is a fundamental limit in the detector background, which is most prominent for the case when a hole is created in the K shell of the silicon and the accompanying emission of its characteristic radiation leaves the crystal. Consequently an escape peak at an energy 1.74 keV less than the full energy peak can be observed. For germanium detectors (two escape peaks for each line) the effect complicates the spectra in the conventional energy region for XRF so much that they are only used for special applications.

The intrinsic **efficiency** for Si(Li) detector systems (Figure 8.5) is governed at the low-energy side mainly by the thickness of the Be detector entrance window on the cryostat vacuum enclosure, the contact layer and the Si dead layer (also responsible for the discontinuity at the absorption-edge energy of 1.84 keV), and at the high-energy side by the thickness of the crystal itself. Between 3 and 20 keV the efficiency is close to 1 for many conventional Si(Li) detectors. Owing to the strong decrease for lower energies, Na is usually quoted as the lowest atomic number detectable by EDXRF.

The energy resolution of a semiconductor detector is determined by electronic noise but also by statistical fluctuations in the number of electron–hole pairs generated by monochromatic radiation. The electronic noise contribution is determined by the input amplifier stage and the detector leakage current, and therefore it is essential to operate the detector at low temperatures (contact to liquid N_2, with a temperature of 77 K). The statistical spread is affected by the average energy required to produce an electron–hole pair. Usually the (FWHM) – as a measure for the detector quality – is given for the energy of Mn Kα (5.89 keV) and ranges for a modern Si(Li) detector between 130 and 180 eV. Another possibility to cool the crystal is the application of the Peltier effect, resulting in small, lightweight detectors (no Dewar for the storage of the

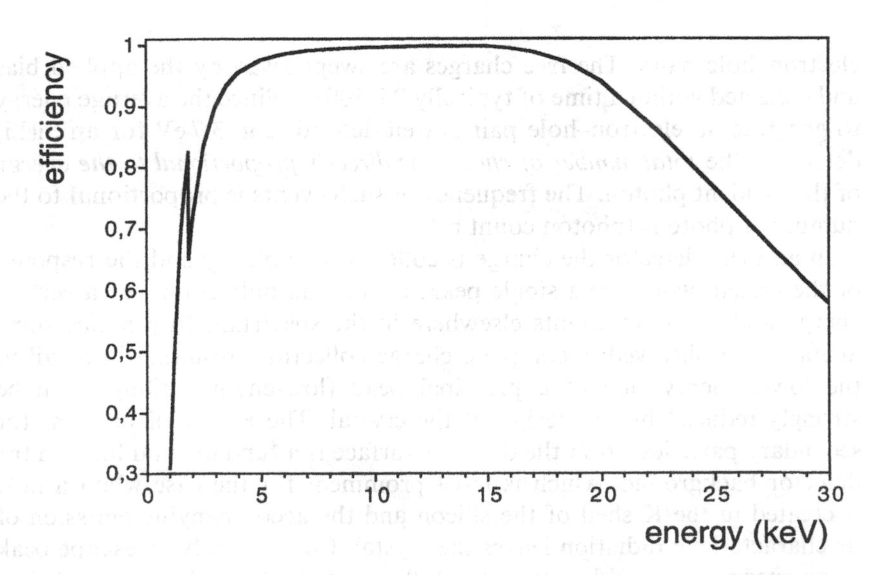

Figure 8.5 Calculated intrinsic efficiency for an Si(Li) detector with a dead layer of 300 nm and a 10 μm thick Be entrance window.

liquid nitrogen) with adequate energy resolution (200 eV). Recently a liquid-helium-cooled detector, operating as a superconducting tunnel junction device, with an energy resolution of 13 eV has been designed, thus coming close to the resolution of wavelength-dispersive spectrometers.

The preamplifier stage integrates each detector charge signal to generate a voltage step proportional to the charge. This is then amplified and shaped in a series of integrating and differentiating stages. Owing to the finite pulse-shaping time, in the range of microseconds, the system will not accept any other incoming signals in the meanwhile (dead time), but extend its measuring time instead. In a further step the height of these signals is digitized as a channel number (analog-to-digital converter, ADC), stored to a memory (multichannel analyzer, MCA) and finally displayed as a spectrum, where the number of counts reflects the respective intensity. In a more modern approach the output signals of the preamplifier are digitized directly, which can increase the throughput of the system significantly.

For high count rates there is an increasing probability that two photons of, for example, a very intense line, are absorbed in the detector crystal within such a short time interval that their charges are not collected as two individual signals with a certain energy, but rather as a single signal with twice the energy (sum peak).

For a pure copper sample, excited by an Rh-anode X-ray tube, a spectrum (Figure 8.6) was collected that exhibits, apart from the dominant

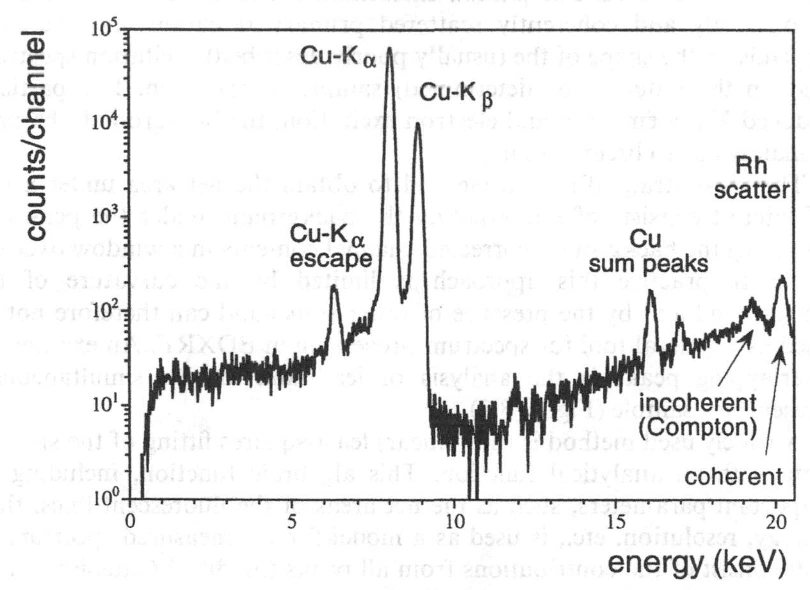

Figure 8.6 Spectrum of a pure copper specimen, collected by an Si(Li) detector (excited by an Rh anode X-ray tube).

Table 8.2 Energies of lines in a spectrum of a pure copper sample (Figure 8.6)

$E_{\mathrm{Cu\,K}\alpha} = 8.04\,\mathrm{keV}$
$E_{\mathrm{Cu\,K}\beta} = 8.91\,\mathrm{keV}$
$E_{\mathrm{sum\,1}} = E_{\mathrm{Cu\,K}\alpha} + E_{\mathrm{Cu\,K}\alpha} = 16.08\,\mathrm{keV}$
$E_{\mathrm{sum\,2}} = E_{\mathrm{Cu\,K}\alpha} + E_{\mathrm{Cu\,K}\beta} = 16.95\,\mathrm{keV}$
$E_{\mathrm{escape}} = E_{\mathrm{Cu\,K}\alpha} - 1.74\,\mathrm{keV} = 6.30\,\mathrm{keV}$
$E_{\mathrm{coherent}} = E_{\mathrm{Rh\,K}\alpha} = 20.20\,\mathrm{keV}$
$E_{\mathrm{Compton}}(90°) = 19.43\,\mathrm{keV}$

characteristic Cu Kα and Cu Kβ lines, an escape peak, sum peaks of the intense Cu lines and the scattered Rh K lines (Table 8.2). The energy shift (Compton shift) of the inelastic scatter peak is directly related to the geometry of the experiment with a scattering angle of about 90°.

Spectrum evaluation. Spectrum evaluation in energy-dispersive XRF is certainly more critical than in WDXRF, because of the relatively low resolution of the solid-state detectors employed. The aim is the extraction of the analytically relevant information (net number of counts under a peak) from experimental spectra.

In EDXRF the characteristic radiation of a particular line can be described in an adequate first-order approximation by a Gaussian function (detector response function). The spectral background results from a variety of processes: For photon excitation, the main contribution is the incoherently and coherently scattered primary radiation and therefore depends on the shape of the (usually poorly described) excitation spectrum and on the (later to be determined) sample composition. For particle-induced X-ray emission and electron excitation, the background observed is mainly due to bremstrahlung.

The most straightforward method to obtain the net area under a line of interest consists of interpolating the background under the peak and summing the background-corrected channel contents in a window over the peak. In practice this approach is limited by the curvature of the background and by the presence of other peaks and can therefore not be used as a general tool for spectrum processing in EDXRF. An example of overlapping peaks is the analysis of lead and arsenic simultaneously present in a sample (Figure 8.7).

A widely used method is (non-linear) least-squares fitting of the spectral data with an analytical function. This algebraic function, including all important parameters, such as the net areas of the fluorescent lines, their energy, resolution, etc., is used as a model for the measured spectrum. It will consist of the contributions from all peaks (modified Gaussian peaks, with corrections for low-energy tailing, escape peaks, etc.) within a certain region of interest and the background (described by, for example, linear

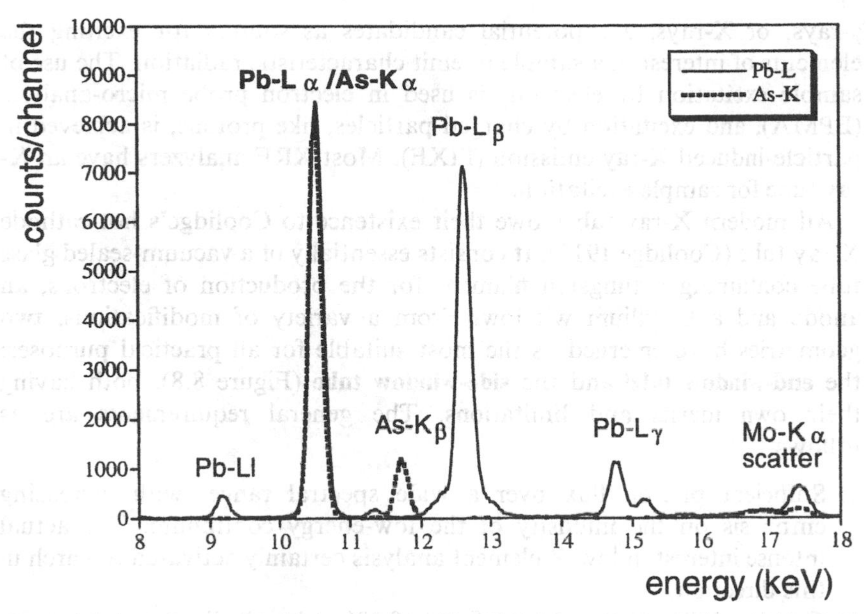

Figure 8.7 A spectrum of As K, overlapped with a Pb L line spectrum, both excited by a Mo X-ray tube, under identical conditions. The energy of As $K\alpha_{1,2}$ (10.53 keV) and of Pb $L\alpha_{1,2}$ (10.55 keV) cannot be separated by an Si(Li) detector.

or exponential polynomials). The optimum values of the parameters are those for which the difference between the model and the measured spectrum is minimal. Unfortunately some of these parameters are non-linear, which places some importance on the minimization procedure (usually the Marquardt algorithm is used).

In another frequently used approach the discrete deconvolution of a spectrum with a so-called top-hat filter suppresses the low-frequency component, i.e. the slowly varying background. A severe distortion of the peaks is introduced. But applying this filter to both the unknown spectrum and well defined, experimentally obtained, reference spectra, a multiple linear least-squares fitting to the filtered spectra will result in the net peak areas of interest. A disadvantage of this method is that reference and unknown spectra should be acquired under preferably identical conditions; especially, energy calibration changes of more than only a few eV can generate large systematic errors.

8.3 X-ray tubes and radioisotope sources

8.3.1 X-ray tubes

A variety of radiation sources of sufficient energy, emitting either particles,

γ-rays, or X-rays, are potential candidates as sources for exciting the elements of interest in a sample to emit characteristic radiation. The use of sample excitation by electrons is used in electron probe micro-analysis (EPMA), and excitation by charged particles, like protons, is achieved in particle-induced X-ray emission (PIXE). Most XRF analyzers have an X-ray tube for sample excitation.

All modern X-ray tubes owe their existence to Coolidge's hot-cathode X-ray tube (Coolidge 1913). It consists essentially of a vacuum-sealed glass tube containing a tungsten filament for the production of electrons, an anode and a beryllium window. From a variety of modifications, two geometries have emerged as the most suitable for all practical purposes: the **end-window tube** and the **side-window tube** (Figure 8.8), both having their own merits and limitations. The general requirements are as follows:

1. Sufficient photon flux over a wide spectral range, with increasing emphasis on the intensity of the low-energy continuum. The actual intense interest in low-Z element analysis certainly activated research in this direction.
2. Good stability of the photon flux (<0.1% at least). Short-term stability is an absolute requirement for obtaining acceptable precision.
3. Tunable tube potential (only for side-window tubes), allowing the creation of the most effective excitation conditions for each element, because the intensity of the analyte lines varies considerably with excitation conditions.
4. Freedom from too many interfering lines from the characteristic spectrum of the anode (scatter peaks).

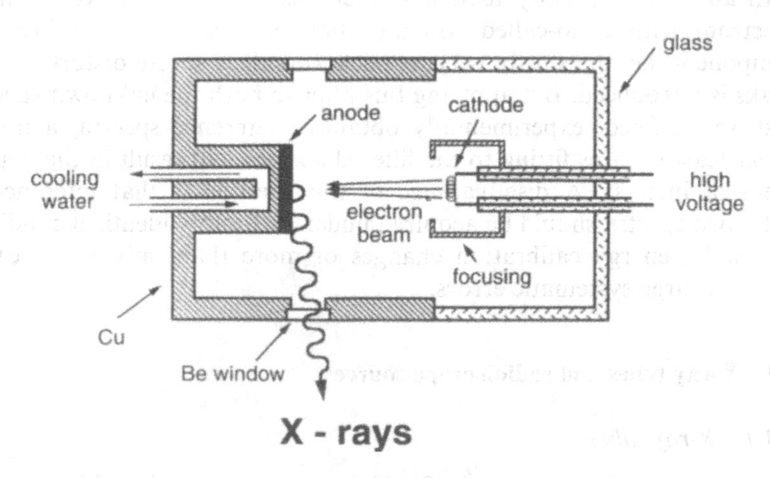

Figure 8.8 Scheme of a side-window X-ray tube.

The working principle is as follows: A low filament voltage (6–14 V) is used to heat the filament wire so that it will thermally emit electrons (cathode). A high voltage, applied between the filament and a target of a pure element, accelerates the electrons, which are focused by the Wehnelt cylinder, to the anode. The deceleration of the electrons in the Coulomb field of atoms of the anode material and inner-shell ionization give rise to the production of continuous (bremstrahlung) and characteristic X-rays emitted from the anode. Because the X-ray tube is under high vacuum, the radiation must pass through a window, usually made of beryllium. The radiation output power is rather poor with respect to the input power and is of the order of 1%, making the device a very inefficient transformer of electron current to electromagnetic radiation. Consequently, for high-power tubes the anode block (copper) has to be cooled by water.

An X-ray tube is characterized by its anode material, its input power (in kW), the range of the high voltage between anode and cathode (10–100 kV), the tube current (in mA), the incident angle of the electron beam on the anode, the take-off angle and the thickness of the beryllium window.

The current can be as high as 300 mA for rotating anodes, or as low as 10 μA in combination with EDXRF systems. Typical anode materials are W, Mo, Cu, Rh, Cr, Sc, Ag and Au. For many applications in XRF the spectral background, caused by the scattered continuous radiation, is the limiting factor. The primary spectrum can be shaped by the insertion of suitable filter materials, suppressing the low-energy continuum and/or reduction of the Kβ line of the anode material.

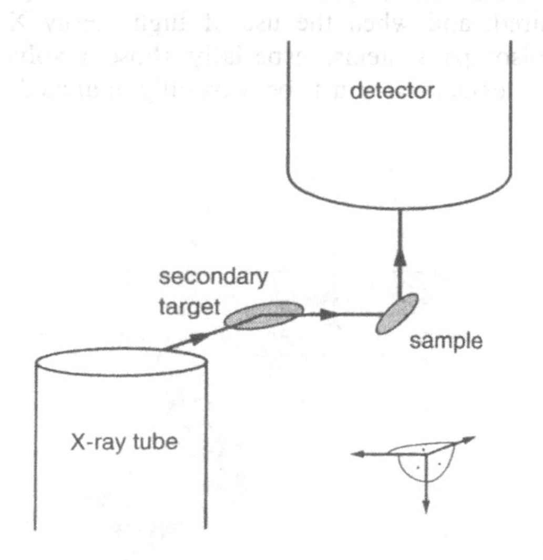

Figure 8.9 Secondary target arrangement (EDXRF).

In a more sophisticated approach the X-ray tube is used to irradiate a suited secondary target material (Figure 8.9). The emerging characteristic radiation is in turn used to excite the sample, with the effect that the spectral background is reduced, because the primary radiation can reach the detector only by double scattering. Furthermore the scattered primary radiation emerging from the secondary target is linearly polarized, if it is at a right angle to the direction of the primary radiation. A further background reduction is gained, exploiting this polarization effect, if the detector is positioned in a triaxial orthogonal geometry. Owing to the substantial excitation efficiency loss, when using a secondary target, high-power X-ray tubes are used and longer counting times are common.

8.3.2 Radioisotope sources

Only a few radioisotopes are frequently used for EDXRF, either as point or annular source (Figure 8.10), but are not used in WDXRF, which is related to the fact that the X-ray yield of radioisotope sources is low, with intensities several orders of magnitude lower than that of X-ray tubes as used in combination with crystal spectrometers.

Radioisotope XRF systems are often tailored to a specific but limited range of applications. They are simpler and often considerably less expensive than analysis systems based on X-ray tubes, but these attributes are often gained at the expense of flexibility. Radioisotope excitation is preferred to X-ray tubes when simplicity, ruggedness, reliability and cost of equipment are important; when minimum size, weight and power consumption are necessary; when a very constant and predictable X-ray output is required; and when the use of high-energy X-rays is advantageous. Radioisotope systems, especially those involving scintillation or proportional detectors, must be carefully matched to the specific application.

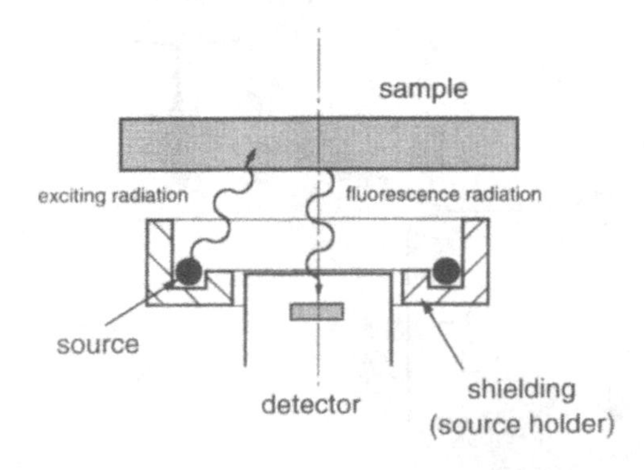

Figure 8.10 Geometry of an EDXRF spectrometer with annular source excitation.

Table 8.3 Typical radioisotope sources used for XRF

Radioisotope	Half-life	Photon energies emitted (keV)	Remarks
^{55}Fe	2.7 y	5.9, 6.5 (Mn K)	Analysis of low-*Z* elements
^{109}Cd	463 d	22.1, 25.0 (Ag K); 88 (γ)	Similar to Mo/Rh K X-ray tube excitation
^{241}Am	432 y	59.6 (γ); 13.9, . . . (Np L)	Dominant line at 59.6 keV
^{57}Co	272 d	14.4, 122, 136 (γ); 6.4, 7.1 (Fe K)	With usual sealing only lines at 122, 136 keV
^{238}Pu	88 y	13.6, . . . (U L)	

The **activity** of radioisotopes is specified in terms of the rate of disintegration of the radioactive atoms, i.e. decays per second or becquerels (Bq). (The becquerel replaces the non-SI unit, the Curie (Ci), which equals 3.7×10^{10} becquerels.) The activity decreases with time from A_0 to $A(t)$ after an elapsed time t:

$$A(t) = A_0 \exp(-0.693t/T_{1/2}) \tag{8.5}$$

where $T_{1/2}$ is the half-life of the radioisotope. The source decays to half of its original emission rate after the time equal to its half-life has passed. The radioisotope source has usually to be replaced after several half-lives. Several sources are listed in Table 8.3.

8.4 Methods of quantitative analysis

The observed photon rate from an analyte element in a specimen is a function of many factors including the concentration (weight fraction) of the element, the matrix (accompanying elements), the specimen type (bulk, powder, liquid, thin-film structure, etc.), size, preparation, geometrical set-up, spectral distribution of the exciting radiation, and the detection system. Theoretical as well as empirical approaches are used to determine concentrations from fluorescent **intensities.**[2]

The theoretical methods are based on mathematical models for the excitation of atoms and subsequent relaxation processes, the absorption of radiation within the specimen, and inter-element effects. Compared to a real set-up, simplifying assumptions are made, for example that the

[2] The term 'intensity' is widely used as a synonym for 'photon count rate' and is then the number of detected photons per second. It depends upon the actual number of emitted photons as well as on the detection efficiency (including the whole spectrometric set-up). The detection efficiency is however generally neglected, because in almost all methods 'count-rate ratios' (relative intensities) are used, where this factor cancels. The count-rate ratio is built from the count rate of the analyzed sample and the count rate of a reference sample for a given analyte line measured under identical experimental conditions.

specimen is perfectly flat and homogeneous, and that the incident primary beam is parallel.

An alternative is the empirical parameter method, which employs relatively simple mathematical descriptions of the relationship between photon counts and concentration (calibration curves). The general principle is that the 'ideal' calibration curve is assumed to be a linear function, which is obtained from the (non-linear) experimental relationship by applying a number of corrections. The coefficients, by which the extent of the various corrections is introduced, are called **empirical parameters**. They are determined experimentally from calibration standards or sometimes by theoretical approaches.

Many times in both methods count-rate ratios ('relative intensities') rather than absolute counts are used. By building ratios of the count rate from a line of an element with the one from the same element in another specimen (which can be a standard, containing this element, or a pure element), many geometrical factors, the detection efficiency and the absolute intensity level of the primary radiation cancel.

8.4.1 Fundamental parameter method

The fundamental parameter equations are based on two physical effects: **absorption** of photons and the **excitation/relaxation** processes of atoms (Sherman 1955, Shiraiwa and Fujino 1966, Criss and Birks 1968). Because of the basic physical parameters involved, such as absorption coefficients, transition probabilities and fluorescent yields, this is called the fundamental parameter method.

A simple mathematical model. The primary X-ray spectrum is described by a function $x(E)$ such that $\mathrm{d}n = x(E)\,\mathrm{d}E$ is the number of virtually monochromatic photons of energies within $[E, E + \mathrm{d}E]$ per unit area and time. They impinge on a specimen surface at an angle ψ_1 and penetrate to depth t (Figure 8.11) corresponding to a path of length $t/\sin\psi_1$, equivalent to the area-related mass $\hat{m}/\sin\psi_1$.

There, the remaining

$$\mathrm{d}n = x(E)\,\mathrm{d}E \exp\left(-\frac{\mu\hat{m}}{\sin\psi_1}\right)$$

photons pass through a thin layer $\rho\,\mathrm{d}t = \mathrm{d}\hat{m}$, whereby

$$\mathrm{d}^2 n = x(E)\,\mathrm{d}E \exp\left(-\frac{\mu\hat{m}}{\sin\psi_1}\right)\frac{\mu\,\mathrm{d}\hat{m}}{\sin\psi_1}$$

are absorbed.

Of the $\mathrm{d}^2 n$ photons, a fraction $w_i\mu_i/\mu$ is absorbed by element i, a fraction $(\tau/\mu)_i$ by photoeffect rather than by scattering, and a fraction $\tau_K/\tau =$

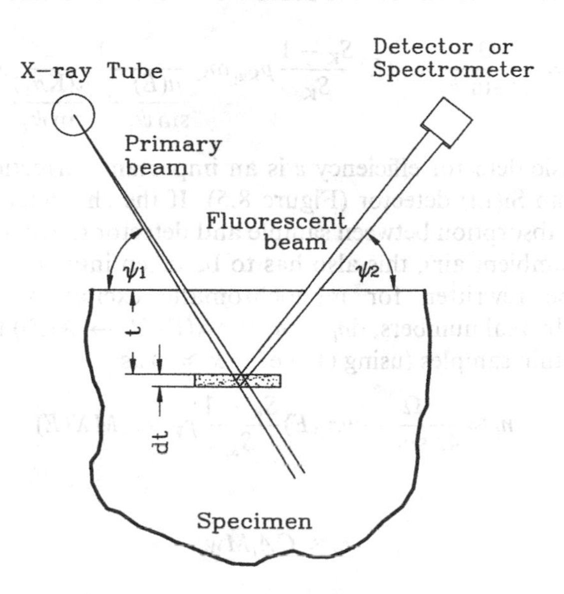

Figure 8.11 Definition of geometry for the fundamental parameter method.

$(S_K - 1)/S_K$ in the K shell. S_K is called the **absorption-edge jump** and defined as the ratio of the mass photoabsorption coefficient on the high- and low-energy sides, respectively, of the K-edge discontinuity. We assume that a $K\alpha_1$ line is the selected analyte line. Relaxation occurs with probability $p_{K\alpha_1}$ by $K \leftarrow L_{III}$, and photoemission (rather than an Auger effect) occurs with probability ω_K. From the point of emission the fluorescent $K\alpha_1$ photon is emitted with a probability $(\Omega/4)\pi$ into the direction of the detector. A factor $\exp(-\mu(K\alpha_1)\hat{m}/\sin\psi_2)$ corrects for absorption on the path to the specimen surface at an angle ψ_2.

Altogether, the number of fluorescent photons at the position of a detector at the specimen surface is

$$\mathrm{d}^2 n = \frac{\Omega}{4\pi \sin\psi_1} w_i \tau_i(E) \frac{S_K - 1}{S_K} p_{K\alpha_1} \omega_K \exp\left[-\hat{m}\left(\frac{\mu(E)}{\sin\psi_1} + \frac{\mu(K\alpha_1)}{\sin\psi_2}\right)\right] x(E)\,\mathrm{d}E\,\mathrm{d}\hat{m}$$

This can now be integrated over t or $\hat{m}$ $\left(\int_{t=0}^{t=D} \mathrm{d}^2 n = \int_{\hat{m}=0}^{\hat{m}=\hat{M}} \mathrm{d}^2 n\right)$

$$\mathrm{d}n_{K\alpha_1} = \frac{\Omega}{4\pi \sin\psi_1} w_i \tau_i(E) \frac{S_K - 1}{S_K} p_{K\alpha_1} \omega_K \frac{1 - \exp\left[-\hat{M}\left(\dfrac{\mu(E)}{\sin\psi_1} + \dfrac{\mu(K\alpha_1)}{\sin\psi_2}\right)\right]}{\dfrac{\mu(E)}{\sin\psi_1} + \dfrac{\mu(K\alpha_1)}{\sin\psi_2}} x(E)\,\mathrm{d}E \tag{8.6}$$

For bulk material $\left(\int_{\hat{m}=0}^{\hat{m}=\infty} \mathrm{d}^2 n\right)$

$$\mathrm{d}n_{\mathrm{K}\alpha_1} = \frac{\Omega}{4\pi \sin \psi_1} w_i \tau_i(E) \frac{S_\mathrm{K} - 1}{S_\mathrm{K}} p_{\mathrm{K}\alpha_1} \omega_\mathrm{K} \frac{1}{\dfrac{\mu(E)}{\sin \psi_1} + \dfrac{\mu(\mathrm{K}\alpha_1)}{\sin \psi_2}} x(E)\,\mathrm{d}E \quad (8.7)$$

The intrinsic detector efficiency ε is an important correction factor and is given for an Si(Li) detector (Figure 8.5). If the characteristic radiation suffers from absorption between sample and detector (particularly for light elements in ambient air), this also has to be taken into account. Equation (8.6) can be rewritten for monochromatic excitation (by replacing differentials by real numbers, $\mathrm{d}n_i \rightarrow n_i$ and $x(E)\,\mathrm{d}E \rightarrow X(E)$) for the special case of very thin samples (using $(1 - \mathrm{e}^{-x})/x \approx 1$) as

$$n_i \approx \frac{\Omega}{4\pi \sin \psi_1} w_i \tau_i(E) \frac{S_\mathrm{K} - 1}{S_\mathrm{K}} p_{\mathrm{K}\alpha_1} \omega_\mathrm{K} \hat{M} X(E) \quad (8.8)$$

hence

$$n_i \approx C A_i \hat{M} w_i \quad (8.9)$$

where

$$C = \frac{\Omega}{4\pi \sin \psi_1} X(E) \quad \text{and} \quad A_i = \tau_i(E) \frac{S_\mathrm{K} - 1}{S_\mathrm{K}} p_{\mathrm{K}\alpha_1} \omega_\mathrm{K}$$

Here C includes all element-independent factors (which are generally not known in detail) and A_i all element-specific constants (which are quite well known). There is no absorption correction needed and n_i is independent of the concentrations w_j of other elements.

By building ratios n_i/n_j, the unknown factors C and $\hat{M}$ cancel,

$$\frac{n_i}{n_j} = \frac{A_i\, w_i}{A_j\, w_j} \quad (8.10)$$

An important application of this method is the quantification in TXRF (section 8.7.1).

Secondary excitation. Sometimes fluorescent photons have sufficient energy to excite fluorescent radiation of other atoms in the specimen. This effect is called **secondary excitation** and can be a major contribution to the observed photons (Figure 8.12). Even a second level of interaction can be observed, when a secondary photon excites a tertiary photon (higher levels of interaction have no practical importance). In the range of transition elements these effects are strong for elements differing by 2 in atomic numbers. Stainless steel (Cr–Fe–Ni) is the usual example for demonstrating secondary excitation (Fe → Cr, Ni → Cr and Ni → Fe) and tertiary excitation (Ni → Fe → Cr). The actual contributions from such effects are often around 5–10%; in a few cases they are up to 30% and higher.

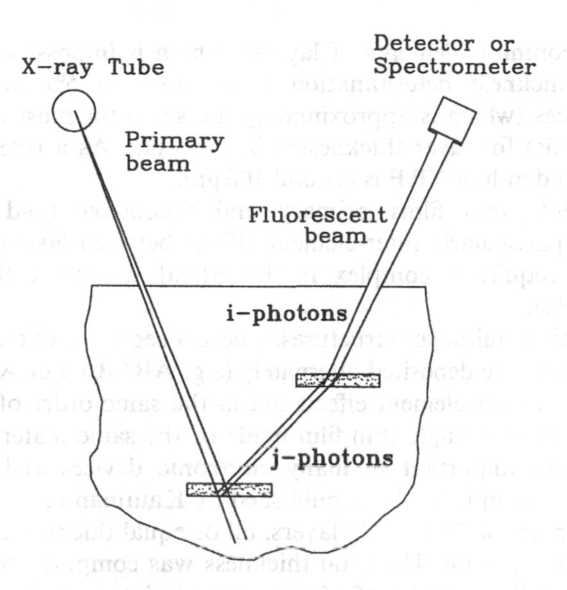

Figure 8.12 Secondary excitation where element j can excite element i.

For bulk material mathematical models for secondary excitation are normally included in commercial fundamental parameter computer algorithms, but tertiary excitation is rather complex and requires considerable computing times. If the energy E_j of photons emitted from element j is higher than the absorption-edge energy $E_{edge,i}$ for the analyte line of element i, the contribution of this j line to the observed i line is

$$n_{i,sec} = \frac{\Omega}{4\pi \sin\psi_1} w_i w_j \frac{1}{2} \frac{S_i - 1}{S_i} \frac{S_j - 1}{S_j} p_i p_j \omega_i \omega_j \int_{E_{edge,j}}^{E_{max}} \frac{\tau_i(E_j)\tau_j(E)x(E)}{\frac{\mu(E)}{\sin\psi_1} + \frac{\mu(E_i)}{\sin\psi_2}}$$
$$\times \left[\frac{\sin\psi_1}{\mu(E)} \ln\left(1 + \frac{\mu(E)}{\mu(E_j)\sin\psi_1}\right) + \frac{\sin\psi_2}{\mu(E_i)} \ln\left(1 + \frac{\mu(E_i)}{\mu(E_j)\sin\psi_2}\right)\right] dE \quad (8.11)$$

The total contribution of secondary excitation is the sum of the contribution of all lines j from all elements with $E_j > E_{edge,i}$. Note that no line can excite another line of the same subshell of a given atom, but for example the K lines of an atom can excite its own L lines.

Applications of the fundamental parameter method. In general, agreement of quantitative XRF by fundamental parameters (K-line analysis) with other methods is better than 1% except for light elements, and even better than 0.1% when similar standards are used.

Thin-film thicknesses can be determined simultaneously with element analysis (even for multiple thin-film structures, however with restrictions regarding common elements of layers), which is impossible by any other method. Thickness determination is accurate to 5% under favorable circumstances (which is approximately the same for most other routinely used methods) for layer thicknesses below 1 μm. As a rule of thumb the information depth of XRF is around 100 μm.

In multiple thin films primary and secondary (and higher-order) excitation, particularly inter-element effects between layers (and with the substrate), require a complex mathematical treatment (Mantler 1987, de Boer 1990).

In repetitive multilayer structures, where a sequence of a few layer types (generally two) are deposited alternately (e.g. ABABAB or ABCABCABC), the combined inter-element effects are in the same order of magnitude as in a homogeneous single thin film made of the same material. Such layer structures are important in many electronic devices and as multilayer mirrors. An example has been published by Kaufmann *et al.* (1994), where a set of 40 pairs of Pd and Co layers, all of equal thickness, was deposited on a silicon substrate. The total thickness was computed by fundamental parameter methods to be 107.7 nm compared to a reference of 105 nm obtained during manufacturing.

8.4.2 Empirical parameter method

Instead of developing mathematical/physical models to describe the excitation process and the possible events in a specimen we can simply compare the observed count rates from the analyte specimen with the count rates of reference specimens having known compositions.

In the case of binaries this is very simple. Because the two concentrations are related by $w_1 + w_2 = 1$, the count rates $n(w_1, w_2)$ of each reference specimen can be drawn as a function of w_1 (or w_2) and the data points connected by a smooth curve, which is called a **calibration curve**. Any flexible, simple mathematical function can be used to fit the curve; for example

$$n_1 = N_1 \frac{w_1}{1 + \alpha(1 - w_1)} \quad \text{or equivalently} \quad w_1 = \frac{n_1}{N_1}(1 + \alpha w_2)$$

for element 1. The constant α is determined from the data of reference specimens. Note that the constant N_1 is the count rate for $w_1 = 1$, i.e. the pure-element count rate. Note further that the curve is not linear, except for very small concentrations, where the denominator is not sensitive to variations in the concentration w, or for $\alpha = 0$. Depending upon the sign of α, the curve can be above the linear relationship **(enhancement)** or below **(absorption)**.

For ternary compounds we have three concentrations related by one equation $w_1 + w_2 + w_3 = 1$, hence each n_i becomes a function of two concentrations, and instead of a one-dimensional calibration curve there is a two-dimensional calibration sphere. The above equation expands in the general case to

$$w_i = \frac{n_i}{N_i}\left(1 + \sum_{j \neq i} \alpha_{i,j} w_j\right) \tag{8.12}$$

Each $\alpha_{i,j}$ can be interpreted as the influence of element j upon the counts of element i, and this can be enhancing, for example by low absorption of i's fluorescent radiation, or by secondary excitation, or it can be absorbing due to a strong absorption coefficient.

This type of equation is often called the **α-coefficients equation** and is employed anywhere in spectroscopy including UV and IR. In the case of X-ray analysis a mathematical relationship between the (properly simplified) fundamental parameter equations and α coefficients can be established and 'fundamental α values' computed from theory.

In practice the agreement between the α-coefficients equation with the 'true' or measured relationship $n_i(w_1, w_2, \ldots)$ is rather poor, if applied to the whole concentration range. Instead, all concentrations of reference and analyte specimens must be within a common, very limited range (generally a few weight per cent, i.e. the unknowns and references must be 'very similar'). In order to reduce statistical errors, the number of standards should be at least twice the number of unknown concentrations. This makes the method suitable for routine applications, where a single set of carefully prepared standards can be employed to analyze a large number of unknowns, but it is less practicable for random-type analyses.

Other empirical parameter approaches. Equation (8.12) can be rewritten by using $R_i = n_i/N_i$:

- Lachance–Traill equation

$$w_i = R_i\left(1 + \sum_{j \neq i} \alpha_{i,j} w_j\right) \tag{8.13}$$

As mentioned above, this is suitable for only a very limited range of concentrations (Lachance and Traill 1966). The following equation can be used to cover a wider range (even the full calibration curve), but requires many more coefficients to be determined from an adequate number of standard specimens (Claisse and Quintin 1967, Lachance and Claisse 1980):

- Claisse–Quintin equation

$$w_i = R_i\left(1 + \sum_{j \neq i} \alpha_{i,j} w_j + \sum_{j \neq i} \sum_{k \neq i,j} \beta_{ijk} w_j w_k\right) \tag{8.14}$$

A different model (Rasberry and Heinrich 1974) that is situated between these two equations, in terms of required standards, is:

- Rasberry–Heinrich equation

$$w_i = R_i\left(1 + \sum_{j \neq i} \alpha_{i,j} w_j + \sum_{j \neq i} \frac{\beta_{i,j} w_j}{1 + w_i}\right) \tag{8.15}$$

The authors assume that the αs describe the absorption effects while the βs are responsible for correcting for secondary excitation, and that only the α terms or the β terms need to be taken into consideration, depending upon the prevailing type of interaction.

A special model for the analysis of steels has been proposed by the Japanese Standards Association (Japanese Industrial Standards Method, JIS). In this case iron is not measured. The concentrations of the constituents are obtained from their raw-intensity values I_i and correction terms similar to those in the Lachance–Traill model:

- JIS equation

$$w_i = (b_0 + b_1 I_i + b_2 I_i^2)\left(1 + \sum_{j \neq i} \alpha_{ij} w_i\right) \tag{8.16}$$

Except for the JIS method, all the above models use relative intensities R_i and therefore a pure element standard for each unknown element. It is, however, possible to treat the pure-element count rates as additional unknowns in the equations above and determine them from the standards.

All above αs and βs are claimed by their authors as being somehow related to fundamental parameters and physical effects, such as absorption or secondary excitation. While there undoubtedly exists a (rather loose) theoretical link, there is much reason to contest an overly detailed interpretation of real data in this way, particularly when the coefficients are determined by least-squares fits from standards.

8.5 Scattered radiation

The effects of elastic (coherent) and inelastic (incoherent, Compton) scattering are usually much weaker than photoabsorption (Figure 8.1). Nevertheless, usually the primary radiation of the excitation source will be scattered by the sample, sample holder, etc., into the detector. Furthermore in the case of tube excitation the spectral background is predominantly caused by the scattered continuum radiation.

Elastic scattering for a single electron can be deduced from classical electrodynamics, but due to its coherent nature the form factor has to be

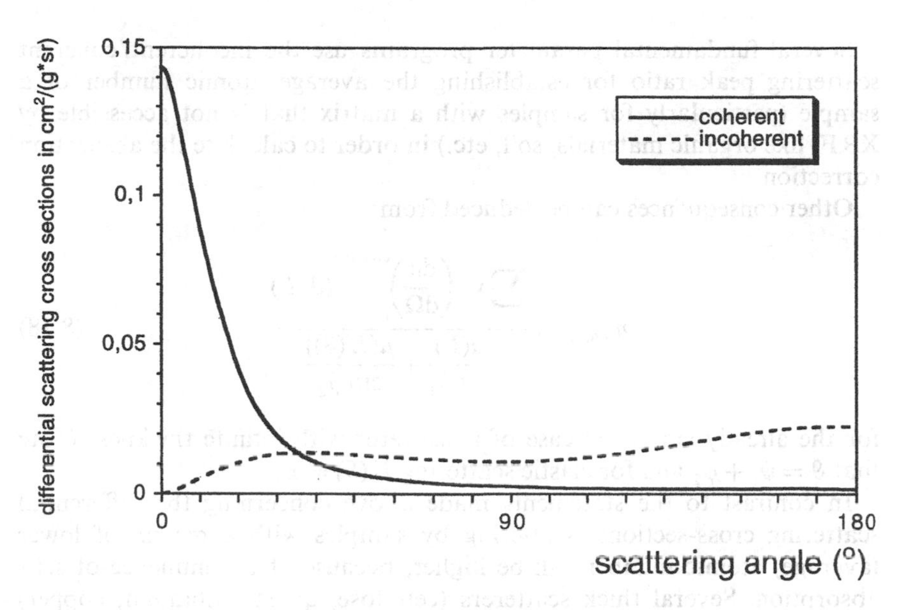

Figure 8.13 Differential coherent and incoherent scattering cross-sections for an incident energy of Mo Kα scattered by carbon.

introduced to account for all electrons of an atom (Hubbell *et al.* 1975, Szaloki 1996). The differential scattering cross-section, which describes this kind of interaction, depends on the energy of the primary radiation, the atomic number of the scattering atom and the scattering angle, because of its inherent anisotropy (Figure 8.13). It shows a pronounced maximum for forward scattering and depends for these small angles on the atomic number like Z^2. As a general statement elastic scattering increases with the atomic number.

Inelastic scattering is relatively independent of the atomic number of the scattering atom, but the observed energy shift $\Delta E = [E - E'(\vartheta)]$ of the Compton peak depends on the scattering angle:

$$E'(\vartheta) = \frac{E}{1 + (E/511)(1 - \cos\vartheta)} \tag{8.17}$$

with E, E' in keV.

In order to make any statements of practical use about the behavior of realistic scatterers, one always has to consider the above-mentioned differential scattering cross-sections, but also the self-absorption. For thin samples it turns out that inelastic scattering is proportional to the thickness of a sample, independent of the atomic number. That is, the Compton peak can be used for the determination of the area-related mass $m/A = \rho t$.

Several fundamental parameter programs use the incoherent/coherent scattering peak ratio for establishing the average atomic number of a sample (particularly for samples with a matrix that is not accessible by XRF, like organic materials, soil, etc.) in order to calculate the absorption correction.

Other consequences can be deduced from:

$$n_{\text{coh, inc}} \propto \frac{\sum_i w_i \left(\frac{d\sigma}{d\Omega}\right)_i^{\text{coh, inc}} (\vartheta, E)}{\frac{\mu(E)}{\sin\psi_1} + \frac{\mu(E'(\vartheta))}{\sin\psi_2}} \tag{8.18}$$

for the already simplified case of a scatterer with infinite thickness (note that $\vartheta = \psi_1 + \psi_2$ and for elastic scattering $E'(\vartheta) = E$).

In contrast to the statements made above concerning the differential scattering cross-sections, scattering by samples with a matrix of lower (average) atomic number will be higher, because of the influence of self-absorption. Several thick scatterers (cellulose, quartz, titanium, copper) have been used to establish the data contained in Figure 8.14, demon-

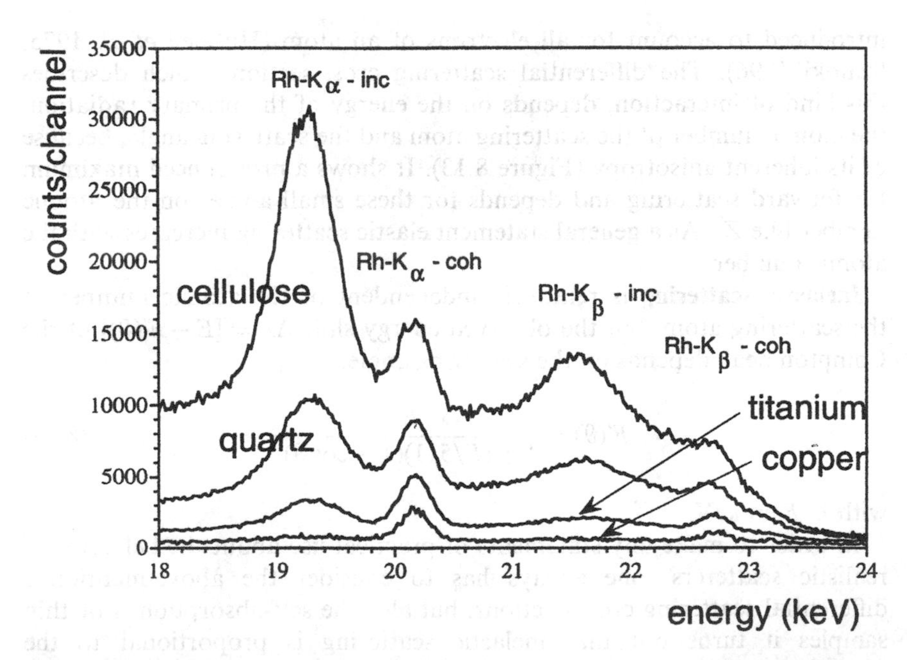

Figure 8.14 The Rh Kα, Rh Kβ region of rhodium X-ray tube spectra, scattered by cellulose, quartz, titanium and copper (incoherent (Compton) and coherent peaks). The absolute height as well as the ratio of incoherent to coherent peaks depend on the scatterers' mean atomic number.

strating that with increasing atomic number scattering effects get weaker and the ratio of coherent to incoherent scattering intensity will change, favoring elastic scattering.

Concerning the energy dependence, the elastic scattering contribution will dominate if the electrons of the scatterer can be treated as well-bound to the atom (lower energies), but for higher energies the inelastic contribution will reach a constant value, outweighing the decreasing elastic contribution, and even become more important than the photoeffect toward the end of the X-ray region (Figure 8.1).

8.6 Electron probe micro-analysis

Electron probe micro-analysis (EPMA) and scanning electron microscopy (SEM) use identical instruments in their basic constructive principles, but adopted to different purposes. In both cases a narrow electron beam (of typically <1 μm diameter) is directed onto a specimen and periodically deflected in order to hit a grid of points on its surface. The kinetic energy of the electrons can be set to values between a few keV and around 50 keV, typically to 20 keV.

Several kinds of interaction between the electron and the sample atoms are possible. The electron can be scattered by an atomic nucleus, which is an acceleration of a charged particle and associated with the emission of a photon. This is the same mechanism as excitation of bremstrahlung in an X-ray tube. The second possibility is the collision with an electron of an atom (inelastic electron scattering), which may lead to ionization and emission of fluorescent radiation. This is the intended interaction in electron microprobe analysis and allows chemical analysis in the same way as photon-excited XRF. Because the electron beam scans the surface of the specimen, two-dimensional maps of chemical information (element distribution maps) can be obtained.

The process is also associated with the emission of photo- and Auger electrons (secondary electrons). Some of them escape from the specimen and are subject to interaction with atomic clusters shaping the specimen's surface (shielding effects). Their observation (without accurately determining their energies) is the main task of 'microscopy', i.e. computing the image of the specimen surface from the shielding information. Other processes are elastic scattering and diffraction by the lattice (which are observed in the transmission electron microscope).

The detection systems for the X-rays in EPMA are either solid-state Si(Li) detectors or crystal spectrometers. The advantages of solid-state detectors are their acceptance of a larger fraction of the emitted X-rays and their easier installation in the limited space of the high-vacuum chamber; they also acquire an interpretable spectrum in a much shorter time. Crystal spectrometers have a better energy (or wavelength) resolution, particularly at lower energies, and are often built in focusing geometry. This helps to collect a larger solid angle of the emitted fluorescent radiation; in this case it is required to vary also the crystal–detector distance while varying the diffraction angle Θ–2Θ in order to maintain the focusing geometry. This distance is often used as an indicator of the observed wavelength rather than the 2Θ value in conventional spectrometers. The same types of crystals are used in EPMA as in conventional XRF, including multilayers for light-element analysis.

Detection of long wavelengths with Si(Li) detectors is difficult because of significant absorption in the beryllium window. The situation can be improved by 'windowless detectors' with special provisions to remove the window during the measurement. This is possible, because the vacuum in the EPMA chamber is normally at least as good as (normally much better than) that of an Si(Li) detector. The disadvantage is that residual gas atoms in the chamber are trapped by the cold crystal surface, forming a dirt layer on the crystal surface. This can be avoided by extremely thin protective windows, which need not be light-tight or mechanically stable.

Spectra from electron excitation exhibit a higher background level than the ones from photon excitation. This is due to the generation of bremstrahlung by the electrons, which is generally the case with excitation by charged particles (e.g. by α-particles or protons). The deterioration of the peak-to-background ratio also affects the detection limits.

Sample preparation is rather important. The specimens must be electrically conductive and have a low vapor pressure (i.e. sustain high vacuum). The surface must be clean, smooth and flat (polished).

8.7 Other XRF techniques

8.7.1 Total reflection X-ray fluorescence analysis

Total reflection X-ray fluorescence analysis (TXRF) is basically an energy-dispersive analytical technique in a special excitation geometry (Figure 8.15). This geometry is achieved by adjusting the sample carrier, not inclined under 45° to the incident beam, as for standard EDXRF, but with angles of about 1 mrad (0.06°) to the primary beam. The incident beam thus impinges at angles below the critical angle of (external) total reflection[3] for X-rays onto the surface of a plane smooth polished reflector.

[3] The term 'external total reflection' refers to the fact that the refractive index n for X-rays is slightly smaller than 1. Therefore this effect can only be observed if the radiation is coming from vacuum (air) and hits the interface of a denser medium (reflector). It is the small size (10^{-6}) of this deviation from 1 that causes the minute critical angle for the total reflection phenomenon.

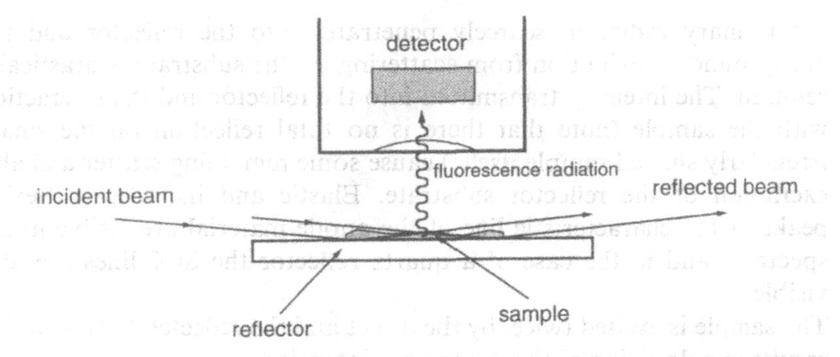

Figure 8.15 Scheme of total reflection X-ray fluorescence (TXRF).

Usually the liquid sample, with a volume of only 1–100 μL, is pipetted in the center of this surface and the droplet will cover an area of a few millimeters in diameter. As a result of the drying process where the liquid part of the sample is evaporated, the residual is irregularly distributed on the reflector (within the above stated diameter), forming a very thin sample (Yoneda and Horiuchi 1971, Aiginger and Wobrauschek 1974, Wobrauschek and Aiginger 1975, Knoth and Schwenke 1978, Schwenke and Knoth 1982).

Some of the reflector materials that are in use are: quartz glass (most common), silicon, germanium, glassy carbon, niobium, boron nitride and (as cheapest material) Plexiglas. The requirements for the reflector are: no interfering fluorescence or diffraction lines, high purity, chemical resistance, hardness, machineability for polishing and an acceptable price. The surface must be flat and the mean roughness in the range of nanometers. Usually they are disk-shaped, with 30 mm diameter and 3–5 mm thickness, but also squares of 30 mm side length and rectangular types are in use.

The simplified equation (valid above the highest K absorption edge of the reflector material) for the critical angle of total reflection φ_{crit} (in mrad) depends on the energy E (in keV) of the incident photons and the density ρ (in g/cm^3) of the reflector material:

$$\varphi_{crit} = \frac{20.3}{E}\sqrt{\rho} \tag{8.19}$$

For example, for incident Mo Kα (17.5 keV) radiation and quartz glass as reflector, the critical angle calculates as 1.7 mrad (=0.1°).

TXRF offers the following advantages as compared to standard EDXRF:

1. If the adjusted incident angle of the instrument is below the critical angle φ_{crit}, almost 100% of the incident photons are totally reflected, so

the primary radiation scarcely penetrates into the reflector and the background contribution from scattering on the substrate is drastically reduced. The intensity transmitted into the reflector and the interaction with the sample (note that there is no total reflection on the small, irregularly shaped sample itself!) cause some remaining scatter and also excitation of the reflector substrate. Elastic and inelastic scattering peaks of the characteristic line of the anode material are visible in the spectrum and in the case of a quartz reflector the Si K lines are also visible.
2. The sample is excited twice, by the direct and the reflected beams, which results in a doubling of the fluorescent intensity.
3. Because of the geometry it is possible to position the detector close to the surface of the reflector where in the center the sample is located. This results in a large solid angle for the detection of the fluorescence signal.

Consequently the peak-to-background ratio in TXRF is high and the detection limits are drastically improved by several orders of magnitude as compared to conventional EDXRF.

As excitation source, high-power X-ray (diffraction) tubes with a line focus $8 \times 0.4\,mm^2$ or long fine focus $12 \times 0.4\,mm^2$ are in use. The most versatile anode material is Mo, and the power applied is up to 3 kW. Depending on the elements of interest, other anodes, e.g. Cr, Cu or W, might be preferred. Rotating anodes of Cu and Mo, with up to 18 kW, are suitable for special applications.

The insertion of spectral modification devices in the beam path of the primary radiation improves the background. A cut-off reflector, acting as bandpass filter, suppresses high-energy photons and improves in particular the background in the low-energy region. A multilayer monochromator suppresses in the ideal case all photons, except the ones with a certain energy, usually the most intense characteristic line of the anode material.

The preferred types of samples are either aqueous or acidic solutions (Figure 8.16). With special sample preparation techniques the pg/g concentration level can be reached. Standard procedures for sample preparation in TXRF are described in several publications (e.g. Prange 1989, Klockenkämper 1996). There are no corrections for absorption or secondary excitation necessary due to the sample formation in a very thin layer. In any case the addition of an internal standard of known concentration is essential for the quantification (typical elements, preferably not present in the sample are Co, Ga, Ge, Y, . . .). Rewardingly the calibration curves are linear over several orders of magnitude and therefore the calculations for converting the measured intensities to concentrations are simple and can be based on experimentally or theoretically determined relative sensitivity curves $S_{rel}(Z)$ as a function of the

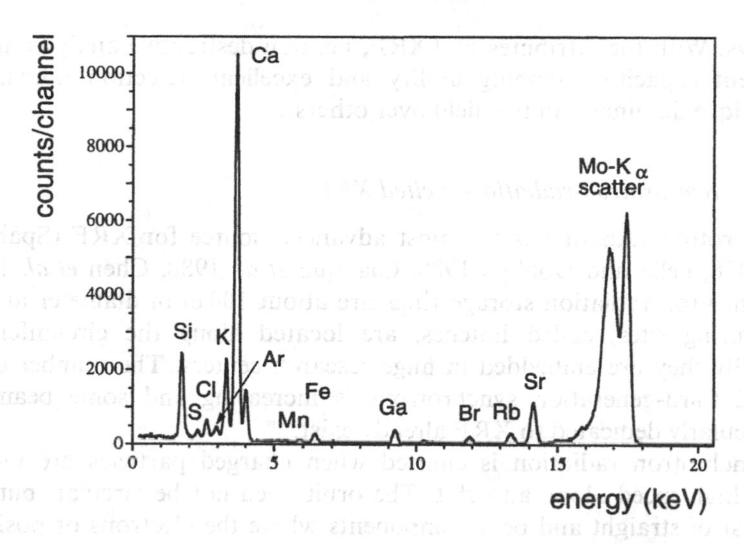

Figure 8.16 Spectrum of a 3 μL mineral water sample, spiked with 1 ng/μL Ga as internal standard element. Excitation in TXRF geometry with a multilayer monochromator by a Mo X-ray tube (50 kV, 10 mA, 1000 s measuring time). (See section 8.8.3 also.)

atomic number Z for all elements in respect of the internal standard element. Rearranging equation (8.10) and defining $S_{rel} = A_i/A_{st}$ the concentration w_i of an element i can be calculated by

$$w_i = \frac{n_i}{n_{st}} \frac{1}{S_{rel}} w_{st} \tag{8.20}$$

Note that n_{st} and w_{st} are the intensity and the concentration of the internal standard element.

For routine applications there are TXRF spectrometers commercially available from several suppliers with full comfort of sample changers and optimized handling for the operator. Also budget-priced modular designs that can be attached to existing standard EDXRF equipment are on the market.

The angular dependence of intensities in the regime of total reflection can be used to investigate surface impurities, thin near-surface layers, and even molecules absorbed on flat surfaces. From these angle-dependent intensity profiles the composition, thickness and density of layers can be obtained. It is the low penetration depth of the primary beam at total reflection that enables also the non-destructive in-depth examination of concentration profiles in the range of 1–500 nm (de Boer 1991).

One of the important applications showing the analytical power of TXRF is the analysis of metal impurities on the surface of Si wafers for the semiconductor industry. The sample is the wafer itself with its polished plane surface having already the quality required for total reflection of

X-rays. With the attributes of TXRF, i.e. non-destructive analysis, multi-element capacity, mapping ability and excellent detection limits, this technique dominates in this field over others.

8.7.2 Synchrotron-radiation-excited XRF

Synchrotron radiation is the most advanced source for XRF (Sparks *et al.* 1978, Pella and Dobbyn 1988, Giauque *et al.* 1988, Chen *et al.* 1990). Synchrotron radiation storage rings are about 100 m in diameter and the measuring sites, called hutches, are located along the circumference. Usually they are embedded in huge research centers. The number of so-called third-generation synchrotrons is increasing and some beamlines particularly dedicated to XRF already exist.

Synchrotron radiation is emitted when charged particles are moving with high speed along an orbit. The orbit need not be circular, but may consist of straight and bent components where the electrons or positrons are traveling in ultra-high vacuum with almost the speed of light and are held in the orbit by magnetic fields. The radiation losses (= synchrotron radiation) of the electron/positron bunches are compensated by radio-frequency power supply in synchronous pulses. The radiation is emitted with high intensity from a small spot, tangential to the orbit.

The features of synchrotron radiation are ideal for XRF: natural collimation in the vertical, high brilliance, a continuous spectrum over the entire range from eV to several hundred keV, and linear polarization in the plane of the orbit (can be used for background reduction). Insertion devices like wigglers and undulators can improve even further some parameters of synchrotron radiation in comparison to the dipole bending magnet.

By use of monochromator crystals, tuning to each desired energy of the synchrotron radiation spectrum can be achieved. Double monochromator units using, for example, an Si (111) crystal transmit a beam with an extremely sharp energy resolution in the eV range. Consequently the excitation close to the absorption edge of an element of interest can favor it before others. In turn characteristic lines of elements can be suppressed, by choosing an energy below the absorption edge. X-ray optical units, like bent mirrors or crystals, can focus the photon beam to micrometer-sized spots for perfect position-resolved analysis. Similar focusing effects are achieved with Fresnel lenses or capillaries to perform micro-XRF of, for example, inclusions. With conventional excitation sources usually the introduction of monochromators, etc., decreases the number of primary photons too much, whereas with synchrotron radiation excitation there are enough photons available to obtain analytical results in a very short time. So elemental mapping across a sample is an application.

Very promising is the combination of synchrotron radiation with TXRF (Iida *et al.* 1985). Figure 8.17 shows some possible geometrical arrangements of

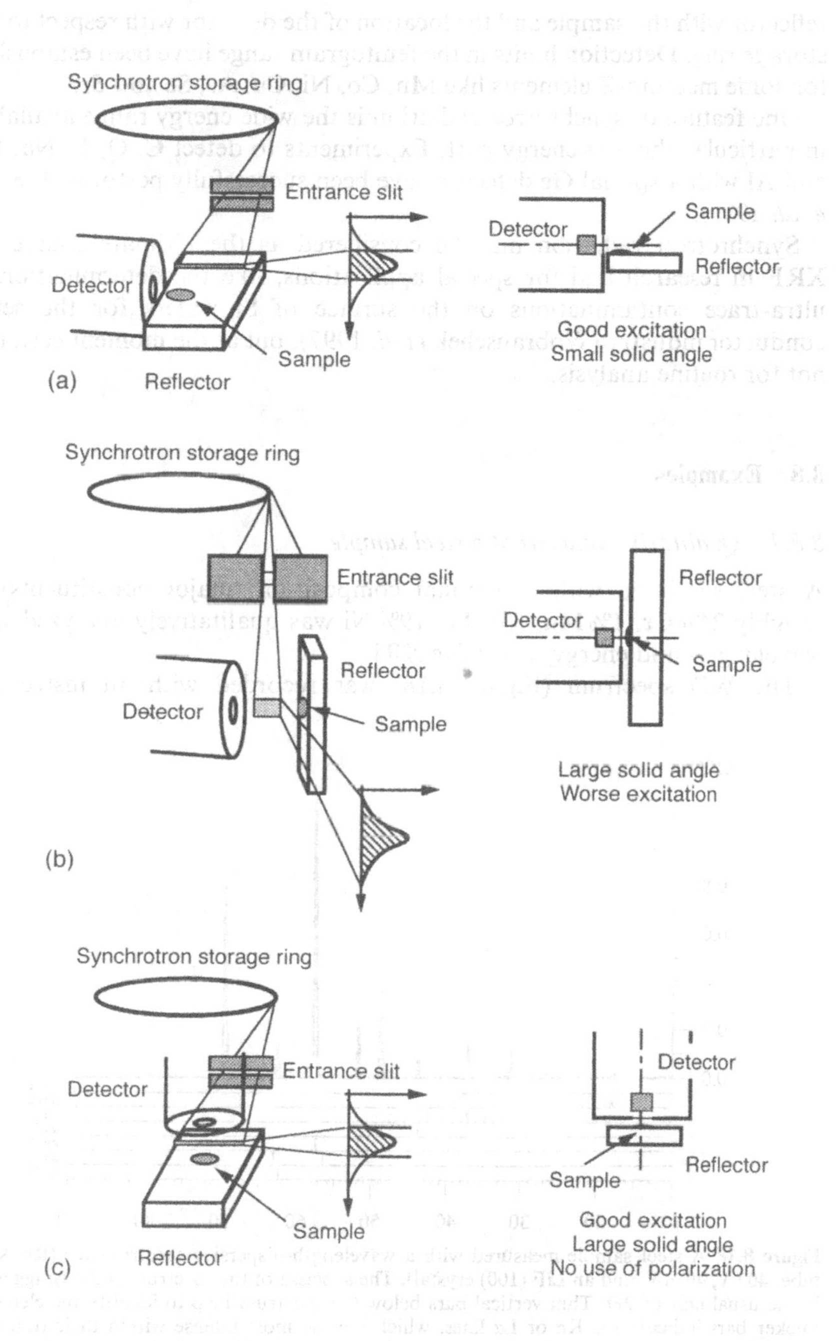

Figure 8.17 Three possible geometrical arrangements of synchrotron-radiation-excited TXRF.

reflector with the sample and the location of the detector with respect to the storage ring. Detection limits in the femtogram range have been established for some medium-*Z* elements like Mn, Co, Ni, Cu, As, Se and Sr.

One feature of synchrotron radiation is the wide energy range available, in particular the low-energy part. Experiments to detect C, O, F, Na, Mg and Al with a special Ge detector have been successfully performed (Streli *et al.* 1997).

Synchrotron radiation may be considered as the ultimate source for XRF in research and for special applications, like the determination of ultra-trace contaminations on the surface of Si wafers for the semiconductor industry (Wobrauschek *et al.* 1997), but at the moment certainly not for routine analysis.

8.8 Examples

8.8.1 Qualitative analysis of a steel sample

A steel specimen with a nominal composition (major constituents) of roughly 24% Cr, 1% Mn, 53% Fe, 19% Ni was qualitatively analyzed with wavelength- and energy-dispersive XRF.

The WD spectrum (Figure 8.18) was recorded with an instrument

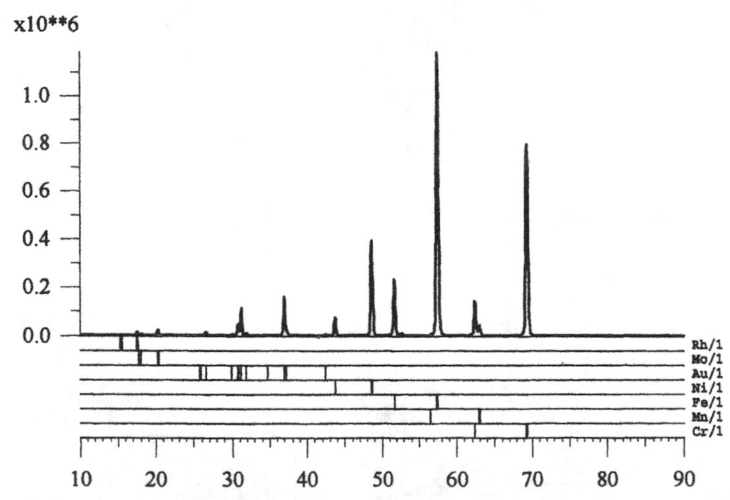

Figure 8.18 A steel sample measured with a wavelength-dispersive spectrometer (Rh X-ray tube, 40 kV, 40 mA; and an LiF (100) crystal). The abscissa of the spectrum is the Bragg angle in the usual unit of 2Θ. Thin vertical bars below the spectrum help to identify the elements; thicker bars indicate the $K\alpha$ or $L\alpha$ lines, which are the most intense within their transition series. In most cases, particularly at longer wavelengths, doublets, like $\alpha_{1,2}$, are not resolved and appear as single lines. The suffix '/1' after the element names refers to the order of reflection, i.e. the *n* in Bragg's equation.

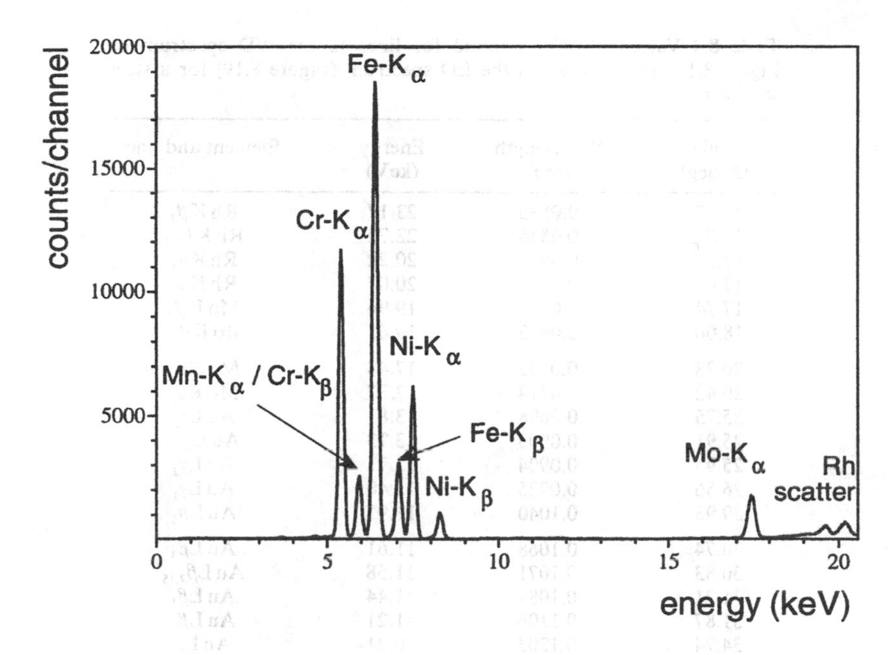

Figure 8.19 The same steel sample (as of Figure 8.18) was measured with an energy-dispersive spectrometer (Rh X-ray tube, 30 kV, 0.20 mA, and an Si(Li) detector). The abscissa of the spectrum is in keV.

equipped with an Rh-target X-ray tube as a source for primary radiation (the applied tube voltage was 40 kV and the tube current 40 mA), an LiF (100) analyzer crystal ($2d = 0.4027$ nm) and a scintillation counter for actual photon detection. An angular range of 2Θ was chosen from 10° to 90° and stepped through in increments of $\Delta 2\Theta = 0.05°$ with a photon counting time of 2 s at each angular position.

For the ED measurement (Figure 8.19) the sample was also excited with an Rh X-ray tube, operated at 30 kV. The tube current was as low as 0.20 mA, which was already sufficient for the optimum throughput of this spectrometer, equipped with an Si(Li) detector.

The spectra contain lines from Cr, Mn, Fe, Ni and Mo, which are constituents of the specimen. Gold lines in the WD spectrum come from the specimen holder (a thin gold coating shields practically all emission lines from the instrument; if gold must be analyzed, specimen holders with different coatings are used). Rhodium lines are part of the tube spectrum, scattered by the specimen into the spectrometer. There is no simple way with this set-up to determine Rh in a specimen. The lines are listed in Table 8.4.

Table 8.4 Values of 2Θ as found for lines in the WD spectrum of Figure 8.18, and energies in the ED spectrum (Figure 8.19) for a steel specimen

Angle 2Θ (deg)	Wavelength (nm)	Energy (keV)	Element and line
15.27	0.0535	23.17	Rh $K\beta_2$
15.57	0.0546	22.72	Rh $K\beta_{1,3}$
17.51	0.0613	20.22	Rh $K\alpha_1$
17.64	0.0618	20.07	Rh $K\alpha_2$
17.74	0.0621	19.96	Mo $K\beta_2$
18.06	0.0632	19.61	Mo $K\beta_{1,3}$
20.28	0.0709	17.48	Mo $K\alpha_1$
20.42	0.0714	17.37	Mo $K\alpha_2$
25.75	0.0898	13.81	Au $L\gamma_3$
25.91	0.0903	13.73	Au $L\gamma_{2,6}$
25.95	0.0904	13.71	Au $L\gamma_2$
26.56	0.0925	13.40	Au $L\gamma_1$
29.93	0.1040	11.92	Au $L\beta_5$
30.74	0.1068	11.61	Au $L\beta_3$
30.83	0.1071	11.58	Au $L\beta_{2,15}$
31.21	0.1084	11.44	Au $L\beta_1$
31.87	0.1106	11.21	Au $L\beta_4$
34.74	0.1203	10.31	Au Ln
36.96	0.1277	9.71	Au $L\alpha_1$
37.36	0.1290	9.61	Au $L\alpha_2$
40.23	0.1385	8.95	Au Ll
43.70	0.1499	8.27	Ni $K\beta$
48.59	0.1658	7.48	Ni $K\alpha$
51.69	0.1756	7.06	Fe $K\beta$
56.62	0.1922	6.49	Mn $K\beta$
57.49	0.1938	6.40	Fe $K\alpha$
62.30	0.2048	5.95	Cr $K\beta$
63.00	0.2105	5.89	Mn $K\alpha$
69.36	0.2292	5.41	Cr $K\alpha$

8.8.2 *Alpha coefficients for the ternary compound Cr–Fe–Ni*

The following example is based on theoretical data computed by fundamental parameter algorithms including primary and secondary excitation. This allows one to 'make' samples of any composition independent of practical difficulties and avoids the demonstrated behavior from being obscured by statistical errors. Table 8.5 shows widely varying concentrations w of the three elements Cr, Fe and Ni. Hence the basic rule, that 'standards must be similar to unknown', is (purposely) violated and a certain error must be expected, if the method is applied to *all* specimens. The Lachance–Traill α-coefficient method (equation (8.13)) is used.

For each element in a ternary alloy two α coefficients must be computed, e.g. $\alpha_{Fe \leftarrow Cr}$ and $\alpha_{Fe \leftarrow Ni}$ for iron (note that no α_{ii} exists). They can be obtained

Table 8.5 Concentration values (weight per cent) and intensity data for Cr–Fe–Ni ternary alloys. R are the relative intensities (counts from alloy divided by counts from pure-element standard), computed for an Rh-anode end-window X-ray tube with a 76 μm Be window, operated at 45 kV, $\psi_1 = 63.5°$, $\psi_2 = 45.0°$

w_{Cr} (%)	w_{Fe} (%)	w_{Ni} (%)	$R_{Cr\,K\alpha_1}$	$R_{Fe\,K\alpha_1}$	$R_{Ni\,K\alpha_1}$
10.00	80.00	10.00	1.57E-01	0.66118	3.67E-02
10.00	70.00	20.00	1.50E-01	0.59707	7.88E-02
20.00	60.00	20.00	2.70E-01	0.43402	8.04E-02
20.00	50.00	30.00	2.62E-01	0.37580	1.30E-01
20.00	40.00	40.00	2.56E-01	0.31438	1.89E-01
60.00	30.00	10.00	6.54E-01	0.13336	4.04E-02
10.00	20.00	70.00	1.34E-01	0.22417	4.43E-01
80.00	10.00	10.00	8.22E-01	0.03780	4.21E-02

from any two standard samples (from which two equations with two unknowns are derived), or from three or more samples (this requires a least-squares fitting algorithm). In fact, as many standards as possible should be used in a real environment, where statistical errors must be taken into account.

We choose the standards with 20–60–20 and 10–70–20 per cent of Cr–Fe–Ni (concentrations must be normalized to 1) and obtain for Fe

$$\frac{w_{Fe}}{R_{Fe}} = 1 + \alpha_{Fe\leftarrow Ni} w_{Ni} + \alpha_{Fe\leftarrow Cr} w_{Cr}$$

which leads to the two equations (for 60 and 70% Fe, respectively):

$$\frac{0.60}{0.43402} = 1 + \alpha_{Fe\leftarrow Ni} 0.20 + \alpha_{Fe\leftarrow Cr} 0.20$$

$$\frac{0.70}{0.59707} = 1 + \alpha_{Fe\leftarrow Ni} 0.20 + \alpha_{Fe\leftarrow Cr} 0.10$$

For the other coefficients we proceed similarly. The matrix of solutions is:

$$\alpha_{Cr\leftarrow Fe} = -0.7407 \qquad \alpha_{Cr\leftarrow Ni} = 0.9259$$
$$\alpha_{Fe\leftarrow Cr} = 2.1003 \qquad \alpha_{Fe\leftarrow Ni} = -0.18815$$
$$\alpha_{Ni\leftarrow Cr} = 1.1481 \qquad \alpha_{Ni\leftarrow Fe} = 1.9857$$

If one defines

$$R_{Fe,corrected}/w_{Fe} = R_{Fe}(1 + \alpha_{Fe\leftarrow Ni} w_{Ni} + \alpha_{Fe\leftarrow Cr} w_{Cr})$$

the usual way to check this result visually is to draw R and $R_{corrected}$ versus w for all standards, one graph for each element. All data points $\{w, R_{corrected}\}$ must be on a straight line after a successful correction (Figure 8.20). Deviations result from statistical and systematic errors in the case

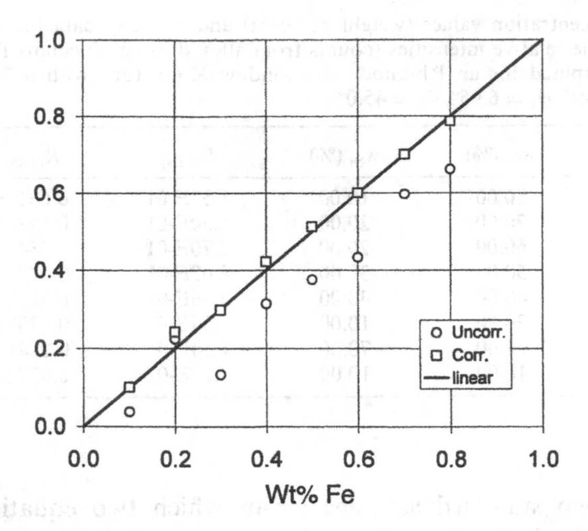

Figure 8.20 Corrected and uncorrected count-rate ratios of iron for a hypothetical ternary compound Cr–Fe–Ni as an example for the Lachance–Traill α-coefficient method.

of an experiment, and if the condition 'similar composition of all standards' is violated, as in our example.

The final step is to set up the system of equations to determine the concentrations of an unknown specimen:

$$-\frac{1}{R_{\mathrm{Cr}}}w_{\mathrm{Cr}} + \alpha_{\mathrm{Cr}\leftarrow\mathrm{Fe}}w_{\mathrm{Fe}} + \alpha_{\mathrm{Cr}\leftarrow\mathrm{Ni}}w_{\mathrm{Ni}} = -1$$

$$\alpha_{\mathrm{Fe}\leftarrow\mathrm{Cr}}w_{\mathrm{Cr}} - \frac{1}{R_{\mathrm{Fe}}}w_{\mathrm{Fe}} + \alpha_{\mathrm{Fe}\leftarrow\mathrm{Ni}}w_{\mathrm{Ni}} = -1$$

$$\alpha_{\mathrm{Ni}\leftarrow\mathrm{Cr}}w_{\mathrm{Cr}} + \alpha_{\mathrm{Ni}\leftarrow\mathrm{Fe}}w_{\mathrm{Fe}} - \frac{1}{R_{\mathrm{Ni}}}w_{\mathrm{Ni}} = -1$$

8.8.3 Trace elements in a mineral water

As an example for trace-element determination, a commercially available mineral water was selected (the sample was taken from a 1 L bottle). The measurements with conventional EDXRF did not contain any useful information about elements to be determined (related to the fact that the main constituent, H_2O, is responsible for a high spectral background, and consequently poor detection limits). As an alternative the mineral water sample was measured with TXRF, providing spectra with several elements of interest superimposed on a low background. A gallium solution was added to the mineral water as internal standard. Only 3–5 μL of the sample

AXIL IBM-PC V3.5
Spectrum: A:\MIN-R54.SPE 1000 s

ChiSqr = 1.8

Line	Ener. (KeV)	Peak area		st.dev.	Chi_sq
S -Ka	2.307	7453.	±	124.	4.48
Ar-Ka	2.957	566.	±	100.	1.41
K -Ka	3.313	80806.	±	316.	1.37
Ca-Ka	3.691	331754.	±	571.	3.37
Mn-Ka	5.895	312.	±	52.	1.39
Fe-Ka	6.399	409.	±	60.	.76
Ga-Ka	9.243	10605.	±	136.	.56
Br-Ka	11.908	2376.	±	104.	.73
Rb-Ka	13.375	4738.	±	136.	.70
Sr-Ka	14.142	17738.	±	205.	1.04

Figure 8.21 The AXIL print-out of calculated net peak areas of a mineral water sample measured in TXRF geometry.

were pipetted onto the sample carrier and the matrix (water) was evaporated by drying under an infrared lamp. The residue was excited by monochromatized Mo Kα radiation. The spectrum (Figure 8.16) contains characteristic lines of Si (from the quartz reflector used as sample support), S, Cl, Ar (from the ambient air), K, Ca, Mn, Fe, Ga (added as internal standard), Br, Rb, Sr and scatter peaks. Their Kα net peak areas are given in Figure 8.21 and were obtained by a (non-linear least-squares fitting)

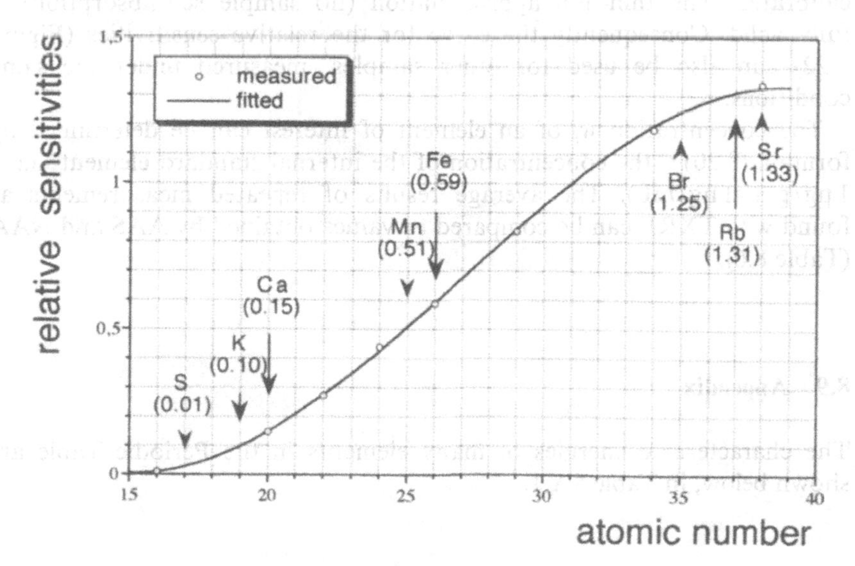

Figure 8.22 Calibration curve (relative sensitivities) for a TXRF system, with the element Ga as internal standard.

Table 8.6 Comparison of concentration values (mineral water sample) as obtained by different analytical methods*

Element	TXRF (mg/kg)	Certified value (mg/L)	ICP-OES (mg/L)	NAA (mg/kg)
Li	–	0.77	0.86	–
Na	–	480	449	434
Mg	–	95.1	112	85.6
Al	–	0.01	<0.3	–
S (as SO_4)	64.3 ± 3.3	89.2	85	–
Cl	–	47.5	–	40
K	80.2 ± 3.9	75	92	80
Ca	229.4 ± 10.6	219.5	224	–
Mn	0.04 ± 0.01	0.25	0.03	0.03
Fe	0.06 ± 0.01	0.05	<0.05	–
Br	0.2 ± 0.01	0.09	–	0.23
Rb	0.34 ± 0.01	–	–	0.69
Sr	1.28 ± 0.03	1.8	1.7	–
I	–	0.03	–	0.03
Ba	–	–	0.11	–

* TXRF, total reflection X-ray fluorescence analysis; ICP-OES, inductively coupled plasma optical emission spectrometry; NAA, neutron activation analysis.

spectrum evaluation program (AXIL), resolving any peak overlaps and subtracting the background.

By means of suitable artificial calibration standards, the system was calibrated. The thin-film approximation (no sample self-absorption) is fully valid. Consequently the curve for the relative sensitivities (Figure 8.22) can also be used for other samples, measured under the same conditions.

The concentration w_i of an element of interest can be determined by formula (8.20) – the concentration of the internal standard element Ga is 1 μg/g (=1 ng/mL). The average results of repeated measurements as found with TXRF can be compared to values obtained by AAS and NAA (Table 8.6).

8.9 Appendix

The characteristic energies of many elements in the Periodic Table are shown below, in Table 8A.1.

Table 8A.1 X-ray emission lines in (keV)

Z	Element	$K\alpha_1$	$K\alpha_2$	$K\beta_1$	$L\alpha_1$	$L\alpha_2$	$L\beta_1$	$L\beta_2$	$L\gamma_1$
3	Li	0.0543							
4	Be	0.1085							
5	B	0.1833							
6	C	0.277							
7	N	0.3924							
8	O	0.5249							
9	F	0.6768							
10	Ne	0.8486	0.8486						
11	Na	1.04098	1.04098	1.0711					
12	Mg	1.25360	1.25360	1.3022					
13	Al	1.48670	1.48627	1.55745					
14	Si	1.73998	1.73938	1.83594					
15	P	2.0137	2.0127	2.1391					
16	S	2.30784	2.30664	2.46404					
17	Cl	2.62239	2.62078	2.8156					
18	Ar	2.95770	2.95563	3.1905					
19	K	3.3138	3.3111	3.5896					
20	Ca	3.69168	3.68809	4.0127	0.3413	0.3413	0.3449		
21	Sc	4.0906	4.0861	4.4605	0.3954	0.3954	0.3996		
22	Ti	4.51084	4.50486	4.93181	0.4522	0.4522	0.4584		
23	V	4.95220	4.94464	5.42729	0.5113	0.5113	0.5192		
24	Cr	5.41472	5.405509	5.94671	0.5728	0.5728	0.5828		
25	Mn	5.89875	5.88765	6.49045	0.6374	0.6374	0.6488		
26	Fe	6.40384	6.39084	7.05798	0.7050	0.7050	0.7185		
27	Co	6.93032	6.91530	7.64943	0.7762	0.7762	0.7914		
28	Ni	7.47815	7.46089	8.26466	0.8515	0.8515	0.8688		
29	Cu	8.04778	8.02783	8.90529	0.9297	0.9297	0.9498		
30	Zn	8.63886	8.61578	9.5720	1.0117	1.0117	1.0347		
31	Ga	9.25174	9.22482	10.2642	1.09792	1.09792	1.1248		
32	Ge	9.88642	9.85532	10.9821	1.18800	1.18800	1.2185		
33	As	10.54372	10.50799	11.7262	1.2820	1.2820	1.3170		
34	Se	11.2224	11.1814	12.4959	1.37910	1.37910	1.41923		
35	Br	11.9242	11.8776	13.2914	1.48043	1.48043	1.52590		
36	Kr	12.649	12.598	14.112	1.5860	1.5860	1.6366		
37	Rb	13.3953	13.3358	14.9613	1.69413	1.69256	1.75217		
38	Sr	14.1650	14.0979	15.8357	1.80656	1.80474	1.87172		
39	Y	14.9584	14.8829	16.7378	1.92256	1.92047	1.99584		
40	Zr	15.7751	15.6909	17.6678	2.04236	2.0399	2.1244	2.2194	2.3027
41	Nb	16.6151	16.5210	18.6225	2.16589	2.1630	2.2574	2.3670	2.4618
42	Mo	17.47934	17.3743	19.6083	2.29316	2.28985	2.39481	2.5183	2.6235
43	Tc	18.3671	18.2508	20.619	2.4240	–	2.5368	–	–
44	Ru	19.2792	19.1504	21.6568	2.55855	2.55431	2.68323	2.8360	2.9645
45	Rh	20.2161	20.0737	22.7236	2.69674	2.69205	2.83441	3.0013	3.1438
46	Pd	21.1771	21.0201	23.8187	2.83861	2.83325	2.99022	3.17179	3.3287
47	Ag	22.16292	21.9903	24.9424	2.98431	2.97821	3.15094	3.34781	3.51959
48	Cd	23.1736	22.9841	26.0955	3.13373	3.12691	3.31657	3.52812	3.71686
49	In	24.2097	24.0020	27.2759	3.28694	3.27929	3.48721	3.71381	3.92081
50	Sn	25.2713	25.0440	28.4860	3.44398	3.43542	3.66280	3.90486	4.13112

Table 8A.1 (Continued)

Z	Element	$K\alpha_1$	$K\alpha_2$	$K\beta_1$	$L\alpha_1$	$L\alpha_2$	$L\beta_1$	$L\beta_2$	$L\gamma_1$
51	Sb	26.3591	26.1108	29.7256	3.60472	3.59532	3.84357	4.10078	4.34779
52	Te	27.4723	27.2017	30.9957	3.76933	3.7588	4.02958	4.3017	4.5709
53	I	28.6120	28.3172	32.2947	3.93765	3.92604	4.22072	4.5075	4.8009
54	Xe	29.779	29.458	33.624	4.1099	–	–	–	–
55	Cs	30.9728	30.6251	34.9869	4.2865	4.2722	4.6198	4.9359	5.2804
56	Ba	32.1936	31.8171	36.3782	4.46626	4.45090	4.82753	5.1565	5.5311
57	La	33.4418	33.0341	37.8010	4.65097	4.63423	5.0421	5.3835	5.7885
58	Ce	34.7197	34.2789	39.2573	4.8402	4.8230	5.2622	5.6134	6.052
59	Pr	36.0263	35.5502	40.7482	5.0337	5.0135	5.4889	5.850	6.3221
60	Nd	37.3610	36.8474	42.2713	5.2304	5.2077	5.7216	6.0894	6.6021
61	Pm	38.7247	38.1712	43.826	5.4325	5.4078	5.961	6.339	6.892
62	Sm	40.1181	39.5224	45.413	5.6361	5.6090	6.2051	6.586	7.178
63	Eu	41.5422	40.9019	47.0379	5.8457	5.8166	6.4564	6.8432	7.4803
64	Gd	42.9962	42.3089	48.697	6.0572	6.0250	6.7132	7.1028	7.7858
65	Tb	44.4816	43.7441	50.382	6.2728	6.2380	6.978	7.3667	8.102
66	Dy	45.9984	45.2078	52.119	6.4952	6.4577	7.2477	7.6357	8.4188
67	Ho	47.5467	46.6997	53.877	6.7198	6.6795	7.5253	7.911	8.747
68	Er	49.1277	48.2211	55.681	6.9487	6.9050	7.8109	8.1890	9.089
69	Tm	50.7416	49.7726	57.517	7.1799	7.1331	8.101	8.468	9.426
70	Yb	52.3889	51.3540	59.37	7.4156	7.3673	8.4018	8.7S88	9.7801
71	Lu	54.0698	52.9650	61.283	7.6555	7.6049	8.7090	9.0489	10.1434
72	Hf	55.7902	54.6114	63.234	7.8990	7.8446	9.0227	9.3473	10.5158
73	Ta	57.532	56.277	65.223	8.1461	8.0879	9.3431	9.6518	10.8952
74	W	59.31824	57.9817	67.2443	8.3976	8.3352	9.67235	9.9615	11.2859
75	Re	61.1403	59.7179	69.310	8.6525	8.5862	10.0100	10.2752	11.6854
76	Os	63.0005	61.4867	71.413	8.9117	8.8410	10.3553	10.5985	12.0953
77	Ir	64.8956	63.2867	73.5608	9.1751	9.0995	10.7083	10.9203	12.5126
78	Pt	66.832	65.112	75.748	9.4423	9.3618	11.0707	11.2505	12.9420
79	Au	68.8037	66.9895	77.984	9.7133	9.6280	11.4423	11.5847	13.3817
80	Hg	70.819	68.895	80.253	9.9888	9.8976	11.8226	11.9241	13.8301
81	Tl	72.8715	70.8319	82.576	10.2685	10.1728	12.2133	12.2715	14.2915
82	Pb	74.9694	72.8042	84.936	10.5515	10.4495	12.6137	12.6226	14.7644
83	Bi	77.1079	74.8148	87.343	10.8388	10.73091	13.0235	12.9799	15.2477
84	Po	79.290	76.862	89.80	11.1308	11.0158	13.447	13.3404	15.744
85	At	81.52	78.95	92.30	11.4268	11.3048	13.876	–	16.251
86	Rn	83.78	81.07	94.87	11.7270	11.5979	14.316	–	16.770
87	Fr	86.10	83.23	97.47	12.0313	11.8950	14.770	14.45	17.303
88	Ra	88.47	85.43	100.13	12.3397	12.1962	15.2358	14.8414	17.849
89	Ac	90.884	87.67	102.85	12.6520	12.5008	15.713	–	18.408
90	Th	93.350	89.953	105.609	12.9687	12.8096	16.2022	15.6237	18.9825
91	Pa	95.868	92.287	108.427	13.2907	13.1222	16.702	16.024	19.568
92	U	98.439	94.665	111.300	13.6147	13.4388	17.2200	16.4283	20.1671
93	Np	–	–	–	13.9441	13.7597	17.7502	16.8400	20.7848
94	Pu	–	–	–	14.2786	14.0842	18.2937	17.2553	21.4173
95	Am	–	–	–	14.6172	14.4119	18.8520	17.6765	22.0652

* Values after Bearden (1967), available under http://xray.uu.se/.

References

Aiginger, H. and Wobrauschek, P. (1974) *Nucl. Instrum. Methods* **114**, 157.
Chen, J.R., Chao, E.T., Minkin, J.A., Back, J.M., Jones, K.W., Rivers, M.L. and Sutton, S.R. (1990) *Nucl. Instrum. Methods* **B49**, 533.
Claisse, F. and Quintin, M. (1967) *Can. Spectrosc.* **12**, 129.
Coolidge, W.D. (1913) *Phys. Rev.* **2**, 409.
Criss, J. and Birks, L. (1968) *Anal. Chem.* **40**, 1080.
de Boer, D.K.G. (1990) *X-Ray Spectrom.* **19**, 145–154.
de Boer, D.K.G. (1991) *Phys. Rev.* **B44**, 498.
de Boer, D.K.G. (1991) *Spectrochim. Acta* **46B**, 1433.
de Boer, D.K.G. and van den Hoggenhof, W.W. (1991) *Spectrochim. Acta* **46B**, 1323.
Giauque, R.D., Thompson, A.C., Underwood, J.H., Wu, Y., Jones, K.W. and Rivers, M.L. (1988) *Anal. Chem.* **60**, 855.
Huang, T.C., Fung, A. and White, R.L. (1989) *X-Ray Spectrom.* **18**, 53–56.
Hubbell, J.H., Veigele, W.J., Briggs, E.A., Brown, R.T., Cromer, D.T. and Howerton, R.J. (1975) *J. Phys. Chem. Ref. Data* **4**, 471.
Iida, A., Sakurai, K., Matsushita, T. and Gohshi, Y. (1985) *Nucl. Instrum. Methods* **228**, 556.
Kaufmann, M., Mantler, M. and Weber, F. (1994) *Adv. X-ray Anal.* **37**, 205–212.
Klockenkämper, R. (1996) *Total-Reflection X-ray Fluorescence Analysis*, John Wiley, New York.
Knoth, J. and Schwenke, H. (1978) *Fresenius Z. Anal. Chem.* **291**, 200.
Lachance, G.R. and Claisse, F. (1980) *X-Ray Spectrom.* **23**, 87.
Lachance, G.R. and Traill, R.J. (1966) *Can. Spectrosc.* **11**, 43.
Love, G. and Scott, V.D. (1987) *Inst. Phys. Conf. Ser.* **90**, 349–353.
Mantler, M. (1987) *Prog. Crystal Growth Charact.* **14**, 213–261.
Moseley, H.G.J. (1913) *Philos. Mag.* **26**, 1024–1034.
Moseley, H.G.J. (1914) *Philos. Mag.* **27**, 703–713.
Nicolosi, J.A., Groven, J.P. and Merlo, D. (1987) *Adv. X-ray Anal.* **30**, 183–192.
Pella, P.A. and Dobbyn, R.C. (1988) *Anal. Chem.* **60**, 684.
Prange, A. (1989) *Spectrochim. Acta* **44B**, 437.
Rasberry, S.R. and Heinrich, K.F.J. (1974) *Anal. Chem.* **46**, 81.
Schwenke, H. and Knoth, J. (1982) *Nucl. Instrum. Methods* **193**, 239.
Sherman, J. (1955) *Spectrochim. Acta* **7**, 283.
Shiraiwa, T. and Fujino, N. (1966) *Jpn. J. Appl. Phys.* **5**, 886.
Sparks, C.J., Raman, S., Ricci, E., Gentry, R.V. and Krause, M.O. (1978) *Phys. Rev. Lett.* **40**, 507.
Streli, Ch., Wobrauschek, P., Bauer, V., Kregsamer, P., Görgl, R., Pianetta, P., Ryon, R., Pahlke, S. and Fabry, L. (1997) *Spectrochim. Acta* **B52**, 861–872.
Szaloki, I. (1996) *X-Ray Spectrom.* **25**, 21–28.
van Eenbergen, A. and Volbert, B. (1987) *Adv. X-ray Anal.* **30**, 201–211.
Wobrauschek, P. and Aiginger, H. (1975) *Anal. Chem.* **47**, 852.
Wobrauschek, P., Görgl, R., Kregsamer, P., Streli, Ch., Pahlke, S., Fabry, L., Haller, M., Knöchel, A. and Radtke M. (1997) *Spectrochim. Acta* **B52**, 901–906.
Yoneda, Y. and Horiuchi, T. (1971) *Rev. Sci. Instr.* **42**, 1069.

9 Analysis of ions using high-performance liquid chromatography

S. LEVIN

9.1 What is ion chromatography?

Ion chromatography (IC) is an analytical technique for the separation and determination of ionic solutes from various origins, such as environmental water (Roessner *et al.* 1987, Tanaka *et al.* 1987, Ammann and Ruettimann 1995, Fung and Dao 1995, Singh *et al.* 1996, Singh and Abbas 1996), water in industrial processes (Cox *et al.* 1987, Jackson 1995, Jackson *et al.* 1995), water in the metal industry (Barkley *et al.* 1992), industrial waste water (Mosko 1984, Nonomura 1987, Nonomura and Hobo 1989, Voloschik *et al.* 1994), samples from biological systems (Porter *et al.* 1985, Hallstrom *et al.* 1989, Boermans 1990, Wolf *et al.* 1992), pharmaceutical samples (Jenke 1988, Hallstrom *et al.* 1989, Saari and Anderson 1991, Den-Hartigh *et al.* 1993, Riley and Nowotnik 1994) and food (Jancar *et al.* 1984, Saari and Anderson 1991, Morawski *et al.* 1993, Santillana *et al.* 1993), etc. IC can be classified as a liquid chromatographic method, in which a liquid permeates through a porous solid stationary phase and elutes the sample components into a flow-through detector. The stationary phase is usually in the form of small-diameter (5–10 μm) uniform particles, packed into a cylindrical column. The column is constructed from a rigid material (such as stainless steel or plastic) and is generally 5–30 cm long and the internal diameter is in the range of 4–9 mm. A high-pressure pump is required to force the mobile phase through the column at typical flow rates of 1–2 mL/min. The sample to be separated is introduced into the system by an injection device, manual or automatic, prior to the column. The detector usually contains a low-volume cell through which the mobile phase passes carrying the separated sample components. A chromatographic system is shown in Figure 9.1.

Any chromatographic system of the type shown in Figure 9.1 can be divided into instrumentation and chemistry components. The instrumentation components are the pump, injector, detector and data station, whereas the chemical components are the mobile phase and the stationary phase. The instrumentation in ion chromatography is typical to high-

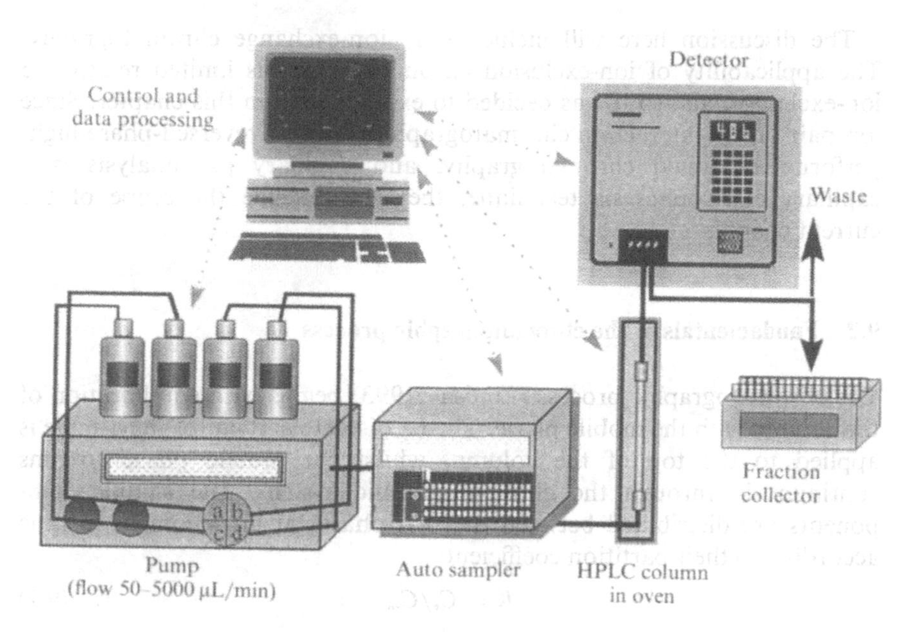

Figure 9.1 Schematic representation of a high-performance liquid chromatographic system.

performance liquid chromatography (HPLC) and the chemistry components are the ones that determine that this mode of HPLC is dedicated to analysis of ions. As in any chromatographic mode, the composition of the mobile phase provides the chemical environment for the interaction of the solutes with the specific stationary phase. Separation can be achieved by controlling and manipulating these interactions, which affect the relative retention times of the various sample components. The types of solutes that can be determined using ion chromatographic techniques are the following:

1. Inorganic ions such as Cl^-, Br^-, SO_4^{2-}, etc.
2. Inorganic cations, including alkali-metal, alkaline-earth, transition-metal and rare-earth ions, but not non-charged metal complexes.
3. Organic acids, including carboxylic, sulfonic, phosphonic acids, etc.
4. Organic bases, including amines.
5. Ionic organometallic compounds.

The liquid chromatographic techniques applicable to the separations described above are termed as the following:

- Ion-exchange chromatography (Haddad and Jackson 1990)
- Ion-exclusion chromatography (Haddad and Jackson 1990)
- Ion-pair (ion-interaction) chromatography (Haddad and Jackson 1990)
- Capillary electrophoresis (Guzman 1993)

The discussion here will include only ion-exchange chromatography. The applicability of ion-exclusion chromatography is limited relative to ion-exchange, and so it was decided to exclude it from this chapter. Since ion-pair or ion-interaction chromatography is in fact reversed-phase high-performance liquid chromatography, and capillary ion analysis is a capillary electrophoresis technique, they are outside the scope of the current chapter.

9.2 Fundamentals of the chromatographic process

The chromatographic process (Lindsay 1993) begins with equilibration of the column with the mobile phase. Then a discrete amount of the sample is applied to the top of the column, whilst the mobile phase streams continuously through the chromatographic system. The sample components are distributed between the two phases at any moment in time according to their partition coefficient:

$$K = C_s/C_m \tag{9.1}$$

where C_s and C_m are the concentrations of the solute in the stationary and mobile phases, respectively, at any given moment. The rate of travel is determined by the average velocity of the mobile phase, the volume ratio of the mobile and stationary phases, and the distribution coefficient of the particular component. If all the chromatographic parameters remain constant, the only factor that changes from one sample component to the other is the distribution ratio, which causes them to migrate at different rates and to reach the detector at different times, the retention times. The detector senses the sample components and produces the **chromatogram**: a graph showing the detector signal as a function of time. The process and a chromatogram are shown in Figure 9.2.

9.2.1 The chromatogram

The chromatogram shown in Figure 9.2 displays three peaks: two retained peaks, one for each sample component (A and B), and the first peak, the so-called 'solvent' or 'injection' peak. The first peak contains the solvent in which the sample was dissolved, together with components that eluted through the column with no interaction with the stationary phase. The time that it takes for the non-retained components to be eluted through the column is known as the **void time**, t_0. Similarly, the time at which each retained component appears in the chromatogram is known as the **retention time**, t_R. The values of t_0 and t_R depend on the flow rate of the mobile phase, and the physical dimensions of the column and the connecting tubing. Therefore it is more conventional to use the

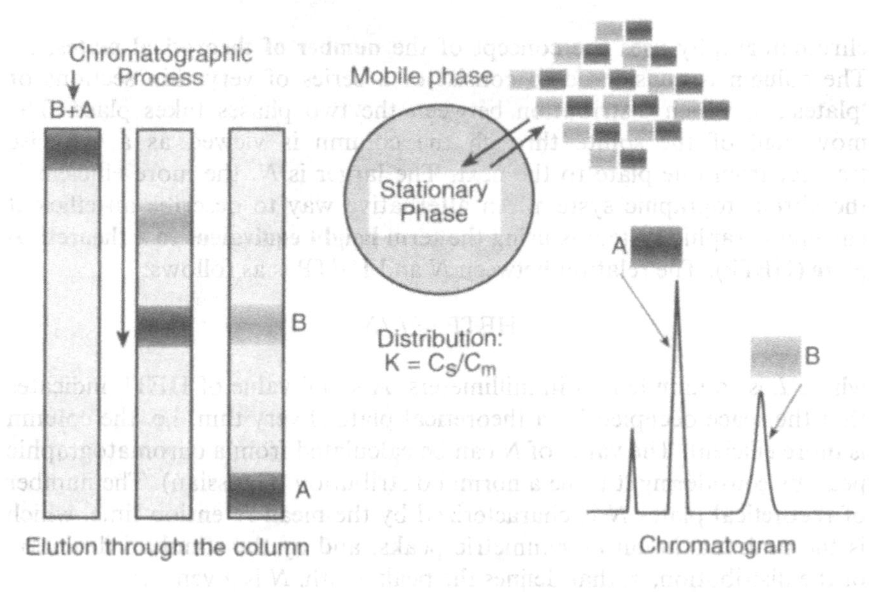

Figure 9.2 Schematic illustration of the separation of two components by IC.

retention factor k', which normalizes the extent of retention by these parameters:

$$k' = \frac{t_R - t_0}{t_0} \tag{9.2}$$

The thermodynamic basis of the retention factor is the ratio between the amounts of the solute in the stationary and mobile phases at any given moment. This ratio is in fact the **capacity ratio** k', which can be described as the following:

$$k' = Q_s/Q_m = (V_s/V_m) \times (C_s/C_m) = \Phi K \tag{9.3}$$

where Q_s and Q_m are the amounts of the solute in the stationary phase and mobile phase respectively; V_s and V_m are the volumes of the stationary and mobile phase respectively; and C_s and C_m are as in equation (9.1).

9.2.2 *Some fundamental chromatographic concepts*

These are described in Lindsay (1993). The sample components begin their migration through the chromatographic column at the column head and they all occupy the same volume, the **injection volume**. As they traverse the column, they form separate zones that migrate at different velocities. Each of these zones is broadened during this migration due to diffusion and mass transfer effects, as the population of molecules migrate through a tortuous path, while distributing between the two phases. The theory of chromatography uses the concept of the **number of theoretical plates**, N. The column is considered to consist of a series of very thin sections or 'plates', in which distribution between the two phases takes place. The movement of the solute through the column is viewed as a stepwise transfer from one plate to the next. The larger is N, the more efficient is the chromatographic system. An alternative way to describe an efficient chromatographic system is using the term **height equivalent to a theoretical plate** (HETP). The relation between N and HETP is as follows:

$$\text{HETP} = L/N \tag{9.4}$$

where L is column length in millimeters. A small value of HETP indicates that the space occupied by a theoretical plate is very thin, i.e. the column is more efficient. The value of N can be calculated from a chromatographic peak by considering it to be a normal distribution (Gaussian). The number of theoretical plates N is characterized by the mean retention time, which is the peak maximum in symmetric peaks, and by the standard deviation of the distribution, σ, that defines the peak width. N is given by:

$$N = (t_R/\sigma)^2 \tag{9.5}$$

From this it can be seen that a large value of N indicates that the standard deviation of the molecules' population remains small at high retention times. By using the Gaussian distribution, and defining the terms $w_{1/2} = 2.345\sigma$ and $w_{base} = 4\sigma$, N can be calculated by the following relationships:

$$N = 5.54(t_R/w_{1/2})^2 = 16(t_R/w_{base})^2 \tag{9.6}$$

These expressions are used routinely as quantitative criteria for column efficiency, as part of system suitability tests in HPLC. The value of N is meaningful only for $k' > 2$, for a number of reasons, which will not be discussed here, as it is outside the scope of this chapter.

The separation between two adjacent peaks is determined by the **resolution factor** R_s, which depends on the differences between their retention times and on the widths at the base of these two peaks:

$$R_s = \frac{(t_{R,2} - t_{R,1})}{0.5(w_{base,2} + w_{base,1})} \tag{9.7}$$

Although the relationships introduced here are valid only for Gaussian peaks, they are usually applied also to non-symmetrical peaks for convenience. Since the values of N and R_s can be used as quantitative criteria for system performance, they fulfill an important role in establishing a newly developed or applied chromatographic method.

9.3 Principles of the separation

The principles of separation in HPLC are described in Haddad and Jackson (1990). In general, the mechanism of interaction of the solutes with the stationary phase determines the classification of the mode of liquid chromatography. In ion chromatography the basic interaction is ionic. The stationary phase is charged due to fixed anions or cations, which are electroneutralized by labile counter-ions of the opposite charge. The counter-ions can be exchanged by other ions either from the mobile phase or from the sample; hence the name **ion-exchange chromatography**.

Figure 9.3 illustrates the principle of retention by exchange of anions in anion-exchange chromatography, and equation (9.8) below describes it as an equilibrium. The functional groups on the stationary phase's surface are fixed positively charged species (M^+). At equilibrium these positively charged functional groups are neutralized by the counter-ions from the running mobile phase (C^-). In the second and third steps, the anionic sample components (A^-) enter the column and distribute between the stationary and mobile phases by displacing the counter-ions, and being displaced by the mobile phase ions back and forth. The distribution equilibrium is determined by the competition between the sample components and the counter-ions, or the anions of the mobile phase on the charged sites of the stationary phase. The process can be described as:

$$M^+C^- + A^- \rightarrow M^+A^- + C^- \tag{9.8}$$

The electroneutrality of the solution and the stationary phase's surface must be maintained during the ion-exchange process. Therefore, the

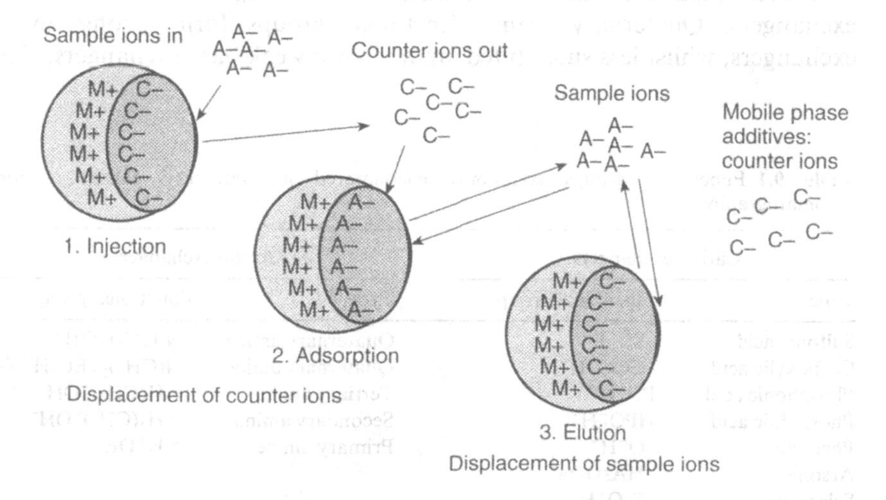

Figure 9.3 Schematic illustration of the retention principle of anion-exchange chromatography.

exchange is stoichiometric, so that a single monovalent anion A^- displaces a single monovalent counter-ion C^-. The process of cation retention is similar, only the stationary phase is negatively charged and the counter-ions are positively charged.

9.4 Types of stationary phases

Ion exchangers are characterized both by the nature of the matrix used as a support and by the nature of the ionic functional groups on the surface. Table 9.1 shows the types of functional groups commonly encountered in ion chromatography.

9.4.1 Functional groups

Ion exchangers' functional groups can function as such only when they are ionized, and therefore they are classified into **strong exchangers** and **weak exchangers** according to their ionization degree over the entire pH range. The strong functional groups are ionized over a wide pH range, in contrast to the weak functional groups, which are ionized over a limited pH range. Sulfonic acid cation exchangers are strong acid types, whilst the remaining cation exchangers' functional groups in Table 9.1 are weak. The weak acidic functional group requires the use of pH higher than its pK_a in order to retain cations. For example, a carboxylic functional group such as resin–COOH will be able to retain cation only in its resin–COO^- form, which exists mainly at pH above its pK_a.

Similarly, anion exchangers are classified as **strong base** and **weak base** exchangers. Quaternary amine functional groups form strong anion exchangers, whilst less substituted amines form weak base exchangers. The

Table 9.1 Functional groups commonly encountered in stationary phases of ion chromatography

Cation exchangers		Anion exchangers	
Type	Functional group	Type	Functional group
Sulfonic acid	$-SO_3^- H^+$	Quaternary amine	$-N(CH_3)_3^+ OH^-$
Carboxylic acid	$-COO^- H^+$	Quaternary amine	$-N(CH_3)_2\ (EtOH)^+ O$
Phosphonic acid	$PO_3^- H^+$	Tertiary amine	$-NH(CH_3)_2^+ OH^-$
Phosphinic acid	$HPO_2^- H^+$	Secondary amine	$-NH_2(CH_3)_2^+ OH^-$
Phenolic	$-O^- H^+$	Primary amine	$-NH_3^+ OH^-$
Arsonic	$-HAsO_3^- H^+$		
Selenonic	$-SeO_3^- H^+$		

strong base will be positively charged over a wide pH range, and therefore will be able to function as an anion exchanger even in basic pH, in contrast to the weak anion exchangers. A weak anion exchanger such as resin–NH_2 requires pH sufficiently low enough (pH < 6–7) to protonate the amine group into resin–NH_3^+. Most of ion chromatography uses strong anion exchanger (SAX) or strong cation exchangers (SCX), either on silica or polymeric-based stationary phases.

9.4.2 *Matrices*

The types of matrices used as support for stationary phases in ion chromatography can be divided into three: silica-based, synthetic organic polymers, and hydrous oxides (Haddad *et al.* 1985, Riley and Nowotnik 1994).

Silica-based ion exchangers. There are two distinct groups of silica-based materials (Yamamoto *et al.* 1984, Heping *et al.* 1990, Hirayama *et al.* 1993). One group includes **functionalized silica**, where a functional group is chemically bound directly to the silica particle. The second group is **polymer-coated silica**, in which the silica particles are first coated with a layer of polymer, such as polystyrene, silicone or fluorocarbon, and this layer is then functionalized. The main advantage of such particles over the total polymeric ones are the faster diffusion of the solutes throughout the thin layer of the polymer, which leads to a better mass transfer between the two phases, the stationary and the mobile. Improved mass transfer leads to better efficiency of the separation.

The functionalized silica-based ion exchangers are produced by chemically bonding quaternary amines to form strong anion exchangers and alkylsulfonates to form strong cation exchangers. Their capacity is usually moderate to high, requiring either ultraviolet–visible (UV-VIS) detection or conductivity with suppression (see section 9.5.1). The polymer-coated silica matrices have low capacities, and therefore they are suitable for non-suppressed ion chromatography (see section 9.5.1).

The most important advantage of silica-based stationary phases is the better chromatographic efficiency, stability and durability under high pressures. A serious drawback of these stationary phases is the limited pH range over which the columns can be operated, 2 < pH < 7, due to hydrolysis of the silica. This pH limitation is typical to another silica-based mode of HPLC, reversed-phase HPLC, which is based on silica gel substituted with hydrocarbons on the surface. Another drawback of the silica-based particles is the affinity of the bare silica and the free silanols on its surface to metal ions with high charge density, such as transition metals. Those are irreversibly adsorbed on the surface, causing interference with the analysis.

Polymer-based ion-exchangers. Polymeric supports for ion-exchange chromatography (Chauret 1989, Miura and Fritz 1989, Heard and Talmadge 1992, Morris and Fritz 1994) are called **resins**. These materials are produced by chemical derivatization of synthetic organic polymers, and they are the most widely used types of ion exchangers. These resins are manufactured by first synthesizing a polymer with suitable physical and chemical properties, and then they are further reacted to introduce the ionic functional groups. Most ion-exchange resins are made from copolymers consisting of styrene and divinylbenzene (PS–DVB), and some consist of copolymers of divinylbenzene and acrylic or methacrylic acid (PMMA).

The fact that a low degree of functionalization is required for ion chromatography implies that a significant proportion of the surface area of the resin exists as a neutral polymer, mainly aromatic moieties. It can therefore be expected that some of the hydrophobic character of the original polymer will be retained, and surface adsorption effects will contribute to the retention of organic ions. These effects are the reason for differences in ion-exchange selectivity between resin- and silica-based ion exchangers with the same functional groups, and the need for eluents containing polar organic solvents to control the selectivity in polymeric stationary phases.

The prime advantage of resin-based ion exchangers is their tolerance towards eluents and samples with extreme pH values, between 0 and 14, in contrast to the silica-based stationary phases, whose pH limits are 2–7. This wide range of pH values enables the exploitation of selectivity effects of multi-charged or weakly ionizable solutes.

The polymeric resins are subject to pressure limitations, because they are relatively soft materials; as a result, the column lengths and flow rates are limited. Macroporous (up to 1000 Å pores) resins are relatively more rigid and stable, and therefore they can be used in long columns and at higher flow rates.

Hydrous oxide. Minerals, such as aluminosilicates, alumina, silica or zirconia, can act as ion exchangers because of their skeleton or matrix material, which carries an excess charge that can be neutralized by mobile counter-ions. For example, anion and cation separations were done using a mixed-bed alumina–silica column (Schmitt and Pietrzyk 1985, Brown and Pietrzyk 1989). The hydrated metal oxide can act as either acid or base, and hence can be used as either a cation or an anion exchanger depending on the pH used in the separation:

$$=\text{M–O–H} \rightarrow =\text{M–O}^- + \text{H}^+ \quad (9.9)$$

$$=\text{M–O–H} \rightarrow =\text{M}^+ + \text{OH}^- \quad (9.10)$$

The pH values at which the reactions in equations (9.9) and (9.10) occur are dependent on the type of hydrous oxide under consideration. The matrix is a cation exchanger at low pH values and an anion exchanger at high pH values and it has an **isoelectric pH point**, depending on the surface chemistry and the type of buffer that is used to maintain the pH. For example, the isoelectric point for silica is 2, and for alumina it can be 3.5 in citrate buffer and 9.2 in carbonate buffer. The pH is therefore a powerful selectivity-controlling parameter in the hydrous oxide stationary phases.

9.4.3 *Characteristics – ion capacity*

The ion capacity of the ion exchanger is determined by the number of functional groups per unit weight of the stationary phase (Foley and Haddad 1986). A typical ion-exchange capacity in IC is 10–100 mequiv/g. The most commonly used unit is milliequivalents of charge per gram of dry packing, or milliequivalents per milliliter of wet packing. In the second case it is customary to state the type of counter-ion present in the stationary phase, since it affects the degree of swelling of the packing and hence its volume. The ion-exchange capacity of a stationary phase determines the concentrations of competing ions used in the mobile phase for elution. High-capacity stationary phases generally require the use of more concentrated mobile phases. Such conditions are problematic when conductometric detectors are used, because they cannot function well with high salt concentrations.

9.4.4 *Characteristics – swelling*

Organic stationary phases consist of crosslinked polymeric chains containing ionic functional groups. When such materials come into contact with water, they tend to swell, with swelling pressures up to 300 atmospheres with high ion-exchange capacities (Kato *et al.* 1983). The higher the ionic capacity and the lower the crosslinking, the more sensitive the polymer is to swelling. The content of the mobile phase is very significant to the effect of swelling. Macroporous resins with high cross-linking and small ion-exchange capacities are commonly used as stationary phases for high-performance ion chromatography, because they swell less.

9.4.5 *Characteristics – selectivity*

The relative affinities of different counter-ions to the stationary phase show considerable variation with the type of ion exchanger and the conditions under which it is used. There are cases where a simple ion-exchange mechanism may not be the sole retention mechanism, such as cases where ion-exclusion effects exist, or cases where there is adsorption to the stationary-phase matrix rather than to the functional groups. However, it is still possible to provide approximate guidelines for the relative affinities of the ion exchangers for different ions. The properties of the solute ions, the mobile-phase ion and the counter-ions that affect the extent of the ionic interactions are the following:

- The charge and the size of the solvated ion
- The degree of crosslinking of the ion-exchange polymers
- The polarizability of the solute ion
- The ion-exchange capacity of the stationary phase
- The type of functional group on the stationary phase
- The extent of interactions with the stationary-phase matrix of the support.

As a rule, an increase of the charge density (charge/solvated size) of the ion enhances its affinity for the stationary phase. Higher charge with smaller solvated ion radius results in higher retention due to higher Coulombic interactions. This trend becomes more pronounced in more diluted mobile phases.

The order of relative affinities of cations to strong acid cation-exchange stationary phases are generally in the following order (Jenke, 1984):

$Pu^{4+} \gg$

$La^{3+} > Ce^{3+} > Pr^{3+} > Eu^{3+} > Y^{3+} > Sc^{3+} > Al^{3+} \gg$

$Ba^{2+} > Pb^{2+} > Sr^{2+} > Ca^{2+} > Ni^{2+} > Cd^{2+} > Cu^{2+} > Co^{2+} > Zn^{2+} > Mg^{2+} > UO_2^{2+} \gg$

$Tl^{+} > Ag^{+} > Cs^{+} > Rb^{+} > K^{+} > NH_4^{+} > Na^{+} > H^{+} > Li^{+} >$

From this series it can be concluded that cation-exchange mobile phases of 0.1 M KCl are stronger than those containing 0.1 M NaCl, provided that all other parameters are identical.

The order of relative affinity of anions to strong base anion-exchanger stationary phases follows the general order (Jenke 1984):

$$\text{citrate} > \text{salicylate} > ClO^- > SCN^- > I^- > S_2O_3^{2-} > WO_4^{2-} > MoO_4^{2-} > CrO_4^{2-}$$
$$> SO_4^{2-} > SO_3^{2-} > HPO_4^{2-} > NO_3^- > Br^- > NO_2^- > CN^- > Cl^- > HCO_3^-$$
$$> H_2PO_4^- > CH_3COO^- > IO_3^- > HCOO^- > BrO_3^- > ClO_3^- > F^- > OH^-$$

Higher degree of crosslinking results in ion-exclusion effects, i.e. exclusion of ions with higher solvated radii from the stationary phase's pores. Since these ions are also less retained, they migrate faster than the smaller, more charged ones, which can enter the small pores. Ions with high charges and small radius are polarizable and therefore are retained longer.

The effects of the last two properties in the above list are hard to predict.

Therefore, it is not possible to provide clear-cut guidelines, regarding the control of the separation, based on these two properties.

9.5 Properties of mobile phases

Elution strength of the mobile phase is controlled by changing ionic strength, pH or type of anions. The mobile phases used in IC are typically aqueous salt solutions, which can be classified into groups of similar characteristics as follows:

- Compatibility with the detection mode – suppressed or non-suppressed.
- Nature of the competing ion
- Concentration of the competing ion
- Mobile phase's pH
- Buffering capacity of the mobile phase
- Ability to complex the ionic sample components
- Organic modifiers

9.5.1 Compatibility with the detection mode – suppressed or non-suppressed

The detection mode that is used is the major factor that determines the types of mobile phases suitable for the desired separation (Dugay *et al.* 1995). The detector signal obtained by the background, i.e. the mobile phase itself, must not be too high, otherwise it would be difficult to obtain linearity, wide dynamic range and stability of the baseline. When high sensitivity is needed, highly responding mobile phase (highly conducting in conductivity detector and highly absorbing in UV-VIS detector) will render it impossible to be used. If a highly conducting mobile phase is the only option for a particular separation, or high sensitivity is a must, the mobile phase should be selected so that its conductivity will be suppressed by a suppressor. This is a device that is installed between the column outlet and the detector to remove ions from the mobile phase, and hence reduce its conductivity.

9.5.2 Nature of the competing ion

In qualitative terms, the mobile-phase characteristics that influence solute retention are the relative affinities of the sample ions and the mobile phase's competing ions (Riley and Nowotnik 1994). The affinity of the mobile-phase ions to the stationary phase is governed by the same factors that affect the affinity of the solute ions, i.e. charge density, degree of hydration, polarizability, etc. (see section 9.4.5). Mobile-phase ions of high affinity to the stationary phase comprise strong eluents, which will reduce the interactions of the sample ions with the stationary phase, and will result in lower retention times.

9.5.3 Concentration of the competing ion

The concentration of the counter-ion in the mobile phase affects the retention of the sample ions as well as its charge density (Jenke 1994). Higher concentrations result in stronger competition, and displacement of the sample ions from the stationary phase, hence lower retention. The effect of concentration on the competition between the solute and the mobile-phase ions is much more pronounced for singly charged ions than for doubly charged ions, although the latter is a stronger eluent.

It is therefore most convenient to choose the type of mobile phase by initially selecting the appropriate charge. The next step will be considering additional effects on selectivity such as size, polarizability, etc., within the group of mobile-phase salts having the desired charge. The last consideration will be manipulating the mobile-phase salt concentration to produce the required separation.

9.5.4 Mobile phase's pH

The mobile phase's pH is a key parameter in determining its characteristics, as it influences the charges on both the mobile-phase ions and the solute ions (Jenke 1994). The effect of pH is particularly important when weak acids or bases are used in the mobile phase, since the pH affects their ionization. The charge on an acidic anion increases with pH, so the eluting power of weak acid eluents increases with pH until the acid is completely dissociated. The opposite trend occurs for weak bases in the mobile phase. With decreasing pH, a higher degree of protonation occurs and the mobile phase becomes a stronger eluent.

Similarly, the degree of ionization of solute ions that are derived from weak acids or bases will be pH-dependent. In this case, increased solute charge will increase its affinity to the functional groups on the stationary phase, hence increase their retention. Examples of solutes showing these effects are F^-, CO_3^{2-}, PO_4^{3-}, SiO_3^{2-}, CN^- and amines. When these ions are present in mixtures with other ions that show no pH dependence, the control of the mobile phase's pH becomes an important variable to be manipulated in the optimization of the separation.

9.5.5 Buffering capacity of the mobile phase

Since both mobile-phase and solute ions can be affected by the pH, the buffering capacity of the mobile phase is very important (Jenke 1984), and should be maintained high. Polyprotic solute ions' retention can be significantly changed with pH, as their charge can increase from singly to doubly and triply charged. In such cases it is very important to make sure that the mobile-phase pH is kept constant, using high-capacity buffers, so that the solutes themselves will not affect the pH.

9.5.6 *Ability to complex the ionic sample components*

When separations of metallic ions are considered, the ability of the mobile phase's salts to complex them is a very important variable (Yamamoto *et al.* 1984). The complexing agent forms complexes with the metal ions that may change their original charge and degree of ionization. The new species now have different retention times, and therefore separation is effected. The degree of complexation depends on the concentration of the complexing agent as well as on the pH of the mobile phase.

9.5.7 *Organic modifiers*

Water-miscible organic solvents (Dumont and Fritz 1995), such as methanol, ethanol, glycerol, acetonitrile and acetone, are sometimes used as additives to the mobile phase for ion-exchange separations. These solvents can affect a variety of parameters related to the separation process, such as affinity of organic ions to the stationary phase, the degree of complexation when such a process occurs, or the degree of ionization of weak acidic and basic ions either in the mobile or the stationary phases or in the samples.

9.6 Ion suppression in ion chromatography

Suppression in ion chromatography (Dugay *et al.* 1995, Singh *et al.* 1996, Singh and Abbas 1996) is needed when conductivity detectors are used and the mobile phase is intensively conducting, saturating the detector's response. A device, called the **suppressor**, is inserted between the ion-exchange separator column and the detector. The suppressor modifies both the mobile phase and the separated solutes coming out of the separator column, so that the mobile phase's conductance is reduced and that of the solutes is enhanced; hence the detectability of the solutes is improved. The suppressor requires a regenerant (or scavenger) solution to enable it to operate for extended periods.

The simplest means to accomplish suppression of an acidic mobile phase is to pass it through a cation-exchange column in the hydrogen form. The simplest example for the function of a suppressor is the case of Cl^- ion as a solute eluted by an eluent that is composed of $NaHCO_3$ (Senior 1990, Edgell *et al.* 1994). The eluent reaction in the suppressor is given in equation (9.11) and the reaction of the solute with the suppressor is given in equation (9.12):

$$\text{resin–H}^+ + \text{Na}^+\text{HCO}_3^{2-} \rightarrow \text{resin–Na}^+ + \text{H}_2\text{CO}_3 \quad (9.11)$$

$$\text{resin–H}^+ + \text{Na}^+\text{Cl}^- \rightarrow \text{resin–Na}^+ + \text{HCl} \quad (9.12)$$

The combined result of these two processes is that the mobile phase's conductance is reduced greatly whilst the conductance of the sample ions is enhanced by the replacement of sodium ions ($50\,S\,cm^2/equiv$) with hydroxonium ions ($350\,S\,cm^2/equiv$). The detectability of the solute is therefore enhanced.

A similar procedure can be applied to cation-exchange chromatography, when the suppressor is an anion-exchange column in the OH^- form, which provides hydroxyl ions to the stream. A simple example would be the separation of Na^+ ions using HCl in the mobile phase (Singh *et al.* 1996, Singh and Abbas 1996). The processes of suppression are shown in equations (9.13) and (9.14).

$$\text{resin–OH}^- + \text{H}^+\text{Cl}^- \rightarrow \text{resin–Cl}^- + \text{H}_2\text{O} \quad (9.13)$$

$$\text{resin–OH}^- + \text{Na}^+\text{Cl}^- \rightarrow \text{resin–Cl}^- + \text{Na}^+\text{OH}^- \quad (9.14)$$

The eluent is converted into water whilst the conductance of the sample band is increased due to replacement of the Cl^- ions ($76\,S\,cm^2/equiv$) by OH^- ions ($198\,S\,cm^2/equiv$).

Table 9.2 describes a few mobile phases with which ion suppression is used.

9.6.1 Mobile phases for non-suppressed ion-exchange chromatography

Non-suppressed IC methods are characterized by the wide range of eluents used (Haddad *et al.* 1985, Dugay *et al.* 1995). Table 9.3 shows the types of eluents employed for the separation of anions and cations.

Eluents for anions.

Aromatic carboxylic acids and their salts. Salts of aromatic carboxylic acids (Diop *et al.* 1986), such as those shown in Figure 9.4, are the most widely employed eluent species in the separation of anions by non-suppressed IC. They have low conductance; therefore, when used in dilute solutions, they provide eluents with low background conductance. The aromatic moiety is an intense UV chromophore, so aromatic acid salts are also ideal for indirect spectrophotometric detection, in which the chromophore causes the detection of non-absorbing solutes (see section 9.7.3 for details). All of these acids are relatively weak, and therefore they have buffering action. Also, since many of them are polyprotic, they can provide the buffering action over a relatively wide range of pH values (Jenke 1994).

Mobile phases prepared from aromatic carboxylate salts are prepared very simply by mixing the acids with the appropriate amounts of lithium

Table 9.2 Eluents used for suppressed ion chromatography

Eluent	Competing ion in eluent	Suppressor form	Products of suppressor reaction	Ref.
Anion-exchange eluents				
$Na_2B_4O_7$	$B_4O_7^{2-}$	S–H^+	S–Na^+ + H_3BO_3	Weinberg (1994)
NaOH	OH^-	S–H^+	S–Na^+ + H_2O	Heitkemper *et al.* (1994)
Na_2CO_3	CO_3^{2-}	S–H^+	S–Na^+ + H_2CO_3	Nonomura (1987)
$NaHCO_3/Na_2CO_3$	HCO_3^{2-}/CO_3^{2-}	S–H^+	S–Na^+ + H_2CO_3	Senior (1990), Edgell *et al.* (1994)
Na(amino acids)	AA–COO^-	S–H^+	S–Na^+ + AA	Zolotov *et al.* (1987)
K_2-ethylenediamine-*N*,*N'*-diacetate (EDDA)	$EDDA^{2-}$	S–Cu^{2+}	S–K^+ + CuEDDA	Sato and Miyanaga (1989)
Na[amine–SO_3^-]	amine–SO_3^-	S–H^+	S–Na^+ + amine$^+$–SO_3^-	Ivey (1984)
Cation-exchange eluents				
HCl	H^+	S–OH^-	S–Cl^- + H_2O	Singh *et al.* (1996), Singh and Abbas (1996)
HNO_3	H^+	S–OH^-		Takeuchi *et al.* (1988)
HNO_3 + $Zn(NO_3)_2$	H^+, Zn^{2+}	S–OH^-	S–NO_3^- + $Zn(OH)_{2(s)}$ + H_2O	Wimberley (1981)
$Ba(NO_3)$ or $Pb(NO_3)$	Ba^{2+} or Pb^{2+}	S–SO_4^{2-}	S–NO_3^- + $BaSO_{4(s)}$ or $Pb(SO_4)_{(s)}$	Nordmeyer *et al.* (1980)

Table 9.3 Eluents used in non-suppressed IC

Anions	Cations
Aromatic carboxylic acids and salts	Inorganic acids
Aliphatic carboxylic acids and salts	Organic bases
Sulfonic acids and salts	Complexing agents
Potassium hydroxide	Inorganic eluents
Polyol–borate complexes	
EDTA	
Inorganic eluents	

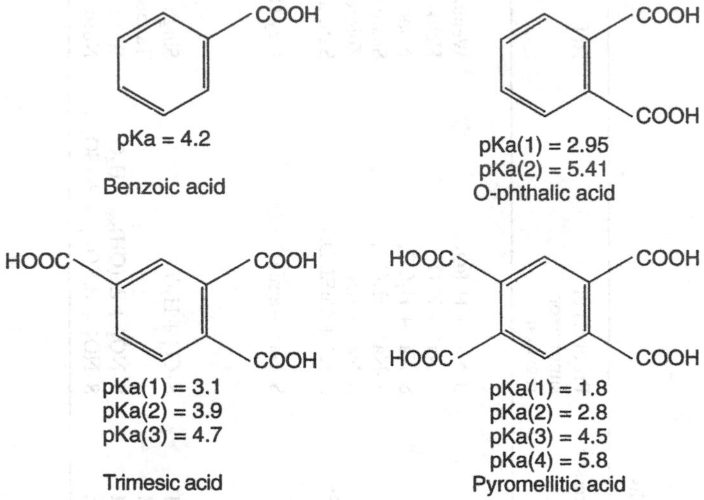

Figure 9.4 Aromatic carboxylic acids used in the mobile phase.

hydroxide, which is less conducting than Na^+ or K^+. When high pH is needed to increase the retention of weak acidic solutes, a borate buffer is used to raise the pH instead of the LiOH.

Chromatograms of phthalate salts used in the mobile phase for the separation of anions are shown in Figure 9.5. Five different columns were evaluated, using phthalate salt in the mobile phase, optimized for each particular column: column (a) is silica-based, column (b) is polyvinyl-aromatic-based, column (c) is polystyrene–divinylbenzene-based, and columns (d) and (e) are based on polymethacrylate.

Aliphatic carboxylic acids. Mobile phases prepared from salts of aliphatic carboxylic acids have been employed widely in non-suppressed IC (Fritz *et al.* 1984, Okada and Kuwamoto 1984). Citric (Nesternko

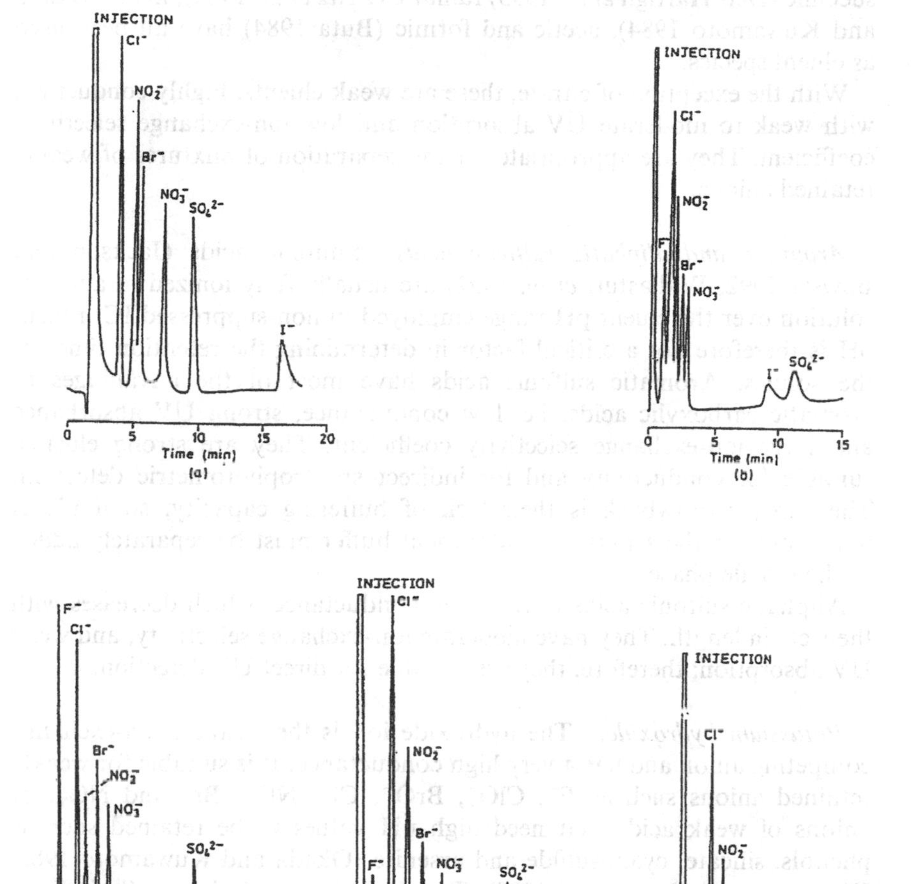

Figure 9.5 Chromatograms obtained with optimized phthalate eluents, using the following columns: (a) Vydac 302 IC 4.6, (b) Interaction ION-100, (c) Hamilton PRP-X100, (d) Bio-Gel TSK IC-Anion-PW, and (e) Waters IC Pak A. Conditions are as follows. Eluents, potassium hydrogenphthalate solutions, at the following concentrations and pH values: (a) 3 mM, pH = 5.3; (b) 1 mM, pH = 4.1; (c) 1 mM, pH = 5.5; (d) 1 mM, pH 5.3; (e) 1 mM, pH 7.0. Flow-rates (mL/min): (a) 2.0; (b) 1.0; (c) 2.0; (d) 1.2; (e) 1.2. Sample: 10 μL of a mixture containing 100 ppm each of the indicated anions. (Reprinted from Haddad *et al.* (1985), figure 2, with kind permission of Elsevier Science (The Netherlands).)

1991, Alumaa and Pentsuk 1994), tartaric (Alumaa and Pentsuk 1994), succinic (Den-Hartigh *et al.* 1993) fumaric (Fritz *et al.* 1984), malic (Okada and Kuwamoto 1984), acetic and formic (Buta 1984) have all been used as eluent species.

With the exception of citrate, these are weak eluents, highly conducting, with weak to moderate UV absorption and low ion-exchange selectivity coefficient. They are appropriate for the separation of mixtures of weakly retained anions.

Aromatic and aliphatic sulfonic acids. Sulfonic acids (Jackson and Bowser 1992, Widiastuti *et al.* 1992) are usually fully ionized in aqueous solution over the eluent pH range employed in non-suppressed IC. Eluent pH is therefore not a critical factor in determining the retention times of the solutes. Aromatic sulfonic acids have most of the advantages of aromatic carboxylic acids, i.e. low conductance, strong UV absorbance and large ion-exchange selectivity coefficient. They are strong eluents, suitable for conductivity and for indirect spectrophotometric detection. Their major drawback is their lack of buffering capacity, so if pH is important for the separation, additional buffer must be separately added to the mobile phase.

Aliphatic sulfonic acids have higher conductance, which decreases with their chain length. They have moderate ion-exchange selectivity, and weak UV absorption; therefore, they are suitable for direct UV detection.

Potassium hydroxide. The hydroxide ion is the weakest ion-exchange competing anion and has a very high conductance. It is suitable for weakly retained anions such as F^-, ClO_3^-, BrO_3^-, Cl^-, NO_2^-, Br^- and NO_3^-, or anions of weak acids that need high pH values to be retained such as phenols, silicate, cyanosulfide and arsenite (Okada and Kuwamoto 1985, Hirayama and Kuwamoto 1988). The detection mode is usually indirect conductivity, in which the mobile phase has high conductivity and causes the detection of non-conducting solutes, which appear as negative peaks (see section 9.7.1 for details).

Polyol–borate complexes. It is well known that both boric acid or borate form neutral or anionic complexes (Jackson and Bowser 1992) with polyhydroxy compounds such as mannitol, glucose, fructose, xylose, glycerol, sorbitol, sucrose or maltose or acidic compounds such as gluconic, tartaric, glucuronic and galacturonic. The complex with the gluconic acid is the most widely used, as can be seen in Figure 9.6

Ethylenediaminetetraacetic acid (EDTA). EDTA can be used as an aliphatic polycarboxylic eluent for anions as well as a strong complexing agent for polyvalent metallic cations (Yamamoto *et al.* 1984, Morawski *et*

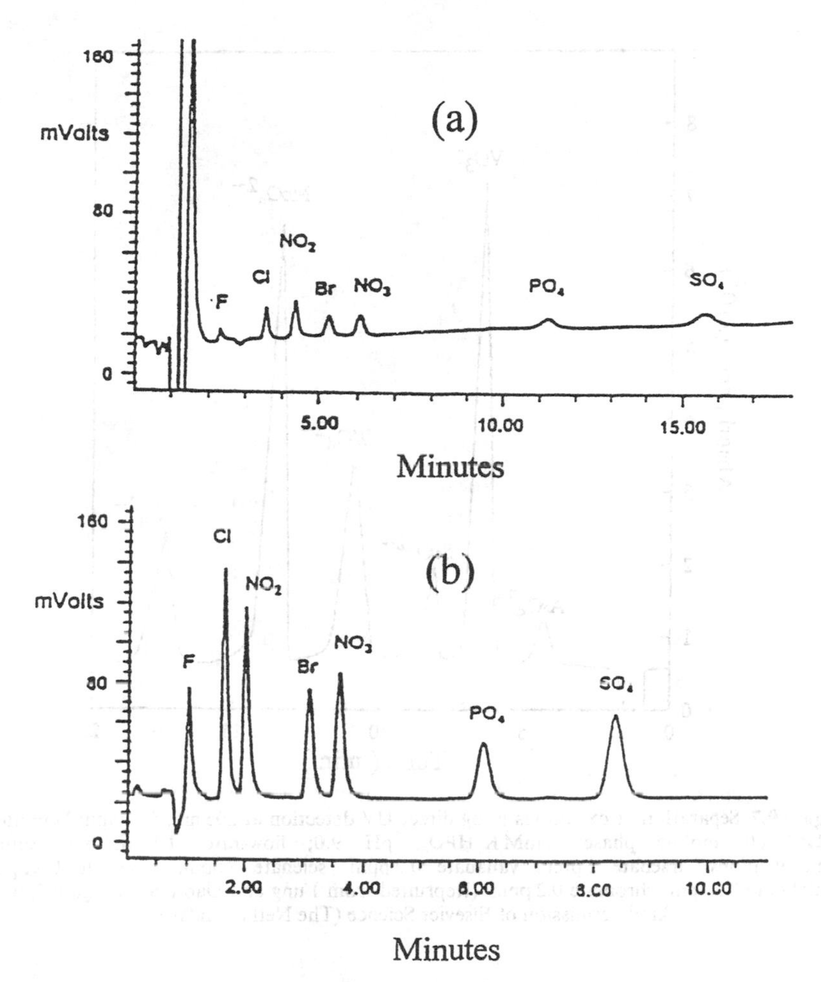

Figure 9.6 Chromatograms of low-level anion standards obtained using conductivity detection: (a) non-suppressed; (b) suppressed. Conditions: a Waters IC Pak Anion HR column; flow-rate 1.0 mL/min; injection volume 100 μL; solutes 0.1–0.6 ppm. Eluents: (a) borate–gluconate, 35°C; (b) bicarbonate–carbonate with Dionex Anion Micromembrane suppressor (AMMS). (Reprinted from Jackson *et al.* (1995), figures 1 and 2, with kind permission of Elsevier Science NL, Sara Burgerhartstraat 25, 1055 KV Amsterdam, The Netherlands.)

al. 1993, Scully *et al.* 1993). The majority of its applications involve the second property, the complexation capability.

Inorganic salts. Inorganic anions such as Cl^-, SO_4^{2-} or PO_4^{3-} can be used as strong eluents (Maruo *et al.* 1989), but due to their high conductance direct conductivity detection cannot be used. Other modes of detection can be UV absorption, refractive index, electrochemical and

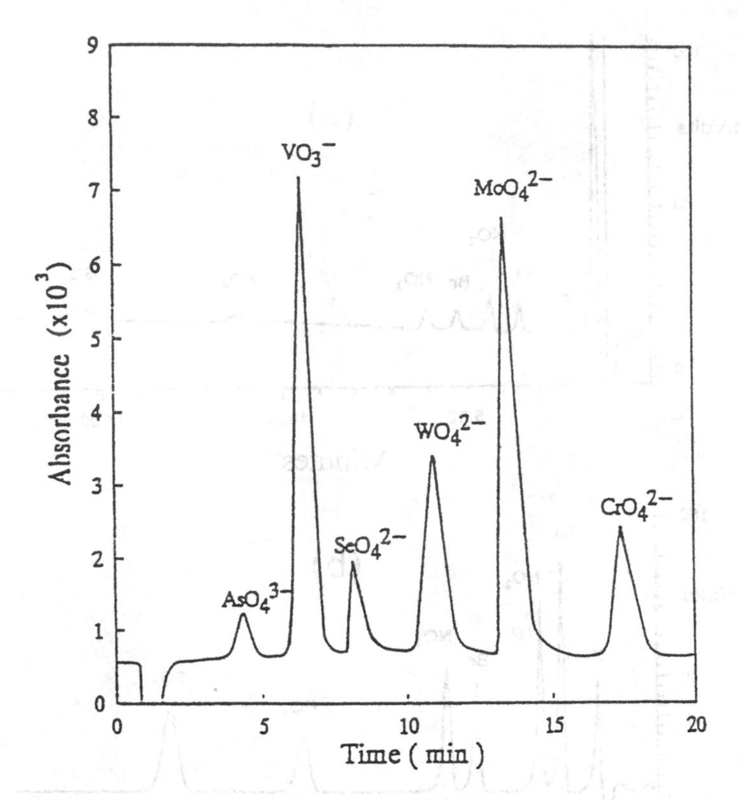

Figure 9.7 Separation of oxyanions using direct UV detection at 205 nm. Column: Hamilton PRP-X100; mobile phase: 3 mM K_2HPO_4, pH 9.0; flow-rate: 1.5 mL/min. Solute concentrations: arsenate 1 ppm, vanadate 0.2 ppm, selenate 2 ppm, tungstate 0.2 ppm, molybdate 0.2 ppm, chromate 0.2 ppm. (Reprinted from Fung and Dao (1995), figure 7, with kind permission of Elsevier Science (The Netherlands).)

post-column reaction. An example of the separation of oxyanions using phosphate buffer as the eluent and a direct UV detection of the anions is presented in Figure 9.7.

Eluents for cations.

Inorganic acids. Dilute solutions of inorganic acids, such as nitric acid, are the most popular eluents for the separation of alkali-metal cations and amines by non-suppressed IC. The eluent strength is determined solely by its pH. The hydroxonium ion is an effective competing cation for these solutes and the very high conductance of the mobile phase enables a sensitive indirect conductivity detection (see section 9.7.1 for details). An example of the separation of cations using nitric acid as the eluent and indirect conductivity detection is presented in Figure 9.8.

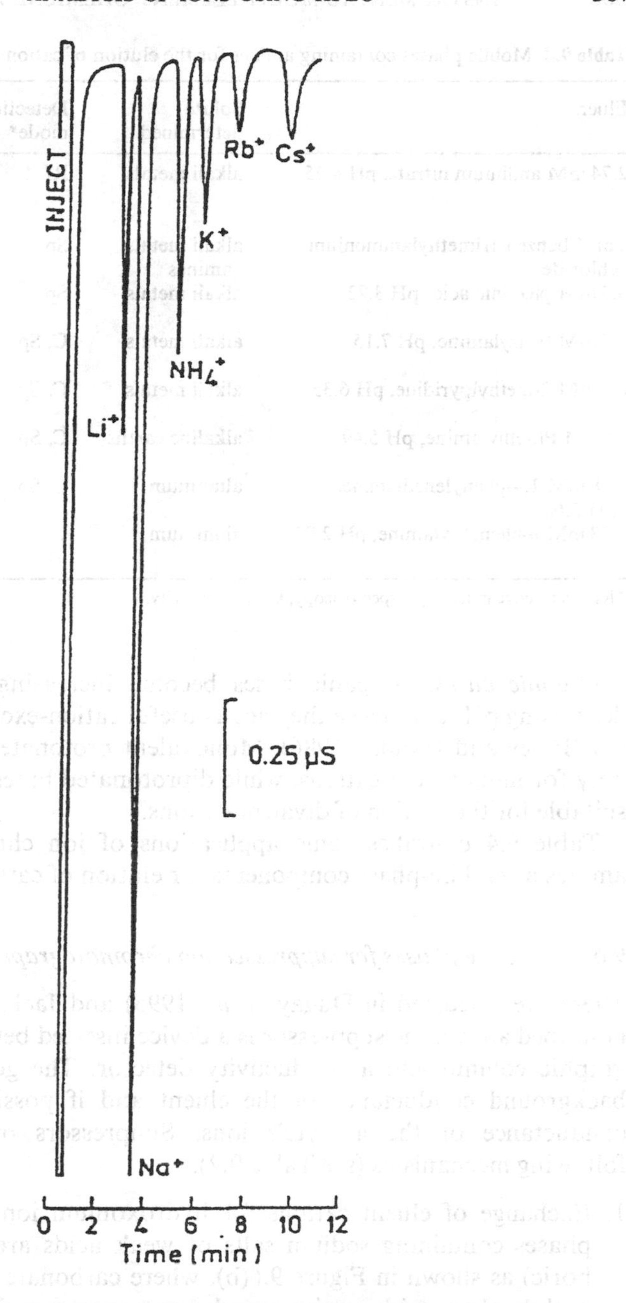

Figure 9.8 Typical indirect conductivity detection of inorganic cations using nitric acid as eluent. Column: Waters IC Pak C; eluent: 2.0 mM nitric acid (pH 2.70); flow-rate: 1.0 mL/min. Solutes: injection volume 8 µL, lithium 0.1 ppm, sodium 0.3 ppm, ammonium 0.2 ppm, potassium 0.4 ppm, rubidium 1.7 ppm and cesium 1.8 ppm. (Reprinted from Foley and Haddad (1986), figure 3, with kind permission of Elsevier Science (The Netherlands).)

Table 9.4 Mobile phases containing amines for the elution of cations

Eluent	Solute determined	Detection mode*	Ref.
2.74 mM anilinium nitrate, pH 4.65	alkali metals	RI	Haddad and Heckenberg (1982)
1 mM benzenetrimethylammonium chloride	alkali metals, amines	Sp	McAleese (1987)
0.5 mM picolinic acid, pH 3.22	alkali metals	Sp	Foley and Haddad (1986)
0.5 mM benzylamine, pH 7.15	alkali metals	C, Sp	Foley and Haddad (1986)
0.5 mM 2-methylpyridine, pH 6.35	alkali metals	C, Sp	Haddad and Foley (1989)
10 mM Ph-ethylamine, pH 5.49	alkaline earths	C, Sp	Haddad and Foley (1989)
5.48 mM 1,4-phenylenediamine, pH 2.68	aluminum	C, Sp	Foley and Haddad (1986)
8.23 mM *p*-phenethylamine, pH 2.94	aluminum	C	Fortier and Fritz (1985)

* RI – refractive index; Sp – spectroscopy; C – conductivity.

Organic bases. Organic bases become increasingly protonated with decreasing pH, and hence they act as useful cation-exchange eluents at low pH (Foley and Haddad 1986). Monovalent protonated bases are effective only for monovalent cations, while diprotonated bases are generally more suitable for the elution of divalent cations.

Table 9.4 describes some applications of ion chromatography using amines as mobile-phase components for elution of cations.

9.6.2 Mobile phases for suppressed ion chromatography

These are discussed in Dugay *et al.* (1995) and Jackson *et al.* (1995). As described above, the suppressor is a device inserted between the chromatographic column and a conductivity detector. The goal is to reduce the background conductance of the eluent and if possible to enhance the conductance of the analyte's ions. Suppressors operate through the following mechanisms (see Table 9.2):

1. Exchange of eluent cations for hydroxonium ions, for which mobile phases containing sodium salts of weak acids are suitable (carbonic, boric) as shown in Figure 9.6(b), where carbonate buffer is used in the mobile phase with a micromembrane suppressor for the separation of inorganic anions.
2. Exchange of eluent anions for hydroxide ions, for which nitrate or chloride salts are suitable (Wimberley 1981).
3. Complete removal of the eluent ions by precipitation, such as removal

of Ba and Pb ions by precipitation with SO_4^{2-} (Nordmeyer *et al.* 1980).

4. Reduction of the ionic charges in the mobile phase by complexing them with Cu^{2+}or other complexant ions. Mobile phases suitable for these suppressors should contain chelates (Sato and Miyanaga 1989).

9.7 Detection in ion chromatography

The following detection methods are available with ion-exchange chromatography:

- Conductivity detection
- Electrochemical (amperometric or coulometric) detection
- Potentiometric detection
- Spectroscopic detection
- Post-column reaction detection

9.7.1 Conductivity detection

Conductivity detection has two major advantages for inorganic ion analysis. First, all the ions are electrically conducting, so that the detector should be universal in response; and secondly, the detectors are relatively simple to construct and operate. Conductivity detection will be discussed here in terms of principle of operation and performance characteristics, modes of detection, cell design, post-column signal enhancement, i.e. suppression, and applications.

Principle of operation. A schematic illustration of the flow cell in a conductivity detector is presented in Figure 9.9. The mobile phase eluting through the detector is in fact a conducting electrolyte. It flows through two electrodes across which potential is applied. The more current conducted by the solution, the higher is the electrical conductivity. The conductance of a solution is determined by several factors, including the ionic strength and type of species in the solution, as well as the temperature. The specific conductance depends on the cross-sectional area (cm^2) of the electrodes inserted into the solution, and L (cm), the distance between them. The conductance is increased for cells in which the electrodes are large in surface area and are close together.

The conductance is subject to activity effects such as ion–ion interactions, and so it can become non-linear at high ionic strength. Since the conductance of the solution results from both the anions and cations of the electrolyte, it is calculated for the individual anions and cations in solution. Most of the common cations and anions have equivalent ionic conductance of 30–100 $S\,cm^2/equiv$. The most conducting cation is the

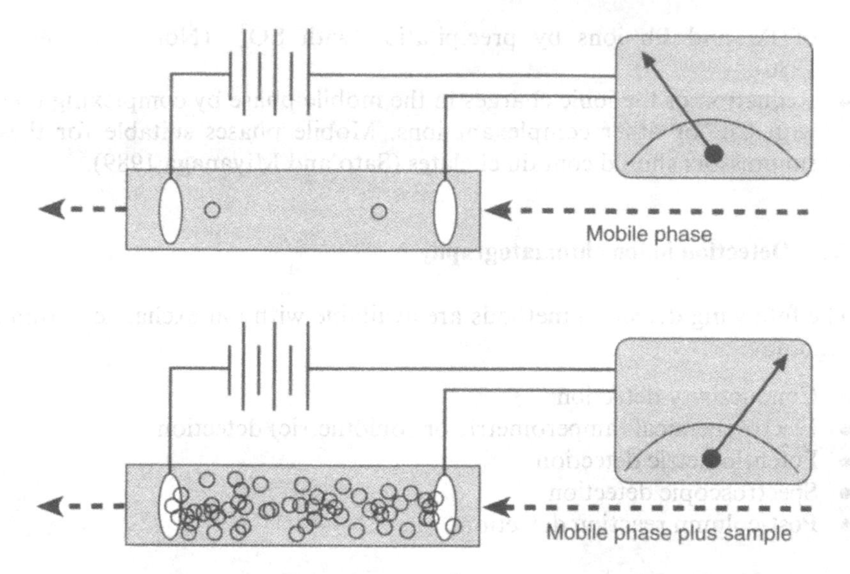

Figure 9.9 Schematic illustration of the flow cell in a conductivity detector.

hydroxonium ion and the most conducting anion is the hydroxyl ion; their values are 350 and 198 $S\,cm^2/equiv$ respectively. The conductance of an ion increases with its charge density and decreases with its viscosity. When a strongly eluting mobile phase with multiply charged ions is needed, the ions can exert high background; therefore, large ions such as phthalate, citrate, or trimesate are used in such cases.

A sensitive detection can be obtained as long as there is a considerable difference in the ionic conductance of the solute and the mobile phase's ions. This difference can be positive or negative, depending on whether the eluent ions are strongly or weakly conducting. If the ionic conductance of the eluent ions is low, then an increase in conductance occurs when higher conducting ionic solutes enter the detection cell, and the peaks appear positive. In general this detection mode is referred to as **direct**. On the other hand, when the mobile phase ions are highly conducting, a decrease in conductance occurs when the less conducting ionic solutes enter the detection cell, and peaks appear negative. This mode of detection is referred to as **indirect**.

Direct conductivity detection is used for most IC methods involving the separation of anions. Eluents for non-suppressed IC are prepared from salts such as potassium hydrogenphthalate (Haddad *et al.* 1985) (see Figure 9.5) or sodium benzoate (Juergens 1990), which are anions with moderately low conductance. Similarly, direct conductivity detection can be used for the separation of cations, using eluents containing organic bases (Foley and Haddad 1986).

Indirect conductivity detection can be applied to anions using hydroxide eluents (Okada and Kuwamoto 1985) and to cations using mineral acid eluents (Maruo *et al.* 1989), as shown in Figure 9.8.

9.7.2 *Electrochemical detection*

The term 'electrochemical detection' (Kawasaki *et al.* 1990, Chadha and Lawrence 1991, Leubolt and Klein 1993) is applied loosely to describe a range of detection techniques involving the application of electrical oxidation–reduction potential via suitable electrodes to a sample solution, containing oxidizable or reducible solutes. The resulting current is measured as a function of time. Electrochemical detection has been applied in situations where extreme sensitivity or selectivity is required. Most commonly the electrochemical detector has been operated in tandem with a conductivity detector, which acts as a universal detector that gives a more general sample analysis. A schematic illustration of the flow cell in an electrochemical detector is presented in Figure 9.10.

Voltammetry. Voltammetry is a well-established technique in which a changing potential is applied to a working electrode with respect to a reference electrode (Jandik *et al.* 1988). The current resulting from the reaction of analyzed species at the working electrode is measured. The key factor is that the applied potential is varied over the course of the measurement.

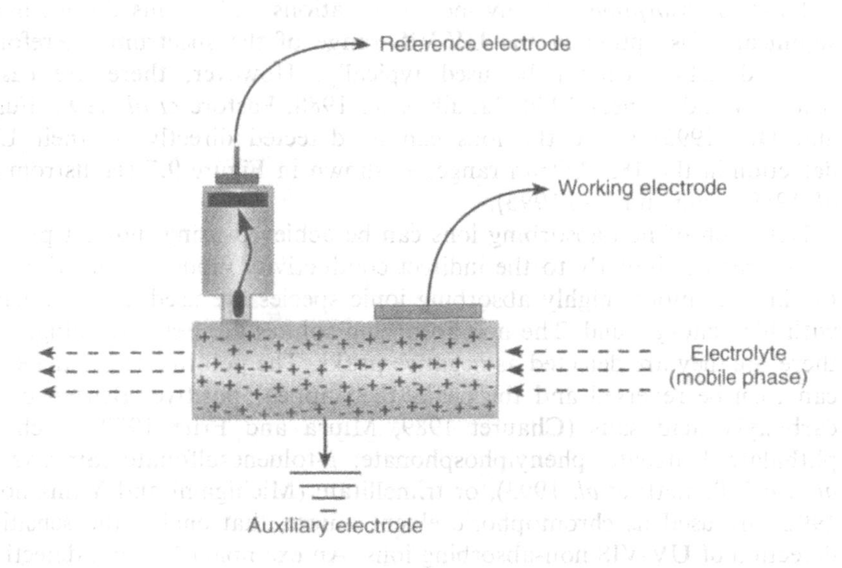

Figure 9.10 Schematic illustration of the flow cell in an electrochemical detector.

Amperometry and coulometry. These techniques are covered by Jandik *et al.* (1988), Ito and Sunahara (1990), Wagner and McGarrity (1992) and Liu and Wang (1993). The term 'amperometry' describes the technique in which a fixed potential is applied to a working electrode with respect to a reference electrode. The working electrode is located in the flow cell through which the mobile phase passes, and the current resulting from the oxidation–reduction reactions occurring at the working electrode is measured. The analyte to be detected undergoes a Faradaic reaction if the applied potential has appropriate polarity and magnitude. When the reaction is incomplete, causing only a fraction of the total analyte to react, the detection mode is termed **amperometry**, while when the working electrode has larger surface area and the reaction is complete, the mode is called **coulometry**.

9.7.3 Spectroscopic methods

Spectroscopic methods of detection are very common in ion chromatography and are second only to conductivity detection in their abundance. This mode of detection can be divided into two major categories: molecular and atomic spectroscopy. Molecular spectroscopy includes methods such as UV-VIS absorption, refractive index, fluorescence and phosphorescence. Atomic spectroscopy includes flame atomic absorption, flame atomic emission and plasma atomic emission.

Molecular spectroscopy.

UV-VIS absorption. Many inorganic cations and anions do not have significant absorption in the UV-VIS range of the spectrum; therefore, direct detection cannot be used typically. However, there are cases (Morrow and Minear 1984, Jandik *et al.* 1988, Pastore *et al.* 1992, Fung and Dao 1995) where the ions can be detected directly by their UV detection in the 185–220 nm range, as shown in Figure 9.7 (Hallstrom *et al.* 1989, Fung and Dao 1995).

Detection of non-absorbing ions can be achieved using indirect photometric mode, similarly to the indirect conductivity mode of detection. In the indirect mode, highly absorbing ionic species are used as the eluents with high background. The non-absorbing solutes are less absorbing, and therefore they are detected as negative peaks. The polarity of the detector can then be reversed and the peaks then appear positive. Benzenepolycarboxylic acid salts (Chauret 1989, Miura and Fritz 1989), such as phthalate, benzoate, phenylphosphonate, *p*-toluenesulfonate (Siriraks *et al.* 1987, Pianetti *et al.* 1993), or trimellitate (Michigami and Yamamoto 1992) are used as chromophoric eluent anions that enable the sensitive detection of UV-VIS non-absorbing ions. An example of such a detection mode is shown in Figure 9.11. A negative chromatogram of inorganic

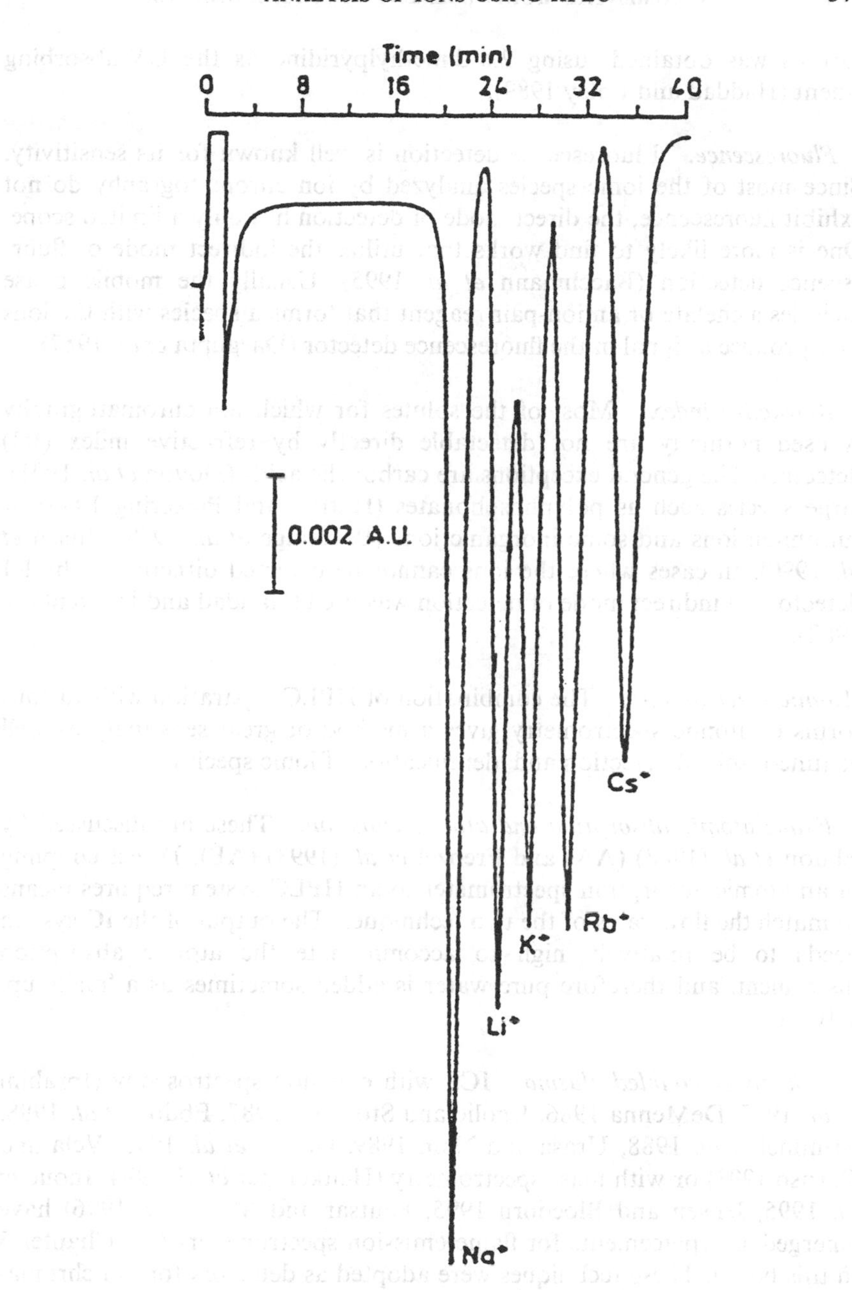

Figure 9.11 Chromatogram obtained with 0.2 mM 2,6-dimethylpyridine at pH 6.35 as the eluent, using indirect UV detection at 269 nm. Solutes: 15 μL of 2×10^{-5} M of each of the indicated ions. (Reprinted from Haddad and Foley (1989), figure 2a, with kind permission of Elsevier Science (The Netherlands).)

cations was obtained, using 2,6-dimethylpyridine as the UV-absorbing eluent (Haddad and Foley 1989).

Fluorescence. Fluorescence detection is well known for its sensitivity. Since most of the ionic species analyzed by ion chromatography do not exhibit fluorescence, the direct mode of detection has only a limited scope. One is more likely to find works that utilize the indirect mode of fluorescence detection (Baechmann *et al.* 1995). Usually the mobile phase includes a chelate or an ion-pair reagent that forms a species with the ions that produce a signal in the fluorescence detector (Dasgupta *et al.* 1987).

Refractive index. Most of the solutes for which ion chromatography is used normally are not detectable directly by refractive index (RI) detectors. The general exceptions are carboxylic acids (Doyon *et al.* 1991), large species such as polyphosphonates (Hatton and Pickering 1993) or sulfonium ions and some inorganic ions (Beveridge *et al.* 1988, Nitsch *et al.* 1990). In cases where the ions cannot be detected directly by the RI detector, an indirect mode of detection was used (Haddad and Heckenberg 1982).

Atomic spectroscopy. The combination of HPLC separation with various forms of atomic spectrometry gives a method of great sensitivity as well as time-resolved detection and identification of ionic species.

Flame atomic absorption and atomic emission. These are discussed by Ebdon *et al.* (1988) (AA) and Frenzel *et al.* (1993) (AE). Direct coupling of an atomic absorption spectrometer to an HPLC system requires means to match the flow-rates of the two techniques. The output of the IC system needs to be relatively high to accommodate the atomic absorption instrument, and therefore pure water is added sometimes as a 'make-up' solvent.

Inductively coupled plasma. ICP with emission spectroscopy (Ibrahim *et al.* 1985, DeMenna 1986, Irgolic and Stockton 1987, Ebdon *et al.* 1988, Frimmel *et al.* 1988, Urasa and Nam 1989, Garcia *et al.* 1993, Vela and Caruso 1993) or with mass spectrometry (Heitkemper *et al.* 1994, Inoue *et al.* 1995, Jensen and Bloedorn 1995, Pantsar and Manninen 1996) have emerged as replacements for flame emission spectrometers (see Chapter 5 in this book). These techniques were adopted as detectors for ion chromatography in recent years to combine the powers of both technologies. The introduction of HPLC coupled directly to ICP MS led to the possibility to analyze various species of one element, i.e. to speciation analysis (Seubert 1995).

The coupling of ion chromatography (IC) with ICP MS enabled the detection of ultra-trace amounts of cationic impurities in the presence of high amounts of matrices, thanks to the separation capabilities of chromatography. Examples can be found in the semiconductor field, where such analyses have been carried out on matrices of Mo, W, Re, As and P (Seubert 1995).

9.7.4 Post-column reaction

Detection by post-column reaction (PCR) involves the chemical reaction of the solutes just as they emerge from the column, on the fly, prior to their introduction to the detector. The main goal of such a procedure is to enhance selectivity and specificity to solutes of small quantities in the presence of large quantities of interferences in the sample matrix. Some of the post-column reagents are ammonium molybdate (Jones *et al.* 1991), 4-(2-pyridylazo)resorcinol (Siriraks *et al.* 1987, Janvion *et al.* 1995, Bruzzoniti *et al.* 1996, Janos *et al.* 1996, Xu *et al.* 1996), *py*,6-dicarboxylic acid (Ehrling *et al.* 1996), phenylfluorone (Sun *et al.* 1995), 2-(5-bromo-2-pyridylazo)-5-(diethylamino)phenol (Papoff *et al.* 1995), etc.

9.8 Applications – summary

Ion chromatography was introduced in the mid-1970s, and since that time it has become widespread at a phenomenal rate. The reasons for that are the capabilities of determining organic and inorganic ions in complex mixtures and matrices. One of the major reasons for the remarkable growth of IC over the past two decades has been its ability to provide rapid and simple solutions to a wide variety of analytical problems. Most of the large volume of publications in the field present practical applications of the technique. The applications can be divided into the following categories: environmental, industrial, chemicals, foods, plants, clinical, pharmaceutical, metallurgical solutions and treated water. Environmental applications include rain water, sea water and brines, surface and underground water, air and aerosols, soil and geological materials. Industrial and chemicals applications include waste water and effluents, water within the industrial process, organic compounds, pulp and paper liquors, acids and bases, detergents and polymers, fuels and oils, explosives and photographic solutions. Food and plants applications include foods, beverages and plant products. Clinical and pharmaceutical applications include physiological fluids and pharmaceutical formulations. Applications to metallic solutions are used in metal plating and metallurgical processing. Treated water includes drinking and purified water.

References

Alumaa, P. and Pentsuk, J. (1994), *Chromatographia* **38**, 566–570.
Ammann, A.A. and Ruettimann, T.B. (1995), *J. Chromatogr., A* **706**, 259–269.
Baechmann, K., Roeder, A. and Haag, I. (1995), *Atmos. Environ.* **29**(2), 175–177.
Barkley, D.J., Bennett, L.A., Charbonneau, J.R. and Pokrajac, L.A. (1992), *J. Chromatogr.* **606**, 195–201.
Beveridge, A., Pickering, W.F. and Slavek, J. (1988), *Talanta* **35**, 307–310.
Boermans, H.J. (1990), *Am. J. Vet. Res.* **51**, Mar, 491–495.
Brown, D.M. and Pietrzyk, D.J. (1989), *J. Chromatogr.* **466**, 291–300.
Bruzzoniti, M.C., Mentasti, E., Sarzanini, C., Braglia, M., Cocito, G. and Kraus, J. (1996), *Anal. Chim. Acta* **322**, 49–54.
Buta, J.G. (1984), *J. Chromatogr.* **295**(2), 506–509.
Chadha, K. and Lawrence, F. (1991), *Int. J. Environ. Anal. Chem.* **44**, 197–202.
Chauret, H.J. (1989), *J. Chromatogr.* **469**, 329–338.
Cox, D., Jandik P. and Jones, W. (1987), *Pulp Pap. Can.* **88**, Sep, 90–93.
Dasgupta, P.K., Soroka, K. and Vithanage, R.S. (1987), *J. Liq. Chromatogr.* **10**, Nov, 3287–3319.
DeMenna, G.J. (1986), *Chromatogr. Int.* **1986**, Sep, 16–19.
Den-Hartigh, J., Langebroek, R. and Vermeij, P. (1993), *J. Pharm. Biomed. Anal.* **11**(10), 977–983.
Diop, A., Jardy, A., Caude, M. and Rosset, R. (1986), *Analusis* **14**(2), 67–73.
Doyon, G., Gaudreau, G., St, G.D., Beaulieu, Y. and Randall, C.J. (1991), *Can. Inst. Food Sci. Technol. J.* **24**, 87–94.
Dugay, J., Jardy, A. and Doury-Berthod, M. (1995), *Analusis* **23**(5), 196–212.
Dumont, P.J. and Fritz, J.S. (1995), *J. Chromatogr.* **706**(1–2), 149–158.
Ebdon, L., Hill, S., Walton, A.P. and Ward, R.W. (1988), *Analyst* **113**, 1159–1165.
Edgell, K.W., Longbottom, J.E. and Pfaff, J.D. (1994), *J. Aoac Int.* **77**, Sep-Oct, 1253–1263.
Ehrling, C., Schmidt, U. and Liebscher, H. (1996), *Fresenius' J. Anal. Chem.* **354**, Apr, 870–873.
Foley, R.C.L. and Haddad, P.R. (1986), *J. Chromatogr.* **366**, 13–26.
Fortier, N.E. and Fritz, J.S. (1985), *Talanta* **32**(11), 1047–1050.
Frenzel, W., Schepers, D. and Schulze, G. (1993), *Anal. Chim. Acta* **277**, 103–111.
Frimmel, F.H., Grenz, R. and Kordik, E. (1988), *Fresenius' Z. Anal. Chem.* **331**, Jul, 253–259.
Fritz, J.S., DuVal, D.L. and Barron, R.E. (1984), *Anal. Chem.* **56**(7), 1177–1182.
Fung, Y.S. and Dao, K.L. (1995), *Anal. Chim. Acta* **300**, 207–214.
Garcia, A.J.I., Sanz, M.A. and Ebdon, L. (1993), *Anal. Chim. Acta* **283**, 261–271.
Guzman, N.A. (1993), *Capillary Electrophoresis Technology*, Marcel Dekker, New York.
Haddad, P.R. and Foley, R.C. (1989), *Anal. Chem.* **61**(13), 1435–1441.
Haddad, P.R. and Heckenberg, A.L. (1982), *J. Chromatogr.* **252**, 177–184.
Haddad, P.R. and Jackson, P.E. (1990), *Ion Chromatography – Principles and Applications*, Elsevier, Amsterdam.
Haddad, P.R., Jackson, P.E. and Heckenberg, A.L. (1985), *J. Chromatogr.* **346**, 139–148.
Hallstrom, A., Carlsson, A., Hillered, L. and Ungerstedt, U. (1989), *J. Pharmacol. Methods* **22**, 113–124.
Hatton, D. and Pickering, W.F. (1993), *Talanta* **40**, 307–311.
Heard, J. and Talmadge, K. (1992), *Int. Lab.* **22**, Jul-Aug, (7 (*Int. Chromatogr. Lab.*, **10**)), 6, 8, 10.
Heitkemper, D.T., Kaine, L.A., Jackson, D.S. and Wolnik, K.A. (1994), *J. Chromatogr. A* **671**, 101–108.
Heping, W., Pacakova, V., Stulik, K. and Barth, T. (1990), *J. Chromatogr.* **519**, 244–249.
Hirayama, N. and Kuwamoto, T. (1988), *J. Chromatogr.* **447**, 323–328.
Hirayama, N., Maruo, M. and Kuwamoto, T. (1993), *J. Chromatogr.* **639**(2), 333–337.
Ibrahim, M., Nisamaneepong, W., Haas, D.L. and Caruso, J.A. (1985), *Spectrochim. Acta, B*,
Inoue, Y., Sakai, T. and Kumagai, H. (1995), *J. Chromatogr., A* **706**, 127–136.

Irgolic, K.J. and Stockton, R.A. (1987), *Mar. Chem.* **22**, Dec, 265–278.
Ito, K. and Sunahara, H. (1990), *J. Chromatogr.* **502**, 121–129.
Ivey, J.P. (1984), *J. Chromatogr.* **287**, 128–132.
Jackson, P.E. (1995), *J. Chromatogr., A* **693**, 155–161.
Jackson, P.E. and Bowser, T. (1992), *J. Chromatogr.* **602**, 33–41.
Jackson, P.E., Romano, J.P. and Wildman, B.J. (1995), *J. Chromatogr., A* **706**, 3–12.
Jancar, J.C., Constant, M.D. and Herwig, W.C. (1984), *J. Am. Soc. Brew. Chem.* **42**, 90–93.
Jandik, P., Haddad, P.R. and Sturrock, P.E. (1988), *CRC Crit. Rev. Anal. Chem.* **20**, 1–74.
Janos, P., Chroma, H. and Kuban, V. (1996), *Fresenius' J. Anal. Chem.* **355**, 135–140.
Janvion, P., Motellier, S. and Pitsch, H. (1995), *J. Chromatogr., A* **715**, 105–115.
Jenke, D.R. (1988), *J. Chromatogr.* **437**, 231–237.
Jenke, D.R. (1994), *Anal. Chem.* **66**, 4466–4470.
Jenke, D.R. and Pagenkopf, G.K. (1984), *Anal. Chem.* **56**(1), 88–91.
Jensen, D. and Bloedorn, W. (1995), *Git Fachz. Lab.* **39**, Jul, 657–658.
Jones, P., Stanley, R. and Barnett, N. (1991), *Anal. Chim. Acta* **12**, 539–544.
Juergens, U. (1990), *LaborPraxis* **14**, 127–128.
Kato, Y., Kitamura, T. and Hashimoto, T. (1983), *J. Chromatogr.* **268**(3), 425–436.
Kawasaki, N., Ishigami, A., Tanimoto, T. and Tanaka, A. (1990), *J. Chromatogr.* **503**, 237–243.
Leubolt, R. and Klein, H. (1993), *J. Chromatogr.* **640**, 271–277.
Lindsay, S. (1993), *High Performance Liquid Chromatography*, Wiley, Chichester.
Liu, Y.B. and Wang, Q. (1993), *J. Chromatogr.* **644**, 73–82.
Maruo, M., Hirayama, N. and Kuwamoto, T. (1989), *J. Chromatogr.* **481**, 315–322.
McAleese, D.L. (1987), *Anal. Chem.* **59**, 541–543.
Michigami, Y. and Yamamoto, Y. (1992), *J. Chromatogr.* **623**(1), 148–152.
Miura, Y. and Fritz, J.S. (1989), *J. Chromatogr.* **482**(1), 155–163.
Morawski, J., Alden, P. and Sims, A. (1993), *J. Chromatogr.* **640**(1–2), 359–364.
Morris, J. and Fritz, J.S. (1994), *LC-GC -Int.* **7**(1), 43–47.
Morrow, C. M. and Minear, R.A. (1984), *Water Res.* **18**, 1165–1168.
Mosko, J.A. (1984), *Anal. Chem.* **56**, Apr, 629–633.
Nesternko, P.N. (1991), *J. High-Resolut. Chromatogr.* **14**(11), 767–768.
Nitsch, A., Kalcher, K. and Posch, U. (1990), *Fresenius' J. Anal. Chem.* **338**, Nov, 618–621.
Nonomura, M. (1987), *Anal. Chem.* **59**, 2073–2076.
Nonomura, M. and Hobo, T. (1989), *J. Chromatogr.* **465**, 395–401.
Nordmeyer, F.R., Hansen, L.D., Eatough, D.J., Rollins, D.K. and Lamb, J.D. (1980), *Anal. Chem.* **52**, 852–856.
Okada, T. and Kuwamoto, T. (1984), *J. Chromatogr.* **284**, 149–156.
Okada, T. and Kuwamoto, T. (1985), *Anal. Chem.* **57**(4), 829–833.
Pantsar, K.M. and Manninen, P.K.G. (1996), *Anal. Chim. Acta* **318**, 335–343.
Papoff, P., Ceccarini, A. and Carnevali, P. (1995), *J. Chromatogr., A* **706**, 43–54.
Pastore, P., Boaretto, A., Lavagnini, I. and Diop, A. (1992), *J. Chromatogr.* **591**, 219–224.
Pianetti, G.A., Moreira, D.C.L.M., Chaminade, P., Baillet, A., Baylocq, F.D. and Mahuzier, G. (1993), *Anal. Chim. Acta* **284**, 291–299.
Porter, D.H., Lin, M. and Wagner, C. (1985), *Anal. Biochem.* **151**, Dec, 299–303.
Riley, A.L.M. and Nowotnik, D.P. (1994), *J. Liq. Chromatogr.* **17**(3), 533–548.
Roessner, B., Behnert, J. and Kipplinger, A. (1987), *Fresenius' Z. Anal Chem.* **327**, Jul, 698–700.
Saari, N.R. and Anderson, J.M.J. (1991), *J. Chromatogr.* **549**, 257–264.
Santillana, M.I., Ruiz, E., Nieto, M.T. and De, A.M. (1993), *J. Liq. Chromatogr.*, 16, May, 1561–1571.
Sato, H. and Miyanaga, A. (1989), *Anal. Chem.* **61**, 122–125.
Schmitt, G.L. and Pietrzyk, D.J. (1985), *Anal. Chem.* **57**(12), 2247–2253.
Scully, H.S., Brumback, L.C. and Kelly, R.G. (1993), *J. Chromatogr.* **640**, 345–350.
Senior, J.P. (1990), *Anal. Proc. (Lond.)* **27**(5), 116.
Seubert, A. (1995), *Git Fachz. Lab.* **39**, Jun, 531–532.
Singh, R.P. and Abbas, N.M. (1996), *J. Chromatogr.* **733**(1–2), 93–99.
Singh, R.P., Abbas, N.M. and Smesko, S.A. (1996), *J. Chromatogr., A* **733**, 73–91.
Siriraks, A., Girard, J.E. and Buell, P.E. (1987), *Anal. Chem.* **59**, 2665–2669.

Sun, Q., Wang, H.T. and Mou, S.F. (1995), *J. Chromatogr., A* **708**, 99–104.
Takeuchi, T., Suzuki, E. and Ishii, D. (1988), *Chromatographia* **25**, Jul, 582–584.
Tanaka, T., Higashi, K., Kawahara, A., Wakida, S., Yamane, M. and Hiiro, K. (1987), *Bunseki Kagaku* **36**, Nov, 647–651.
Urasa, I.T. and Nam, S.H. (1989), *J. Chromatogr. Sci.* **27**, Jan, 30–37.
Vela, N.P. and Caruso, J.A. (1993), *J. Anal. At. Spectrom.* **8**, Sep, 787–794.
Voloschik, I.N., Litvina, M.L. and Rudenko, B.A. (1994), *J. Chromatogr., A* **671**, 249–252.
Wagner, H.P. and McGarrity, M.J. (1992), *J. Am. Soc. Brew. Chem.* **50**, 1–3.
Weinberg, H. (1994), J. *Chromatogr., A* **671**, 141–149.
Widiastuti, R., Haddad, P.R. and Jackson, P.E. (1992), *J. Chromatogr.* **602**(1–2), 43–50.
Wimberley, J. W. (1981), *Anal. Chem.* **53**, 2137–2138.
Wolf, W.R., LaCroix, D.E. and Slagt, M.E. (1992), *Anal. Lett.* **25**, Nov, 2165–2174.
Xu, Y., Sun, J. and Zheng, Z.M. (1996), *Sepu* **14**, Mar, 137–139.
Yamamoto, M., Yamamoto, H., Yamamoto, Y., Matsushita, S., Baba, N. and Ikushige, T. (1984), *Anal. Chem.* **56**, 832–834.
Zolotov, Y.A., Shpigun, O.A., Pazukhina, Y.E. and Voloshik, I.N. (1987), *Int. J. Environ. Anal. Chem.* **31**, 99–105.

10 Scattering methods

E. RAUHALA

10.1 Introduction

Characterization of materials by scattering methods is based on the properties of the elastic collision between two charged atomic particles. In the ion beam analysis of materials, energetic ions from a particle accelerator are steered to strike the surface of a solid sample. Most of the incident ions penetrate the sample, gradually slowing down due to the energy exchange between the ion and the atoms and electrons of the medium. In these processes, the atoms of the solid are excited and ionized and the charge state of the ion is modified. A few of the ions experience encounters with the atoms close enough to produce significant change in the ion trajectories.

When a close collision takes place, different interactions between the ion, atom and nucleus may occur. In an **elastic collision**, the kinetic energy and momentum of the incident particle are conserved. They are divided between the colliding particles, but no internal processes are affected. The geometry of the elastic collision is shown in Figure 10.1. In addition to the Coulomb potential scattering, nuclear and resonance scattering may contribute.

In analytical applications involving scattering techniques, either the scattered ions or the recoiled target atoms are detected and analyzed. In the former case, the projectile is lighter than the target particle and it is usually detected in a backward scattering direction. This is the basic principle of the **ion backscattering method**, schematically illustrated in Figure 10.2. In the latter case, the projectile is the heavier particle and the lighter recoiling target atom is detected in the forward direction. This, in turn, is the principle of the **elastic recoil method**, shown in Figure 10.3. In this simple arrangement, an auxiliary foil is used to stop the scattered ions. In the case of a thin self-supporting sample foil, the recoils may penetrate through the sample and an alternative method is to detect the recoils behind the sample.

The backscattering and recoil processes are thus two facets of the same elastic collision. Qualitative basics of the kinematics of the collision may be readily examined. For example, heavier ions cannot scatter backwards

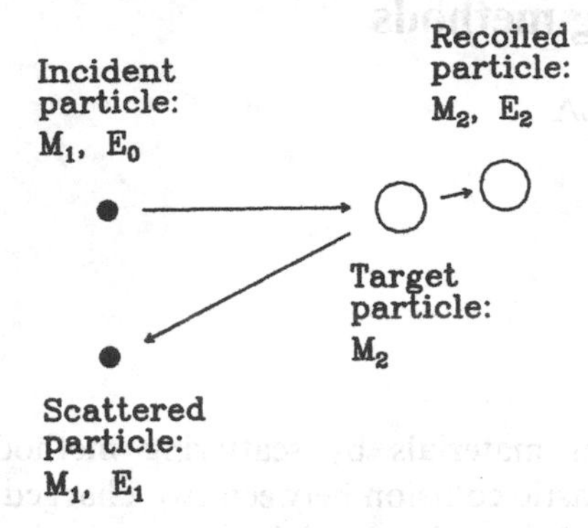

Figure 10.1 The elastic collision of charged particles. The projectile mass M_1 is here assumed smaller than the mass of the target atom M_2. For an almost head-on collision, the incident particle is scattered in the backward direction and the target particle is recoiled in the forward direction.

from a lighter target and target atoms cannot recoil in the backward direction. Some other simple rules may also be understood intuitively: the heavier the target or the smaller the backscattering angle, the higher the scattered ion energy. A similar rule applies for the recoils: the heavier the projectile or the smaller the deviation of the recoil angle from the incident ion direction, the larger the recoiled atom energy.

The two analytical techniques are commonly abbreviated as RBS (**Rutherford backscattering**) and ERD (**elastic recoil detection** or forward

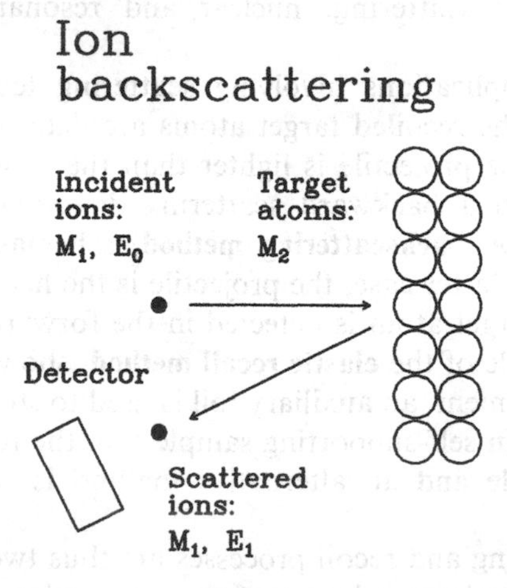

Figure 10.2 The elastic backscattering method. Light incident ions are scattered from heavy target atoms.

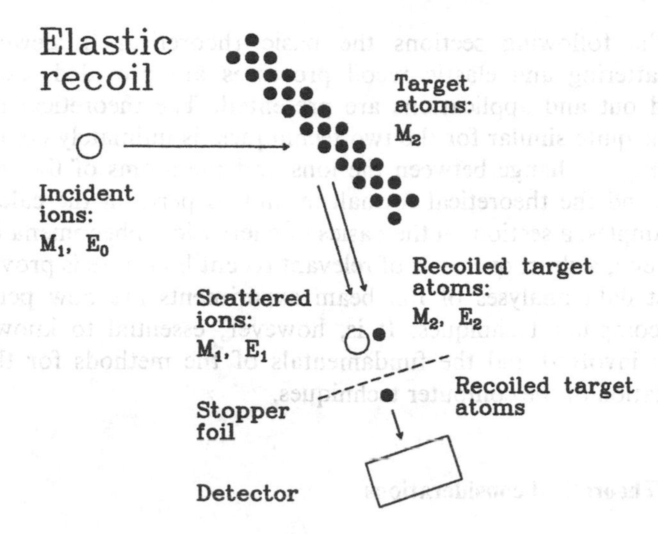

Figure 10.3 The elastic recoil method. Both the recoiled light target atoms and forward-scattered heavy ions are emitted from the target. The scattered ions are obstructed from entering the detector by using a stopper foil.

recoil scattering), respectively. In this context, elastic scattering is referred to as Rutherford scattering for historical reasons. The name is thus considered to include also cases where, in addition to the Coulomb potential, other forms of potentials contribute.

Neither ion backscattering nor elastic recoil detection are novel techniques. They have both been used for decades and are well established as analytical tools. On the other hand, new developments in the methodology are still often published. There is an overwhelming amount of studies published on the application of these techniques. By using a computer database (*Current Contents*, Institute for Scientific Information), covering publications in 1994 and 1995 and keywords 'Rutherford backscattering' and 'Elastic recoil', about 640 and 113 references, respectively, were found. For comparison, the keyword 'PIXE' gave about 245 entries during the same time interval. Judging from these numbers, ion backscattering is by far the most widely used single analytical technique in the field of ion beam analysis.

Textbooks and extensive reviews have been published about the techniques. The basic reference on backscattering is Chu *et al.* (1978). An extended recent treatment of the elastic recoil method is Tirira *et al.* (1996). Excellent handbook articles on both RBS (Leavitt *et al.* 1995) and ERD (Barbour and Doyle 1995) have also been published recently. The present chapter focuses on the basics of the techniques and on practical examples and applications. For a more general treatment of the techniques and for other references, the reader is referred to the books above or to other reviews (e.g. Rauhala 1992, 1994).

In the following sections the basic theoretical framework of ion backscattering and elastic recoil processes are provided, examples are worked out and applications are presented. The theoretical framework, which is quite similar for the two techniques, is intimately connected with the energy exchange between the ions and the atoms of the medium. To understand the theoretical formalism and to perform the calculations in the examples, a section on the basics of energy loss phenomena is included. At the end, a short overview of relevant recent literature is provided.

Most data analyses of ion beam experiments are now performed by using computer techniques. It is, however, essential to know the basic physics involved and the fundamentals of the methods for the effective exploitation of the computer techniques.

10.2 Theoretical considerations

With RBS and ERD techniques it is possible to gain knowledge on the elemental composition and to obtain quantitative and depth information about the sample under study. The capabilities of both methods depend, among other factors, on the sample and the elements itself. Backscattering is sensitive to heavy elements but cannot distinguish them from each other. The sensitivity of light-element detection is poor, especially in a heavier matrix. Conversely, the recoil method is used to detect sample elements lighter than the projectile. The two techniques are thus often used to supplement each other. No information about the chemical structure or the chemical surroundings of the atoms studied can be derived by either of the two methods.

The main analytical characteristics of the techniques – the capability of mass, quantitative and depth analysis – are based on the energy exchange between the colliding atomic particles and electrons, and the probability of elastic scattering. These phenomena are treated theoretically in terms of collision kinematics, scattering cross-section and stopping power.

10.2.1 Kinematics, ion backscattering and elastic recoil processes

Figure 10.4 shows a schematic of a typical backscattering event. The angles of beam incidence Θ_1 and exit Θ_2 are measured with respect to the direction of the normal to the sample surface. In the simplest case, illustrated here, all directions defined are coplanar.

The ions are incident from the right with an energy of E_0. Part of the beam is scattered at the surface of the sample through a scattering angle of Θ to the detector. Most of the beam penetrates the surface. Some ions are scattered at some depth x_1 below the surface through the same angle Θ and reach the sample surface and the detector. Some more ions scatter in

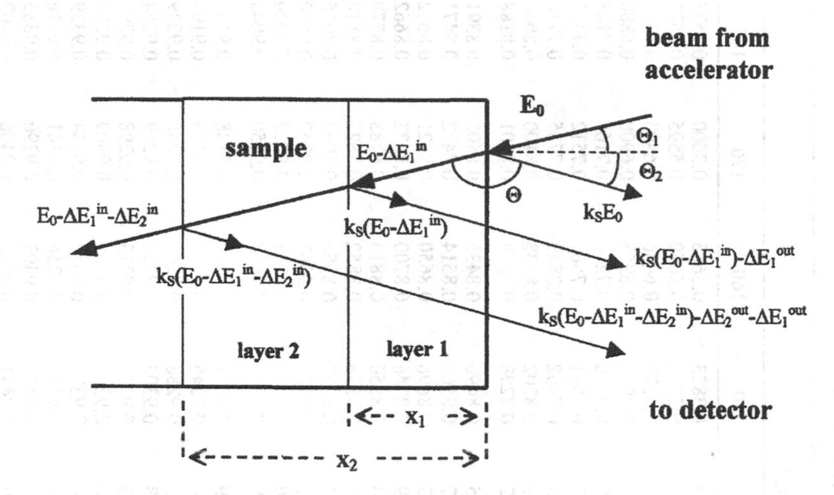

Figure 10.4 A schematic of the ion backscattering process at the surface and at depths x_1 and x_2 beneath the surface of an elemental sample.

the same way at other depths x_2, etc. The ions scattering from below the surface suffer energy losses depending on the depth of the collision.

The ratio of the incident beam energy to that of the beam scattered at the surface is called the **kinematic factor**,

$$k_S = \frac{E_1}{E_0} = \left(\frac{\sqrt{1 - x^2 \sin^2 \Theta} + x \cos \Theta}{1 + x}\right)^2 \tag{10.1}$$

where the subscript in k_S refers to the scattered particles (as opposed to recoiled particles) and x is the mass ratio $x = M_2/M_2$. Tables 10.1 and 10.2 give the kinematic factors for backscattering of ^1H and ^4He ions by the elements through angles of 90–180°. Given the experimental conditions, equation (10.1) yields the energy E_1 of the particles scattered from the surface of an elemental sample. In the important special case of large backscattering angles, when $\Theta \approx 180°$, equation (10.1) reduces to:

$$k_S = \frac{E_1}{E_0} \approx \left(\frac{1 - x}{1 + x}\right)^2 = \left(\frac{M_2 - M_1}{M_2 + M_1}\right)^2 \quad \text{for } \Theta \approx 180° \tag{10.2}$$

Example 10.1

Can gold and tantalum signals be resolved in the backscattering spectrum taken with 2.0 MeV ^4He ions?

Assume a scattering angle of close to 180° (the optimum mass resolution). Differentiating E_1 in equation (10.2) with respect to M_2 one obtains:

Table 10.1 Kinematic factors for proton backscattering by the elements through scattering angles of 90° to 180°

Z_2	Atom	Scattering angle (deg)									
		90	100	110	120	130	140	150	160	170	180
2	He	0.5977	0.5461	0.5004	0.4611	0.4285	0.4023	0.3823	0.3683	0.3600	0.3573
3	Li	0.7464	0.7093	0.6751	0.6445	0.6182	0.5963	0.5792	0.5669	0.5595	0.5571
4	Be	0.7988	0.7682	0.7397	0.7139	0.6913	0.6725	0.6576	0.6486	0.6403	0.6381
5	B	0.8295	0.8029	0.7780	0.7553	0.7354	0.7187	0.7054	0.6958	0.6900	0.6880
6	C	0.8452	0.8208	0.7979	0.7769	0.7585	0.7430	0.7306	0.7216	0.7161	0.7143
7	N	0.8658	0.8443	0.8241	0.8055	0.7891	0.7752	0.7641	0.7561	0.7512	0.7495
8	O	0.8815	0.8624	0.8442	0.8276	0.8128	0.8003	0.7902	0.7829	0.7785	0.7770
9	F	0.8992	0.8828	0.8672	0.8527	0.8399	0.8290	0.8202	0.8138	0.8100	0.8086
10	Ne	0.9049	0.8893	0.8745	0.8608	0.8486	0.8382	0.8298	0.8238	0.8201	0.8188
11	Na	0.9160	0.9022	0.8889	0.8767	0.8658	0.8565	0.8490	0.8435	0.8402	0.8391
12	Mg	0.9204	0.9072	0.8946	0.8830	0.8726	0.8637	0.8566	0.8514	0.8482	0.8471
13	Al	0.9280	0.9160	0.9046	0.8939	0.8845	0.8763	0.8698	0.8650	0.8621	0.8612
14	Si	0.9307	0.9192	0.9081	0.8979	0.8887	0.8809	0.8746	0.8700	0.8672	0.8662
15	P	0.9370	0.9264	0.9163	0.9070	0.8986	0.8914	0.8856	0.8814	0.8788	0.8779
16	S	0.9391	0.9289	0.9191	0.9100	0.9019	0.8949	0.8893	0.8852	0.8827	0.8818
17	Cl	0.9447	0.9354	0.9265	0.9182	0.9108	0.9044	0.8993	0.8956	0.8933	0.8925
18	Ar	0.9508	0.9425	0.9345	0.9271	0.9204	0.9147	0.9101	0.9067	0.9047	0.9040
19	K	0.9497	0.9413	0.9331	0.9256	0.9188	0.9130	0.9083	0.9048	0.9027	0.9020
20	Ca	0.9509	0.9427	0.9347	0.9273	0.9207	0.9150	0.9104	0.9070	0.9050	0.9043
21	Sc	0.9561	0.9487	0.9416	0.9349	0.9290	0.9239	0.9197	0.9167	0.9148	0.9142
22	Ti	0.9588	0.9518	0.9451	0.9388	0.9332	0.9284	0.9245	0.9216	0.9199	0.9193
23	V	0.9612	0.9546	0.9483	0.9424	0.9371	0.9325	0.9288	0.9261	0.9245	0.9239
24	Cr	0.9620	0.9555	0.9493	0.9435	0.9383	0.9338	0.9302	0.9276	0.9259	0.9254
25	Mn	0.9640	0.9578	0.9519	0.9464	0.9415	0.9373	0.9338	0.9313	0.9298	0.9292
26	Fe	0.9645	0.9585	0.9527	0.9473	0.9424	0.9382	0.9349	0.9324	0.9309	0.9304
27	Co	0.9664	0.9607	0.9551	0.9500	0.9454	0.9414	0.9382	0.9358	0.9344	0.9339
28	Ni	0.9662	0.9605	0.9550	0.9498	0.9452	0.9412	0.9379	0.9356	0.9341	0.9336
29	Cu	0.9688	0.9635	0.9583	0.9535	0.9492	0.9455	0.9425	0.9403	0.9390	0.9385
30	Zn	0.9696	0.9645	0.9595	0.9548	0.9506	0.9470	0.9441	0.9419	0.9406	0.9402

31	Ga	0.9715	0.9666	0.9619	0.9576	0.9536	0.9502	0.9475	0.9455	0.9442	0.9438
32	Ge	0.9726	0.9679	0.9634	0.9592	0.9554	0.9521	0.9495	0.9476	0.9464	0.9460
33	As	0.9735	0.9689	0.9645	0.9604	0.9568	0.9536	0.9510	0.9492	0.9480	0.9476
34	Se	0.9748	0.9705	0.9663	0.9624	0.9589	0.9559	0.9535	0.9517	0.9506	0.9502
35	Br	0.9751	0.9708	0.9667	0.9629	0.9594	0.9564	0.9540	0.9522	0.9512	0.9508
36	Kr	0.9762	0.9722	0.9682	0.9646	0.9613	0.9584	0.9561	0.9544	0.9534	0.9530
37	Rb	0.9767	0.9727	0.9688	0.9652	0.9620	0.9592	0.9569	0.9553	0.9543	0.9539
38	Sr	0.9773	0.9734	0.9696	0.9661	0.9629	0.9602	0.9580	0.9564	0.9554	0.9550
39	Y	0.9776	0.9737	0.9700	0.9666	0.9634	0.9608	0.9586	0.9570	0.9560	0.9557
40	Zr	0.9781	0.9744	0.9708	0.9674	0.9644	0.9617	0.9596	0.9580	0.9571	0.9568
41	Nb	0.9785	0.9749	0.9713	0.9680	0.9650	0.9624	0.9603	0.9588	0.9579	0.9575
42	Mo	0.9792	0.9756	0.9722	0.9690	0.9661	0.9636	0.9616	0.9601	0.9592	0.9589
43	Tc	0.9798	0.9764	0.9730	0.9699	0.9671	0.9647	0.9627	0.9613	0.9604	0.9601
44	Ru	0.9803	0.9769	0.9736	0.9705	0.9678	0.9654	0.9635	0.9621	0.9612	0.9609
45	Rh	0.9806	0.9773	0.9741	0.9710	0.9683	0.9660	0.9641	0.9627	0.9619	0.9616
46	Pd	0.9812	0.9780	0.9749	0.9720	0.9694	0.9671	0.9653	0.9639	0.9631	0.9628
47	Ag	0.9815	0.9783	0.9752	0.9724	0.9698	0.9675	0.9657	0.9644	0.9636	0.9633
48	Cd	0.9822	0.9792	0.9762	0.9735	0.9710	0.9688	0.9671	0.9658	0.9650	0.9648
49	In	0.9826	0.9796	0.9767	0.9740	0.9716	0.9695	0.9678	0.9665	0.9658	0.9655
50	Sn	0.9832	0.9803	0.9775	0.9748	0.9725	0.9705	0.9688	0.9676	0.9669	0.9666
51	Sb	0.9836	0.9808	0.9780	0.9755	0.9732	0.9712	0.9696	0.9684	0.9677	0.9674
52	Te	0.9843	0.9816	0.9790	0.9766	0.9744	0.9725	0.9710	0.9698	0.9691	0.9689
53	I	0.9842	0.9815	0.9789	0.9765	0.9742	0.9723	0.9708	0.9697	0.9690	0.9687
54	Xe	0.9848	0.9821	0.9796	0.9772	0.9751	0.9733	0.9718	0.9707	0.9700	0.9698
55	Cs	0.9849	0.9824	0.9799	0.9775	0.9754	0.9736	0.9721	0.9710	0.9703	0.9701
56	Ba	0.9854	0.9829	0.9805	0.9782	0.9762	0.9744	0.9730	0.9719	0.9713	0.9711
57	La	0.9856	0.9831	0.9807	0.9785	0.9764	0.9747	0.9733	0.9722	0.9716	0.9714
58	Ce	0.9857	0.9833	0.9809	0.9787	0.9766	0.9749	0.9735	0.9725	0.9719	0.9716
59	Pr	0.9858	0.9834	0.9810	0.9788	0.9768	0.9751	0.9737	0.9726	0.9720	0.9718
60	Nd	0.9861	0.9837	0.9814	0.9793	0.9773	0.9756	0.9743	0.9733	0.9726	0.9724
61	Pm	0.9864	0.9840	0.9818	0.9796	0.9777	0.9761	0.9747	0.9738	0.9732	0.9729
62	Sm	0.9867	0.9844	0.9822	0.9801	0.9782	0.9766	0.9753	0.9743	0.9737	0.9735
63	Eu	0.9868	0.9846	0.9824	0.9803	0.9784	0.9768	0.9756	0.9746	0.9740	0.9738
64	Gd	0.9873	0.9851	0.9829	0.9810	0.9792	0.9776	0.9764	0.9754	0.9749	0.9747

Table 10.1 (Continued)

Z_2	Atom	Scattering angle (deg)									
		90	100	110	120	130	140	150	160	170	180
65	Tb	0.9874	0.9852	0.9831	0.9812	0.9794	0.9778	0.9766	0.9757	0.9751	0.9750
66	Dy	0.9877	0.9855	0.9835	0.9816	0.9798	0.9783	0.9771	0.9762	0.9757	0.9755
67	Ho	0.9879	0.9858	0.9837	0.9818	0.9801	0.9786	0.9775	0.9766	0.9760	0.9759
68	Er	0.9880	0.9860	0.9840	0.9821	0.9804	0.9789	0.9778	0.9769	0.9764	0.9762
69	Tm	0.9881	0.9861	0.9841	0.9823	0.9806	0.9791	0.9780	0.9771	0.9766	0.9764
70	Yb	0.9884	0.9864	0.9845	0.9827	0.9810	0.9796	0.9785	0.9777	0.9771	0.9770
71	Lu	0.9885	0.9866	0.9847	0.9829	0.9813	0.9799	0.9787	0.9779	0.9774	0.9772
72	Hf	0.9888	0.9868	0.9850	0.9832	0.9816	0.9803	0.9791	0.9783	0.9778	0.9777
73	Ta	0.9889	0.9870	0.9852	0.9834	0.9819	0.9805	0.9794	0.9786	0.9781	0.9780
74	W	0.9891	0.9872	0.9854	0.9837	0.9822	0.9808	0.9797	0.9790	0.9785	0.9783
75	Re	0.9892	0.9874	0.9856	0.9839	0.9824	0.9811	0.9800	0.9792	0.9787	0.9786
76	Os	0.9895	0.9876	0.9859	0.9842	0.9827	0.9815	0.9804	0.9797	0.9792	0.9790
77	Ir	0.9896	0.9878	0.9860	0.9844	0.9829	0.9816	0.9806	0.9799	0.9794	0.9792
78	Pt	0.9897	0.9879	0.9862	0.9846	0.9832	0.9819	0.9809	0.9802	0.9797	0.9795
79	Au	0.9898	0.9881	0.9864	0.9848	0.9833	0.9821	0.9811	0.9803	0.9799	0.9797
80	Hg	0.9900	0.9883	0.9866	0.9850	0.9836	0.9824	0.9814	0.9807	0.9803	0.9801
81	Tl	0.9902	0.9885	0.9869	0.9853	0.9839	0.9827	0.9818	0.9811	0.9806	0.9805
82	Pb	0.9903	0.9886	0.9870	0.9855	0.9841	0.9830	0.9820	0.9813	0.9809	0.9807
83	Bi	0.9904	0.9887	0.9871	0.9856	0.9843	0.9831	0.9822	0.9815	0.9810	0.9809
84	Po	0.9904	0.9888	0.9872	0.9857	0.9844	0.9832	0.9822	0.9816	0.9811	0.9810
85	At	0.9904	0.9888	0.9872	0.9857	0.9844	0.9832	0.9822	0.9816	0.9811	0.9810
86	Rn	0.9910	0.9894	0.9879	0.9865	0.9852	0.9841	0.9832	0.9825	0.9821	0.9820
87	Fr	0.9910	0.9894	0.9879	0.9865	0.9853	0.9842	0.9833	0.9826	0.9822	0.9821
88	Ra	0.9911	0.9896	0.9881	0.9867	0.9855	0.9844	0.9835	0.9828	0.9825	0.9823
89	Ac	0.9912	0.9896	0.9882	0.9868	0.9855	0.9844	0.9836	0.9829	0.9825	0.9824
90	Th	0.9913	0.9899	0.9884	0.9871	0.9858	0.9848	0.9839	0.9833	0.9829	0.9828
91	Pa	0.9913	0.9898	0.9884	0.9870	0.9858	0.9847	0.9838	0.9832	0.9828	0.9827
92	U	0.9916	0.9901	0.9887	0.9874	0.9862	0.9852	0.9843	0.9837	0.9833	0.9832

Table 10.2 Kinematic factors for ^{4}He ion backscattering by the elements through scattering angles of 90° to 180°

Z_2	Atom	Scattering angle (deg)									
		90	100	110	120	130	140	150	160	170	180
3	Li	0.2684	0.2101	0.1663	0.1343	0.1114	0.0953	0.0844	0.0773	0.0733	0.0720
4	Be	0.3849	0.3241	0.2747	0.2356	0.2056	0.1832	0.1671	0.1564	0.1502	0.1482
5	B	0.4596	0.4002	0.3502	0.3093	0.2768	0.2519	0.2335	0.2209	0.2136	0.2112
6	C	0.5001	0.4424	0.3929	0.3518	0.3187	0.2929	0.2736	0.2604	0.2526	0.2501
7	N	0.5555	0.5009	0.4532	0.4127	0.3795	0.3532	0.3333	0.3194	0.3113	0.3086
8	O	0.5998	0.5483	0.5027	0.4635	0.4309	0.4047	0.3848	0.3708	0.3625	0.3597
9	F	0.6520	0.6050	0.5627	0.5258	0.4946	0.4693	0.4498	0.4360	0.4278	0.4250
10	Ne	0.6690	0.6236	0.5826	0.5466	0.5162	0.4913	0.4721	0.4584	0.4503	0.4476
11	Na	0.7034	0.6615	0.6233	0.5896	0.5607	0.5369	0.5185	0.5053	0.4974	0.4948
12	Mg	0.7173	0.6769	0.6399	0.6071	0.5790	0.5558	0.5377	0.5248	0.5171	0.5145
13	Al	0.7416	0.7040	0.6693	0.6384	0.6117	0.5896	0.5724	0.5600	0.5525	0.5500
14	Si	0.7505	0.7139	0.6802	0.6500	0.6239	0.6022	0.5852	0.5731	0.5657	0.5633
15	P	0.7711	0.7370	0.7054	0.6770	0.6523	0.6318	0.6156	0.6040	0.5970	0.5946
16	S	0.7780	0.7448	0.7139	0.6861	0.6620	0.6418	0.6260	0.6145	0.6077	0.6054
17	Cl	0.7971	0.7663	0.7375	0.7115	0.6889	0.6699	0.6549	0.6441	0.6376	0.6354
18	Ar	0.8179	0.7898	0.7634	0.7395	0.7186	0.7010	0.6871	0.6770	0.6709	0.6689
19	K	0.8143	0.7857	0.7589	0.7347	0.7135	0.6956	0.6815	0.6713	0.6651	0.6631
20	Ca	0.8184	0.7904	0.7641	0.7403	0.7194	0.7019	0.6880	0.6779	0.6718	0.6698
21	Sc	0.8365	0.8109	0.7869	0.7650	0.7457	0.7295	0,7166	0.7073	0.7016	0.6997
22	Ti	0.8458	0.8215	0.7986	0.7778	0.7594	0.7439	0.7315	0.7226	0.7171	0.7153
23	V	0.8543	0.8312	0.8095	0.7896	0.7720	0.7572	0.7454	0.7368	0.7316	0.7298
24	Cr	0.8570	0.8344	0.8130	0.7934	0.7761	0.7615	0.7498	0.7414	0.7362	0.7345
25	Mn	0.8642	0.8425	0.8221	0.8033	0.7867	0.7727	0.7615	0.7534	0.7485	0.7468
26	Fe	0.8662	0.8449	0.8247	0.8062	0.7898	0.7760	0.7649	0.7569	0.7520	0.7504
27	Co	0.8728	0.8524	0.8331	0.8154	0.7997	0.7864	0.7758	0.7681	0.7634	0.7618
28	Ni	0.8724	0.8519	0.8325	0.8147	0.7990	0.7857	0.7750	0.7673	0.7626	0.7610
29	Cu	0.8815	0.8624	0.8442	0.8276	0.8128	0.8003	0.7902	0.7829	0.7785	0.7770
30	Zn	0.8846	0.8660	0.8483	0.8320	0.8175	0.8053	0.7955	0.7883	0.7840	0.7825

Table 10.2 (Continued)

Z_2	Atom	Scattering angle (deg)									
		90	100	110	120	130	140	150	160	170	180
31	Ga	0.8914	0.8738	0.8570	0.8416	0.8279	0.8163	0.8069	0.8001	0.7960	0.7946
32	Ge	0.8955	0.8785	0.8623	0.8474	0.8341	0.8229	0.8138	0.8072	0.8032	0.8019
33	As	0.8986	0.8820	0.8663	0.8518	0.8389	0.8279	0.8191	0.8127	0.8087	0.8074
34	Se	0.9035	0.8877	0.8727	0.8588	0.8464	0.8359	0.8275	0.8213	0.8176	0.8163
35	Br	0.9046	0.8890	0.8741	0.8604	0.8481	0.8377	0.8294	0.8233	0.8195	0.8183
36	Kr	0.9088	0.8939	0.8796	0.8664	0.8546	0.8446	0.8366	0.8307	0.8272	0.8260
37	Rb	0.9105	0.8958	0.8818	0.8688	0.8573	0.8474	0.8395	0.8338	0.8302	0.8291
38	Sr	0.9126	0.8982	0.8845	0.8718	0.8605	0.8509	0.8431	0.8375	0.8340	0.8329
39	Y	0.9138	0.8996	0.8861	0.8736	0.8624	0.8529	0.8452	0.8396	0.8362	0.8351
40	Zr	0.9159	0.9021	0.8888	0.8766	0.8657	0.8563	0.8489	0.8434	0.8401	0.8389
41	Nb	0.9174	0.9038	0.8907	0.8787	0.8679	0.8588	0.8514	0.8460	0.8427	0.8416
42	Mo	0.9199	0.9067	0.8940	0.8823	0.8718	0.8629	0.8557	0.8505	0.8473	0.8462
43	Tc	0.9223	0.9094	0.8971	0.8857	0.8755	0.8669	0.8599	0.8548	0.8516	0.8506
44	Ru	0.9238	0.9112	0.8991	0.8879	0.8779	0.8694	0.8625	0.8575	0.8545	0.8534
45	Rh	0.9251	0.9127	0.9008	0.8898	0.8800	0.8716	0.8648	0.8599	0.8569	0.8559
46	Pd	0.9275	0.9154	0.9039	0.8932	0.8837	0.8755	0.8690	0.8642	0.8612	0.8602
47	Ag	0.9284	0.9165	0.9052	0.8946	0.8852	0.8771	0.8706	0.8659	0.8630	0.8620
48	Cd	0.9312	0.9198	0.9088	0.8986	0.8895	0.8818	0.8755	0.8709	0.8681	0.8672
49	In	0.9326	0.9214	0.9106	0.9007	0.8917	0.8841	0.8780	0.8735	0.8707	0.8698
50	Sn	0.9348	0.9239	0.9134	0.9037	0.8951	0.8877	0.8817	0.8773	0.8747	0.8738
51	Sb	0.9363	0.9257	0.9155	0.9060	0.8976	0.8903	0.8845	0.8802	0.8776	0.8767
52	Te	0.9392	0.9290	0.9192	0.9102	0.9020	0.8951	0.8895	0.8854	0.8829	0.8820
53	I	0.9388	0.9286	0.9188	0.9097	0.9015	0.8945	0.8889	0.8848	0.8823	0.8814
54	Xe	0.9408	0.9309	0.9214	0.9126	0.9047	0.8979	0.8924	0.8884	0.8860	0.8852
55	Cs	0.9415	0.9317	0.9223	0.9136	0.9058	0.8991	0.8937	0.8897	0.8873	0.8865
56	Ba	0.9434	0.9339	0.9247	0.9163	0.9087	0.9022	0.8969	0.8931	0.8907	0.8899
57	La	0.9440	0.9346	0.9256	0.9172	0.9096	0.9032	0.8980	0.8942	0.8919	0.8911
58	Ce	0.9445	0.9351	0.9262	0.9178	0.9104	0.9040	0.8989	0.8951	0.8928	0.8920
59	Pr	0.9448	0.9355	0.9266	0.9183	0.9109	0.9045	0.8994	0.8956	0.8933	0.8926

60	Nd	0.9460	0.9369	0.9282	0.9201	0.9128	0.9066	0.9016	0.8979	0.8957	0.8949
61	Pm	0.9470	0.9381	0.9295	0.9215	0.9144	0.9083	0.9034	0.8997	0.8975	0.8968
62	Sm	0.9481	0.9394	0.9310	0.9232	0.9162	0.9102	0.9054	0.9019	0.8997	0.8990
63	Eu	0.9487	0.9400	0.9317	0.9240	0.9171	0.9111	0.9064	0.9028	0.9007	0.9000
64	Gd	0.9504	0.9420	0.9339	0.9265	0.9198	0.9140	0.9094	0.9060	0.9039	0.9032
65	Tb	0.9509	0.9426	0.9346	0.9272	0.9206	0.9149	0.9103	0.9069	0.9048	0.9041
66	Dy	0.9519	0.9438	0.9360	0.9288	0.9222	0.9167	0.9122	0.9089	0.9068	0.9062
67	Ho	0.9526	0.9446	0.9369	0.9298	0.9233	0.9178	0.9134	0.9101	0.9081	0.9075
68	Er	0.9533	0.9454	0.9378	0.9307	0.9244	0.9189	0.9145	0.9113	0.9094	0.9087
69	Tm	0.9537	0.9459	0.9384	0.9314	0.9251	0.9197	0.9154	0.9122	0.9102	0.9096
70	Yb	0.9548	0.9471	0.9398	0.9329	0.9268	0.9215	0.9173	0.9142	0.9123	0.9116
71	Lu	0.9553	0.9477	0.9404	0.9337	0.9276	0.9224	0.9182	0.9151	0.9132	0.9125
72	Hf	0.9561	0.9487	0.9416	0.9349	0.9290	0.9238	0.9197	0.9167	0.9148	0.9142
73	Ta	0.9567	0.9494	0.9423	0.9358	0.9299	0.9248	0.9207	0.9178	0.9159	0.9153
74	W	0.9574	0.9502	0.9432	0.9386	0.9310	0.9260	0.9220	0.9190	0.9172	0.9166
75	Re	0.9579	0.9508	0.9439	0.9375	0.9318	0.9269	0.9229	0.9200	0.9182	0.9176
76	Os	0.9588	0.9518	0.9451	0.9388	0.9332	0.9284	0.9245	0.9216	0.9198	0.9193
77	Ir	0.9592	0.9523	0.9456	0.9394	0.9339	0.9291	0.9252	0.9224	0.9206	0.9201
78	Pt	0.9598	0.9530	0.9464	0.9403	0.9348	0.9301	0.9263	0.9235	0.9218	0.9212
79	Au	0.9602	0.9534	0.9469	0.9408	0.9354	0.9307	0.9270	0.9242	0.9225	0.9219
80	Hg	0.9609	0.9542	0.9478	0.9419	0.9365	0.9319	0.9282	0.9255	0.9238	0.9233
81	Tl	0.9616	0.9551	0.9488	0.9429	0.9377	0.9332	0.9295	0.9268	0.9252	0.9246
82	Pb	0.9621	0.9557	0.9495	0.9437	0.9385	0.9340	0.9304	0.9278	0.9262	0.9256
83	Bi	0.9624	0.9560	0.9499	0.9442	0.9390	0.9346	0.9310	0.9284	0.9268	0.9262
84	Po	0.9626	0.9562	0.9501	0.9444	0.9393	0.9349	0.9313	0.9287	0.9271	0.9266
85	At	0.9626	0.9562	0.9501	0.9444	0.9393	0.9349	0.9313	0.9287	0.9271	0.9266
86	Rn	0.9646	0.9586	0.9528	0.9473	0.9425	0.9383	0.9349	0.9324	0.9309	0.9304
87	Fr	0.9647	0.9587	0.9530	0.9476	0.9427	0.9386	0.9352	0.9327	0.9312	0.9307
88	Ra	0.9652	0.9593	0.9536	0.9482	0.9435	0.9394	0.9360	0.9336	0.9321	0.9316
89	Ac	0.9653	0.9595	0.9538	0.9485	0.9437	0.9396	0.9363	0.9339	0.9324	0.9319
90	Th	0.9661	0.9603	0.9547	0.9496	0.9449	0.9409	0.9376	0.9353	0.9338	0.9333
91	Pa	0.9659	0.9601	0.9546	0.9493	0.9447	0.9406	0.9374	0.9350	0.9335	0.9330
92	U	0.9669	0.9613	0.9559	0.9508	0.9462	0.9423	0.9392	0.9368	0.9354	0.9349

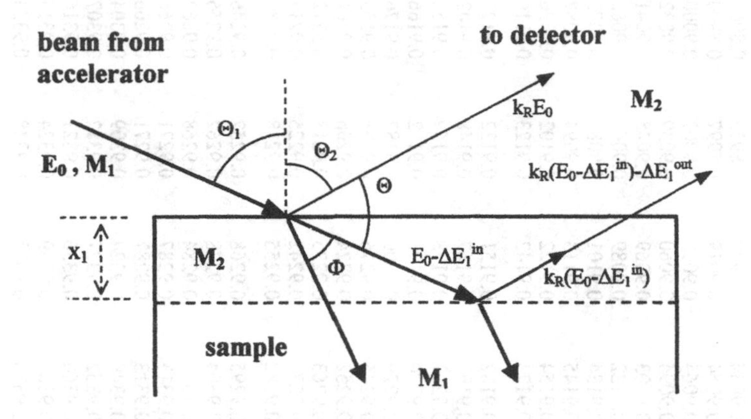

Figure 10.5 A schematic of the elastic recoil process at the surface of an elemental sample.

$$dE_1 = 4E_0M_1\frac{M_2 - M_1}{(M_2 + M_1)^3}dM_2 \tag{10.3}$$

where dM_1 is the **mass resolution** and dE_1 is the **energy resolution** of the experimental system (detector). For the ^1H and ^4He ions at 2.0 MeV, the energy resolution dE_1 in a backscattering experiment is typically of the order of 15 keV. The mass difference (in atomic mass units, u) for Au and Ta is $dM_2 = M_{Au} - M_{Ta} = 197u - 181u = 16u$. Substituting these quantities in equation (10.3) we obtain for the energy difference of the signals $dE_1 \approx 13$ keV, which is not adequate for resolving the Au and Ta signals. Increasing the beam energy to 2.5 MeV would solve the problem.

In the case of the elastic recoil process, shown schematically in Figure 10.5, the kinematic factor gives the ratio of the incident beam energy E_0 to that of the recoiled target atoms E_2:

$$k_R = \frac{E_2}{E_0} = \frac{4M_1M_2}{(M_1 + M_2)^2}\cos^2\Theta \tag{10.4}$$

The subscript in k_R now refers to the recoiled particles. The recoil angle Θ is the angle between the direction of the detector and the incident beam direction, as defined in Figure 10.5. The other angles and directions are analogous to those in Figure 10.4.

10.2.2 Scattering cross-sections of ion backscattering and elastic recoil processes

The probability of scattering and hence the sensitivities of the techniques are a consequence of the scattering cross-section. When the Rutherford theory of scattering may be assumed valid, the cross-section can, in principle, be computed exactly by the classical Coulomb scattering theory. Often, however, the ions penetrate the Coulomb barrier and the scattering must be treated by using wave mechanics as a combination of the Coulomb, nuclear potential and resonance scattering contributions. In practice, the non-Rutherford scattering cross-sections are based on experimental data. The non-Rutherford threshold energies can be predicted for different ion–target combinations and energies.

Rutherford scattering. The Rutherford formulas for the **scattering** and **recoil cross-sections** in the laboratory frame of reference are:

$$\left(\frac{d\sigma}{d\Omega}\right)_{RBS} \approx \left(\frac{Z_1 Z_2 e^2}{16\pi\varepsilon_0 E}\right)^2 \left[\sin^{-4}\left(\frac{\Theta}{2}\right) - 2x^2 + \ldots\right] \quad (10.5)$$

$$\left(\frac{d\sigma}{d\Omega}\right)_{ERD} = \left(\frac{Z_1 Z_2 e^2}{8\pi\varepsilon_0 E}\right)^2 \frac{1}{\cos^3\Theta}(1+x)^2 \quad (10.6)$$

where $x = M_1/M_2$, e is the electron charge and E is the collision energy. Equation (10.5) is an approximation from a series expansion, valid for $x \ll 1$. The first omitted term is of the order of x^4. Table 10.3 presents the Rutherford scattering cross-sections of the elements for ions $Z_1 = 1$–8 at an energy of 1.0 MeV, scattered through an angle of 170°.

For both of the techniques, the Rutherford cross-section is found to be inversely proportional to the square of the collision energy and approximately proportional to the squares of the atomic numbers of the particles. Further, the backscattering yield increases rapidly with decreasing scattering angle, while the yield of the recoiled particles increases with increasing recoil angle.

Non-Rutherford scattering. As the incident ion energy approaches the Coulomb barrier energy of the target nucleus, new scattering mechanisms begin to compete with pure Coulomb scattering. Scattering from the nuclear potential produces gradual modifications to the scattering probability while abrupt changes are found at energies coinciding with the excited states of the compound nucleus formed in the collision. The total **non-Rutherford scattering** is a combination of the three contributions: Rutherford, nuclear potential and resonance scattering. Even at low energies, the nuclear charge is screened by the atomic electrons and a small screening correction (of the order of a few per cent) should be applied to the Rutherford cross-section (see e.g. Rauhala 1992).

The enhancement of the cross-section relative to Rutherford, often encountered when light elements are bombarded by light projectiles (^1H, ^4He) of sufficient energy, is now routinely employed to improve the detection sensitivity of light elements. In heavy-ion RBS and ERD the

Table 10.3 Rutherford backscattering cross-sections (b/sr) (1 b = 10^{-24} cm^2) of the elements for ions Z_1 = 1–8 at an energy of 1.0 MeV. A scattering angle of 170° is assumed. For other energies E (in MeV) multiply the cross-sections by $1/E^2$

		Ions							
Z_2	Atom	^{1}H	^{4}He	^{7}Li	^{9}Be	^{11}B	^{12}C	^{14}N	^{16}O
1	H	0	0	0	0	0	0	0	0
2	He	0.00463	0	0	0	0	0	0	0
3	Li	0.01135	0.02130	0	0	0	0	0	0
4	Be	0.02054	0.05459	0.02995	0	0	0	0	0
5	B	0.03233	0.09838	0.10047	0.05005	0	0	0	0
6	C	0.04671	0.15023	0.18694	0.14726	0.03106	0.00001	0	0
7	N	0.06382	0.21803	0.32810	0.35877	0.24000	0.16800	0	0
8	O	0.08355	0.29655	0.49730	0.63406	0.59206	0.59009	0.23119	0
9	F	0.10598	0.38982	0.71834	1.031	1.188	1.403	1.108	1.5909
10	Ne	0.13093	0.48629	0.91863	1.358	1.638	2.001	1.760	1.188
11	Na	0.1586	0.5993	1.182	1.833	2.380	3.059	3.121	2.752
12	Mg	0.1888	0.7179	1.437	2.265	3.012	3.931	4.188	3.952
13	Al	0.2217	0.8513	1.743	2.818	3.882	5.182	5.864	6.050
14	Si	0.2572	0.9907	2.044	3.331	4.640	6.238	7.189	7.610
15	P	0.2954	1.146	2.402	3.979	5.671	7.773	9.239	10.271
16	S	0.3362	1.306	2.752	4.582	6.576	9.005	10.872	12.255
17	Cl	0.3796	1.483	3.163	5.334	7.784	10.769	13.335	15.533
18	Ar	0.4258	1.672	3.607	6.154	9.121	12.737	16.131	19.330
19	K	0.4744	1.861	4.008	6.825	10.091	14.070	17.756	21.184
20	Ca	0.5256	2.064	4.455	7.602	11.273	15.745	19.951	23.924
21	Sc	0.5797	2.285	4.974	8.563	12.842	18.057	23.251	28.438
22	Ti	0.6363	2.512	5.492	9.491	14.306	20.176	26.166	32.283
23	V	0.6955	2.750	6.032	10.460	15.836	22.391	29.213	36.304
24	Cr	0.7573	2.996	6.578	11.419	17.309	24.494	32.014	39.872
25	Mn	0.8212	3.255	7.165	12.469	18.964	26.885	35.298	44.198
26	Fe	0.8889	3.522	7.758	13.510	20.566	29.172	38.348	48.088
27	Co	0.9586	3.802	8.393	14.648	22.358	31.766	41.914	52.794
28	Ni	1.031	4.088	9.024	15.747	24.032	34.140	45.036	56.709
29	Cu	1.106	4.392	9.721	17.010	26.051	37.084	49.154	62.247
30	Zn	1.184	4.702	10.417	18.244	27.973	39.846	52.894	67.105

31	Ga	1.264	5.025	11.154	19.570	30.077	42.901	57.129	72.745
32	Ge	1.347	5.357	11.904	20.907	32.174	45.926	61.265	78.173
33	As	1.432	5.699	12.674	22.277	34.313	49.008	65.459	83.648
34	Se	1.521	6.053	13.477	23.715	36.582	52.292	69.981	89.630
35	Br	1.611	6.415	14.287	25.146	38.802	55.474	74.270	95.169
36	Kr	1.705	6.790	15.136	26.665	41.193	58.932	79.020	101.437
37	Rb	1.801	7.174	15.997	28.192	43.571	62.351	83.654	107.460
38	Sr	1.899	7.568	16.884	29.768	46.032	65.893	88.470	113.740
39	Y	2.001	7.973	17.791	31.375	48.531	69.481	93.325	120.039
40	Zr	2.105	8.389	18.726	33.038	51.130	73.224	98.419	126.692
41	Nb	2.211	8.815	19.683	34.735	53.775	77.027	103.579	133.405
42	Mo	2.320	9.252	20.669	36.493	56.528	80.999	109.003	140.517
43	Tc	2.432	9.700	21.679	38.292	59.347	85.064	114.554	147.793
44	Ru	2.547	10.158	22.708	40.120	62.202	89.172	120.140	155.079
45	Rh	2.664	10.626	23.760	41.988	65.116	93.365	125.835	162.499
46	Pd	2.784	11.105	24.843	43.918	68.142	97.731	131.805	170.336
47	Ag	2.906	11.594	25.941	45.866	71.178	102.097	137.727	178.042
48	Cd	3.031	12.096	27.074	47.891	74.361	106.695	144.034	186.349
49	In	3.159	12.606	28.223	49.934	77.553	111.293	150.292	194.524
50	Sn	3.289	13.128	29.401	52.033	80.846	116.043	156.788	203.052
51	Sb	3.422	13.660	30.599	54.166	84.183	120.854	163.347	211.638
52	Te	3.557	14.203	31.829	56.366	87.644	125.857	170.218	220.701
53	I	3.695	14.755	33.063	58.548	91.033	130.719	176.781	229.191
54	Xe	3.836	15.319	34.336	60.819	94.593	135.857	183.808	238.421
55	Cs	3.980	15.892	35.624	63.106	98.162	140.992	190.783	247.510
56	Ba	4.126	16.477	36.945	65.459	101.852	146.316	198.061	257.064
57	La	4.274	17.071	38.280	67.831	105.552	151.639	205.294	266.489
58	Ce	4.426	17.676	39.639	70.242	109.312	157.047	212.634	276.047
59	Pr	4.580	18.291	41.020	72.692	113.129	162.535	220.078	285.731
60	Nd	4.736	18.918	42.431	75.205	117.061	168.202	227.806	295.845
61	Pm	4.895	19.555	43.865	77.755	121.048	173.944	235.627	306.068
62	Sm	5.057	20.202	45.324	80.351	125.110	179.799	243.609	316.514
63	Eu	5.222	20.860	46.802	82.976	129.207	185.695	251.623	326.962
64	Gd	5.389	21.529	48.313	85.671	133.433	191.794	259.964	337.917

Table 10.3 (Continued)

Z_2	Atom	Ions							
		^{1}H	^{4}He	^{7}Li	^{9}Be	^{11}B	^{12}C	^{14}N	^{16}O
65	Tb	5.559	22.208	49.839	88.381	137.664	197.882	268.240	348.710
66	Dy	5.731	22.898	51.393	91.147	141.991	204.118	276.742	359.838
67	Ho	5.906	23.598	52.968	93.946	146.365	210.416	285.315	371.034
68	Er	6.084	24.308	54.566	96.787	150.804	216.807	294.012	382.390
69	Tm	6.264	25.029	56.186	99.666	155.298	223.275	302.806	393.860
70	Yb	6.447	25.761	57.836	102.603	159.895	229.902	311.845	405.698
71	Lu	6.632	26.503	59.505	105.568	164.525	236.567	320.910	417.528
72	Hf	6.820	27.256	61.200	108.585	169.244	243.366	330.177	429.650
73	Ta	7.011	28.019	62.917	111.637	174.013	250.234	339.525	441.858
74	W	7.205	28.793	64.659	114.734	178.854	257.206	349.020	454.267
75	Re	7.401	29.577	66.423	117.870	183.753	264.261	358.620	466.804
76	Os	7.599	30.372	68.214	121.058	188.740	271.447	368.419	479.627
77	Ir	7.801	31.177	70.025	124.276	193.766	278.683	378.262	492.475
78	Pt	8.005	31.993	71.861	127.541	198.870	286.033	388.271	505.554
79	Au	8.211	32.819	73.719	130.843	204.026	293.456	398.367	518.731
80	Hg	8.420	33.656	75.604	134.196	209.270	301.012	408.663	532.195
81	Tl	8.632	34.504	77.513	137.592	214.581	308.664	419.091	545.835
82	Pb	8.847	35.362	79.443	141.025	219.946	316.391	429.611	559.580
83	Bi	9.064	36.230	81.396	144.495	225.365	324.191	440.221	573.427
84	Po	9.283	37.109	83.371	148.003	230.841	332.071	450.931	587.393
85	At	9.506	37.997	85.368	151.548	236.370	340.025	461.732	601.462
86	Rn	9.731	38.900	87.408	155.194	242.102	348.309	473.097	616.442
87	Fr	9.958	39.810	89.454	158.828	247.776	356.474	484.198	630.919
88	Ra	10.189	40.731	91.527	162.514	253.537	364.771	495.495	645.680
89	Ac	10.422	41.662	93.621	166.234	259.342	373.127	506.854	660.496
90	Th	10.657	42.605	95.744	170.013	265.256	381.649	518.474	675.705
91	Pa	10.895	43.557	97.882	173.807	271.172	390.159	530.026	690.748
92	U	11.136	44.521	100.056	177.679	277.238	398.905	541.969	706.403

cross-sections fall off rapidly above the non-Rutherford threshold, rendering analytical work impracticable.

The **energy thresholds** of non-Rutherford scattering for different projectile–target combinations can be found in the literature (Nurmela *et al.* 1998, Bozoian 1993, Räisänen *et al.* 1994, Rauhala and Räisänen 1994, Räisänen and Rauhala 1995). For example, the non-Rutherford effects must be considered in RBS analysis of carbon by protons above $E_H = 0.3$ MeV, by ^{4}He above 2.2 MeV and in ERD analysis of hydrogen above $E_{He} = 1.3$ MeV. In general, the cross-sections are based on experimental data; therefore a considerable amount of experiments have been performed (for reviews, see e.g. Rauhala 1992, Leavitt *et al.* 1996).

10.2.3 Energy loss phenomena

Basic definitions. In materials analysis using beams of charged atomic particles, knowledge of slowing down of ions traversing matter is of fundamental importance. Depth analysis is based directly on the energy lost by the probing particles and energy loss calculations are needed to obtain quantitative and compositional information from experimental data.

The **stopping power** $S = \mathrm{d}E/\mathrm{d}x$ is defined in terms of the amount of energy lost ΔE per unit distance Δx traversed by the ion (Figure 10.6):

$$S = \frac{\mathrm{d}E}{\mathrm{d}x} = \lim_{\Delta x \to 0} \frac{\Delta E}{\Delta x} \tag{10.7}$$

The **stopping cross-section** ε is the stopping power divided by the atomic density N:

$$\varepsilon = \frac{1}{N} \frac{\mathrm{d}E}{\mathrm{d}x} \tag{10.8}$$

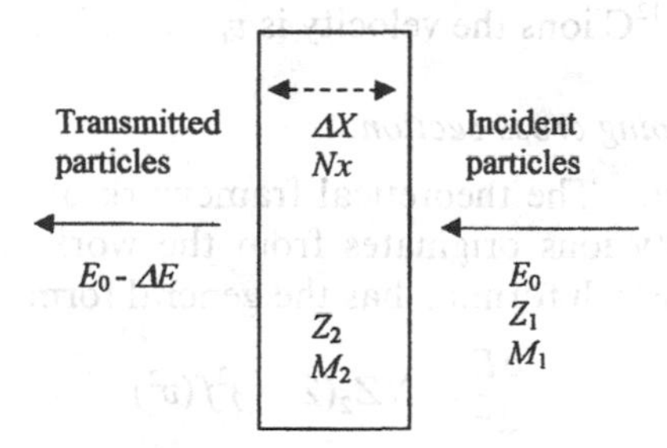

Figure 10.6 Energy loss ΔE of charged particles with atomic number Z_1 and mass M_1 in penetrating through a layer of thickness Δx. The layer consists of a material with elements of mass M_2, corresponding to an areal density of Nx.

The unit for the stopping power is, for example, keV/μm and for the stopping cross-section, eV cm^2/10^{15} atoms.

The physics of energy loss phenomena is very complex, involving many kinds of interactions between the projectile ion, target nuclei and target electrons. The relative importance of the various interaction processes between the ion and the target medium depends mostly on the ion velocity and on the charges of the ion and target atoms. At ion velocities v significantly lower than the Bohr velocity v_0 of the atomic electrons (see Example 10.2), the ion is assumed to carry its electrons and tends to neutralize by electron capture. At these velocities elastic collisions with the target nuclei, i.e. the **nuclear energy loss**, dominate. As the ion velocity is increased, the **electronic energy loss**, i.e. inelastic collisions with the atomic electrons, soon becomes the main interaction. The **total energy loss** is obtained as a sum of the nuclear and electronic contributions. At higher velocities $v \gg v_0$ the charge state of the ion increases and finally it becomes fully stripped of its electrons.

At ion velocities relevant to RBS or ERD analysis, the nuclear energy loss may be ignored and the energy loss calculations are based on formulations of the electronic energy loss.

Example 10.2
Bohr velocity v_0

The ion velocity in **Bohr velocity** units is

$$\frac{v}{v_0} = 6.325\sqrt{E} \tag{10.9}$$

where E should be given in energy-per-mass units MeV/u (u is the projectile mass in atomic mass units). For protons the Bohr velocity thus corresponds to an energy of:

$$E = 1/(6.325)^2 \text{ MeV/u} = 0.025 \text{ MeV/u} = 25 \text{ keV}$$

The velocity of 2.0 MeV ^{4}He is $v_{He} = 4.47$ times the Bohr velocity, and for 20 MeV ^{12}C ions the velocity is $v_C = 8.17 v_0$.

Calculation of stopping cross-section.

Elemental targets. The theoretical framework of the electronic energy loss of high-velocity ions originates from the works of Bohr, Bethe and Bloch. The Bethe–Bloch formula has the general form:

$$\frac{dE}{dx} = NZ_2(Z_1e^2)^2 f(v^2) \tag{10.10}$$

where e is the electron charge and $f(v^2)$ is a function depending only on the target and the projectile velocity, not on the type of the projectile.

To extrapolate the theoretical treatment from the high-energy region to the intermediate region where the ion is only partially stripped, the concept of **effective charge** has been introduced. As a consequence of equation (10.10), the effective charge also relates the energy loss of different ions: the ratio of heavy-ion energy loss to light-ion energy loss is obtained from the square of the effective charge. Based on both theoretical considerations and the analysis of experimental energy loss data, analytical formulations of the effective charge and the energy loss for all kinds of projectile ions in all elemental targets have been derived.

Proton stopping cross-sections ε_H above 25 keV may be obtained from an equation modified from the Bethe–Bloch formula (Ziegler *et al.* 1985, Ziegler and Biersack 1991):

$$\begin{aligned} \varepsilon_H &= \frac{\varepsilon_{low}\varepsilon_{high}}{\varepsilon_{low} + \varepsilon_{high}} \\ \varepsilon_{low} &= A_1 E^{A_2} + A_3 E^{A_4} \\ \varepsilon_{high} &= \frac{A_5}{E^{A_6}} \ln\left[\left(\frac{A_7}{E} + A_8 E\right)\right] \end{aligned} \tag{10.11}$$

where E is in keV/u and the stopping cross-sections are obtained in units of eV cm^2/10^{15}atoms. The coefficients A_1–A_8 are given in Table 10.4.

From equation (10.10) the stopping cross-sections of two different projectiles a and b having *the same velocity* v in a medium may be derived:

$$\left(\frac{\varepsilon}{(\gamma Z_1)^2}\right)_a = \left(\frac{\varepsilon}{(\gamma Z_2)^2}\right)_b \tag{10.12}$$

In equation (10.12) γ is the **fractional effective charge**, defined as

$$\gamma = Z_1^*(v, Z_2)/Z_1(v, Z_2)$$

where $Z_1^*(v, Z_2)$ is the effective charge of the ion at velocity v in the medium Z_2. Taking protons as ions a, the stopping cross-section of heavy ions (HI) is then obtained from the stopping of protons (p) *at the same velocity* (assuming $\gamma_p = 1$), by the **heavy-ion scaling rule**:

$$\varepsilon_{HI} = \varepsilon_p (Z_{HI})^2 (\gamma_{HI})^2 \tag{10.13}$$

The higher the ion energy, the closer γ approaches unity. When $v \gg v_0$ the ion is assumed to be fully stripped and $\gamma = 1$ (Example 10.3).

For the heavy-ion fractional effective charge γ, many different Z_1- and Z_2-dependent formulations have been proposed by various authors in the literature (see Rauhala 1995). Semi-empirical parametrizations based on an extensive amount of experimental data were presented, for example, by Ziegler (1980) and Ziegler *et al.* (1985).

For ^{4}He ions, Ziegler *et al.* (1985) and Ziegler and Manoyan (1988) give the following simple formula for the effective charge:

Table 10.4 Coefficients A_1–A_8 for calculating proton stopping cross-sections ε_H in elements from equations (10.11). The stopping cross-sections are in eV cm^2/10^{15} atoms

Z_2	Atom	A_1	A_2	A_3	A_4	A_5	A_6	A_7	A_8
1	H	0.0121702	0.00533578	1.12874	0.364197	1120.7	1.12128	2477.31	0.009770990
2	He	0.4890013	0.0050512491	0.8613451	0.4674054	745.3815	1.0422672	7988.3889	0.033328667
3	Li	0.8583748	0.0050147482	1.6044494	0.3884424	1337.3032	1.047033	2659.2306	0.018979873
4	Be	0.8781010	0.0051049349	5.4231571	0.2031973	1200.6151	1.0211124	1401.8432	0.038529280
5	B	1.4607952	0.0048835929	2.3380238	0.4424895	1801.2741	1.0352217	1784.1234	0.020239625
6	C	2.10544	0.00490795	2.08723	0.46258	1779.22	1.01472	2324.45	0.020269400
7	N	0.645636	0.00508289	4.09503	0.33879	2938.49	1.04017	2911.08	0.010721900
8	O	0.751093	0.00503003	3.93983	0.346199	2287.85	1.01171	3997.24	0.018426800
9	F	1.30187	0.00514136	3.82737	0.28151	2829.94	1.02762	7831.3	0.020940300
10	Ne	4.7339096	0.0044505735	0.0298622	1.4940358	1825.3641	0.9789632	130.76313	0.021576591
11	Na	6.097248	0.0044291901	3.1929400	0.4576301	1363.3487	0.9518161	2380.6086	0.081834623
12	Mg	14.013106	0.0043645904	2.2641223	0.3632649	2187.3659	0.9909772	6264.8005	0.046200118
13	Al	0.0390926	0.0045416623	6.9692434	0.3297639	1688.3008	0.9594386	1151.9784	0.048981572
14	Si	2.178134	0.0044454523	2.6045162	0.6088463	1550.2068	0.9330245	1703.8459	0.031619771
15	P	17.575478	0.0038345645	0.0786935	1.2388076	2805.9699	0.9728416	1037.5875	0.012878599
16	S	3.1473	0.0044716	4.9747	0.41024	4005.8	1.0011	1898.8	0.007659200
17	Cl	3.3544	0.004474	5.9206	0.41003	4403.9	0.99623	2006.9	0.008231000
18	Ar	2.0378865	0.0044775111	3.0742856	0.5477292	3505.0123	0.9757545	1714.0455	0.011700915
19	K	0.7417072	0.0043051307	1.1514679	0.9508284	917.2098	0.8781994	389.93209	0.189257680
20	Ca	9.1315679	0.0043809163	5.4610696	0.3132704	3891.8065	0.9793344	6267.9299	0.015195719
21	Sc	7.2247467	0.0043718065	6.1016923	0.3751071	2829.2167	0.9521775	6376.1223	0.020398311
22	Ti	0.1469983	0.0048456345	6.3484619	0.4105728	2164.1297	0.9402756	5292.6185	0.050263311
23	V	5.0611377	0.0039867276	2.6173973	0.5795689	2218.8786	0.9236074	6323.0262	0.025669274
24	Cr	0.5326669	0.0042967841	0.3900533	1.2725384	1872.6784	0.9077604	64.16607	0.030106759
25	Mn	0.47669667	0.0043038251	0.3145150	1.3289335	1920.5366	0.9064879	45.57647	0.027469102
26	Fe	0.0274264	0.0035443371	0.0315632	2.1754694	1919.5468	0.9009877	23.90246	0.025362735
27	Co	0.1638265	0.0043042231	0.0734540	1.8591669	1918.4213	0.8967809	27.60974	0.023184259
28	Ni	4.2562307	0.0043736651	1.5605733	0.7206703	1546.8412	0.8795769	302.01726	0.040944280
29	Cu	2.3508344	0.0043236552	2.8820471	0.5011277	1837.724	0.8999210	2376.9522	0.049650048
30	Zn	3.1095234	0.0038454560	0.1147724	1.5037112	2184.6911	0.8930896	67.30570	0.016587522

31	Ga	15.321773	0.0040306222	0.6539050	0.6766839	3001.7127	0.9248419	3344.1849	0.016366484
32	Ge	3.6931934	0.0044813010	8.6080131	0.2763752	2982.6649	0.9276011	3166.5676	0.030873833
33	As	7.1372758	0.0043134079	9.4247048	0.2793700	2725.8278	0.9159744	3166.0773	0.025007912
34	Se	4.8979355	0.0042936681	3.7793113	0.5000385	2824.4987	0.9102752	1282.4213	0.017061405
35	Br	1.3682731	0.0043024195	2.5678732	0.6082169	6907.8291	0.9817031	628.00764	0.006805497
36	Kr	2.2352779	0.0043096475	4.8856225	0.4788315	4972.2755	0.9514553	1185.2258	0.009235595
37	Rb	0.4205605	0.0041168716	0.0169498	2.3615903	2252.7284	0.8919173	39.75195	0.027756682
38	Sr	30.779774	0.0037736081	0.5581330	0.7681590	7113.1556	0.9769719	1604.3936	0.006526788
39	Y	11.575976	0.0042119068	7.0244432	0.3776444	4713.5219	0.9426371	2493.2196	0.011269740
40	Zr	6.2405791	0.0041916428	5.2701225	0.4945333	4234.5506	0.9323158	2063.9198	0.011844234
41	Nb	0.3307332	0.0041243377	1.7246018	1.1062067	1930.1909	0.8690703	27.41631	0.038208323
42	Mo	0.0177470	0.0041715317	0.145851	1.7305221	1803.6188	0.8631518	29.66948	0.032122562
43	Tc	3.7228678	0.0041768103	4.6286038	0.5676889	1678.0247	0.8620204	3093.9512	0.062440177
44	Ru	0.1399838	0.0041328551	0.2557267	1.4241133	1919.2613	0.8632628	72.79694	0.032235102
45	Rh	0.2858978	0.0041385894	0.3130097	1.3423521	1954.8161	0.8617511	115.18178	0.029341706
46	Pd	0.7600193	0.0042179200	3.3859683	0.7628467	1867.3886	0.8580521	69.99413	0.036447779
47	Ag	6.395683	0.0041934620	5.4689075	0.4137814	1712.6134	0.8539727	18493.003	0.056470873
48	Cd	3.471693	0.0041343969	3.2337247	0.6378845	1116.3584	0.8195894	4766.0254	0.117895110
49	In	2.5265128	0.0042282025	4.5319769	0.5356240	1030.8484	0.8165170	16252.232	0.197218350
50	Sn	7.3682953	0.0041006764	4.6791094	0.5142830	1160.0010	0.8245361	17964.821	0.133160090
51	Sb	7.7197216	0.0043879762	3.2419754	0.6843444	1428.1143	0.8339777	1786.6706	0.066512413
52	Te	16.779901	0.0041917673	9.3197716	0.295638	3370.9153	0.9028867	7431.7168	0.026159672
53	I	4.2132343	0.0042097824	4.6753325	0.5794508	3503.9280	0.8926145	1468.8716	0.014359044
54	Xe	4.6188934	0.0042203349	5.8164363	0.5218418	3961.2382	0.9040966	1473.2618	0.014194525
55	Cs	0.1851741	0.0036214664	0.0005878	3.5315422	2931.2998	0.8893605	26.17981	0.026392988
56	Ba	4.8248318	0.0041457784	6.0934255	0.5702562	2300.1078	0.8635875	2980.7187	0.038678811
57	La	0.4985675	0.0041054097	1.9775408	0.9587658	0786.5483	0.7850915	806.59969	0.408823790
58	Ce	3.275439	0.0042177424	5.7680306	0.5405402	6631.2873	0.9428168	744.06608	0.008302589
59	Pr	2.9978278	0.0040901358	4.5298608	0.6202474	2161.1538	0.8566884	1268.5942	0.043030519
60	Nd	2.8701111	0.0040959577	4.2567723	0.6137956	2130.4334	0.8523469	1704.1091	0.039384664
61	Pm	10.852925	0.0041148811	5.8907486	0.4683363	2857.1706	0.8754973	3654.168	0.029955419
62	Sm	3.64072	0.0041782043	4.8742398	0.5786142	1267.6986	0.8221108	3508.1718	0.241737030
63	Eu	17.645466	0.0040991510	6.5855038	0.3273433	3931.3048	0.9075401	5156.6611	0.036278412
64	Gd	7.5308869	0.0040813717	4.9389060	0.5067915	2519.6680	0.8581850	3314.6247	0.030514306

Table 10.4 (Continued)

Z_2	Atom	A_1	A_2	A_3	A_4	A_5	A_6	A_7	A_8
65	Tb	5.4741845	0.0040828895	4.8969573	0.5111306	2340.0736	0.8529648	2342.6752	0.035661774
66	Dy	4.266075	0.0040667385	4.5031787	0.552674	2076.3920	0.8415133	1666.5639	0.040801264
67	Ho	6.8312811	0.0040485776	4.3986952	0.5167493	2003.0028	0.8343741	1410.4455	0.034779520
68	Er	1.2707048	0.0040553442	4.6294611	0.5742751	1626.2816	0.8185828	995.68122	0.055319240
69	Tm	5.7561274	0.0040490505	4.356999	0.5249555	2207.3232	0.8379555	1579.5099	0.027165033
70	Yb	14.127459	0.0040595861	5.8303935	0.3775487	3645.8910	0.8782317	3411.7714	0.016392116
71	Lu	6.6947551	0.0040602520	4.9361227	0.4796132	2719.0292	0.8524863	1885.8388	0.019713210
72	Hf	3.0618944	0.0040511084	3.5802967	0.5908198	2346.0989	0.8371319	1221.9881	0.020071670
73	Ta	10.810811	0.0033007875	1.3767142	0.7651179	2003.7300	0.8226856	1110.5686	0.024957749
74	W	2.7100691	0.0040960825	1.2289456	0.9859815	1232.3872	0.7906638	155.42021	0.047294287
75	Re	0.5234523	0.0040243824	1.4037997	0.8551002	1461.3907	0.7967727	503.34277	0.036789456
76	Os	0.4616005	0.0040202735	1.3014089	0.8704331	1473.5357	0.7968670	443.08542	0.036301488
77	Ir	0.9781437	0.0040374101	2.0126698	0.7225008	1890.8068	0.8174691	930.70144	0.027690448
78	Pt	3.2085673	0.0040510075	3.6657671	0.5361780	3091.1569	0.8560235	1508.1176	0.015401358
79	Au	2.0035097	0.0040430629	7.4882362	0.3560990	4464.3312	0.8883581	3966.5437	0.012838852
80	Hg	15.429952	0.0039432123	1.1237408	0.7070324	4595.7209	0.8843687	1576.4704	0.008853375
81	Tl	3.1512351	0.0040523543	4.0995555	0.5424994	3246.3125	0.8577231	1691.7661	0.015058053
82	Pb	7.1896291	0.0040587571	8.6927070	0.3584227	4760.5609	0.8883332	2888.2709	0.011029181
83	Bi	9.320869	0.0040539730	11.542820	0.3202666	4866.1620	0.8912398	3213.3794	0.011934944
84	Po	29.242217	0.0036194863	0.1686396	1.1226448	5687.9614	0.8981205	1033.2571	0.007130312
85	At	1.8522161	0.0039972862	3.1556025	0.6509577	3754.9715	0.8638291	1602.0163	0.012041676
86	Rn	3.221995	0.0040040926	5.9023588	0.5267790	4040.1546	0.8680370	1658.3527	0.011746940
87	Fr	9.3412359	0.0039660558	7.920988	0.4297687	5180.8957	0.8877259	2173.1554	0.009200702
88	Ra	36.182673	0.0036003237	0.5834109	0.8674703	6990.2108	0.9108200	1417.0974	0.006218743
89	Ac	5.9283892	0.0039694789	6.4082404	0.5212246	4619.5148	0.8808273	2323.5230	0.011627375
90	Th	5.2453649	0.0039744049	6.7968897	0.4854236	4586.3094	0.8779443	2481.5001	0.011282428
91	Pa	33.701736	0.003690143	0.4725717	0.8923500	5295.6866	0.8892973	2053.3026	0.009190849
92	U	2.7588977	0.0039805707	3.2091513	0.6612173	2505.3660	0.8286302	2065.1403	0.022815839

$$\gamma_{He}^2 = 1 - \exp\left(-\sum_{i=0}^{5} a_i (\ln E_{He})^i\right) \tag{10.14}$$

where E_{He} is in keV/u and the fitting constants are $a_i = 0.2865$, 0.1266, −0.001429, 0.02402, −0.01135 and 0.001475 for $i = 0$ to 5, respectively.

For heavy ions ($Z_1 > 3$), Ziegler (1980) presents a simple parametrization for γ in the energy region 0.2 MeV/u < E < 22 MeV/u:

$$\begin{aligned} \gamma_{HI} &= 1 - \exp(-A)[1.034 - 0.177\exp(-0.08114 Z_{HI})] \\ A &= B + 0.0378\sin(B\pi/2) \\ B &= 0.1772\sqrt{E_{HI}}(Z_{HI})^{-2/3} \end{aligned} \tag{10.15}$$

For Li ions ($Z_1 = 3$), a separate parametrization is given:

$$\begin{aligned} \gamma_{Li} &= A\{1 - \exp[-(B + C)]\} \\ A &= 1 + (0.007 + 5 \times 10^{-5} Z_2)\exp[-(7.6 - \ln E_{Li})^2] \\ B &= 0.7138 + 0.002797 E_{Li} \\ C &= 1.348 \times 10^{-6} E_{Li}^2 \end{aligned} \tag{10.16}$$

In equations (10.14)–(10.16) all energies E, E_{HI}, and E_{Li} are in units keV/u and the sine function argument $B\pi/2$ (equation (10.15)) is in absolute angular units (radians).

Other recent formulations of the effective charge, such as, for example, the parametrization of Ziegler *et al.* (1985), applicable to $Z_1 > 2$, $E > 25$ keV/u, require more extensive calculations.

The most elaborate calculations of ion energy loss found in the literature involve either corrections to the simple formulas above or formulations based on different conceptions. The present formulations for the stopping cross-sections usually deviate from those calculations by less than 2% for H and He, and more (usually less than 5–10%) for heavier projectiles.

Example 10.3
Proton and helium-ion stopping power calculations, and the effective charge γ

Quick estimates of high-energy heavy-ion stopping in elements may be calculated from the proton stopping powers, equation (10.13), by taking γ equal to unity.

At the same velocity the heavy-ion and proton energies ($E = \frac{1}{2}mv^2$) are related by:

$$E_{HI} = (m_{HI}/m_p)E_p$$

where m_{HI} and m_p are the heavy-ion and proton masses, respectively. Taking $m_p = 1$ (atomic mass units) we have $E_{HI} = m_{HI}E_p$. Protons at 0.5 MeV thus have the same velocity as 2.0 MeV ^{4}He.

For 2.0 MeV ^{4}He we obtain from equation (10.13): $\varepsilon_{He}(2.0\,MeV) \approx 4\varepsilon_p$ (0.5 MeV). At $E_{He} = 2.0$ MeV, corresponding to $v_{He} = 4.47v_0$ (Example 10.2), this rule is accurate to 3%, the factor $Z_{He}^2\gamma_{He}^2 = 4\gamma_{He}^2$ being actually 3.88 (as calculated from equation (10.14)) instead of 4.

Assuming $\gamma_{He} = 1$ thus leads to the mnemonic: *2 MeV ^{4}He ions lose energy four times more rapidly than ^{1}H ions at 0.5 MeV.*

Example 10.4
Approximate carbon stopping cross-section for ^{7}Li ions at 5.0 MeV

The stopping cross-sections of carbon for ^{7}Li ions may be again estimated from equation (10.13) by assuming $\gamma_{Li} = 1$. As $M_{Li} \approx 7M_H$, the proton energy is $E_H \approx E_{Li}/7 \approx 714$ keV at the same velocity as 5.0 MeV ^{7}Li. Now $Z_{Li}^2\gamma_{Li}^2 = 9$ and we may write equation (10.13) as:

$$\varepsilon_{Li}(5.0\text{ MeV}) \approx 9\varepsilon_H(714\text{ keV})$$

From equations (10.11) the proton stopping cross-section is $\varepsilon_H(714\,\text{keV}) \approx 5.70\,\text{eV cm}^2/10^{15}$ atoms and thus $\varepsilon_{Li}(5.0\,\text{MeV}) \approx 51\,\text{eV cm}^2/10^{15}$ atoms.

Example 10.5
^{7}Li ion stopping in carbon at 5.0 MeV as calculated from equations (10.13) and (10.16)

As before in Example 10.4: $E_{Li} = 5.0$ MeV $= 713$ keV/u ($\approx 5.3v/v_0$). Equation (10.16) yields $\gamma_{Li}(5.0\text{ MeV}) = 0.969$. From equation (10.13) and Example 10.4 we then obtain:

$$\varepsilon_{Li}(5.0\text{ MeV}) = 5.70 \times 9 \times (0.969)^2\,\text{eV cm}^2/10^{15}\text{ atoms}$$
$$= 49.7\text{ cm}^2/10^{15}\text{ atoms}$$

Comparing this value to that obtained in Example 10.4, we observe a difference of 3%.

Multi-elemental targets. A simple linear additivity rule for energy loss in compounds may be adopted on the assumption that the interaction processes between ions and component target elements are independent of the surrounding target atoms. Consider a compound or mixture A_mB_n. Using **Bragg's rule** for the stopping cross-section ε^{AB} of the compound, the stopping may be obtained as a weighted sum of the stopping cross-sections of the component elements as:

$$\varepsilon^{AB} = m\varepsilon^A + n\varepsilon^B \tag{10.17}$$

where $m + n$ is normalized to unity and ε is in units eV cm^2/atom.

Example 10.6
^4He ion stopping at 10 MeV in aluminum oxide Al_2O_3

Equations (10.11), (10.13) and (10.14) give for the helium ions: $\varepsilon^{Al}(10\text{ MeV}) = 17.03\text{ eV cm}^2/10^{15}$ atoms and $\varepsilon^{O}(10\text{ MeV}) = 12.25$ eV $\text{cm}^2/10^{15}$ atoms. For the oxide we then obtain:

$$\varepsilon^{Al_2O_3} = (2\varepsilon^{Al} + 3\varepsilon^{O})/5 = 0.4\varepsilon^{Al} + 0.6\varepsilon^{O} \approx 14.16\text{ eV cm}^2/10^{15}\text{ atoms}$$

10.2.4 Problems

1. The kinematic factors are a consequence of the conservation of energy and momentum in an elastic collision. Derive equations (10.1) and (10.4). Plot k_S and k_R for $M_1 = 4$ and 35 as a function of M_2 and use the plots to study the mass resolution.

2. Study the backscattering process for $M_1 > M_2$. Show that in ERD the incident ions cannot scatter into the detector if $\sin\Theta < M_2/M_1$. In this case the absorber foil can be eliminated.

3. Recall that the scattering cross-section for RBS is proportional to E^{-2} and approximately proportional to Z_1^2, Z_2^2 and $\sin^{-4}(\Theta/2)$. Investigate these dependences by comparing the cross-sections (a) for $Z_1 = 1, 2$ and 8 for a given target element and (b) for light, heavy and medium-heavy target elements and a given incident particle. (c) Calculate the cross-sections for $Z_1 = 2$, $Z_2 = 14$ at incident ion energies of 500, 1000 and 2500 keV. (d) Compare the ^4He ion scattering though angles $\Theta = 90°$, 130° and 170°.

4. Choose several pairs of Z_2, calculate the square of the ratio and compare it to the ratio of the cross-sections from Table 10.3 (see equations (10.5) and (10.6)).

5. Perform similar comparisons as in problem 3 for ERD: (a) take $E_0 = 25$ MeV, $Z_1 = 17$, $\Theta = 30°$ and calculate the cross-sections for elements $Z_2 = 1$ to 10.

6. The non-Rutherford energy threshold due to the nuclear potential scattering contribution can be estimated for ^1H and ^4He ion backscattering ($\Theta > 150°$) as being approximately proportional to Z_2:

$$E_{th,H} = [(M_1 + M_2)/M_2]Z_2/10\text{ MeV} \quad \text{and}$$
$$E_{th,He} = [(M_1 + M_2)/M_2]Z_2/4\text{ MeV}$$

respectively (Bozoian 1993). Study the threshold energy E_{th} as a function of Z_2 in both cases.

7. Use equations (10.13) and (10.14) to calculate the helium-ion stopping cross-sections of aluminum at $E_{He} = 0.5$, 1.5 and 3.0 MeV. Find the corresponding Bohr velocities and verify that the factors $4\gamma_{He}^2$ are 2.88, 3.75 and 3.97, respectively (see Example 10.3).

8. Verify that the assumption $\gamma_{Li} = 1$ leads to approximate ^{7}Li ion stopping cross-sections of carbon at $E_{Li} = 2.0$ and 10.0 MeV as 91 and 32 eV cm^2/10^{15} atoms respectively (see Example 10.4).

9. Compare the approximate ^{7}Li ion stopping cross-sections of carbon at $E_{Li} = 2.0$ and 10.0 MeV (problem 8) to stopping cross-sections obtained from equations (10.13) and (10.16). Verify that the result falls 36% below the approximate value for 2.0 MeV ($3.4v/v_0$), but that the change is less than 1% for 10.0 MeV ($7.6v/v_0$).

10.3 The experimental arrangement

A typical RBS or ERD experiment requires only simple experimental nuclear physics facilities. The accelerator, the scattering chamber, a particle detector, the basic electronics and a multichannel analyzer for spectrum collection and storage constitute the main experimental set-up. Performing the experiment itself – the measurement of an energy spectrum – is rather elementary in comparison to many more complex problems in experimental nuclear physics. In both scattering techniques, however, advanced experimental systems and complicated analysis problems require more expertise. We will focus on the fundamental aspects and examine the standard set-ups.

10.3.1 The incident ion and energy

The choice of the probing ion and incident energy depends on the analytical problem and sample structure. In RBS, ^{4}He ions at about 2.0 MeV energy might be considered as a good alternative to start with; this will give a general view of the sample. If this is not adequate, increasing the energy will improve the mass resolution but decrease the depth resolution. In addition, at some point the non-Rutherford effects will have to be considered. Lower energies or heavier projectiles are needed for better depth resolution. Choosing a heavier projectile and higher energy will increase the mass resolution for heavier sample elements. For the lightest elements, however, ^{1}H ions should be used (see problem 10.2.4.1). For the accessible depth, see problem 10.4.3.2. The detection sensitivity of RBS is poor for light elements, especially in a heavy matrix (see problems 10.2.4.3 and 10.4.3.1). The sensitivity can be improved by using ^{1}H or ^{4}He ions and increasing the energy above the non-Rutherford energy threshold. The non-Rutherford cross-sections have been surveyed, for example, in Rauhala (1992).

The choice of the incident ion and energy is more straightforward in the simple forms of ERD experiments. ^{4}He ions at about 2–3 MeV are typically used for the depth distribution analysis of hydrogen. If heavier elements are to be profiled, heavier ions and higher energies need to be available. When an absorber foil is used, the energy loss of the recoils in the absorber depends on the choice of the ion species and energy. This will affect which sample elements are best analyzed (see e.g. section 10.6). Typically, different ions such as Si, Cl, Ni, I, Au, etc., at energies ranging from about 20 to 200 MeV have been used.

10.3.2 The detector and electronics

A silicon barrier detector or a similar charged-particle detector is regularly used for recording the scattered or recoiled ions. The pulse originating at the detector is then amplified and modified in the preamplifier–amplifier–analog-to-digital converter (ADC) chain. The pulse height is finally converted to energy and an energy spectrum is generated in a multichannel analyzer (MCA). A computer program is often used as an MCA. Auxiliary equipment is often used, such as ratemeters for monitoring the count rate (suitable count rate is of the order of $10^3\,s^{-1}$) and pulse generators to monitor the energy/channel calibration and to correct for the detection dead time.

The thickness of the detector active area should be sufficient for the ions used, e.g. ^{1}H ions of a few MeV require a thickness of a few hundred micrometers. The energy resolution for ^{1}H and ^{4}He ions is of the order of 15 keV, but is significantly worse for heavier ions of high energy, e.g. about 30–60 keV for 10–20 MeV ^{7}Li, or more than 100 keV for 20 MeV ^{16}O. Consequently, for heavy ions, the kinematic mass resolution cannot be fully exploited.

In RBS, the detector is usually placed at a scattering angle of close to 180°. A typical choice is 170°. For ERD, recoil angles of 20–30° and 75–80° incident angles are used. If the detector active area is larger than about 50 mm^2, preventing the incident beam from hitting the detector may limit the scattering angle to a smaller value than otherwise desirable. To define the scattering angle adequately and to avoid the kinematic broadening of the resolution, the detector acceptance angle should not exceed 8–10 msr in RBS. This implies that the beam spot diameter at the target is of the order of 0.5–2.0 mm, target-to-detector distance more than about 50 mm and the detector aperture of the order of a few millimeters in diameter. In ERD, the collimation of the detector is even more crucial. Owing to the small angle between the sample surface and incident beam, the beam spot tends to widen at the sample surface. Therefore, rectangular or curved slits, instead of circular apertures, are often used to define the recoil geometry. A typical solid angle for ERD should not exceed 5 msr.

RBS and ERD spectra can be measured simultaneously or subsequently. This arrangement provides several advantages: a more complete analysis of the sample, determination of the energy/channel calibration, normalization of the beam parameters, etc.

10.3.3 Spectrum measurement

The experimental data needed for the data analysis consist of the energy spectrum and various parameters defining the experimental conditions. A minimum set of parameters required includes the ion species, incident energy, scattering or recoil angle, the angle of incidence and the energy/channel calibration. In ERD, the absorber foil thickness and composition must be known.

The energy/channel calibration, δE, is defined by the settings in the amplifier. Values of 3–5 keV/channel are customary. Calibration δE has to be measured from the spectrum, given either by the RBS channel positions ch_A and ch_B of two known elements A and B or the channels ch_1 and ch_2 of one element at two different incident ion energies E_1 and E_2. Often two suitable elements can be found in the surface layer of the very sample studied. To obtain good accuracy, the masses of the elements or energies should be chosen well apart. The energy/channel calibration is then obtained from:

$$\delta E = \frac{E_0(k_B - k_A)}{ch_B - ch_A} \quad \text{or} \quad \delta E = \frac{k_A(E_2 - E_1)}{ch_2 - ch_1}$$

A special calibration standard with several elements in a thin compound film or the use of radioactive α sources may be advantageous in the energy/channel calibration.

A typical beam current is of the order of 1–100 nA. The number of incident ions, Q, is obtained from the accumulated charge, measured with the charge integrator, by dividing by the elementary charge e. The measurements of the ion current and the accumulated charge are problematic. Simple DC integration yields only relative values for a given sample and energy. This value cannot be used to extrapolate to other samples or energies. Therefore, beam chopper systems or Faraday cups with electron suppressors are used for ion dose measurement. Insulating samples often pose another problem. By evaporating a very thin conducting film on the sample surface, of either a heavy (e.g. Au) or light (C) material, the accumulating beam charge may be grounded to the integrator. Even a soft pencil may sometimes be applied. If this is not feasible, a conducting material (clamps or apertures) should be applied very close (1–2 mm) to the beam spot.

To calibrate the height of the spectrum, the total number of incident ions Q and the solid acceptance angle Ω must be obtained either by direct measurement or by using a suitable reference sample and measuring the relative accumulated charge for the two samples. Often, however, only relative signal heights and areas are needed and the product $Q\Omega$ is irrelevant.

The statistical uncertainty of the number of counts N in the signal of a spectrum is proportional to $\sqrt{N}$ (see problem 10.4.3.1). Consequently, the accuracy of signal height or area increases with the number of counts and thus with the total number of incident ions Q and the measuring time. Depending on the mass of the sample elements, the energy/channel calibration, the beam current and the solid acceptance angle of the detector, the usually adequate signal height of 1000–5000 counts (statistical error of the number of counts in one channel 3.2–1.4%) corresponds to a measuring time from a few minutes to about half an hour.

10.4 Spectrum analysis

10.4.1 Ion backscattering

Signal width and layer thickness.

Single-element layer. Ions backscattered from beneath the surface of a sample have traversed the material twice: first penetrating the layer on the inward path, then back on the outward path after scattering. Considering the layer close to the surface, we can assume a constant value of stopping cross-sections ε_{in} and ε_{out} on both inward and outward paths. From equation (10.8) and using the notation of Figure 10.4, the ion energy loss in the first layer of thickness x_1 may be written as $\Delta E_1^{in} = \varepsilon_{in} N x_1 / \cos\Theta_1$ and $\Delta E_1^{out} = \varepsilon_{out} N x_1 / \cos\Theta_2$. The total energy difference ΔE for ions scattered from atoms at the surface and at depth x is then given by:

$$\Delta E = [\varepsilon_0] N x \tag{10.18}$$

where $[\varepsilon_0]$ is defined as the **stopping cross-section factor**:

$$[\varepsilon_0] = \frac{k_S}{\cos\Theta_1}\varepsilon_{in}(E_0) + \frac{1}{\cos\Theta_2}\varepsilon_{out}(k_S E_0) \tag{10.19}$$

the stopping cross-sections ε_{in} and ε_{out} (equations (10.11)–(10.16)) are taken at some constant energies on the inward and outward paths, for example, at E_0 and $k_S E_0$, respectively, in the **surface energy approximation**. The stopping cross-section factors $[\varepsilon_0]$ for ^4He ions are presented in Table 10.5.

Table 10.5 Stopping cross-section factors $[\varepsilon_0]$ for 1000–3000 keV ^{4}He ions in the surface energy approximation. The scattering angle is assumed as $\Theta = 170°$ and the angles of incidence and exit are $\Theta_1 = \Theta_2 = 5°$. The stopping cross-section factors are in eV cm^2/10^{15} atoms

		^{4}He ion energy (keV)										
Z_2	Atom	1000	1200	1400	1600	1800	2000	2200	2400	2600	2800	3000
3	Li	13.69	14.54	15.28	15.94	16.55	17.11	17.63	18.13	18.59	19.02	19.43
4	Be	28.74	29.29	29.67	29.90	30.04	30.09	30.08	30.02	29.93	29.80	29.65
5	B	38.88	40.33	41.39	42.13	42.60	42.84	42.90	42.79	42.56	42.22	41.80
6	C	44.19	45.69	46.73	47.40	47.77	47.88	47.79	47.52	47.12	46.61	46.03
7	N	55.38	56.82	57.66	58.06	58.09	57.85	57.38	56.72	55.92	55.01	54.02
8	O	59.75	61.05	61.64	61.70	61.38	60.77	59.94	58.96	57.87	56.71	55.50
9	F	56.13	57.88	58.87	59.31	59.32	59.02	58.47	57.74	56.89	55.93	54.92
10	Ne	61.18	62.74	63.41	63.48	63.12	62.47	61.61	60.61	59.51	58.35	57.17
11	Na	85.56	84.59	83.12	81.38	79.49	77.55	75.60	73.66	71.77	69.93	68.15
12	Mg	86.80	86.00	84.54	82.68	80.61	78.44	76.24	74.07	71.96	69.92	67.96
13	Al	85.77	84.40	82.71	80.85	78.92	76.96	75.00	73.08	71.20	69.37	67.61
14	Si	108.5	104.6	100.4	96.37	92.53	88.96	85.64	82.58	79.74	77.10	74.65
15	P	120.2	116.2	111.3	106.2	101.3	96.86	92.75	89.02	85.61	82.51	79.66
16	S	124.4	120.4	115.2	109.6	104.0	98.67	93.72	89.18	85.05	81.31	77.91
17	Cl	147.8	142.4	135.7	128.7	121.8	115.3	109.4	104.0	99.16	94.76	90.79
18	Ar	142.3	137.2	131.0	124.6	118.5	112.8	107.6	102.9	98.65	94.81	91.32
19	K	148.6	144.5	139.6	134.4	129.2	124.1	119.3	114.7	110.3	106.3	102.5
20	Ca	138.5	137.9	135.3	131.7	127.5	123.1	118.6	114.2	110.0	106.1	102.3
21	Sc	160.3	156.3	150.7	144.5	138.0	131.8	125.8	120.3	115.2	110.4	106.1
22	Ti	153.1	149.2	144.3	138.9	133.5	128.3	123.3	118.7	114.4	110.4	106.6
23	V	171.2	167.0	160.8	153.8	146.6	139.8	133.3	127.3	121.8	116.8	112.2
24	Cr	147.5	145.3	141.8	137.6	133.2	128.8	124.4	120.2	116.1	112.3	108.7
25	Mn	146.2	144.5	141.4	137.5	133.3	129.0	124.7	120.5	116.5	112.8	109.2
26	Fe	151.7	148.9	145.0	140.6	136.0	131.4	126.9	122.6	118.5	114.6	111.0
27	Co	147.0	145.3	142.1	138.3	134.1	129.8	125.6	121.5	117.6	113.9	110.4
28	Ni	134.1	134.4	133.2	131.1	128.5	125.5	122.3	119.1	115.9	112.8	109.8
29	Cu	133.8	133.8	132.4	130.1	127.4	124.4	121.3	118.3	115.2	112.3	109.4
30	Zn	137.1	138.1	137.2	135.1	132.4	129.2	125.9	122.5	119.2	116.0	112.8

31	Ga	146.7	146.4	144.0	140.5	136.4	132.2	128.0	124.0	120.1	116.5	113.1
32	Ge	151.1	149.4	146.5	142.9	139.1	135.3	131.5	127.9	124.4	121.1	118.0
33	As	163.4	159.0	153.8	148.4	143.1	138.1	133.3	128.9	124.8	121.0	117.5
34	Se	162.4	158.0	152.9	147.9	143.0	138.4	134.1	130.1	126.3	122.8	119.5
35	Br	179.6	172.2	164.6	157.6	151.4	145.9	141.0	136.6	132.6	128.9	125.5
36	Kr	191.8	184.4	176.2	168.2	160.8	154.2	148.2	142.8	137.9	133.4	129.4
37	Rb	206.1	200.4	193.8	186.8	179.9	173.1	166.7	160.6	154.9	149.6	144.6
38	Sr	222.5	214.9	205.2	195.0	185.2	176.0	167.8	160.4	153.7	147.8	142.5
39	Y	217.7	210.1	201.0	191.6	182.7	174.3	166.7	159.8	153.6	147.9	142.8
40	Zr	230.8	220.9	209.8	199.0	189.0	180.0	171.8	164.5	158.0	152.1	146.8
41	Nb	223.2	217.9	211.2	203.9	196.5	189.3	182.3	175.7	169.6	163.8	158.4
42	Mo	205.8	201.0	194.8	188.2	181.5	174.9	168.6	162.7	157.1	151.8	146.9
43	Tc	218.0	211.2	203.7	196.3	189.0	182.2	175.6	169.5	163.8	158.5	153.5
44	Ru	213.8	210.0	204.4	198.1	191.4	184.8	178.4	172.2	166.4	161.0	155.8
45	Rh	209.8	206.4	201.4	195.5	189.3	183.0	176.9	171.1	165.5	160.2	155.3
46	Pd	204.8	203.2	199.6	194.8	189.4	183.7	178.1	172.6	167.3	162.3	157.5
47	Ag	208.2	206.5	202.1	196.2	189.7	183.1	176.6	170.4	164.5	159.1	154.0
48	Cd	217.3	212.3	206.2	199.5	192.9	186.4	180.1	174.1	168.5	163.2	158.3
49	In	220.3	216.1	210.2	203.5	196.7	190.0	183.5	177.4	171.6	166.2	161.1
50	Sn	226.7	221.2	214.0	206.2	198.5	191.0	183.9	177.3	171.2	165.5	160.2
51	Sb	229.5	222.7	215.5	208.2	200.9	194.0	187.3	181.0	175.1	169.6	164.4
52	Te	239.7	234.2	226.6	218.2	209.7	201.5	193.8	186.5	179.9	173.7	168.0
53	I	256.5	244.3	232.4	221.5	211.7	202.9	195.0	187.9	181.5	175.6	170.2
54	Xe	258.2	247.0	235.6	225.0	215.4	206.7	198.8	191.7	185.3	179.4	174.0
55	Cs	272.0	264.5	255.6	246.4	237.1	228.1	219.6	211.5	204.0	196.9	190.4
56	Ba	284.3	271.7	259.2	247.5	236.8	226.9	217.9	209.7	202.2	195.3	188.9
57	La	284.6	274.8	264.0	253.1	242.5	232.4	223.0	214.2	206.1	198.6	191.7
58	Ce	262.8	249.0	236.5	225.6	216.1	207.8	200.5	194.0	188.0	182.6	177.6
59	Pr	255.1	248.7	241.6	234.1	226.6	219.3	212.3	205.5	199.2	193.2	187.5
60	Nd	253.7	246.8	239.2	231.4	223.8	216.4	209.4	202.8	196.5	190.6	185.1
61	Pm	261.6	254.7	246.3	237.4	228.8	220.5	212.8	205.7	199.0	192.9	187.2
62	Sm	261.9	259.2	254.1	247.7	240.6	233.4	226.1	219.1	212.3	205.8	199.7
63	Eu	232.8	230.9	226.7	221.5	215.8	209.9	204.2	198.7	193.4	188.4	183.6
64	Gd	254.3	248.7	241.4	233.6	225.7	218.2	211.1	204.4	198.2	192.4	187.0

Table 10.5 (Continued)

Z_2	Atom	^{4}He ion energy (keV)										
		1000	1200	1400	1600	1800	2000	2200	2400	2600	2800	3000
65	Tb	240.2	236.3	230.9	224.9	218.6	212.4	206.4	200.6	195.1	189.9	184.9
66	Dy	239.4	236.1	231.3	225.9	220.1	214.2	208.4	202.8	197.3	192.2	187.2
67	Ho	216.1	214.1	210.8	206.7	202.2	197.5	192.9	188.2	183.8	179.4	175.2
68	Er	232.8	231.5	228.4	224.1	219.1	213.7	208.3	202.9	197.7	192.6	187.7
69	Tm	224.2	220.8	216.3	211.3	206.0	200.8	195.7	190.8	186.0	181.5	177.2
70	Yb	234.2	231.1	225.6	219.0	212.2	205.5	199.0	193.0	187.4	182.2	177.4
71	Lu	229.0	224.8	219.1	213.1	207.0	201.1	195.6	190.3	185.4	180.7	176.3
72	Hf	228.9	224.2	218.7	213.0	207.3	201.7	196.4	191.3	186.5	181.9	177.6
73	Ta	228.1	225.5	221.5	216.9	211.9	206.8	201.7	196.8	191.9	187.3	182.9
74	W	231.9	229.9	225.8	220.4	214.4	208.2	202.1	196.1	190.4	184.9	179.8
75	Re	227.4	226.8	224.1	220.1	215.3	210.2	205.0	199.7	194.6	189.7	184.9
76	Os	226.8	226.7	224.3	220.6	216.0	210.9	205.7	200.5	195.5	190.5	185.8
77	Ir	226.7	225.2	222.1	218.0	213.4	208.6	203.7	198.8	194.0	189.4	184.9
78	Pt	227.6	224.9	220.3	214.9	209.4	204.0	198.7	193.8	189.1	184.6	180.4
79	Au	236.2	237.1	234.2	229.2	223.1	216.7	210.2	204.0	198.1	192.6	187.4
80	Hg	244.6	242.1	236.4	229.3	221.9	214.8	208.1	201.9	196.2	190.9	186.0
81	Tl	252.2	247.5	240.8	233.5	226.3	219.5	213.0	207.0	201.4	196.2	191.3
82	Pb	269.9	265.6	258.2	249.5	240.6	231.9	223.7	216.1	209.1	202.7	196.9
83	Bi	290.6	284.0	274.6	264.3	254.0	244.1	234.9	226.4	218.7	211.6	205.2
84	Po	293.6	284.9	273.5	262.0	251.2	241.3	232.5	224.7	217.6	211.2	205.3
85	At	299.0	289.2	277.6	266.1	255.4	245.6	236.7	228.6	221.4	214.7	208.6
86	Rn	304.8	293.8	281.5	269.6	258.6	248.6	239.6	231.5	224.1	217.4	211.3
87	Fr	315.5	305.6	293.1	280.1	267.7	256.3	245.9	236.5	228.1	220.6	213.7
88	Ra	331.4	320.1	305.8	291.0	277.0	264.2	252.9	242.8	233.9	225.9	218.7
89	Ac	346.3	333.3	318.2	303.0	288.8	275.9	264.2	253.8	244.5	236.0	228.4
90	Th	330.0	320.1	307.4	294.0	281.1	269.2	258.3	248.5	239.7	231.6	224.4
91	Pa	322.2	316.0	305.7	293.9	282.1	270.8	260.4	250.9	242.2	234.4	227.2
92	U	299.3	291.5	282.1	272.5	263.3	254.5	246.4	238.8	231.8	225.3	219.2

Example 10.7
Calculation of layer thickness from a helium backscattering spectrum for a thick Al foil

Figure 10.7 presents a backscattering spectrum for 2.0 MeV ^{4}He ions incident on an Al foil on a light substrate. Only the Al signal is shown. The scattering angle is $\Theta = 170°$, with the incident and exit angles $\Theta_1 = \Theta_2 = 5°$.

To find the edge positions and heights of the signal we fit lines to the plateau (line A) and the leading edge (line B). According to Figure 10.7 the position $E_1 = E_{Al} = k_{S,Al}E_0$ (equation (10.1)) is defined as the energy (or channel) of the half height point of line B. The height H_0 is the height of the intersection point of line A and a vertical line at E_1. From a similar geometrical construction at the trailing edge we obtain the signal width $\Delta E = 152\,\text{keV}$ as the difference between the leading and trailing edge positions. Solving for Nx from equation (10.18) we obtain:

$$Nx = \frac{\Delta E}{[\varepsilon_0]}$$

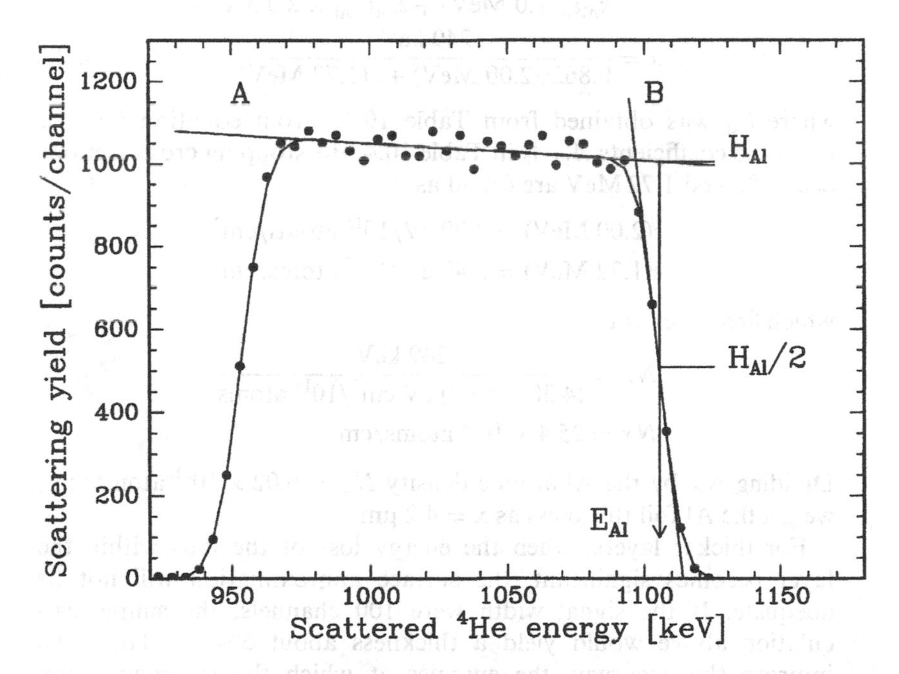

Figure 10.7 A backscattering spectrum for 2.0 MeV ^{4}He ions incident on an aluminum film on a light substrate. The helium-ion dose is 3.6 μC and the detector solid acceptance angle $\Omega = 5.35$ msr.

From Table 10.5 we find $[\varepsilon_0]_{Al} = 76.96 \times 10^{-15}$ eV cm^2/atom for 2.0 MeV ^{4}He. This gives Nx:

$$Nx = \frac{152 \text{ keV}}{76.96 \times 10^{-15} \text{ eV cm}^2/\text{atom}} = 1975 \times 10^{15} \text{ atoms/cm}^2$$

Dividing Nx by the Al atomic density $N_{Al} = 6.02 \times 10^{22}$ atoms/cm^3 we get the Al foil thickness as $x = 328$ nm.

Example 10.8
Calculation of foil thickness from a proton backscattering spectrum for a thick Al foil

Presume the proton incident energy $E_H = 2.0$ MeV and the scattering geometry as in the previous example. Assume the energy width of the signal as $\Delta E = 50 \text{ ch} \times 5 \text{ keV/ch} = 250$ keV. Again from equation (10.18) and using equation (10.19) for the proton stopping cross-section factor we obtain:

$$Nx = \frac{\Delta E}{[\varepsilon_0]}$$

$$Nx = \frac{250 \text{ keV} \times \cos 5°}{k_{Al}\varepsilon_{in}(2.0 \text{ MeV}) + \varepsilon_{out}(k_{Al} \times 2.0 \text{ MeV})}$$

$$Nx = \frac{249 \text{ keV}}{0.862\varepsilon(2.00 \text{ MeV}) + \varepsilon(1.72 \text{ MeV})} \tag{10.20}$$

where k_{Al} was obtained from Table 10.1. From equation (10.11), using the coefficients A_1–A_8 in Table 10.4, the stopping cross-sections ε_0 at 2.00 and 1.72 MeV are found as

$$\varepsilon(2.00 \text{ MeV}) = 4.99 \text{ eV}/10^{15} \text{ atoms/cm}^2$$
$$\varepsilon(1.72 \text{ MeV}) = 5.49 \text{ eV}/10^{15} \text{ atoms/cm}^2$$

which finally leads to:

$$Nx = \frac{249 \text{ keV}}{(4.30 + 5.49) \text{ eV cm}^2/10^{15} \text{ atoms}}$$
$$Nx = 25.4 \times 10^{18} \text{ atoms/cm}^2$$

Dividing Nx by the Al atomic density $N_{Al} = 6.02 \times 10^{22}$ atoms/cm^3 we get the Al foil thickness as $x = 4.2$ µm.

For thicker layers, when the energy loss of the ions within the layer becomes significant, the surface approximation will not be adequate. If the signal width were 100 channels, the simple calculation above would yield a thickness about 5% too large. To improve the accuracy, the energies at which the stopping cross-sections are evaluated should be determined more accurately (see Example 10.14 and problem 10.4.3.6).

Multi-element layer. Extending the treatment to multi-elemental samples is straightforward, but the notation becomes awkward as the number of elements increases. When scattering takes place from target atoms of different atomic species, the energy loss of ions after scattering depends on the target atom.

The important case of binary compounds (or mixtures) A_mB_n, $(m+n=1)$ is elucidated below. In analogy to equation (10.18) the energy widths of the signals corresponding to scattering from atoms A and B are:

$$\begin{aligned} \Delta E_A &= [\varepsilon_0]_A^{AB} Nx \\ \Delta E_B &= [\varepsilon_0]_B^{AB} Nx \end{aligned} \tag{10.21}$$

where the subscripts A and B indicate the atomic species of the target, the superscript AB indicates the medium and the **generalized stopping cross-section factors** are:

$$\begin{aligned} [\varepsilon_0]_A^{AB} &= \frac{k_{S,A}}{\cos\Theta_1}\varepsilon_{in}^{AB}(E_0) + \frac{1}{\cos\Theta_2}\varepsilon_{out}^{AB}(k_{S,A}E_0) \\ [\varepsilon_0]_B^{AB} &= \frac{k_{S,B}}{\cos\Theta_1}\varepsilon_{in}^{AB}(E_0) + \frac{1}{\cos\Theta_2}\varepsilon_{out}^{AB}(k_{S,B}E_0) \end{aligned} \tag{10.22}$$

For example, $\varepsilon_{out}^{AB}(k_{S,B}E_0)$ is the stopping cross-section for ions scattered from atoms B in the compound medium A_mB_n. This must be obtained by using Bragg's rule, i.e. $\varepsilon^{AB} = m\varepsilon^A + n\varepsilon^B$, where ε^A and ε^B should be evaluated at $k_{S,B}E_0$, the energy of incident energy E_0 times the scattering kinematic factor $k_{S,B}$ for target atoms B.

Example 10.9

Calculation of layer thickness for a thick ZnS layer on a glass substrate by 2.0 MeV ^{4}He ion backscattering

In Figure 10.8 the dots represent the experimental spectrum, the solid lines the result of a theoretical simulation (see section 10.5). From the spectrum taken at $\Theta = 170°$ with $\Theta_1 = \Theta_2 = 5°$, we obtain the energy width the Zn signal as: $\Delta E_{Zn}^{ZnS} = 240$ keV. Solving for Nx from equation (10.21):

$$Nx = \frac{\Delta E_{Zn}^{ZnS}}{[\varepsilon_0]_{Zn}^{ZnS}} = \frac{\Delta E_S^{ZnS}}{[\varepsilon_0]_S^{ZnS}}$$

Equation (10.22) yields the stopping cross-section factor for scattering from Zn:

$$[\varepsilon_0]_{Zn}^{ZnS} = \frac{1}{\cos\Theta_1}[k_{Zn}\varepsilon^{ZnS}(E_0) + \varepsilon^{ZnS}(k_{Zn}E_0)]$$

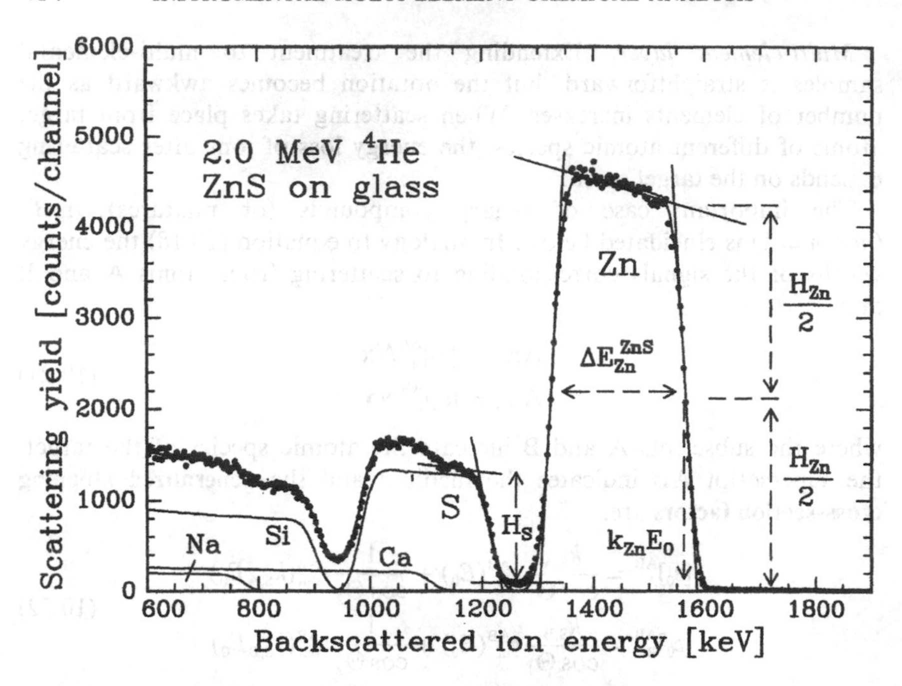

Figure 10.8 A backscattering spectrum for 2.0 MeV ^{4}He ions incident on a ZnS layer on a glass substrate. The dots (•) represent the experimental spectrum, and the solid lines (——) are a result of a theoretical simulation.

where according to equation (10.17), assuming stoichiometric ZnS,

$$\varepsilon^{\mathrm{ZnS}}(E_0) = \frac{\varepsilon^{\mathrm{Zn}}(E_0) + \varepsilon^{\mathrm{S}}(E_0)}{2}$$

$$\varepsilon^{\mathrm{ZnS}}(k_{\mathrm{Zn}}E_0) = \frac{\varepsilon^{\mathrm{Zn}}(k_{\mathrm{Zn}}E_0) + \varepsilon^{\mathrm{S}}(k_{\mathrm{Zn}}E_0)}{2}$$

The stopping cross-sections above are given by equations (10.11), (10.13) and (10.14) (in eV cm^2/10^{15} atoms):

$$\varepsilon^{\mathrm{Zn}}(E_0) = 69.33 \qquad \varepsilon^{\mathrm{S}}(E_0) = 50.57$$

$$\varepsilon^{\mathrm{Zn}}(k_{\mathrm{Zn}}E_0) = 73.68 \qquad \varepsilon^{\mathrm{S}}(k_{\mathrm{Zn}}E_0) = 58.76$$

where $k_{\mathrm{Zn}} = 0.7840$ from Table 10.2 and $E_0 = 2000$ keV. Substituting the numbers, we get for the compound stopping cross-sections: $\varepsilon^{\mathrm{ZnS}}(E_0) = 59.95$ eV cm^2/10^{15} atoms and $\varepsilon^{\mathrm{ZnS}}(k_{\mathrm{Zn}}E_0) = 66.37$ eV cm^2/10^{15} atoms, which yield the stopping cross-section factor for the compound: $[\varepsilon_0]^{\mathrm{ZnS}}_{\mathrm{Zn}} = 113.4$ eV cm^2/10^{15} atoms.

The first equation of the example then finally gives the ZnS layer areal density:

$$Nx = \frac{240\ \text{keV}}{113.4\ \text{eV cm}^2/10^{15}\ \text{atoms}} = 2120 \times 10^{15}\ \text{atoms/cm}^2$$

Multiplying Nx (in 10^{15} atoms/cm^2) by $0.0166M/\rho$, where $M = (M_{Zn} + M_S)/2 = 48.7\,\text{u}$ is the average atomic mass and $\rho = 3.98\,\text{g/cm}^3$ the mass density of ZnS, we obtain the thickness of the ZnS layer in nanometers:

$$x_{ZnS} = 430\ \text{nm}$$

Signal height and sample composition.

Thick-target yield. Assume a sample layer thick enough that the width of the signal in the spectrum is significantly wider than the energy resolution of the system. Suppose, first, that the layer consists of a single element, that the layer is amorphous and homogeneous. The back-scattering yield from the surface of the layer H_0 or the **surface height** of the signal is then given by:

$$H_0 = \frac{\sigma(E_0)Q\Omega\delta E}{[\varepsilon_0]\cos\Theta_1} \tag{10.23}$$

The yield is thus directly proportional to the scattering cross-section $\sigma(E_0)$, the number of incident particles Q, the solid acceptance angle of the detector Ω and the energy width of a channel δE and inversely proportional to the stopping cross-section factor $[\varepsilon_0]$ and the geometrical factor $\cos\Theta_1$. Calculating the surface yield, one evaluates $\sigma(E_0)$ and $[\varepsilon_0]$ at incident energy.

For a multi-element layer, the height of each of the elemental signals increases with the relative amount of atoms of the given atomic species. For two elements A and B in a compound (or mixture) of A_mB_n $(m + n = 1)$, the surface heights $H_{A,0}$ and $H_{B,0}$ corresponding to equation (10.23) are:

$$\begin{aligned} H_{A,0} &= \frac{\sigma_A(E_0)Q\Omega m\delta E}{[\varepsilon_0]_A^{AB}\cos\Theta_1} \\ H_{B,0} &= \frac{\sigma_B(E_0)Q\Omega n\delta E}{[\varepsilon_0]_B^{AB}\cos\Theta_1} \end{aligned} \tag{10.24}$$

Example 10.10

Determination of the stoichiometry of a two-element compound layer by RBS

From equations (10.24) one obtains for the ratio m/n:

$$\frac{m}{n} = \frac{H_{A,0}\sigma_B(E_0)[\varepsilon_0]_A^{AB}}{H_{B,0}\sigma_A(E_0)[\varepsilon_0]_B^{AB}} \tag{10.25}$$

where the ratio of the stopping cross-section factors approaches unity when the atomic numbers Z_A and Z_B approach each other. As a zeroth approximation m/n is thus given by the ratio of the surface heights weighed by the inverse of the squares of the atomic numbers (equation (10.5)):

$$\frac{m}{n} \approx \frac{H_{A,0}}{H_{B,0}} \left(\frac{Z_B}{Z_A}\right)^2 \tag{10.26}$$

From this m/n one then calculates the ratio of the stopping cross-section factors using equations (10.22) and then a better estimate of m/n from equation (10.25).

From Figure 10.8 (Example 10.9) the surface heights of Zn and S are measured as: $H_{Zn} = 4240$ and $H_S = 1220$ counts. Equation (10.26) gives the zeroth approximation for the composition:

$$m/n = (4240/1220) \times (16/30)^2 = 0.989$$

The normalization $m + n = 1$ yields $m = 0.503$ and $n = 0.497$. With these values m and n, calculating the stopping cross-section factors as in Example 10.9 and taking the scattering cross-sections from Table 10.3, equation (10.25) becomes:

$$\frac{m}{n} = \frac{H_{Zn}}{H_S} \frac{\sigma_S(E_0)}{\sigma_{Zn}(E_0)} \frac{[\varepsilon_0]_{Zn}^{ZnS}}{[\varepsilon_0]_S^{ZnS}} = \frac{4240}{1220} \frac{1.306}{4.702} \frac{113.4}{108.9} = 1.005$$

This leads to $m = 0.499$ and $n = 0.501$, a result showing almost no change compared to the values obtained from the zeroth-order approximation, thus verifying the stoichiometric ZnS composition.

Thin-target yield. If the layer thickness corresponds to a signal width of the order of the system energy resolution, the height of the signal is no longer given by equation (10.23). In this case, the height decreases with worsening resolution. The total number of counts below the signal peak or the **signal area** is, however, independent of resolution and energy loss in the medium:

$$A_0 = \sigma(E_0) Q \Omega N x / \cos \Theta_1 \tag{10.27}$$

Example 10.11
Figure 10.9 shows a backscattering spectrum for 2.0 MeV ^{4}He ions incident on a multilayer structure of Bi/Cu/Au on a silicon substrate – find the amount of Bi in the surface layer

As shown by the signals in the spectrum, the bismuth and gold layers are significantly thinner than the copper layer. The signal area of bismuth $A_{Bi} = 63\,250$ counts and the surface height of the copper

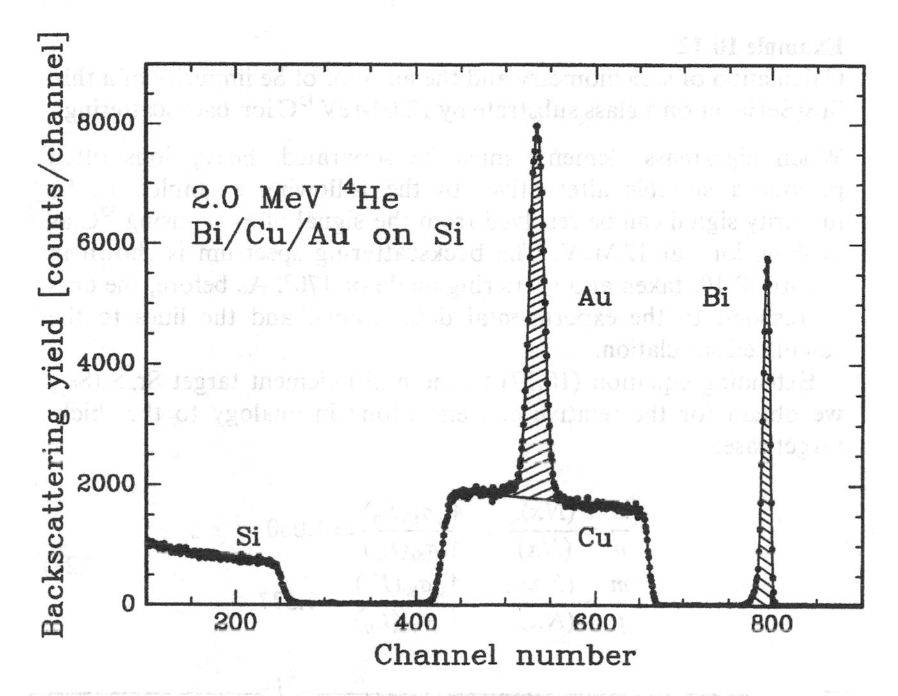

Figure 10.9 A backscattering spectrum for 2.0 MeV ^{4}He ions incident on a multilayer sample of Bi/Cu/Au on a silicon substrate.

signal $H_{Cu} = 1580$ counts are measured from the spectrum. The energy/channel calibration δE was set to 2.36 keV/ch and the scattering angle to $\Theta = 170°$, with the incident and exit angles $\Theta_1 = \Theta_2 = 5°$. The number of incident ions Q and the solid angle of the detector Ω are unknown.

We will calculate the unknown quantities Q and Ω by using the Cu signal as a reference. From equation (10.23) we obtain:

$$Q\Omega = \frac{H_{Cu}[\varepsilon_0]_{Cu} \cos \Theta_1}{\sigma(E_0)_{Cu}\delta E}$$

Solving for Nx from equation (10.27) and substituting $Q\Omega$ from above:

$$(Nx)_{Bi} = \frac{A_{Bi} \cos \Theta_1}{\sigma(E_0)_{Bi} Q\Omega} = \frac{A_{Bi}}{H_{Cu}} \frac{\sigma(E_0)_{Cu}}{\sigma(E_0)_{Bi}} \frac{\delta E}{[\varepsilon_0]_{Cu}} \approx \frac{A_{Bi}}{H_{Cu}} \left(\frac{29}{83}\right)^2 \frac{\delta E}{[\varepsilon_0]_{Cu}}$$

where the last expression follows by approximating the scattering cross-section ratio with the ratio of the squares of the atomic numbers. Table 10.5 gives $[\varepsilon_0]_{Cu}$ as 124.4 eV cm^2/10^{15} atoms. Substituting the numbers yields $(Nx)_{Bi} = 93 \times 10^{15}$ atoms/cm^2 $\sim$ 33 nm.

Example 10.12
Calculation of stoichiometry and the amount of Se impurity of a thin SrS(Se) layer on a glass substrate by 12.0 MeV ^{12}C ion backscattering

When high-mass elements must be separated, heavy ions often provide a suitable alternative. In the following example, the Se impurity signal can be resolved from the signal of Sr by using ^{12}C as probing ions at 12 MeV. The backscattering spectrum is shown in Figure 10.10, taken at a scattering angle of 170°. As before, the dots correspond to the experimental data points, and the lines to the calculated simulation.

Extending equation (10.27) to the multi-element target $Sr_mS_n(Se_p)$ we obtain for the relative concentrations in analogy to the thick-target case:

$$\begin{aligned} \frac{m}{n} &= \frac{(Nx)_{Sr}}{(Nx)_{S}} = \frac{A_{Sr}\sigma_{S}(E_0)}{A_{S}\sigma_{Sr}(E_0)} = 1.060 \\ \frac{m}{p} &= \frac{(Nx)_{Sr}}{(Nx)_{Se}} = \frac{A_{Sr}\sigma_{Se}(E_0)}{A_{Se}\sigma_{Sr}(E_0)} = 7.577 \end{aligned} \tag{10.28}$$

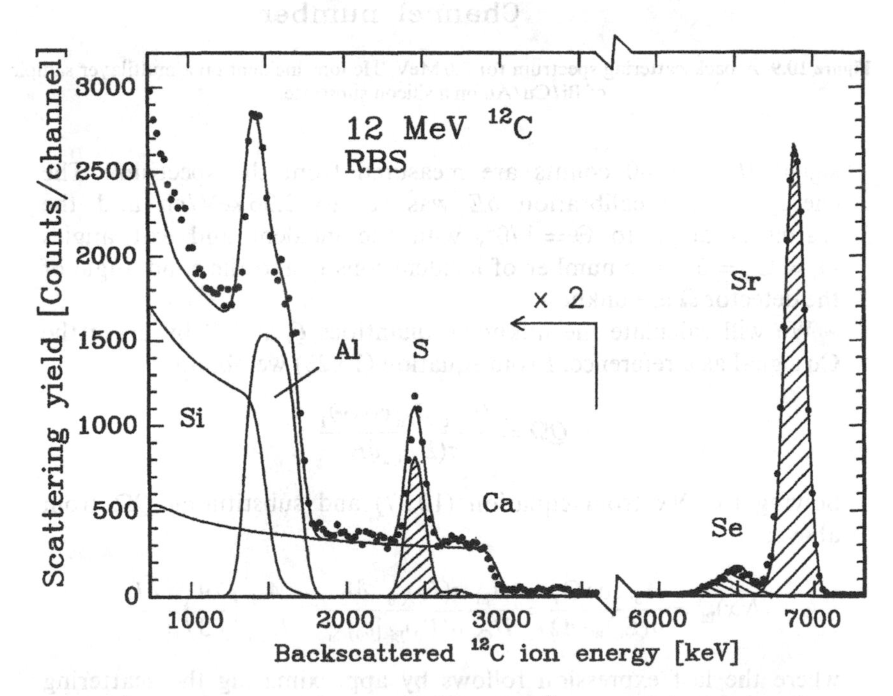

Figure 10.10 A backscattering spectrum for 12.0 MeV ^{12}C ions incident on an SrS film on an aluminum oxide/glass substrate. The film contains a minor impurity of selenium. The counts below 3.6 MeV are multiplied by 2 for clarity. The experimental spectrum (•) and the theoretical simulations (——) are shown.

where the signal areas $A_{Sr} = 19\,000$, $A_S = 2450$ and $A_{Se} = 1990$ have been determined from the spectrum by evaluating the total number of counts under the signal peaks, and the ^{12}C ion scattering cross-sections have been obtained from Table 10.3. For the S signal, the background due to scattering from the glass substrate has been subtracted prior to calculating the peak area.

Taking $m + n + p = 1$ into account we find $m = 0.48$, $n = 0.46$ and $p = 0.06$. The SrS film is thus found to be almost stoichiometric with an Se impurity concentration of about 6 at.%.

To find the film thickness one again needs to know the product $Q\Omega$. This may be obtained by direct measurement or by using a known reference sample, in analogy to Example 10.11.

10.4.2 Elastic recoil detection

Signal width and layer thickness. The depth information in the recoil technique is based on the energy loss of both incident ions and the recoiled particles. The conventional ERD involves the detection of hydrogen isotopes recoiled from the sample by helium-ion bombardment. In addition, an absorber foil is used to stop the incident ions (Figure 10.3). In experiments that make use of high-energy heavy incident ions, the various light or medium-heavy particles recoiling from the sample are detected simultaneously. The absorber foil is then often replaced by elaborate arrangements for particle detection and identification, such as time-of-flight systems, magnetic and electrostatic analyzers, etc. These techniques require complex methods of data analysis and are beyond the scope of the present chapter.

Assume a sample comprising a single-element layer, with a small light-element impurity. Ions heavier than the impurity strike the sample at an angle of Θ_1 with respect to the surface normal and the light ions are emitted at an angle of Θ_2, with the recoil angle defined as $\Theta = \pi - \Theta_1 - \Theta_2$ (Figure 10.5). The depth scale is given by an equation analogous to equation (10.18) for ion backscattering:

$$\Delta E = [\varepsilon_{0,\mathrm{ERD}}]Nx$$
$$[\varepsilon_{0,\mathrm{ERD}}] = \frac{k_R}{\cos\Theta_1}\varepsilon_{\mathrm{in,I}}(E_0) + \frac{1}{\cos\Theta_2}\varepsilon_{\mathrm{out,R}}(k_R E_0) \qquad (10.29)$$

Here $[\varepsilon_{0,\mathrm{ERD}}]$ is the recoil stopping cross-section factor, equivalent to equation (10.19), k_R is the kinematic factor for the elastic recoil process of equation (10.4) and $\varepsilon_{\mathrm{in,I}}(E_0)$ and $\varepsilon_{\mathrm{out,R}}(k_R E_0)$ denote the stopping cross-sections of the **incident ions** on the inward path and **recoiled ions** on the outward path, assuming surface energy approximation.

Example 10.13
Depth scale for a hydrogen ERD spectrum taken with ^{4}He ions incident at 2.0 MeV on a thick carbon sample

Assume a recoil angle of $\Theta = 30°$, with angles of incidence and exit $\Theta_1 = \Theta_2 = 75°$, the corresponding angles with respect to the surface of the sample being 15° (Figure 10.5). For simplicity, assume a minor impurity concentration and ignore the effect of the hydrogen to the energy loss of the ions, thus assuming a pure carbon target in the stopping calculations. The stopping cross-sections $\varepsilon_{in,He}(E_0)$ and $\varepsilon_{out,H}(k_R E_0)$ of equation (10.29), given by equations (10.11), (10.13) and (10.14) are 27.77 and 4.690 eV cm^2/10^{15} atoms, where the recoil kinematic factor is $k_R = 0.482$ and the energy of the recoiled hydrogen from the surface is $k_R E_0 = 964$ keV (equation (10.4)).

Solving for Nx from equation (10.29) and substituting the stopping cross-sections obtained above we have:

$$Nx = \frac{(\cos\Theta_1)\Delta E}{k_R \varepsilon_{in,He}(E_0) + \varepsilon_{out,H}(k_R E_0)}$$

$$= \frac{0.259 \times \Delta E}{(0.482 \times 27.77 + 4.690)\ \text{eV cm}^2/10^{15}\ \text{atoms}}$$

$$= \frac{14.3 \times 10^{15}\ \text{atoms/cm}^2}{\text{keV}} \Delta E$$

$$x = 0.814\ \text{nm} \left(\frac{\Delta E}{\text{keV}}\right)$$

where the last expression follows by dividing by the carbon atomic density $N = 1.76 \times 10^{23}$ atoms/cm^3. Recalling that ΔE is the energy width of the signal in the spectrum, this establishes the depth scale for hydrogen recoils in the surface energy approximation, given the experimental conditions defined above.

For example, hydrogen ions emitting from the sample surface at 764 keV, with an energy 200 keV lower than those recoiled from the sample surface, originate from a depth of about 160 nm.

The calculation above was based on the surface energy approximation and on the assumption of a small hydrogen content. In addition, the effect of the absorber foil was disregarded.

In Example 10.13 the relation between the energy of the hydrogen emitting from the sample surface and the collision depth of the process was introduced. As the absorber foil stops the heavier scattered incident particles, it also slows down the lighter recoiled particles.

The absorber significantly affects the depth scale and the effect cannot be ignored in real data analysis. To take the effect accurately into account, the actual energy dependence of the stopping power would need to be considered. This would involve presuming the depth of the foil to be divided into thin sections and performing a numerical integration of the energy loss over the sections. The procedure can be quite tedious, usually demanding computer methods, especially when multi-elemental absorber foils (Mylar, Havar, etc.) are used. Often such a foil is approximated as an elemental (e.g. carbon) foil and a correction is achieved by adopting some hypothetical mass density. Interpolation tables can also be used (see e.g. Barbour and Doyle 1995). In an approximation, the stopper foil introduces a linear increase to the depth scale of the hydrogen emitting from the sample surface.

In addition to decreasing the recoiled hydrogen energies and increasing the signal widths, an absorber contributes to an increase of the **energy straggling** of the ions (see e.g. Chu *et al.* 1978), deteriorating the energy resolution. This phenomenon is a consequence of the statistical nature of energy loss.

Example 10.14
The effect of the absorber foil on the depth scale of Example 10.13

Assume the sample of Example 10.13, 2.0 MeV incident ^{4}He ions and an Al absorber foil of 8.0 μm thickness, positioned perpendicular to the scattered and recoiled beams in front of the detector, as shown in Figure 10.11. The energy of the ^{4}He ions scattered

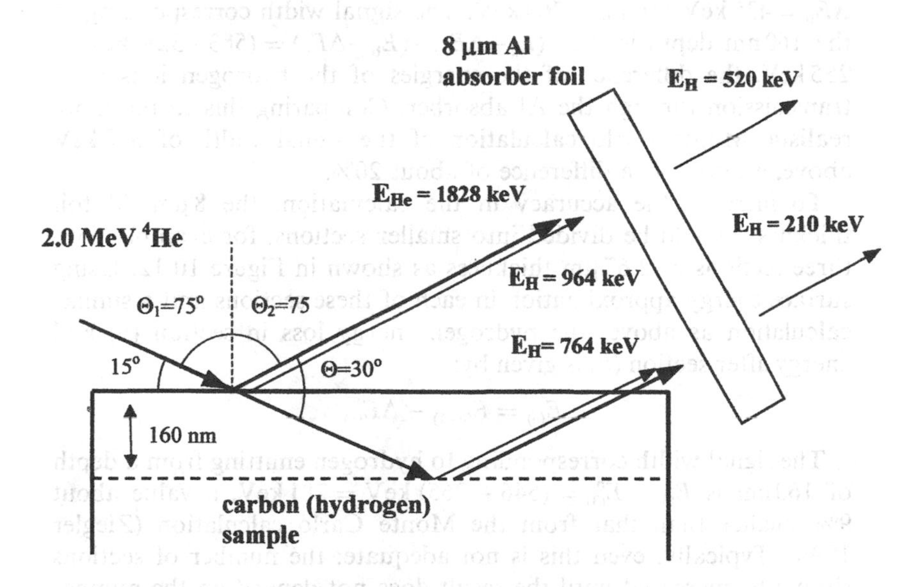

Figure 10.11 A schematic of the kinematics and energy loss of the scattered and recoiled particles from a carbon sample containing a small hydrogen impurity.

at an angle of 30° from the surface of the carbon sample is $E_{1,He} = k_{S,C}E_0 = 1828$ keV (equation (10.1)). These ions are stopped by the Al foil.

The energy of the hydrogen ions recoiling from the surface of the sample through the same angle is $E_{1,H} = k_R E_0 = 964$ keV (equation (10.4)). In the Al absorber foil, these ions suffer an energy loss of about 445 keV (±17 keV, standard deviation of energy straggling), leaving the foil at an average energy of 520 keV, as calculated by a Monte Carlo program (Ziegler 1996). The ions originating from a depth of 160 nm, entering the absorber at an energy of about 764 keV (Example 10.13), lose an energy of about 555 keV (±20 keV) in the foil, leaving the foil at an average energy of 210 keV (Figure 10.11).

The effect of the absorber was thus to extend the energy difference of 200 keV, corresponding to a depth of about 160 nm of the sample, to a signal width of about 310 keV in the ERD spectrum. As an approximation, a linear interpolation or extrapolation to smaller or somewhat larger depths may be performed. Furthermore, the energy of the leading edge is also shifted to a lower energy of 520 keV from 964 keV, corresponding to the energy of the hydrogen recoiled from the sample surface.

Using the surface energy approximation to estimate the absorber effect involves calculating the energy loss of hydrogen in the absorber from $\Delta E = Nx\varepsilon_H(E)$, where E is the energy of the hydrogen entering the foil, x being the Al foil thickness. From equation (10.11) we get $\Delta E_a = 381$ keV for $E_a = 964$ keV and $\Delta E_b = 435$ keV for $E_b = 764$ keV. The signal width corresponding to the 160 nm depth is then $(E_a - \Delta E_a) - (E_b - \Delta E_b) = (583 - 328)$ keV = 255 keV, the difference of the energies of the hydrogen ions after transmission through the Al absorber. Comparing this to the more realistic Monte Carlo calculation of the signal width of 310 keV above, we observe a difference of about 20%.

To increase the accuracy in the calculation, the 8 μm Al foil thickness should be divided into smaller sections; for example, into three sections of 2.67 μm thickness as shown in Figure 10.12. Using surface energy approximation in each of these sections and a similar calculation as above, the hydrogen energy loss in section (n) and energy after section (n) is given by:

$$E_{(n)} = E_{(n-1)} - \Delta E_{(n)}$$

The signal width corresponding to hydrogen emitting from a depth of 160 nm is $E^a_{(3)} - E^b_{(3)} = (546 - 265)$ keV = 281 keV, a value about 9% smaller than that from the Monte Carlo calculation (Ziegler 1996). Typically, even this is not adequate: the number of sections should be increased until the result does not depend on the number

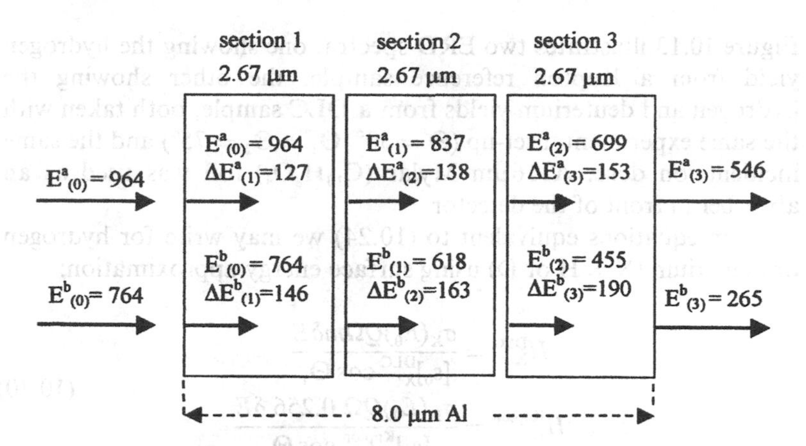

Figure 10.12 The division of the self-supporting 8.0 μm aluminum foil into sections of 2.67 μm thickness. The superscripts a and b refer to particles recoiling from the sample surface and from a depth of 160 nm, respectively. The energies are in keV.

of sections adopted. It may be noted that the simple stopping power calculations adopted here (in section 10.2.3) differ slightly (less than 2%) from those used in Ziegler (1996).

Signal height and amount of material. In hydrogen profiling by elastic recoil spectrometry, the incident helium-ion energy is in practice always above the non-Rutherford energy threshold. This implies that the recoil cross-section cannot be obtained from the Rutherford cross-section of equation (10.6). For heavier ions, however, the Rutherford model can usually be applied, the validity of equation (10.6) depending on the energy of the incident ions and the mass of the target atoms to be detected (Rauhala and Räisänen 1994, Räisänen *et al.* 1994).

The quantitative data analysis of simple ERD experiments is in principle analogous to that by RBS. Equations similar to (10.23), (10.25) and (10.27) may be written also for ERD, where the stopping cross-section factors are replaced by equation (10.29) or its extensions to the multi-element case. When profiling hydrogen with a helium beam, reference samples with known amount of hydrogen are conventionally used, since the cross-sections cannot be calculated. A typical reference sample is Kapton ($C_{22}H_{10}O_5N_2$) with a hydrogen content of 25.6 at.%.

Example 10.15

Calculation of hydrogen and deuterium content at the sample surface from an ERD spectrum for 2.4 MeV ^{4}He ions incident on a diamond-like coating (DLC) film with a small hydrogen impurity

Figure 10.13 illustrates two ERD spectra, one showing the hydrogen yield from a Kapton reference sample, the other showing the hydrogen and deuterium yields from a DLC sample, both taken with the same experimental set-up ($\Theta = 30°$, $\Theta_1 = \Theta_2 = 75°$) and the same incident ion dose. A 16 μm Mylar ($C_{10}H_8O_4$) foil was used as an absorber in front of the detector.

From equations equivalent to (10.24) we may write for hydrogen or deuterium (X = H or D) using surface energy approximation:

$$H_X^{DLC} = \frac{\sigma_X(E_0)Q\Omega m\delta E}{[\varepsilon_0]_X^{DLC}\cos\Theta_1}$$
$$H_H^{Kapton} = \frac{\sigma_H(E_0)Q\Omega\, 0.256\, \delta E}{[\varepsilon_0]_H^{Kapton}\cos\Theta_1} \tag{10.30}$$

where 0.256 is the concentration of hydrogen in Kapton. Taking the ratio of the surface heights of the hydrogen or deuterium signals from DLC and Kapton, H_X^{DLC}/H_H^{Kapton}, the common factors Q, Ω, δE and $\cos\Theta_1$ cancel out. Solving for the hydrogen or deuterium concentration, m, we have:

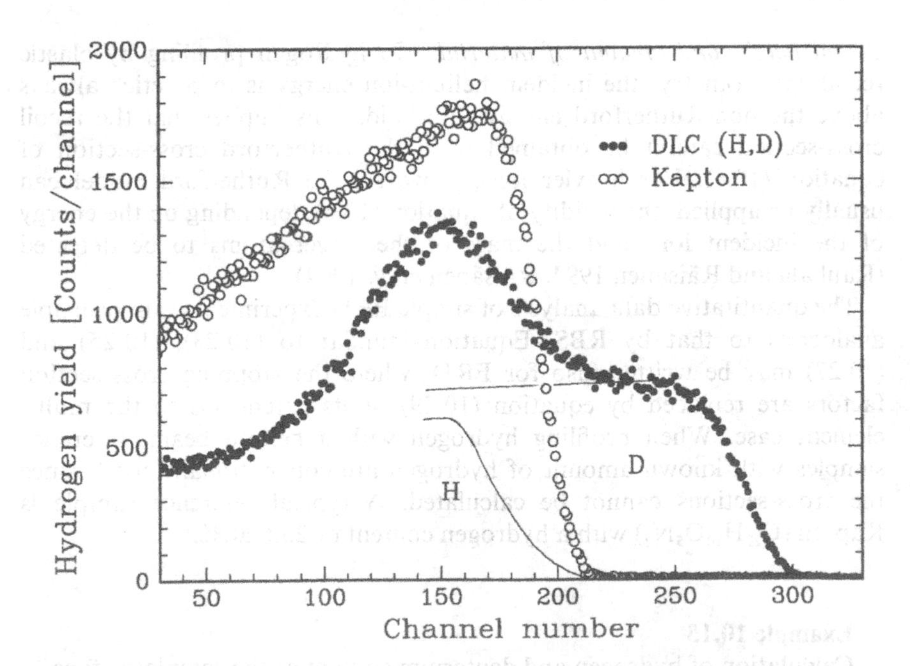

Figure 10.13 Hydrogen and deuterium ERD spectra for 2.0 MeV ^{4}He ions incident on a diamond-like coating material (DLC) (•) and a Kapton foil (o). The hydrogen and deuterium originating from close to the sample surface are approximated by the solid lines.

$$m = \frac{H_X^{DLC}}{H_H^{Kapton}} \frac{\sigma_H(E_0)}{\sigma_X(E_0)} \frac{[\varepsilon_0]_X^{DLC}}{[\varepsilon_0]_H^{Kapton}} 0.256 \approx \frac{\sigma_H(E_0)}{\sigma_X(E_0)} \frac{H_X^{DLC}}{H_H^{Kapton}} 0.256 \qquad (10.31)$$

The last expression is an approximation, assuming the ratio of the recoil stopping cross-section factors as unity.

In Figure 10.13 the hydrogen and deuterium signals are seen to overlap, but approximate surface heights may still be estimated: $H_H^{DLC} = 600$, $H_D^{DLC} = 700$ and $H_H^{Kapton} = 1800$. For hydrogen the recoil cross-section ratio $\sigma_H(E_0)/\sigma_X(E_0)$ equals unity (X = H) and the above approximation yields the hydrogen surface concentration as $m = (600/1800) \times 0.256 = 8.5$ at.%. In order to calculate the deuterium concentration, we need to know the non-Rutherford cross-section ratio. From Baglin *et al.* (1992) and Kellock and Baglin (1993) the cross-sections $\sigma_H = 273$ mb/sr and $\sigma_D = 440$ mb/sr are obtained, giving the ratio $\sigma_H(E_0)/\sigma_D(E_0) = 0.62$. Using X = D and this ratio in equation (10.31) leads to the deuterium concentration of 6 at.%.

Owing to overlapping of the signals and the approximations used, the uncertainty of the results may amount to a few at.%. It should be noted that the absorber does not influence this calculation of impurity content.

10.4.3 Problems

1. Investigate the RBS **detection limit** of a light element in a thick two-component compound or mixture. Assume that a signal of a light element, superposed on a signal of a heavier element may be detected if its surface height is $H_{light} > 3(H_{heavy})^{1/2}$. Take $H_{heavy} = 10^4$ and use equation (10.26) as an approximation, i.e. ignore the effect of the stopping cross-section factors. This simple calculation may be in error by less than about 20%. Verify, for example, that oxygen can be detected in a thick tantalum oxide by 2.0 MeV ^{4}He only if its concentration exceeds 70 at.%.

2. The **accessible depth** in RBS is often defined by using the criterion $E_1 > \frac{1}{4} K_S E_0$, where E_1 and E_0 are the detected and incident energies, respectively. Investigate the accessible depths for ^{4}He projectiles by using Table 10.5. Assume $\Theta = 170°$, incident ^{4}He energies between 1.0 and 3.0 MeV and target elements $Z_2 = 6$, 13, 22, 42, 73 and 79.

3. Using Figure 10.8, find out the layer thickness for the ZnS layer by using the S signal instead of the Zn signal used in Example 10.9.

4. Calculate the Zn and S concentrations from the signal areas by using equation (10.27) for the ZnS sample of Examples 10.9 and 10.10, depicted in Figure 10.8. The total number of counts under the Zn signal

is $A_{Zn} = 340 \times 10^3$. As seen from the theoretical simulation in Figure 10.8, the Ca signal from the glass substrate and the S signal of the film partly overlap. The S signal area is estimated as $A_S = 95 \times 10^3$.

5. The areal densities of the Cu and Au layers in the sample of Figure 10.9 are 4230×10^{15} atoms/cm^2 and 130×10^{15} atoms/cm^2. Try to determine these areal densities from the data given in Example 10.11 and from the spectrum plotted in Figure 10.9. What kind of complications arise in these calculations?
6. Investigate the depth scale of ions penetrating a solid. Find out the dependence of the ion energy and the penetration depth for 2.0 MeV ^{1}H ions in carbon of 20 µm thickness. Assume the thickness divided into small sections and perform numerical integrations. Verify that dividing the 20 µm thickness into six sections results in a total energy loss of the ions of about 1250 keV.
7. Find the areal densities of the components and the total areal density of the SrS layer of Figure 10.10 (Example 10.12). Assume that Q and Ω are known, $Q = 8.8$ µC/e and $\Omega = 5.35$ msr, and use Table 10.3 for the scattering cross-sections. Compare the result to that given in section 10.5.

10.5 Numerical methods

Computer methods for the data analysis of scattering techniques are now widely adopted. In general, the analysis usually consists of simulating the experimental spectra by calculated theoretical spectra, based on the principles outlined above. The goal is a theoretical spectrum closely fitting the experimental one. The simulation typically starts by guessing the sample composition. A theoretical spectrum is then simulated. Successive iterations are used and after each iteration the sample composition is modified to produce a better match between the two spectra. Descriptions of a number of computer programs have been published, many of which can be obtained from the authors (e.g. Doolittle 1985, 1986, Saarilahti and Rauhala 1992, Barradas *et al.* 1997). Various Internet sites provide relevant information for the ion beam analysis community. Sigmabase, for example, at http://ibaserver.physics.isu.edu/sigmabase/ contains data analysis programs for downloading, scattering cross-sections, documents, etc., and links to other nuclear data servers and home pages.

Complex experimental situations cannot be treated adequately without numerical methods. These situations include, for example, samples with many different layers and elements, samples with gradually changing compositions, data from experiments utilizing non-Rutherford cross-sections, data from ERD experiments with complicated detection systems, etc.

Examples of computer data analysis are presented in Figures 10.8 and 10.10. The solid curves 'connecting' the dotted experimental spectra are the theoretical simulations. The quality of the fits reflects the accuracy of analysis. The elemental signals are also shown in both figures. Essentially the same sample parameters were used in the simulations as those procured in Examples 10.9 and 10.12. The signals observable from the glass substrate, depicted in Figure 10.8, correspond to 3 at.% of calcium, 24 at.% of silicon and 10 at.% of sodium. The oxygen signal is not shown.

The analyses above were obtained with a program GISA (Saarilahti and Rauhala 1992) in a few minutes. The input data fed into the program comprise the experimental conditions, the incident ion, the energy, the scattering geometry, the detector energy resolution, the solid acceptance angle, the number of incident ions and the energy/channel calibration. The parameters describing the sample composition were the areal densities and the elemental concentrations of the films. For example, the simulations shown in Figure 10.10 were obtained by using the experimental parameters $\delta E = 23.0$ keV/channel, the energy at zero channel (energy offset) = 160 keV, full width at half-maximum (FWHM) of the detector energy resolution = 90 keV, $\Omega = 5.35$ msr and $Q = 8.8\ \mu C/e$. The areal density of the SrS(Se) layer was 295×10^{15} atoms/cm^2. It could be noted that the two isotopes of sulfur (95% ^{32}S and 4% ^{34}S) are clearly resolved in the theoretical simulation in Figure 10.10.

10.6 Applications to elemental analysis

In the preceding sections the basic principles of the scattering techniques as analytical tools were outlined and a number of practical examples were presented. The capabilities of the scattering techniques as analytical tools, however, extend far beyond these examples. By using channeling techniques, for example, problems such as the characterization of crystal quality, depth profiles and annealing behavior of implantation-induced defects, lattice location of impurities, etc., may be unraveled. In the following, some examples of analytical applications of the techniques published recently are summarized. An extensive review of the current literature on the scattering methods is beyond the scope of the present chapter.

The analytical characteristics of the scattering methods depend on several factors, like the ion species, ion energy, sample matrix, scattering geometry, analysis depth, detector system, etc. Hence, no elementary rules can be given for this topic in a general case. As an example, the depth resolution for 1.0–2.0 MeV ^4He RBS using a standard Si surface barrier detector is typically about 10–30 nm at the sample surface. Isotopic mass resolution extends up to about chlorine. RBS is very sensitive to heavy elements, up to 0.01 atomic layers, but the mass resolution is poor. On the other hand, tens of at.% of light impurities may remain undetected in a heavy matrix (see problem 10.4.3.1). For ERD (H or D profiling by ^4He) a typical depth resolution is of the order of 20–60 nm and the detection sensitivity about 0.5–0.01 at.%. The accessible depths obtained with 2.0 MeV ^4He or ^1H RBS are of the order of 1 and 10 μm, respectively.

Using heavy ions and sophisticated detection systems the mass and depth resolutions are improved considerably. For RBS, an isotopic mass resolution may be achieved even for heavy elements and the depth resolution of a few tenths of a nanometer at the surface may be accomplished. Using high energies and/or heavy ions and replacing the standard Si detectors by more advanced detection systems further extends the analytical capabilities of both RBS and ERD techniques. With high-energy protons or helium projectiles the accessible depths are increased, up to about 50 or 100 μm with 8 MeV ^4He or ^1H, respectively.

The enhanced scattering cross-sections and combinations of various techniques are routinely exploited in ion beam experiments to expand the capabilities of conventional RBS or ERD. An example of an experiment to enhance the detection sensitivity of light elements is described in Tesmer *et al.* (1996). Figure 10.14 illustrates a backscattering spectrum for 3.5 MeV $^4He^+$ incident on a CN_x film on an Si substrate. The CN_x is a technologically interesting material since it has a theoretical bulk modulus comparable with diamond, and may therefore possess a hardness similar to diamond. In characterizing these films, the current experiments involved both high-energy RBS to quantify C and N in the films and ERD to find out the hydrogen content. Figure 10.14 displays the measured spectrum by the + symbols and a simulation using Rutherford cross-sections by a solid line. The inset shows the non-Rutherford simulation (solid line) and the measured C and N signals (+) after background subtraction. The cross-section for C is 5.6 times and for N 2 times the Rutherford cross-section. The simulation corresponds to a sample composition of $C_{0.65}N_{0.35}$ and a film thickness of 44 nm. By using a scattering geometry with $\Theta = 164°$ and a sample tilt of $\Theta_1 = 55°$ (see inset) a depth resolution of 9–10 nm was achieved. The hydrogen content was below 2% as affirmed by an ERD experiment.

To attain the most accurate results, the non-Rutherford cross-sections and stopping powers must be calibrated for the experimental situation used. This may involve careful studies of the calibration standard. Hydrogen may be present in large amounts within the sample, almost without a trace in the RBS spectrum. Second-order effects in the stopping powers, such as bonding effects, may sometimes have a significant

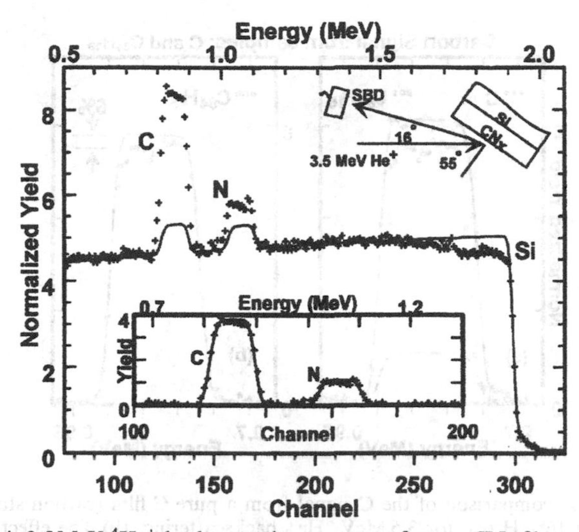

Figure 10.14 A 3.5 MeV $^4He^+$ backscattering spectrum from a CN_x film on an Si substrate. The measured spectrum (+) and a simulation (—) using the Rutherford cross-sections are shown. The sample was tilted 55° relative to the incident beam, $\Theta = 164°$, $\Omega = 6.95$ msr and the incident charge was 20 μC. The inset shows the C and N signals after background subtraction and the simulated spectra using enhanced cross-sections. (Reprinted from *Materials Chemistry and Physics*, **46**, J.R. Tesmer, C.J. Maggiore, M. Nastasi and J.C. Barbour, A review of high energy backscattering spectrometry, pp. 189–197 (1996), with kind permission from Elsevier Science SA, PO Box 564, 1001 Lausanne, Switzerland.)

contribution. Figure 10.15 represents these aspects for the calibration standard used in the experiments of the CN_x films above. In Figure 10.15(a) a spectrum for a pure carbon (+) standard, a $C_{0.84}H_{0.16}$ film, containing 16% hydrogen (o), together with a Rutherford simulation (dashed line) and the non-Rutherford simulation (solid line) are displayed. The heights of the signals from the two films are found to differ significantly. Thus, knowledge of the H content of the standard is important to avoid errors of the order of 10% when calibrating the cross-section. In addition, the stopping power of the film changes with the hydrogen content, which affects the analysis of the film thickness.

Figure 10.15(b) illustrates the possible 5% contribution of chemical bonding (Ziegler and Manoyan 1988, Rauhala 1995) of the carbon–hydrogen samples on the stopping power and thus on the height of the signal. These effects influence both composition and thickness determination in RBS. A similar bonding correction may also affect the analysis of the CN_x films above, and thereby increase the inherent uncertainty in determining the N/C ratio.

Another example of three complementary ion beam techniques is presented by Burkhart and Barbour (1995), who combine Rutherford/non-Rutherford and elastic recoil spectrometries to analyze perovskite-

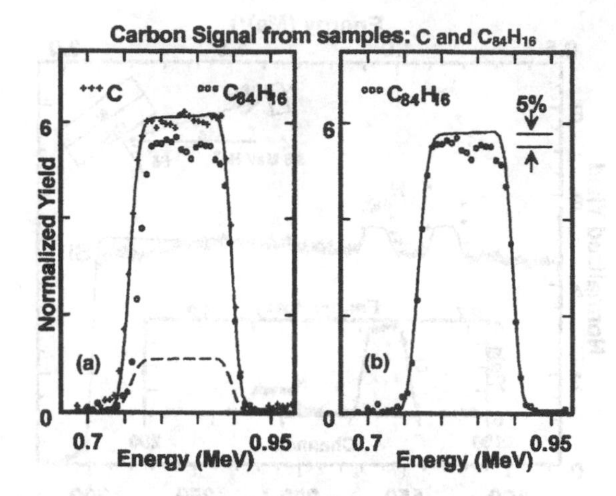

Figure 10.15 (a) A comparison of the C signal from a pure C film (carbon standard) (+) to a film containing 16% H (o), for 3.5 MeV $^4He^+$ backscattering. (b) The effect on the signal height of a 5% correction to the stopping power due to bonding effects. (Reprinted from *Materials Chemistry and Physics*, **46**, J.R. Tesmer, C.J. Maggiore, M. Nastasi and J.C. Barbour, A review of high energy backscattering spectrometry, pp. 189–197 (1996), with kind permission from Elsevier Science SA, PO Box 564, 1001 Lausanne, Switzerland.)

type metal oxides. Figure 10.16 represents backscattering spectra for 2.2 and 8.7 MeV $^4He^+$ ions incident on a $La_xSr_yCoO_z$ film on a $LaAlO_3$ substrate. The low-energy spectrum gave the composition of the heavier elements (Co, Sr and La) and the high-energy spectrum was used to determine the concentrations of the lighter elements. A 30 MeV Si ERD experiment was used to complement the backscattering experiments by revealing the composition of the light elements from H to O. In Figure 10.16 the oxygen signal in the 2.2 MeV spectrum is barely visible, while the signal obtained with 8.7 MeV is 22 times enhanced relative to that calculated by the Rutherford formula. As a result of the analysis, a thickness of 190 nm and a composition of $La_{0.88}Sr_{0.15}CoO_3$ was procured.

Further recent examples of applications of RBS in different fields of materials analysis are, for example, the characterization of standards (Climent-Font *et al.* 1994), an external proton backscattering set-up for the analysis of pigments of paintings (Neelmeijer *et al.* 1995), industrial applications involving the analysis of integrated circuits by micro-beams (Bakhru *et al.* 1995), and the depth profile and dose determination of implanted alloys (Wätjen *et al.* 1996). Resonant or nuclear potential scattering has been exploited by using high-energy ^{4}He ions to analyze silicon carbide (Leavitt *et al.* 1996) and silicon oxide (Watamori *et al.* 1996a,b) films, and many light elements simultaneously (Zhou *et al.* 1995, Markwitz *et al.* 1997).

Heavy-ion ERD is a sensitive technique for the analysis of light elements

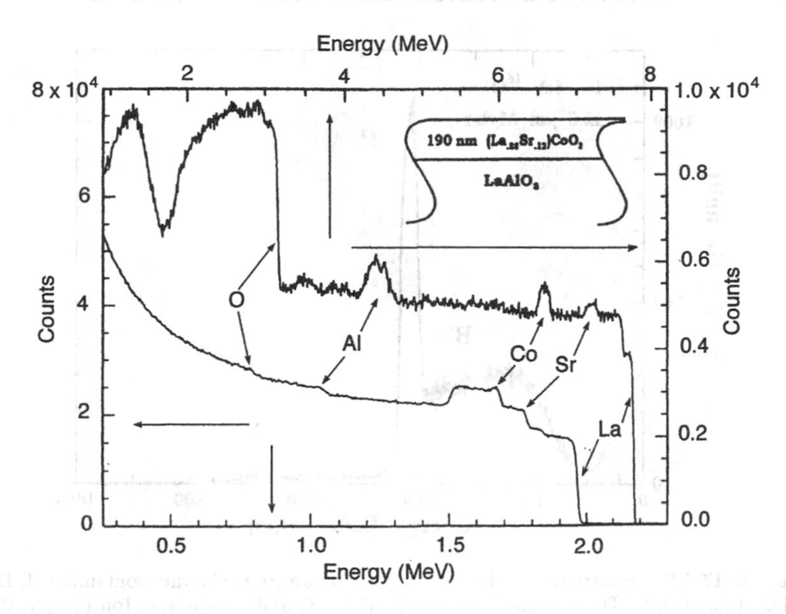

Figure 10.16 Backscattering spectra for 2.2 MeV (bottom) and 8.7 MeV (top) ^{4}He ions incident on an $La_xSr_yCoO_z$ film on a $LaAlO_3$ substrate. The arrows indicate the correct axis for each spectrum. (Reprinted from *Nuclear Instruments and Methods*, **B99**, J.H. Burkhart and J.C. Barbour, Material analysis using combined elastic recoil detection and Rutherford/enhanced Rutherford backscattering spectrometry, pp. 484–487 (1995), with kind permission from Elsevier Science–NL, Sara Burgerhartstraat 25, 1055 KV Amsterdam, The Netherlands.)

such as the isotopes of hydrogen and helium, but also for medium-heavy elements up to a mass number of 100 and above. Goppelt-Langer *et al.* (1996) present ERD profiling of a thin B layer on an iron substrate containing H, D, He and C at the surface by 8.0 and 16.0 MeV ^{16}O and 30 MeV ^{58}Ni. The corresponding ERD spectra are shown in Figures 10.17, 10.18 and 10.19. Conventional experimental set-ups with absorber foils of 6.5–12.5 μm Mylar were used. Simultaneously with ERD, RBS spectra were recorded, showing the signals of the heavier elements present in the sample. The experiments demonstrate that all the elements may be analyzed by a suitable combination of the experimental parameters. In Figure 10.17, taken with 16 MeV ^{16}O, the B layer interferes with H, while D is well separated. Figure 10.18 illustrates the spectrum for the same sample after reducing the energy to 8 MeV. H, D and He are now separated, with the B signal shown at a lower energy. With 30 MeV ^{58}Ni (Figure 10.19), the carbon signal is also visible and all the elements are shown separately. The relative positions of the elemental signals in the spectra are found to depend on the mass and energy of the incident ion and on the thickness of the absorber foil.

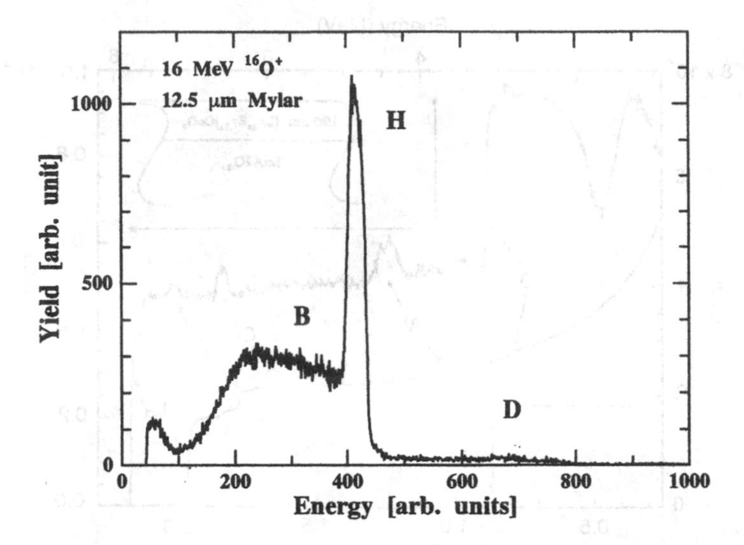

Figure 10.17 ERD spectrum of a thin boron layer on an iron substrate containing H. D, He and C at the surface. The incident beam was 16 MeV ^{16}O and the absorber foil 12.5 μm Mylar. The recoil angle $\Theta = 30°$ and the angle of incidence $\Theta_1 = 75°$. (Reprinted from *Nuclear Instruments and Methods*, **B118**, P. Goppelt-Langer, S. Yamamoto, Y. Aoki, H. Takeshita and H. Naramoto, Light and heavy element profiling using heavy ion beams, pp. 251–255 (1996), with kind permission from Elsevier Science–NL, Sara Burgerhartstraat 25, 1055 KV Amsterdam, The Netherlands.)

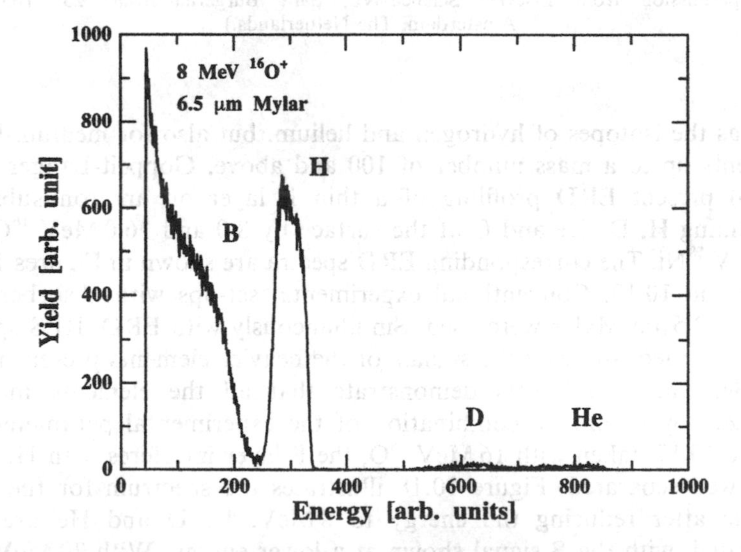

Figure 10.18 ERD spectrum of the sample of Figure 10.17. The incident beam was 8 MeV ^{16}O and the absorber foil 6.5 μm Mylar. The angles are as in Figure 10.17. (Reprinted from *Nuclear Instruments and Methods*, **B118**, P. Goppelt-Langer, S. Yamamoto, Y. Aoki, H. Takeshita and H. Naramoto, Light and heavy element profiling using heavy ion beams, pp. 251–255 (1996), with kind permission from Elsevier Science–NL, Sara Burgerhartstraat 25, 1055 KV Amsterdam, The Netherlands.)

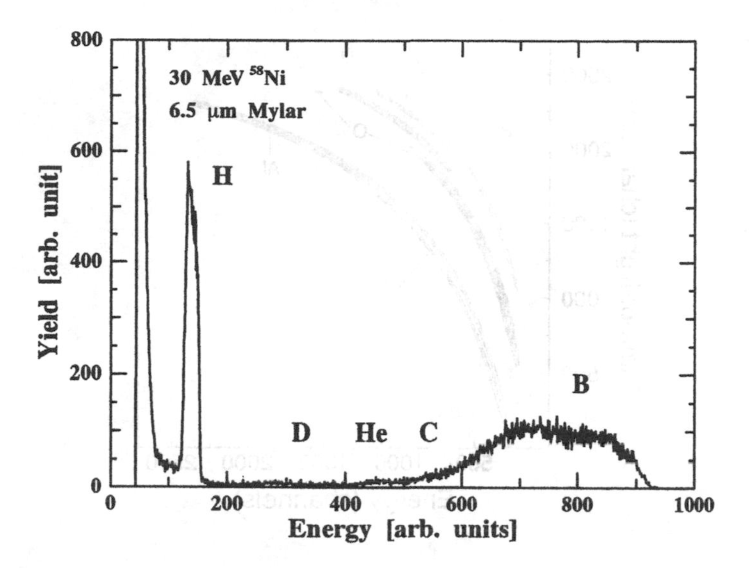

Figure 10.19 ERD spectrum of the sample of Figures 10.17 and 10.18. The incident beam was 30 MeV ^{58}Ni. The absorber and angles are as in Figure 10.18. (Reprinted from *Nuclear Instruments and Methods*, **B118**, P. Goppelt-Langer, S. Yamamoto, Y. Aoki, H. Takeshita and H. Naramoto, Light and heavy element profiling using heavy ion beams, pp. 251–255 (1996), with kind permission from Elsevier Science–NL, Sara Burgerhartstraat 25, 1055 KV Amsterdam, The Netherlands.)

A time-of-flight (TOF) detector system for ERD has recently been described by Jokinen (1997). Other advanced experimental systems in ERD are, for example, spectrometers employing magnetic and electric fields (e.g. Arnoldbik *et al.* 1996) and set-ups with transmission detectors placed in front of the energy detector (ΔE–E detectors) (e.g. Wielunski *et al.* 1997). In these experimental set-ups the absorber is omitted.

In a TOF detector, the flight time over a known flight path and the energy of the recoiling elements from the sample are recorded. These yield information on both the energy and mass of the particles. By using such a detector, a significantly better mass resolution is achieved as compared to conventional ERD analysis.

Figure 10.20 shows the measured energy–TOF data for an AlN thin film on a borosilicate glass (Jokinen *et al.* 1996). The primary ions were 37 MeV $^{197}Au^{7+}$. The signals from boron, carbon, nitrogen, oxygen, aluminum and silicon are all well separated in the figure. From these data the concentration–depth profiles of Figure 10.21 were derived. The hydrogen profile was obtained by using a $^{1}H(^{15}N,\alpha\gamma)^{12}C$ reaction. A stoichiometry of $Al_{0.49}N_{0.44}C_{0.07}$ was obtained and an uncertainty of the concentrations of 5%, due to possible errors in the stopping powers, was estimated.

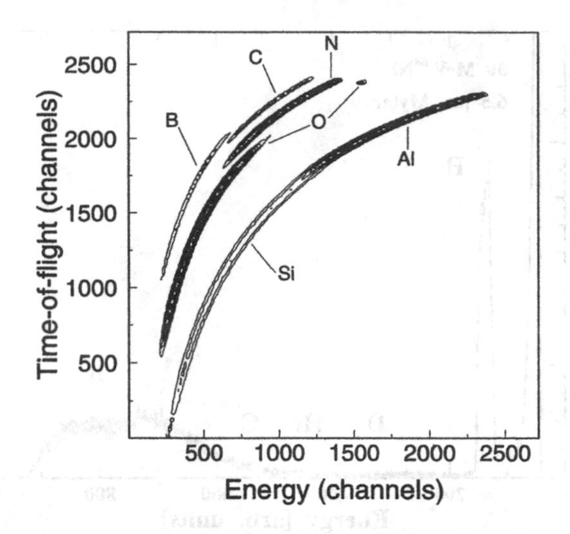

Figure 10.20 An energy–TOF spectrum for 37 MeV $^{197}Au^{7+}$ ions incident on a sample of an AlN film on a borosilicate glass. (Reprinted from *Thin Solid Films*, **289**, J. Jokinen, P. Haussalo, J. Keinonen, M. Ritala, D. Riihelä and M. Leskelä, Analysis of AlN thin films by combining TOF–ERDA and NRB techniques, pp. 159–165 (1996), with kind permission from Elsevier Science SA, PO Box 564, 1001 Lausanne, Switzerland.)

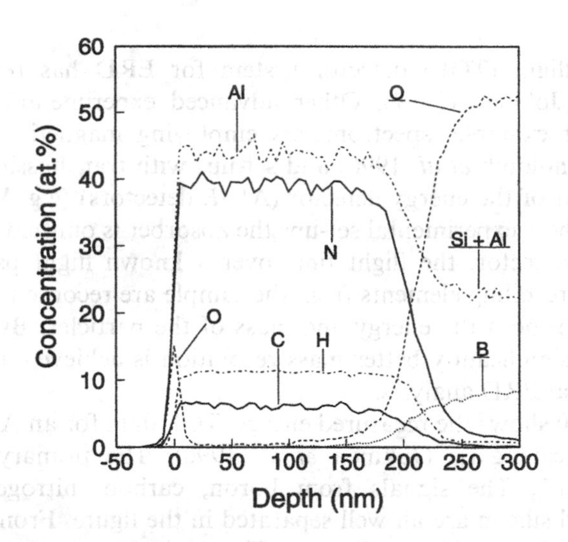

Figure 10.21 The concentration–depth profiles from TOF–ERD measurements as extracted from the data of Figure 10.20. (Reprinted from *Thin Solid Films*, **289**, J. Jokinen, P. Haussalo, J. Keinonen, M. Ritala, D. Riihelä and M. Leskelä, Analysis of AlN thin films by combining TOF–ERDA and NRB techniques, pp. 159–165 (1996), with kind permission from Elsevier Science SA, PO Box 564, 1001 Lausanne, Switzerland.)

High-energy heavy ions render the analysis of heavy constituents in the sample possible. Other recent applications using TOF–ERD systems are, for example, the analysis of $Al_xGa_{1-x}As$ (Walker *et al.* 1994) and $Ba_xSr_{1-x}TiO_3$ (Stannard *et al.* 1995) films by 77 and 98 MeV $^{127}I^{10+}$ ions. Dytlewski *et al.* (1996) and Gujrathi *et al.* (1996) have investigated high-dose metal-implanted germanium and silicon-compound multilayer and graded-index optical coatings, respectively. Even higher energies, up to 200 MeV, and ^{127}I and ^{197}Au ions and various detection techniques have been utilized to study different analytical problems by Siegele *et al.* (1994, 1996), Dollinger *et al.* (1996) and Assmann *et al.* (1996). These authors also investigate the methodology even more generally.

Resonance backscattering by the reaction $^{12}C(\alpha,\alpha)^{12}C$ at $E_C = 4.265$ MeV and ERD by 24 MeV Si ions for the quantification of carbon in SiGeC layers have been compared by Bair *et al.* (1996). Diamond-like carbon films have been analyzed by both RBS and ERD by Hirvonen *et al.* (1996).

Theoretical considerations for the depth resolution calculations in the RBS and ERD techniques and numerical algorithms in the quantification of ERD have been presented by Szilágyi *et al.* (1995) and Bergmaier *et al.* (1995), respectively. Conditions to obtain monolayer resolution in RBS have been studied by Kimura *et al.* (1995), and Wielunski (1996) has investigated the multiple and double scattering contributions to depth resolution and the low-energy background in hydrogen ERD analysis. The modification of the sample by the analysis beam, and hydrogen desorption induced by heavy ions during 130–200 MeV nickel- and iodine-ion ERD analysis has been studied by Behrish *et al.* (1996).

References

Arnoldbik, W.M., Wolfswinkel, W., Inia, D.K., Verleun, V.C.G., Lobner, S., Reinders, J.A., Labohm, F. and Boerma, D.O. (1996), *Nucl. Instrum. Methods* **B118**, 566–572.

Assmann, W., Davies, J.A., Dollinger, G., Forster, J.S., Huber, H., Reichelt, Th. and Siegele, R. (1996), *Nucl. Instrum. Methods* **B118**, 242–250.

Baglin, J.E.E., Kellock, A.J., Crockett, M.A. and Shih, A.H. (1992), *Nucl. Instrum. Methods* **B64**, 469–474.

Bair, A.E., Atzmon, Z., Russell, S.W., Barbour, J.C., Alford, T.L. and Mayer, J.W. (1996), *Nucl. Instrum. Methods* **B118**, 274–277.

Bakhru, H., Nickles, E. and Haberl, A.W. (1995), *Nucl. Instrum. Methods* **B99**, 410–413.

Barbour, J.C. and Doyle, B.L. (1995), in *Handbook of Modern Ion Beam Materials Analysis* (eds Tesmer, J.R. and Nastasi, M.), Materials Research Society, Pittsburgh, pp. 83–138.

Barradas, N.P., Jeynes, C. and Webb, R.P. (1997), *Appl. Phys. Lett.* **71**(2), 291–293.

Behrish, R., Prozesky, V.M., Huber, H. and Assmann, W. (1996), *Nucl. Instrum. Methods* **B118**, 262–267.

Bergmaier, A., Dollinger, G. and Frey, C.M. (1995), *Nucl. Instrum. Methods* **B99**, 488–490.

Bozoian, M. (1993), *Nucl. Instrum. Methods* **B82**, 602–603.

Burkhart, J.H. and Barbour, J.C. (1995), *Nucl. Instrum. Methods* **B99**, 484–487.

Chu, W.K., Mayer, J.W. and Nicolet, M.-A. (1978), *Backscattering Spectrometry*, Academic Press, New York.

Climent-Font, A., Fernández-Jiminéz, M.T., Wätjen U. and Perrière, J. (1994), *Nucl. Instrum. Methods* **A353**, 575–578.
Dollinger, G., Boulouednine, M., Bergmaier, A., Faestermann, T. and Frey, C.M. (1996), *Nucl. Instrum. Methods* **B118**, 291–300.
Doolittle, L.R. (1985), *Nucl. Instrum. Methods* **B9**, 344–351. Program 'RUMP' for the simulation and analysis of RBS and ERD spectra, see http://www. genplot.com
Doolittle, L.R. (1986), *Nucl. Instrum. Methods* **B15**, 227–231.
Dytlewski, N., Evans, P.J., Noorman, J.T., Wielunski, L.S. and Bunder, J. (1996), *Nucl. Instrum. Methods* **B118**, 278–282.
Goppelt-Langer, P., Yamamoto, S., Aoki, Y., Takeshita, H. and Naramoto, H. (1996), *Nucl. Instrum. Methods* **B118**, 251–255.
Gujrathi, S.C., Poitras, D., Klemberg-Sapieha, J.E. and Martinu, L. (1996), *Nucl. Instrum. Methods* **B118**, 560–565.
Hirvonen, J.-P., Koskinen, J., Torri, P., Lappalainen, R. and Anttila, A. (1996), *Nucl. Instrum. Methods* **B118**, 596–601.
Jokinen, J. (1997), *Acta Polytech. Scand.* **212**, 1–38.
Jokinen, J., Haussalo, P., Keinonen, J., Ritala, M., Riihelä, D. and Leskelä, M. (1996), *Thin Solid Films* **289**, 159–165.
Kellock, A.J. and Baglin, J.E.E. (1993), *Nucl. Instrum. Methods* **B79**, 493–497.
Kimura, K., Ohshima, K., Nakajima, N., Fujii, Y., Mannami, M. and Gossmann, H.-J. (1995), *Nucl. Instrum. Methods* **B99**, 472–475.
Leavitt, J.A., McIntyre, L.C., Jr. and Weller, M.R. (1995), in *Handbook of Modern Ion Beam Materials Analysis* (eds Tesmer, J.R. and Nastasi, M.), Materials Research Society, Pittsburgh, pp. 37–81.
Leavitt, J.A., McIntyre, L.C., Ashbaugh, M.D., Cox, R.P., Lin, Z. and Gregory, R.B. (1996), *Nucl. Instrum. Methods* **B118**, 613–616.
Markwitz, A., Ruvalcaba-Sil, J.L. and Demortier, G. (1997), *Nucl. Instrum. Methods* **B122**, 685–688.
Neelmeijer, C., Wagner, W., Schramm, H.P. and Thiel, U. (1995), *Nucl. Instrum. Methods* **B99**, 390–393.
Nurmela, A., Zazubovich, V., Räisänen, J., Rauhala, E. and Lappalainen, R. (1998), *J. Appl. Phys.* submitted.
Rauhala, E. (1992), in *Elemental Analysis by Particle Accelerators* (eds Alfassi, Z.B. and Peisach, M.), CRC Press, Boca Raton, pp. 179–241.
Rauhala, E. (1994), in *Chemical Analysis by Nuclear Methods* (ed. Alfassi, Z.B.), John Wiley, New York, pp. 253–291.
Rauhala, E. (1995), in *Handbook of Modern Ion Beam Materials Analysis* (eds Tesmer, J.R. and Nastasi, M.), Materials Research Society, Pittsburgh, pp. 3–19.
Rauhala, E. and Räisänen, J. (1994), *J. Appl. Phys.* **75**, 642–644.
Räisänen, J. and Rauhala, E. (1995), *J. Appl. Phys.* **77**, 1762–1765.
Räisänen, J., Rauhala, E., Knox, J.M. and Harmon, J.F. (1994), *J. Appl. Phys.* **75**, 3273–3276.
Saarilahti, J. and Rauhala, E. (1992), *Nucl. Instrum. Methods* **B64**, 734–738. Program 'GISA' for the simulation and analysis of RBS spectra, contact Jaakko.Saarilahti@VTT.Fi
Siegele, R., Davies, J.A., Forster, J.S. and Andrews, H.R. (1994), *Nucl. Instrum. Methods* **B90**, 606–610.
Siegele, R., Assmann, W., Davies, J.A. and Forster, J.S. (1996), *Nucl. Instrum. Methods* **B118**, 283–290.
Stannard, W.B., Johnston, P.N., Walker, S.R., Bubb, I.F., Scott, J.F., Cohen, D.D., Dytlewski, N. and Martin, J.W. (1995), *Nucl. Instrum. Methods* **B99**, 447–449.
Szilágyi, E., Pászti, F. and Amsel, G. (1995), *Nucl. Instrum. Methods* B100, 103–121.
Tesmer, J.R., Maggiore, C.J., Nastasi, M. and Barbour, J.C. (1996) *Mater. Chem. Phys.* **46**, 189–197.
Tirira, J., Serryus, Y. and Trocellier, P. (1996), *Forward Recoil Spectrometry*, Plenum, New York.
Walker, S.R., Johnston, P.N., Bubb, I.F., Stannard, W.B., Cohen, D.D., Dytlewski, N., Hult, M., Whitlow, H.J., Zaring, C., Östling, M. and Andersson, M. (1994), *Nucl. Instrum. Methods* **A353**, 563–567.

Watamori, M., Oura, K., Hirao, T. and Sasabe, K. (1996a), *Nucl. Instrum. Methods* **B118**, 228–232.

Watamori, M., Oura, K., Hirao, T. and Sasabe, K. (1996b), *Nucl. Instrum. Methods* **B118**, 233–237.

Wätjen, U., Bax, H. and Räisänen, J. (1996), *Nucl. Instrum. Methods* **B118**, 676–680.

Wielunski, L.S. (1996), *Nucl. Instrum. Methods* **B118**, 256–261.

Wielunski, M., Mayer, M., Behrisch, R., Roth, J. and Scherzer, B.M.U. (1997), *Nucl. Instrum. Methods* **B122**, 113–120.

Zhou, Z.Y., Zhou, Y.Y., Zhang, Y., Xu, W.D., Zhao, G.Q., Tang, J.Y. and Yang, F.J. (1995), *Nucl. Instrum. Methods* **B100**, 524–528.

Ziegler, J.F. (1980), *Handbook of Stopping Cross-Sections for Energetic Ions in All Elements*, Pergamon Press, New York.

Ziegler, J.F. (1996), SRIM-96 computer program (version 96.01), private communication.

Ziegler, J.F. and Biersack, J.P. (1991), TRIM-91 computer program, private communication.

Ziegler, J.F. and Manoyan, J.M. (1988), *Nucl. Instrum. Methods* **B35**, 215–228.

Ziegler, J.F., Biersack, J.P. and Littmark U. (1985), *The Stopping and Range of Ions in Solids*, vol. 1, Pergamon Press, New York.

11 Elemental analysis of surfaces

M. POLAK

11.1 Introduction: overview of surface phenomena and major techniques

Atoms composing the surfaces of solids have different coordination numbers and bonding compared to bulk atoms, often leading to distinct surface structures and composition. Surface reconstructions and atomic redistributions have been observed often in single-crystal solids under clean (ultra-high-vacuum, UHV) conditions. Furthermore, being the first to encounter any interactions with an external environment, the highly reactive surface atoms (especially in a metal) quite rapidly adsorb gas molecules. This can lead to structural reconstruction and compositional redistribution (e.g. of alloy constituents). Thus, the chemical composition of the surface region in a multicomponent material usually differs from the bulk composition as a result of 'surface segregation', either of intrinsic nature, or extrinsically induced by preferential chemical interactions with adsorbed species. The chemical identity and distribution of the superficial layer or contamination on mono- and multi-element solids are of interest as well. Diverse processes can occur at thin or ultra-thin film surfaces and interfaces, such as segregation and interdiffusion in semiconductor heterostructures.

Surface compositional variations, which often extend beneath the outermost surface layer, affect various macroscopic properties of solids. Corrosion resistance, catalytic activity, friction, adhesion, electrical and magnetic properties are just a few examples. Over the years this has motivated the development of a variety of surface-sensitive techniques applied to material issues in metallurgy, thin-film semiconductors, ceramics and polymers. Thus, in addition to basic research on well-defined samples, including mechanisms of gas adsorption or surface segregation and reactions on atomically clean single-crystal surfaces, some techniques have been very useful also in 'real-world' surface analysis, such as contamination identification.

In particular, **Auger electron spectroscopy** (AES), **X-ray photoelectron spectroscopy** (XPS or ESCA) and **secondary-ion mass spectrometry** (SIMS) are the three most widely used techniques for surface chemical analysis, and therefore have been chosen for this chapter. Probably the category of practical applications, which also includes trouble-shooting and quality control, has contributed the most to remarkable developments, experimental and instrumental as well as theoretical, in these methods (Briggs and Seah 1990, 1992).

AES, XPS and SIMS use monoenergetic electrons, photons and ions, respectively, to excite the surface atoms (see Figure 11.1). The experiment is typically carried out under UHV conditions (10^{-9} to 10^{-10} Torr), mainly to avoid surface contamination and modification by residual gases in the test chamber (it sets limits on the analysis acquisition times). Each technique is named after the particle emitted after the excitation, which is then analyzed and detected. In AES and XPS characteristic electron energies furnish absolute identification of all elements except H and He, while in SIMS *all* elements (and isotopes) can be identified according to their atomic mass number. Prominent advantages of these methods include fast multi-element simultaneous analysis via commonly well-separated spectral lines, as well as chemical-state information accessible through line shifts or shape changes (in XPS and AES), and characteristic molecular-cluster mass numbers in SIMS.

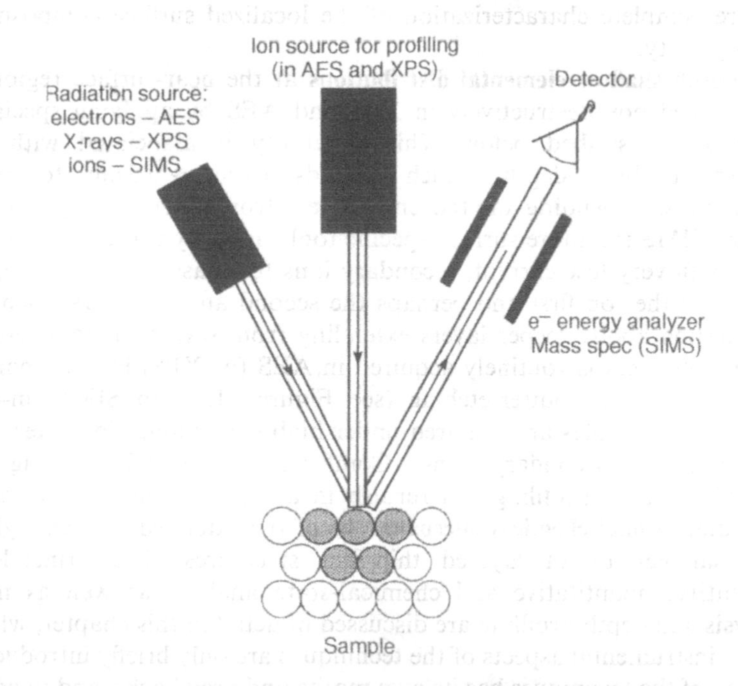

Figure 11.1 Schematic of the basic experimental set-up for surface chemical analyses by AES, XPS or SIMS. The heart of each apparatus is the 'spectrometer': an electron kinetic-energy analyzer ('filter') in AES/XPS, or a 'mass filter' in SIMS. In-depth composition profiling in SIMS does not require an additional ion gun.

Among the three techniques, only SIMS is sensitive enough for trace-element detection (in the ppm–ppb range). However, if a trace analyte originally present in a solid, gas or solution is deposited as a very thin film on a solid surface, AES and XPS, with their intrinsic high absolute detection power, become suitable for trace analysis with good accuracy and precision (Polak 1994).

An important aspect of modern surface analysis to be addressed in this chapter is the capability to analyze spatial chemical heterogeneities in a sample, including both lateral and in-depth compositional gradients. The first capability, **surface microanalysis**, is often used in order to shed light on localized surface-related phenomena, and it can be achieved by using focused beams to excite atoms in a restricted, small area of the sample. This capability is highly important in quality control of modern semiconductor devices with submicrometer defects or surface particulate contamination, for which AES (using state-of-the-art instrumentation) usually surpasses the capabilities of all other techniques (such as electron probe microanalysis, EPMA). Two-dimensional elemental imaging (mapping) obtained, for example, by scanning the primary beam used in the three techniques (electrons, photons or ions) over the surface provides a more complete characterization of the localized surface compositional heterogeneity.

In-depth **shallow elemental distributions** at the near-surface region can be obtained non-destructively in XPS and AES by applying specialized procedures described below. This capability is associated with their inherent probing depth, which extends from ~ 0.5 nm to several nanometers, depending on the emitted electron kinetic energy and the matrix. SIMS is a more surface-specific tool, since, by using a primary-ion beam with very low current, secondary ions for mass analysis are ejected only from the top first and perhaps the second atomic layers. Elemental information about deeper layers extending from several nanometers to a few micrometers is routinely acquired in AES (or XPS) by incorporating simultaneous ion sputter-etching (see Figure 11.1). In SIMS, in-depth composition profiles are acquired under high enough ejection rates of the mass-analyzed secondary ions. Combining elemental imaging with controlled depth profiling can furnish in all three techniques a detailed **three-dimensional chemical microanalysis** of the extended surface region in bulk samples or of layered thin-film structures. The principles of qualitative, quantitative and chemical-state analysis as well as micro-analysis and depth profiling are discussed in detail in this chapter, whereas purely instrumental aspects of the techniques are only briefly introduced.

Each of the techniques has its own merits and drawbacks, and to achieve a more complete and unambiguous solution of a complex surface problem, a combination of two or more complementary techniques is desirable. For instance, the synergistic advantages of XPS and SIMS in organic and polymeric applications, such as characterization of a modified surface or examination of structure–property relationships in biomaterials, can be combined to improve the interpretation of surface chemical features. When atomic structural features are of interest too (e.g. in single-crystal surface adsorption or segregation), one can choose from another class of techniques, such as low-energy electron diffraction (LEED) or scanning tunneling microscopy (STM), in order to elucidate the chemical information achievable by AES, XPS or SIMS. Another surface-specific instrumental technique, ion scattering spectroscopy (ISS), can provide multi-element chemical analysis as well as structural information. However, since ISS is much less commonly used, it is not included in this chapter.

11.2 Auger electron spectroscopy and X-ray photoelectron spectroscopy

11.2.1 General principles

Since their advent (around 1967) AES and XPS have become the most common and powerful tools for surface chemical analysis. Both techniques are based on the energy analysis of the slow electrons that are emitted from atoms excited by an X-ray or electron beam in XPS and AES, respectively. Since in both spectroscopies peak overlap is quite rare (and can be overcome by appropriate deconvolution procedures), simultaneous multi-element analysis is straightforwardly feasible. Other merits include the availability and the relatively easy use of equipment, and the ability to analyze 'real surfaces', without the need for any special sample preparation. Furthermore, vast amounts of support data have been accumulated during the years and are available in various handbooks (Davis *et al.* 1978, Wagner *et al.* 1979), textbooks (Briggs and Seah 1990, 1992) and computerized databases (Yoshitake and Yoshihara 1997, Seah and Gilmore 1996).

Three basic stages are involved in the XPS and AES experiments (see Figure 11.1):

1. Excitation of sample atoms resulting in the emission of characteristic (relatively slow) electrons from the surface region.
2. Kinetic-energy measurements of the emitted electrons by an electrostatic analyzer.
3. Detection and amplification of the electron signals.

Thus, during a continuous irradiation of the sample, an **electron spectrum** is recorded, and of particular interest is the discrete energy distribution of these backwardly emitted electrons.

The well-known fundamental physical phenomena behind the first stage

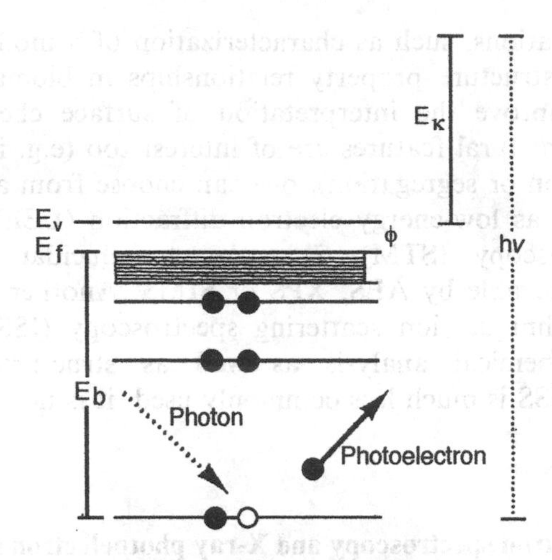

Figure 11.2 The basic principle of photoelectron (photoemission) spectroscopy. The electron binding energies (E_b) are usually referenced to the Fermi level (E_f) of the solid (a metal in this illustration); E_v – the 'vacuum level'.

are the **photoionization effect** in the case of XPS (Figure 11.2) and the **Auger process** in AES (described below). In particular, in XPS the energy of a photon ($h\nu$) is transferred to an electron in an atomic orbital having binding energy E_b. If $h\nu$ exceeds E_b (plus the few eV workfunction, ϕ), the electron is emitted from the solid surface with characteristic kinetic energy

$$E_k = h\nu - E_b - \phi \tag{11.1}$$

Since the energies of the atomic core levels (see Table 11.1) are unique to each element (Figure 11.3), and $h\nu$ is fixed in XPS (typically Mg or Al Kα radiation), measurements of photoelectron kinetic energies constitute a

Table 11.1 Quantum numbers and notations for hydrogen-like wavefunctions in the *j–j* spin–orbit coupling approximation

n	*l*	*j*	Spectroscopic notation	X-ray notation
1	0	1/2	1s	K
2	0	1/2	2s	L_1
2	1	1/2	$2p_{1/2}$	L_2
2	1	3/2	$2p_{3/2}$	L_3
3	0	1/2	3s	M_1
3	1	1/2	$3p_{1/2}$	M_2
3	1	3/2	$3p_{3/2}$	M_3
3	2	3/2	$3d_{3/2}$	M_4
3	2	5/2	$3d_{5/2}$	M_5

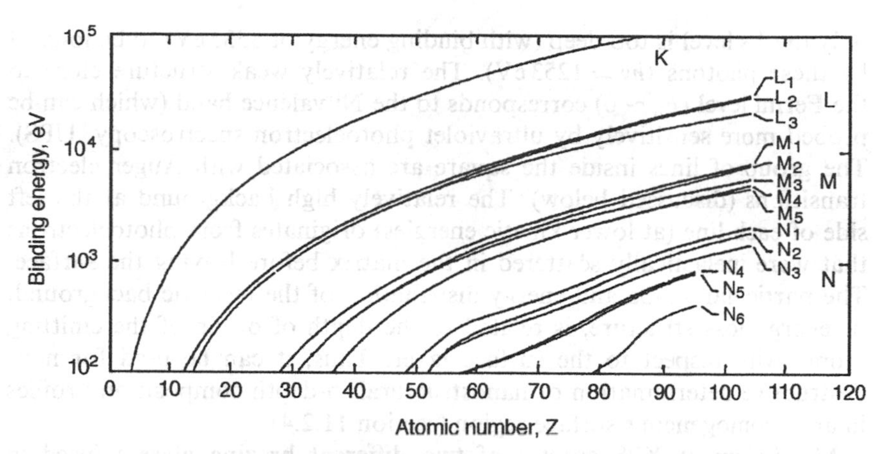

Figure 11.3 The binding-energy variations of the K, L, M and N core-level electrons as function of the elemental atomic number. (Reprinted from Feldman and Mayer (1986), with kind permission.)

straightforward means to identify the elements (with $Z > 2$) composing the surface.

As a first example, the XPS spectrum of pure nickel taken with Mg Kα radiation is shown in Figure 11.4. Among the core levels of this element,

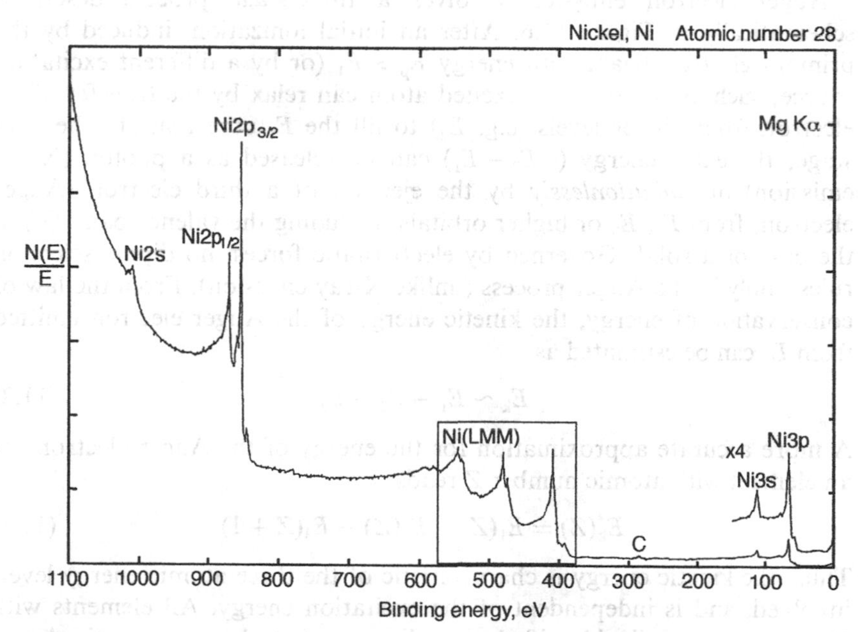

Figure 11.4 The X-ray photoelectron spectrum of nickel. (From Wagner *et al.* (1979), with permission.) Accurate binding-energy determination requires a higher-resolution spectrum (with expanded energy scale).

only the 1 s level is too deep (with binding energy of 8333 eV) to be ionized by these photons ($h\nu = 1253$ eV). The relatively weak structure close to the Fermi level ($E_b \sim 0$) corresponds to the Ni valence band (which can be probed more sensitively by ultraviolet photoelectron spectroscopy, UPS). The group of lines inside the square are associated with Auger electron transitions (discussed below). The relatively high background at the left side of each line (at lower kinetic energies) originates from photoelectrons that were inelastically scattered in the matrix before leaving the surface. The particular shape and energy distribution of the inelastic background, or energy loss structure, is related to the depth of origin of the emitting atoms with respect to the surface layer. Thus, it can be used for non-destructive determination of nanostructural in-depth composition profiles in an inhomogeneous surface region (section 11.2.4).

Multi-element XPS spectra of two different brazing glasses (used in battery fabrication) are shown in Figure 11.5. Since one of them did not satisfy the required brazing conditions, differences in cation composition were suspected, and were clearly revealed by XPS: Glass A contains Ca, B and Al not present in glass B, and the relative concentrations of Ba, Si, K and Na, as reflected by the relative line intensities, are different in the two glass materials (the indium signal came from the foil in which the glass powder had to be embedded for analysis).

Auger electron emission involves a three-stage process described schematically in Figure 11.6. After an initial ionization, induced by the primary electron beam with energy $E_p > E_1$ (or by a different excitation source, such as X-rays), the excited atom can relax by the transfer of an electron from higher levels (e.g. E_2) to fill the E_1 core hole. In the final stage, the extra energy ($\sim E_2 - E_1$) can be released as a photon (X-ray emission) or *radiationlessly* by the ejection of a third electron (Auger electron) from E_2, E_3 or higher orbitals, including the valence band (V) in the case of a solid. Governed by electrostatic forces, no dipole selection rules apply in the Auger process (unlike X-ray emission). From the law of conservation of energy, the kinetic energy of the Auger electron emitted from E_3 can be estimated as

$$E_k \sim E_1 - E_2 - E_3 \tag{11.2}$$

A more accurate approximation for the energy of the Auger electron for an element with atomic number Z reads

$$E_k(Z) = E_1(Z) - E_2(Z) - E_3(Z+1) \tag{11.3}$$

Thus, the kinetic energy is characteristic of the three atomic energy levels involved, and is independent of the excitation energy. All elements with $Z > 2$ can be easily identified according to several characteristic Auger emission lines (Figure 11.7). For example, according to Table 11.1, six KLL transitions are expected to occur (KL_1L_1, KL_1L_2, KL_1L_3, KL_2L_3,

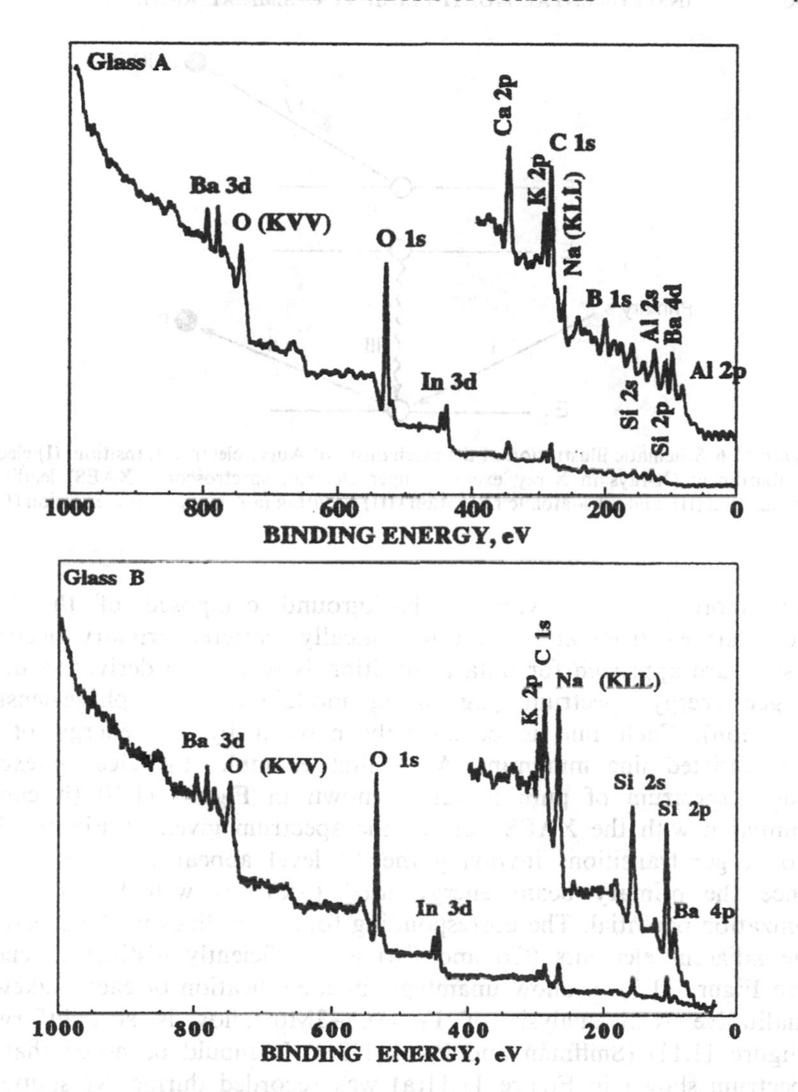

Figure 11.5 Application of XPS to the bulk composition analysis of two different glass materials. Each survey spectrum was recorded after short argon sputtering aimed to remove surface contamination and modified layers. (Courtesy of Analytical Services, The Laboratory of Surface Science, Ben-Gurion University of the Negev, Beer-Sheeva, Israel.)

KL_2L_2 and KL_3L_3). However, pure j–j coupling holds only for $Z > 80$, pure L–S coupling for $Z < 20$ (also six transitions), whereas for the rest of the elements nine KLL lines are observed ('intermediate coupling').

The general spectrum of electrons emitted from an electron-bombarded solid (Figure 11.8) is shown schematically in Figure 11.9. Since the Auger signals are relatively weak (but sharp) and superimposed on the intense

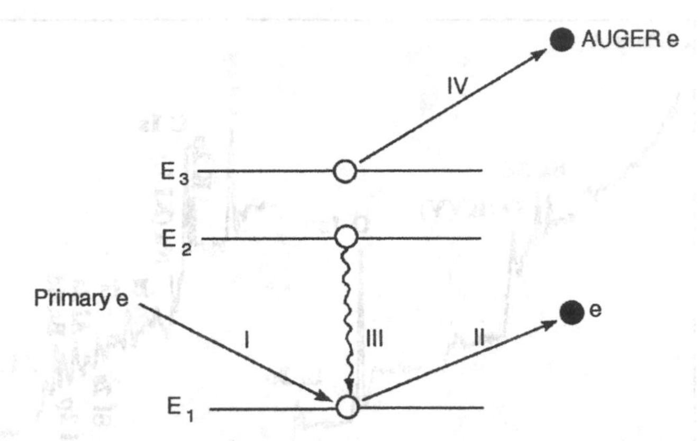

Figure 11.6 Schematic illustration of the mechanism of Auger electron transition: (I) electron bombardment (X-rays in X-ray excited Auger electron spectroscopy, XAES) leading to ionization (II), and intra-atomic relaxation (III) resulting in Auger electron emission (IV).

(but more gradually varying) background composed of the 'true' secondary electrons and of the inelastically scattered primary electrons, a standard approach for data acquisition is to get the derivative of the Auger energy spectrum (e.g. using modulation and phase-sensitive detection). Each line is customarily marked by the energy of the differentiated line minimum. As a first example, the electron-excited Auger spectrum of pure nickel is shown in Figure 11.10 (it can be compared with the XAES part of the spectrum given in Figure 11.4). No Auger transitions involving the 1s level appear in the spectrum, since the primary beam energy used (3 kV) is well below the 1s ionization potential. The corresponding four main lines in the spectra of the adjacent elements (Co and Cu) are sufficiently distinct in energy (see Figure 11.7) to allow unambiguous identification of each. Likewise, qualitative AES analysis of Fe–18Cr–3Mo alloy is straightforward (Figure 11.11) (Shiffman and Polak 1986). It should be noted that the spectrum shown in Figure 11.11(a) was recorded during Ar sputtering under relatively clean UHV background, revealing excess Cr at the surface region, whereas the spectrum acquired with some residual CO in the test chamber shows Cr depletion (Figure 11.11b), as compared to the spectrum of the *in-situ* scribed alloy representing the bulk composition (not shown). This demonstrates the problem of **preferential sputtering** in surface and interface analysis of multi-element solids (see section 11.2.4) and the resultant uncertainties in quantitative information. The surface analysis by AES of insulating materials is complicated by electron-beam-induced charging effects. Efficient means of charge compensation in oxides have been discussed recently (Guo *et al.* 1997).

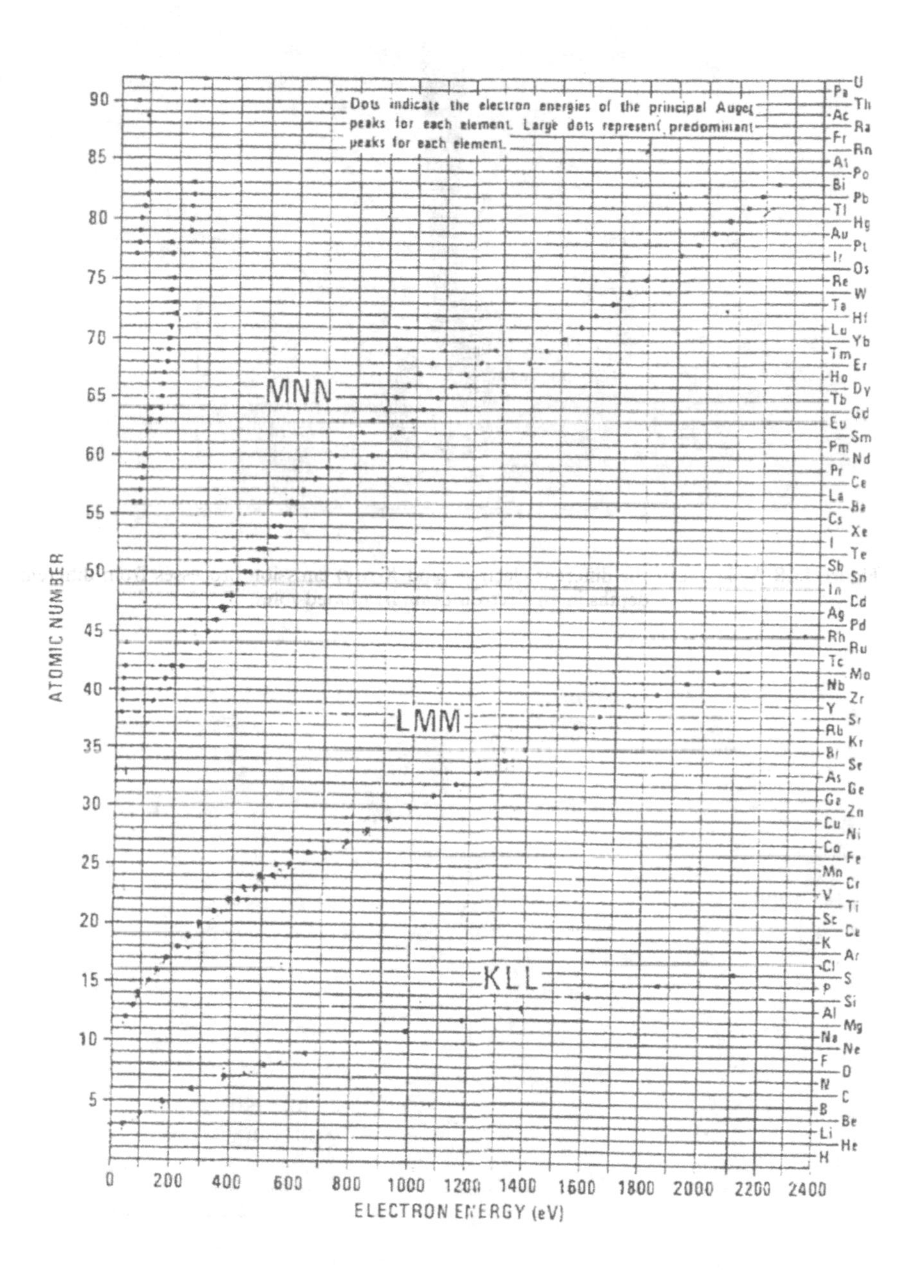

Figure 11.7 The Auger electron energies of the elements (from Davis *et al.* (1978), with permission.) The notations KLL, LMM etc. refer to the three shells involved in the Auger transition (E_1, E_2 and E_3, respectively, in Figure 11.6).

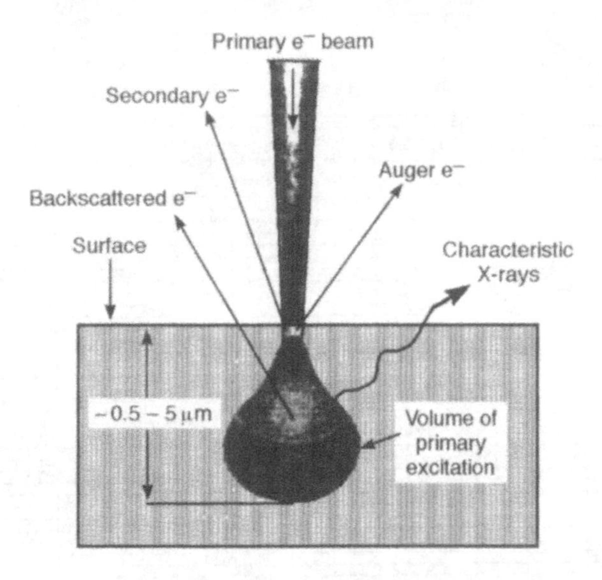

Figure 11.8 A sketch of the different electron (and X-ray) emission processes from different depths in an electron-beam irradiated solid.

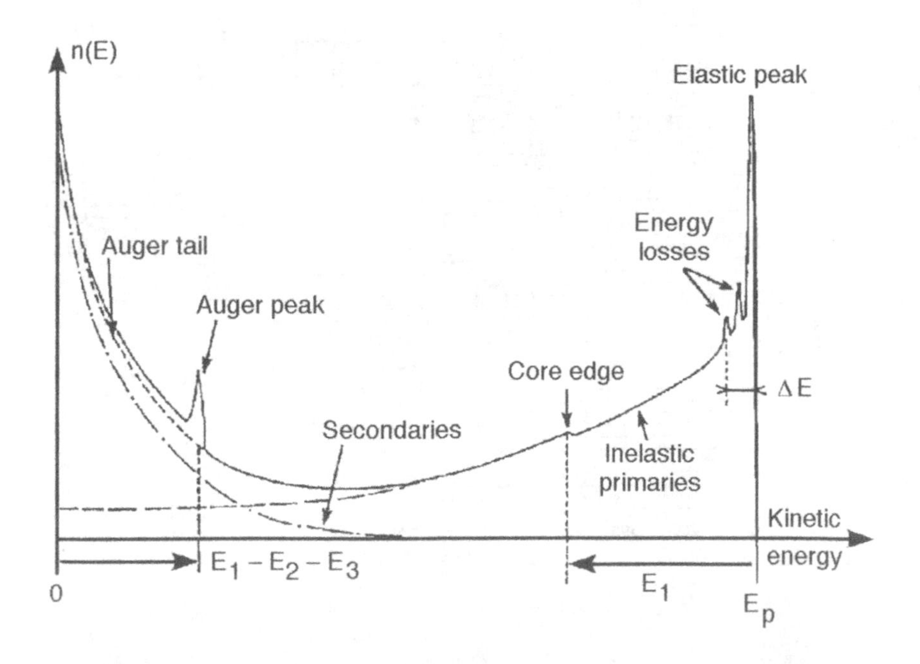

Figure 11.9 Schematic illustration of the various contributions to the energy spectrum of electrons emitted from a solid surface subjected to electron-beam irradiation (with energy E_p). (Reproduced with permission from Langeron (1989).)

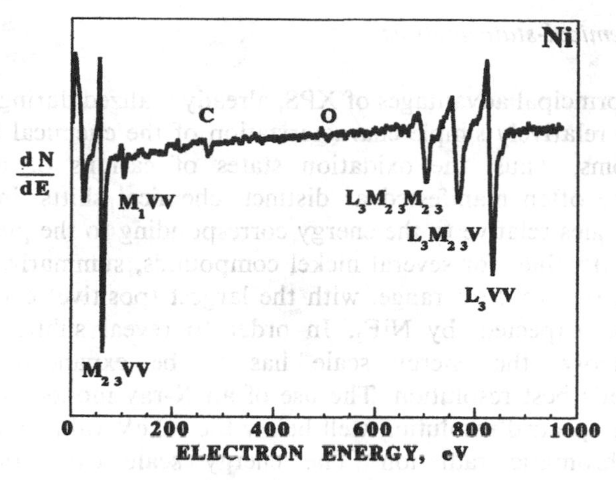

Figure 11.10 The Auger electron differentiated spectrum of nickel. The carbon and oxygen contamination after the Ar sputter cleaning are due to residual gases in the test chamber.

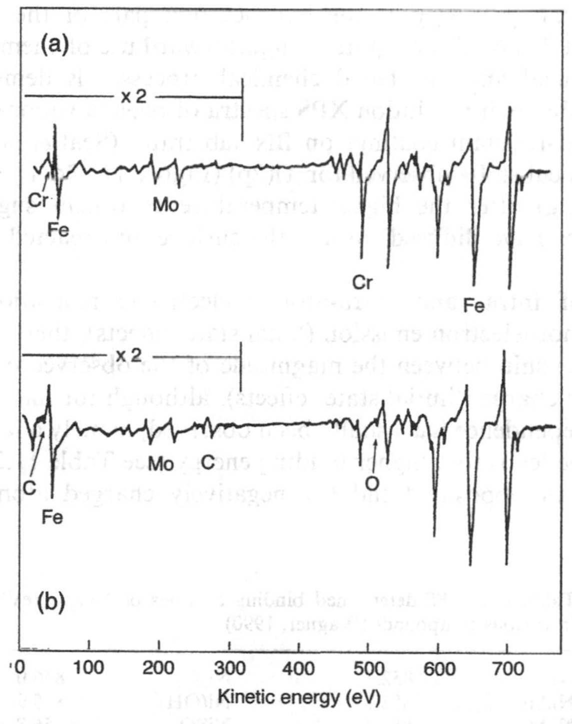

Figure 11.11 Auger electron spectra of Fe–18Cr–3Mo(100) surface acquired during Ar sputtering at 780 K: (a) under relatively clean UHV background, and (b) with some residual CO (primary e^- beam – 3 kV). (Reproduced with permission from Shiffman and Polak (1986).)

11.2.2 Chemical-state analysis

One of the principal advantages of XPS, already realized during its earliest days, is the relatively simple characterization of the chemical bonding of surface atoms. Thus, the oxidation states of cations or anions, for example, are often manifested as distinct 'chemical shifts' in core-level binding energies relative to the energy corresponding to the pure element. The Ni($2p_{3/2}$) values for several nickel compounds, summarized in Table 11.2, are in the 0.5–5 eV range, with the largest (positive) chemical shift exhibited, as expected, by NiF_2. In order to reveal subtle changes in binding energy, the energy scale has to be expanded using the spectrometer's best resolution. The use of an X-ray monochromator can improve the spectral resolution well below the ~1 eV value obtained with non-monochromatic radiation. The energy scale calibration of the spectrometer is usually based on the Au($4f_{7/2}$) line fixed at its known binding energy of 84.0 eV. Positive charging shifts of binding energies induced by the photoionization of insulating sample surfaces can be removed by a combination of electron-flood-gun adjustments and fixing the C(1 s) binding energy of the hydrocarbon part of the adventitious carbon line at 284.6 eV. The quite straightforward use of chemical shifts to unravel surfacial and interfacial chemical processes is demonstrated in Figure 11.12 by high-resolution XPS spectra of plasma vapor-deposited Ti and TiN wear-resistant coatings on BN substrates (Seal *et al.* 1997). The distinct chemical shifts observed for Ti(2p) (Figure 11.12d–f) and for B(1s) (Figure 11.12g) after the higher-temperature treatment suggest that B and N species have diffused toward the surface and reacted with Ti and oxygen.

Because of intra- and extra-atomic electronic relaxation processes during the photoelectron emission ('final state' effects), there is usually no simple relationship between the magnitude of the observed chemical shift and the ionic charge ('initial-state' effects), although for many elements a monotonic dependence has usually been observed, namely a larger positive atomic charge leads to a higher binding energy (see Table 11.2 and Figure 11.12), and an opposite trend for negatively charged atoms. Binding-

Table 11.2 XPS-determined binding energies of Ni $2p_{3/2}$ (eV) in various compounds (Wagner, 1990)

Ni	852.7	Ni_2O_3	856.0
NiAu	852.1	$Ni(OH)_2$	855.9
$NiAl_3$	853.7	$NiSO_4$	856.8
Ni_2P	853.1	$NiCl_2$	856.7
$Ni(CO)_4$	854.4	NiF_2	857.4
NiO	854.4		

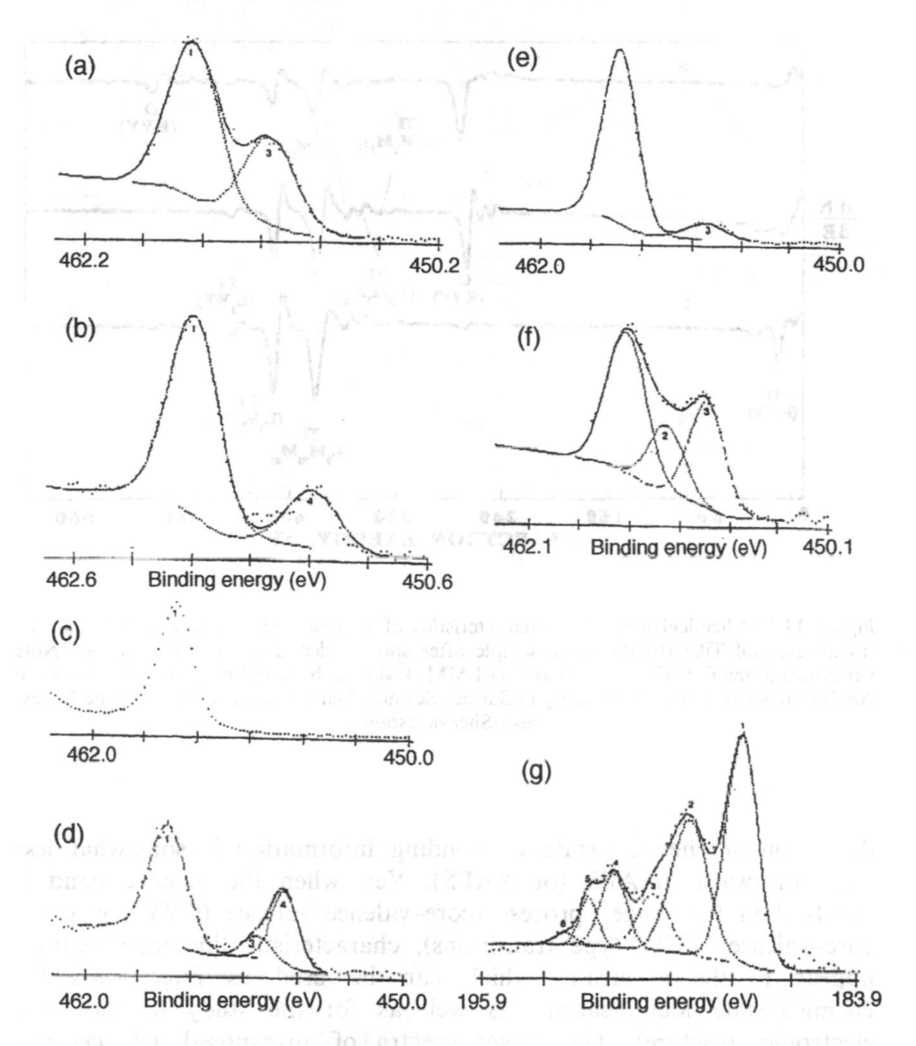

Figure 11.12 High resolution XPS Ti(2p) deconvoluted spectra: (a) TiN standard, (b) TiB_2 standard, (c) and (d) PVD Ti on BN substrate (treated at 1000 and 1400°C), (e) and (f) PVD TiN on BN (treated at 1000 and 1400°C). The B(1s) spectrum of Ti + BN treated at 1400°C is shown in (g). Possible peak identification: Ti(2p) – (1) TiO_2, (2) TiN_x, (3) TiN and (4) TiB_2; B(1s) – (1) TiB_2, (2) TiB_xN_y, (3) BN, (4) BN_yO_z, (5) B–O and (6) B_2O_3. (Reprinted with permission from Seal *et al.* (1997), copyright 1997 American Vacuum Society.)

energy data compilations for different compounds and bonding states of all elements detected by XPS are available in the literature (Wagner 1990, Yoshitake and Yoshihara 1997).

Owing to the more complex nature of the Auger process, leaving two holes on the emitting atom (and hence more extra-atomic relaxation compared to the one-hole situation in photoelectron emission), the

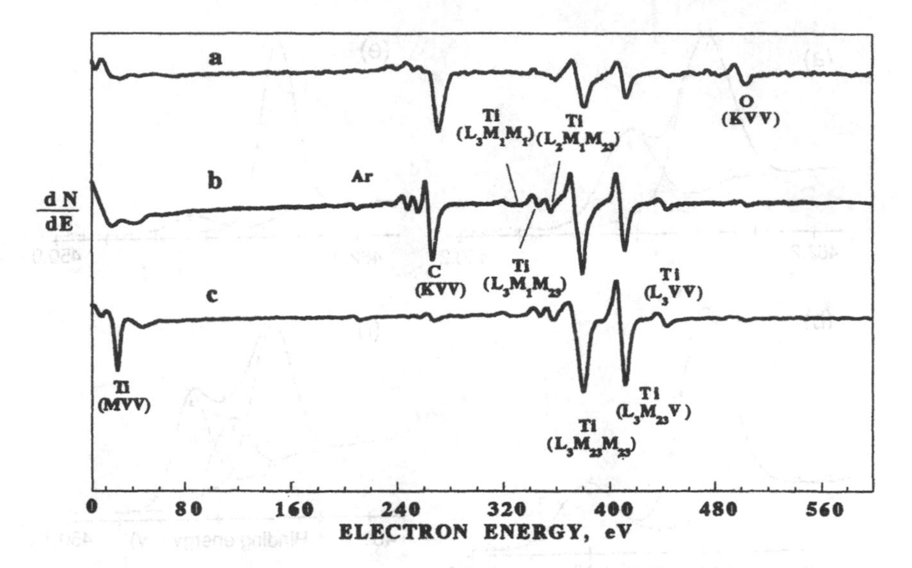

Figure 11.13 Chemical-state AES characteristics of carbon and titanium in TiC and Ti: (a) air-exposed TiC; (b) the same sample after sputter cleaning; (c) clean pure Ti. Note variations in the C KVV (a vs. b) and Ti LMM, LMV (a, b vs. c) line shapes. (Courtesy of Analytical Services, The Laboratory of Surface Science, Ben-Gurion University of the Negev, Beer-Sheeva, Israel.)

deduction of chemical-state or bonding information is somewhat less straightforward in AES (or XAES). Yet, when the valence band is involved in the Auger process (core–valence–valence (CVV) or core–core–valence (CCV) type transitions), characteristic lineshape features appear in the spectrum, which can be used as fingerprints for chemical-state identification (as well as for the study of the local electronic structure). The Auger spectra of air-exposed TiC ceramic recorded before and after exposing the bulk material by sputter-etching several nanometers from the surface region are given in Figure 11.13(a) and (b), respectively. Besides clear changes in composition, as reflected by the respective Ti/C intensity ratios, the shape of the carbon KVV spectrum of the carbidic phase (Figure 11.13b) differs considerably from the one given in Figure 11.13(a), which is typical of a contamination overlayer, probably originating from residual hydrocarbons in the UHV test chamber (compare also the Ti lineshape variations measured for pure Ti, Figure 11.13c). Likewise, the high-resolution carbon KVV spectrum of a diamond thin film (Figure 11.14) reveals distinct features compared to spectra of low-quality diamond or graphite.

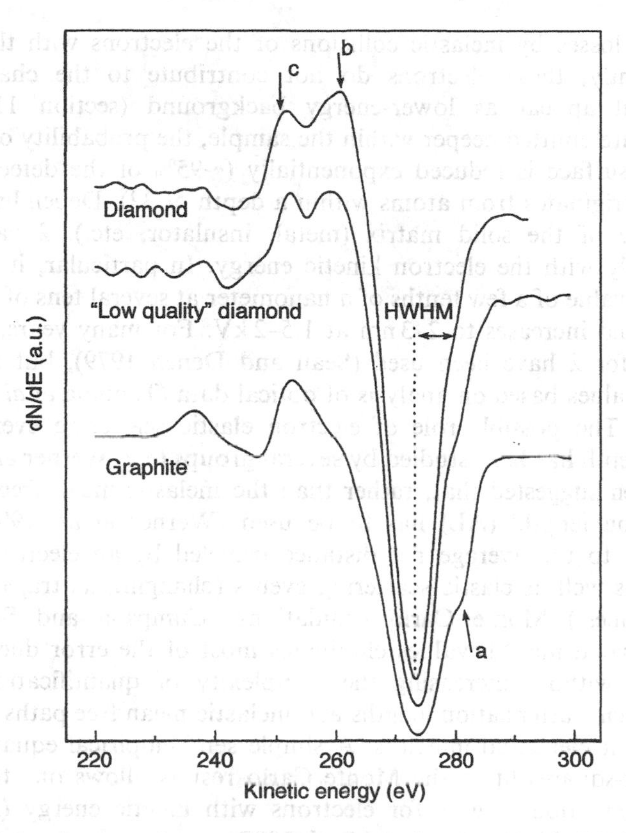

Figure 11.14 Auger spectra characteristic of a good quality diamond thin film, a low-quality diamond and graphite. Feature b is characteristic to diamond, and feature a is the self convolution of occupied p states in the valence band of sp^2 atoms and is absent in the diamond spectrum. These features enable the determination of the ratio of diamond to non-diamond carbon in the surface region. (Reprinted with permission from Guttierez *et al.* 1997, copyright 1997 American Vacuum Society.)

11.2.3 The probing depth

A key factor in the high sensitivity of electron spectroscopy to the near-surface atoms, namely its shallow probing depth, is the use of slow electrons with relatively short inelastic mean free path (IMFP or λ in the range 0.5–3 nm) as information carriers. Assuming **Beer's law** behavior, the residual number of electrons (N) left with their initial energy after traveling along a distance d in a solid ('straight-line approximation') is expressed by

$$N = N_0 \exp(-d/\lambda) \tag{11.4}$$

where N_0 is the initial number of electrons, such as that produced inside the solid by the photoelectric or the Auger process. This attenuation is due to energy losses by inelastic collisions of the electrons with the matrix. Consequently, these electrons do not contribute to the characteristic signal, but appear as lower-energy background (section 11.2.1). As electrons are emitted deeper within the sample, the probability of escaping from the surface is reduced exponentially (~95% of the detected signal intensity originates from atoms within a depth of 3λ). Depending also on the nature of the solid matrix (metal, insulator, etc.), λ values vary significantly with the electron kinetic energy. In particular, it exhibits a minimum value of a few tenths of a nanometer at several tens of eV kinetic energy,[1] and increases to 2–3 nm at 1.5–2 kV. For many years, empirical formulas for λ have been used (Seah and Dench 1979), but now more accurate values based on analysis of optical data (Tanuma *et al.* 1991) are available. The possible role of electron elastic scattering events in the probing depth has been studied by several groups (e.g. Werner *et al.* 1991). It has been suggested that, rather than the inelastic mean free path, the 'attenuation length' (AL) has to be used (Werner *et al.* 1991). AL is equivalent to the average net distance traveled by an electron between inelastic as well as elastic scattering events (changing its trajectory from straight lines.) Monte Carlo simulations (Cumpson and Seah 1997) showed that using AL values eliminates most of the error due to elastic scattering without increasing the complexity of quantification. Figure 11.15 presents attenuation lengths and inelastic mean free paths calculated for different elemental matrices. A simple semi-empirical equation based on a least-squares fit to the Monte Carlo results allows one to estimate easily attenuation lengths for electrons with kinetic energy (E_k) in the range 50–2000 eV (Cumpson and Seah 1997):

$$\lambda_{AL} = 0.316a^{3/2}\{E_k/[Z^{0.45}(\ln E/27 + 3)] + 4\}\ \text{nm} \qquad (11.5)$$

where Z is the average atomic number of the matrix and a is the lattice parameter, which can be estimated from

$$a = 10^8(\mu/\rho N_{av})^{1/3}\ \text{nm}$$

with μ the average atomic mass of the matrix (g) and ρ its density (kg m^{-3}). λ_{AL} values calculated in this way can be used in the quantification equations given below. Jablonski and Powell recently introduced another approach for evaluation of correction parameters for elastic scattering effects in XPS and AES (Jablonsky and Powell 1997).

[1] For the highest surface specificity, lines around this energy range of several tens of eV should be analyzed (it can be readily achieved in photoemission with a tunable excitation source, i.e. synchrotron radiation).

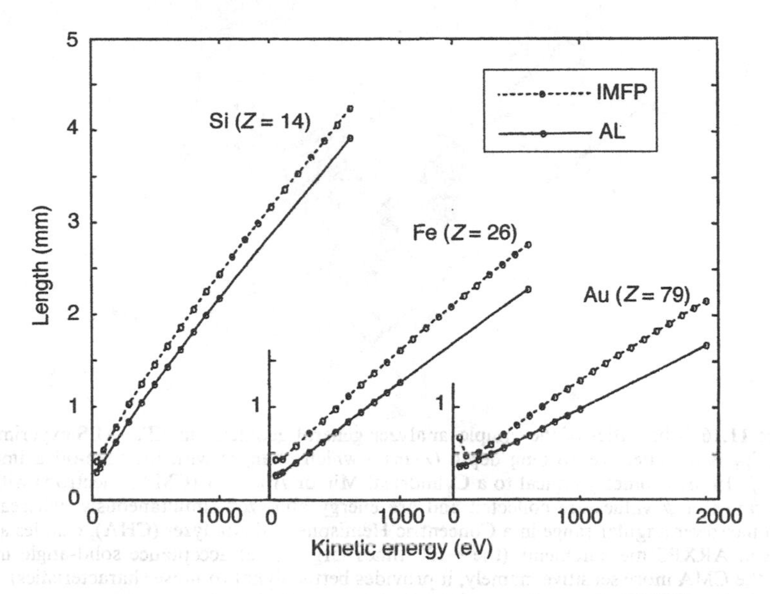

Figure 11.15 Attenuation length (AL) and inelastic mean free path (IMFP) calculated as function of the electron kinetic energy in three materials. (Reproduced with permission from Cumpson and Seah (1997).)

11.2.4 Quantification

Quantitative analysis by XPS or AES, namely the determination of the near-surface elemental composition from measured line intensities, still involves significant inaccuracies related to uncertainties in the basic physical processes and in instrumental factors. Furthermore, modifications in chemical composition and bonding in a multi-element solid material usually are not confined to the outermost surface layer (monolayer), but extend to depths of several nanometers **(shallow composition gradients)** or even much more **(extended composition gradients)**. Surface segregation phenomena, air oxidation products, corroded metals or multilayer electronic materials are just a few examples for in-depth non-uniform distributions. Thus, a major complication in quantification stems from the fact that the measured signal is a superposition of different contributions from a region several atomic layers thick. Only when certain procedures are strictly employed, can the information furnished by the two techniques be considered as being beyond semi-quantitative.

Shallow composition gradients and thin overlayers. Analogously to most other spectrometric techniques, the photoelectron or Auger line intensity can be expressed by means of a product of several terms related to the excitation process, to electron transport in the solid matrix and its

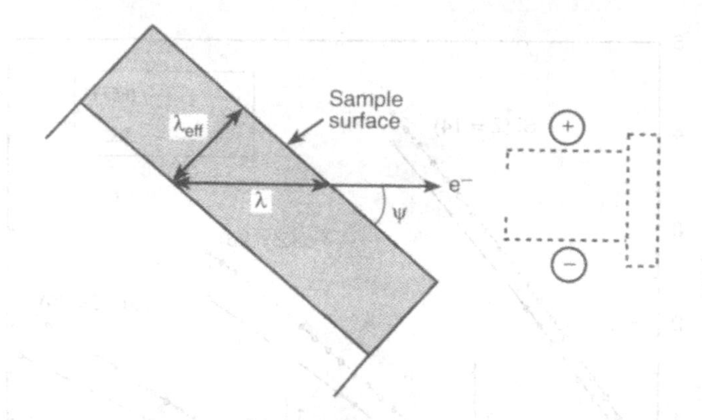

Figure 11.16 Schematics of the sample–analyzer general geometry in XPS/AES experiments. Here λ_{eff} is the effective probing depth ($\lambda \sin \psi$) which changes with the take-off azimuthal angle ψ. In the geometry typical to a Cylindrical Mirror Analyzer (CMA), electrons within a wide range of ψ values are collected and are energy analyzed simultaneously, whereas the much narrower angular range in a Concentric Hemispherical Analyzer (CHA), enables angle-resolved ARXPS measurements (the ~100 times larger total acceptance solid-angle makes the CMA more sensitive, namely, it provides better signal-to-noise characteristics).

emission outside, and to instrumental factors (see Figures 11.1 and 11.16). In XPS, the general expression for the detected line intensity of an element with inhomogeneous depth distribution per unit volume $n(z)$ can be written with some simplifications as

$$I = \Phi \sigma \phi T D \int_0^\infty n(z) \exp(-z/\lambda_{AL} \sin \psi)\, dz \qquad (11.6)$$

where Φ is the X-ray flux on the sample (replaced by I_p, the primary electron beam current, in AES); σ is the total photoionization cross-section (the corresponding term in AES is the product of three terms, (i) the cross-section for direct ionization by the primary electrons, (ii) the primary electron backscattering factor (see Figure 11.8), which amplifies I_p and increases with the matrix atomic number (Briggs and Seah 1990, Leveque and Bonnet 1995), and (iii) the probability for Auger process following ionization); ϕ is an angular factor dependent upon the angle between the photon and detected electron directions (irrelevant in AES); T = transmission efficiency, related to characteristics of the electron energy analyzer; and D = the electron detector efficiency.

Because some of these factors are not well known, and since the line intensity (I) represents integration over several nanometers, quantitative determination of an unknown inhomogeneous $n(z)$ distribution from the intensity of a single line is actually impossible. The latter problem has been recently demonstrated (Figure 11.17) by Tougaard, who introduced a new method for determination of near-surface nanostructures by the

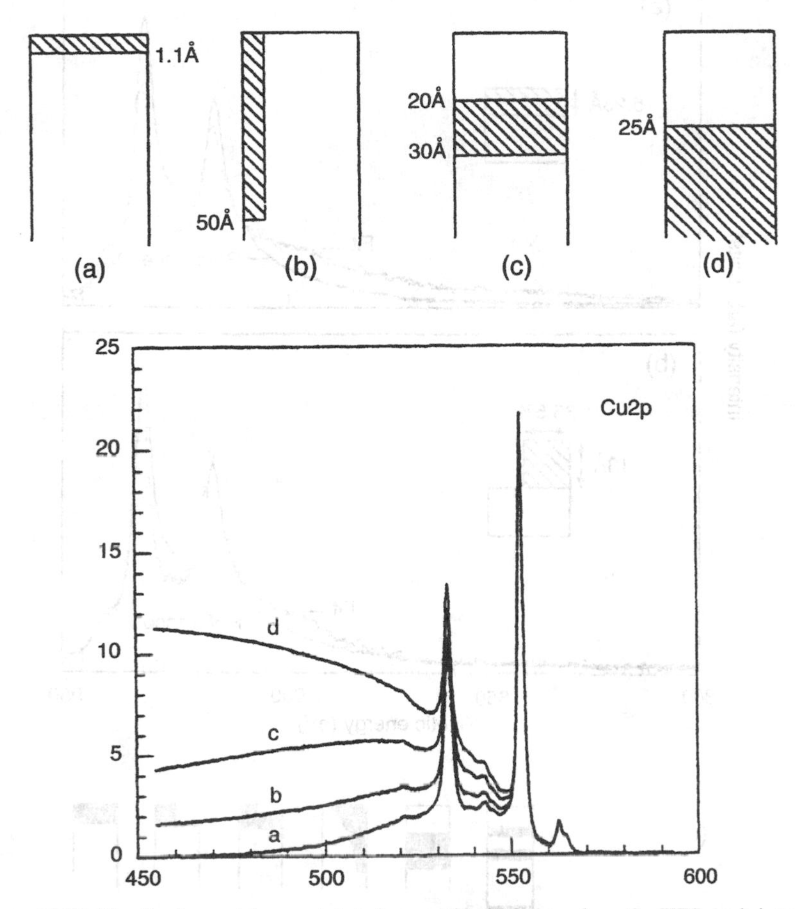

Figure 11.17 The fundamental uncertainty in quantification based on the XPS peak intensity alone: all four hypothetical distinct structures of copper on/in gold (a–d, top) should give the same Cu 2p peak intensity (a–d, bottom). The line shape at kinetic energies below the peak, however, depends strongly on the in-depth elemental distribution, and can be used to unravel it. (Reproduced with permission from Tougaard (1996), copyright 1996 American Vacuum Society.)

analysis of the XPS peak intensity as well as its inelastic background shape in a wide (~100 eV) energy range below it (Tougaard 1996b). The inherent shallow probing depth of XPS/AES (see section 11.2.3) obviously limits the information depth obtainable ($d \lesssim 3\lambda$), but even for ultrathin films one often wants to know the layer thicknesses or the distribution of elements and chemical states as functions of depth. As an illustration, Figure 11.18 shows Pt(4d) spectra for deposited Pt on Si(111) analyzed according to two different growth models (Tougaard 1996a). Clearly, the island model with the indicated thickness and coverage gives the overall best fit regarding lineshape and intensity (all other structures in Figure 11.18(c)

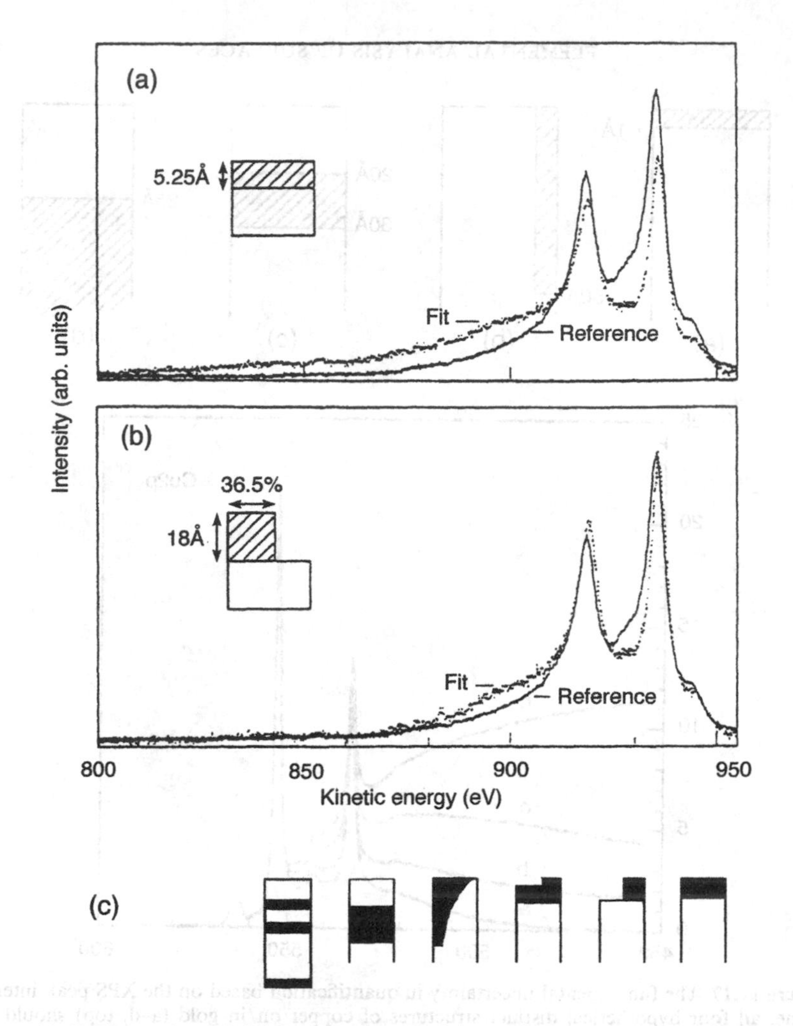

Figure 11.18 Illustration of the Tougaard method for surface nanostructure determination by XPS peak shape analysis for Pt growth on Si(111). The background corrected Pt(4d) spectrum is fitted (with the indicated thickness and coverage parameters) to the reference spectrum from pure Pt: (a) layer-by-layer growth mode (rectangular overlayer); (b) a single island model; (c) all near-surface nanostructures that can be treated by the software package. (Reprinted from Tougaard (1996a) with kind permission of Elsevier Science-NL.)

gave inferior fits (Tougaard 1996a)). This relatively new approach for quantitative XPS and AES is sensitive on the ~1–10 nm scale, and it seems to reduce the uncertainty to a level of 10–20% (Tougaard 1996b).

Another non-destructive method that can provide information on the shallow depth distribution of atoms is based on the angular dependence of XPS intensities (ARXPS) obtained by tilting the sample with respect to the analyzer entrance, thereby changing the angle between the surface and the electron take-off angle, namely, the effective probing depth (see Figure 11.16). Enhanced sensitivity for surface elements can be achieved by using low electron take-off angles ('grazing-angle'). Several algorithms for deriving concentration–depth profiles from ARXPS (or ARAES) measurements have been published (Tyler *et al.* 1989). The use of the 'regularization method' is exemplified in Figure 11.19 for a carbon-contaminated thin silicon oxide layer on Si, and in Figure 11.20 for near-surface variations in the composition of a fluorine-containing polyether urethane sample (Tyler *et al.* 1989). The latter profile indicates that the top 1.5 nm layer of the sample is enriched in the polyether segments (O and C). It should be noted that significant depth-resolution limitations (and concentration uncertainties) are still inherent in ARXPS analysis due to the limited extent of useful information present in such data (Cumpson 1995), so the achievable depth resolution is governed by signal-to-noise ratio, and not by the number of emission angles.

Prior knowledge of the nanostructure form, such as in the simple case of a single homogeneous overlayer on a substrate, is very useful and makes

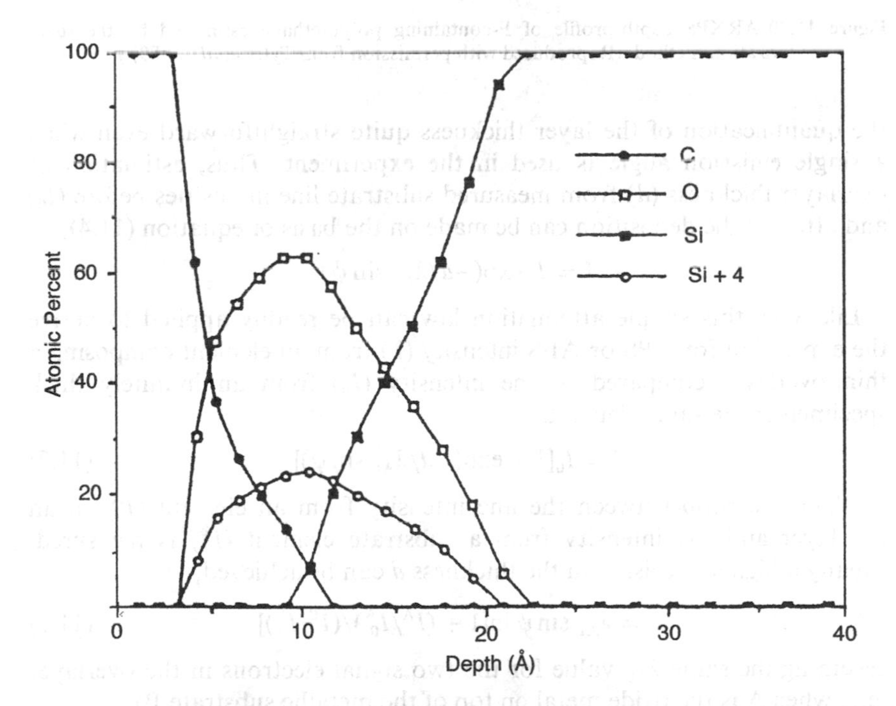

Figure 11.19 ARXPS depth profile of a carbon contaminated thin oxide layer on elemental Si. Signal intensities were collected at three take-off angles: 90°, 35° and 10° (the counts for Si^0, for example, were 803, 407 and 16, respectively). (Reproduced with permission from Tyler *et al.* (1989).)

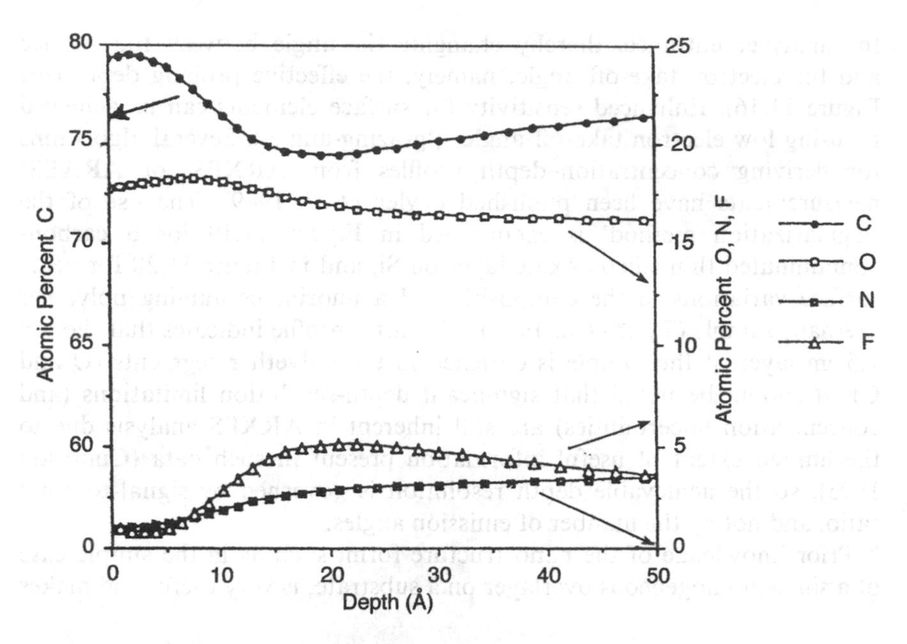

Figure 11.20 ARXPS depth-profile of F-containing polyurethane estimated by the regularization method. (Reproduced with permission from Tyler *et al.* (1989).)

the quantification of the layer thickness quite straightforward even when a single emission angle is used in the experiment. Thus, estimation of overlayer thickness (d) from measured substrate line intensities before (I_0) and after (I) the deposition can be made on the basis of equation (11.4),

$$I = I_0 \exp(-d/\lambda_{AL} \sin \psi)$$

Likewise, this simple attenuation law can be readily applied to derive the expression for XPS or AES intensity (I) from an element composing a thin overlayer compared to the intensity (I_0) from an infinitely thick specimen of the same element,

$$I = I_0[1 - \exp(-d/\lambda_{AL} \sin \psi)] \tag{11.7}$$

When the *ratio* between the line intensity from an element (I^A) in an overlayer and the intensity from a substrate element (I^B) is measured, usually a higher precision in the thickness d can be achieved,

$$d = \lambda_{AL} \sin \psi \ln[1 + (I^A/I_0^A)/(I^B/I_0^B)] \tag{11.8}$$

assuming the same λ_{AL} value for the two signal electrons in the overlayer (e.g. when A is the oxide metal on top of the metallic substrate B).

As mentioned in section 11.2.3, in order to get more accurate results the attenuation length, including inelastic as well as elastic scattering effects, has been used in the above equations. Since they were derived for a *flat* substrate supporting a uniform overlayer, efforts have been made to incorporate effects of roughness or curvature in the XPS analysis (Chatelier *et al.* 1997, Gunter *et al.* 1997).

Homogeneous in-depth atomic distribution. In the quite uncommon case of an *homogeneous* surface region (at least to a depth of $\sim 3\lambda$), the detected XPS signal intensity can be readily derived from equation (11.6):

$$I = n\Phi\sigma\phi\lambda_{AL}\sin\psi TD \tag{11.9}$$

where n is the number of atoms of the element of interest per unit volume. The problem of several unknown parameters in this equation can be partially overcome by taking relative intensities of spectral lines corresponding to two elements (A, B). The relative atomic fraction is then expressed by means of 'atomic sensitivity factors', S:

$$x_A/x_B = (I_A/I_B)(S_B/S_A) \tag{11.10}$$

Various approaches have been used to evaluate the elemental sensitivity factors in XPS (Tougaard and Jansson 1992). They can be calculated from theoretical photoelectric cross-sections, estimated λ (or λ_{AL}) and the spectrometer transmission function (all being energy-dependent) (Wagner *et al.* 1979), or they can be determined empirically (Wagner 1983) and normalized relative to the intensity of a particularly strong line, such as F(1s) (with $S = 1$). Cancellation of errors from uncertainties in equation (11.9) can be achieved when intensities of spectral lines from pure-element standards (I°) are used as the sensitivity factors. For the general case of a multi-element sample, the atomic fraction of a particular element (A) then reads (Briggs and Seah 1990)

$$x_A = \frac{I_A/I^\circ_A}{\sum_B I_B/I^\circ_B} \tag{11.11}$$

where the summation involves line intensities from all elements.

To go beyond this rough, first-order approximation in quantitative XPS (or AES) analysis, it is necessary to include calculated 'matrix correction factors' giving the second-order approximation,

$$x_A = \frac{I_A I^\circ_A}{\sum_B F_{BA} I_B/I^\circ_B} \tag{11.12}$$

where, in XPS, the matrix-dependent factor F consists of contributions from *atomic density* and *attenuation length* differences, while in AES it includes the relative (A vs. B matrices) *electron backscattering* efficiency as well (see Figure 11.8). Matrix factors calculated for many pairs of elements show that this correction is usually significant (up to several tens of per cent) in both XPS and AES (Briggs and Seah 1990, Hall and Morabito 1979). It should be noted that measuring the pure element intensities (I°) in the same spectrometer (and under the same experimental conditions) is preferable to using published values.

As in the quantification of many other 'non-absolute' techniques, relatively higher degree of accuracy can be obtained by the use of locally produced 'real' standards, namely, measuring relative intensities from an adequately prepared sample with a chemical matrix similar to that of the test specimen (a series of such samples can provide a 'calibration curve'). However, preparation of standard samples with accurately known, uniform composition within the outermost 2–3 nm of the surface region is not an easy task. For example, ion bombardment used for cleaning the surface usually leads to preferential removal of atoms of one particular element, and thus to deviation from the bulk composition (see, for example, Figure 11.11). Even when one tries to fracture or scribe the sample with a hard sharp tool (inside the UHV test chamber), it is not certain that the actual bulk composition is exposed without alteration (e.g. due to local surface segregation) associated with the mechanical fracture or deformation. Therefore, the application of the other XPS (or AES) quantification procedures mentioned above is much more common in routine analysis.

Of particular importance in quantitative analysis is the accurate determination of photoelectron or Auger line intensities from a measured spectrum. Therefore, efforts have been made to develop appropriate procedures for removal of the background due to the inelastically scattered electrons, deconvolution of overlapping lines ('curve fitting') and other data processing procedures (Briggs and Seah 1990, 1992, Grant 1989, Powell and Seah 1990). Recently, the validity of traditional methods for background subtraction from XPS lines, namely the Shirley method (Shirley 1972) and the linear subtraction procedure, has been studied by comparison with first-principles intensity calculations (Tougaard and Jansson 1992). In particular, a much higher accuracy has been achieved using a new method based on a detailed description of the physical processes involved.

Although samples are not likely to be homogeneous in the surface region, this convenient assumption is often made in quantitative XPS and AES, yielding, at best, some 'average' composition. Limitations in the accuracy of calculated sensitivity factors used for quantification have been analyzed, with the conclusion that enhancing the knowledge of the actual depth–composition gradients near the surface (see previous section) is of even greater importance (Tougaard 1996b).

Extended composition gradients: ion sputter profiling. For determination of compositional variations between ~3 nm and 1000 nm sputter-etching by inert ions is needed. With this somewhat destructive technique, the sample surface is gradually eroded by 0.5–5 kV ions (see Figure 11.1) while the newly exposed atomic layers are analyzed by AES (or XPS). Approximate sputtering rate calibration can be accomplished by measuring the time it takes to fully sputter, under the same ion-beam parameters, a chemically similar layer of known thickness. Figure 11.21 shows an example for the use of AES and ion sputtering in obtaining

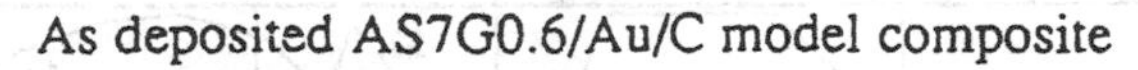
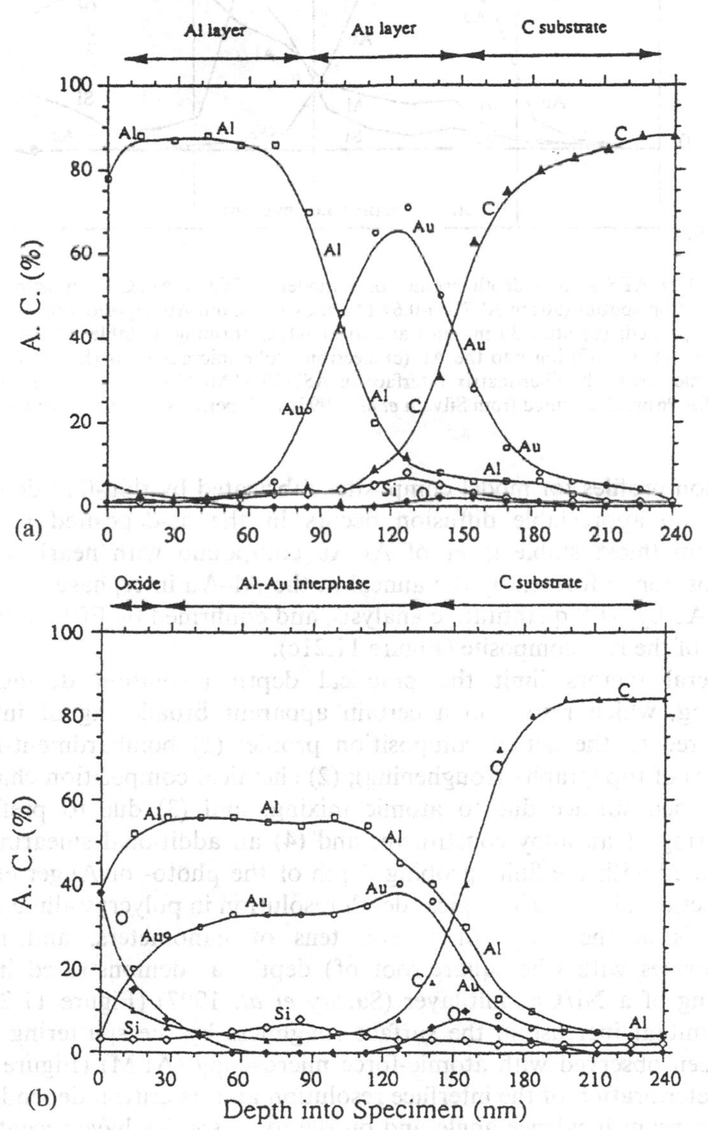

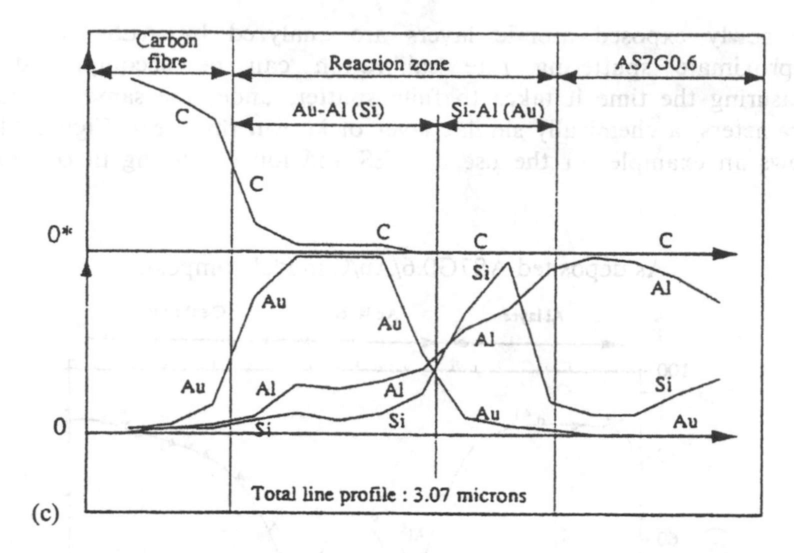

Figure 11.21 AES sputter depth profiles of a model AS7G0.6/Au/C composite fabricated by magnetron sputter (90 nm Al-7%Si-0.6%Mg alloy on 60 nm Au deposited on C substrate): (a) as deposited; (b) after 30 minutes anneal at 600 C forming a stable Al–Au interphase which inhibits C diffusion into the Al; (c) electron probe microanalysis (EPMA) qualitative line profile across the fiber-matrix interface in AS7G0.6/Au/C real composite annealed at 600 C for 96 hr. (Reprinted from Silvain *et al.* (1996) with permission from Elsevier Science.)

diffusion profiles for model composites fabricated by thin-film deposition. While no appreciable diffusion occurs in the as-deposited sample, a ~120 nm thick, stable layer of Al–Au compound with nearly constant composition is formed by the anneal at the Al–Au interphase – estimated as Al_2Au by AES quantitative analysis, and confirmed by EPMA measurements of the real composite (Figure 11.21c).

Several factors limit the practical depth resolution during depth profiling, which results in a certain apparent broadening of interfaces compared to the actual composition profile: (1) bombardment-induced changes of topography (roughening); (2) chemical composition changes of the eroded surface due to atomic mixing, and (3) due to preferential sputtering of an alloy constituent; and (4) an additional smearing effect associated with the finite probing depth of the photo- or Auger electrons (see section 11.2.3). The typical depth resolution in polycrystalline metallic solids is in the range of several tens of nanometers, and it often deteriorates with (the square-root of) depth, as demonstrated in depth profiling of a Ni/Cr multilayer (Satory *et al.* 1997) (Figure 11.22). The concomitant increase in the surface roughness by the sputtering process has been observed with atomic-force microscopy (AFM) (Figure 11.23). The deterioration of the interface resolution and its critical dependence on the ion-beam incidence angle and on the ionic species have recently been

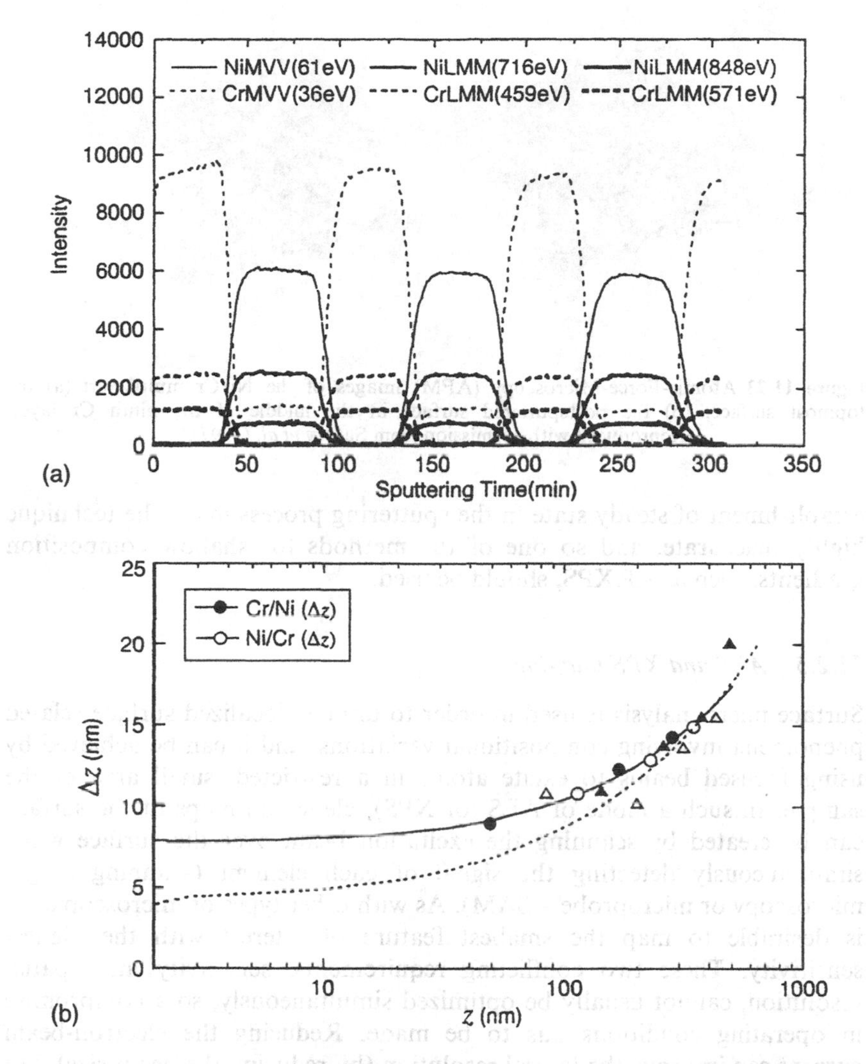

Figure 11.22 Sputter depth profiling of a Ni/Cr multilayer (SRM No. 2135, NIST – 5 × 53 nm Cr/4 × 64 nm Ni on Si substrate): (a) AES intensities vs. sputter time; (b) the depth dependence of the depth-resolution (the distance over which a 16–84% change in signal is measured) for the Cr/Ni and Ni/Cr interfaces in the case of Ni MVV. (Redrawn with permission from Satori *et al.* (1977), copyright 1977 American Vacuum Society.)

demonstrated in another study of Cr/Ni multilayer structures (Moon and Kim 1996). Specific refinements such as sample rotation (Zalar 1986) can improve the depth resolution to 1–2 nm. A thin Ta_2O_5 layer on Ta, was especially developed as a reference material for depth profiling (Certified Reference Material 1983).

For depths less than ~3 nm, the changes that occur during the

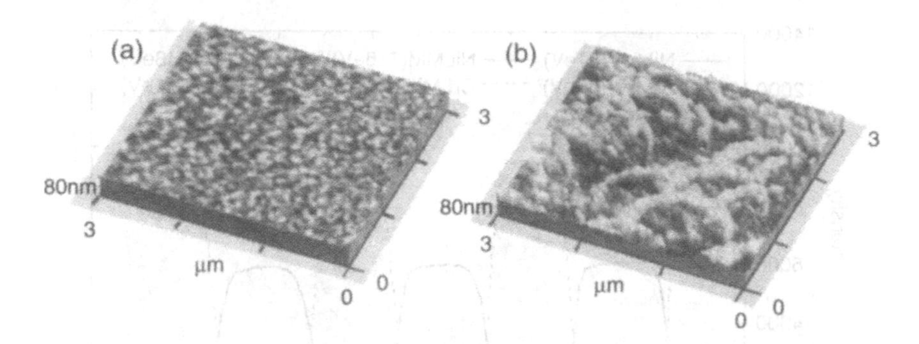

Figure 11.23 Atomic-Force-Microscopy (AFM) images of the Ni/Cr multilayer: (a) the topmost surface; (b) the postsputtered surface in the middle of the ninth Cr layer. (Reproduced with permission from Satory *et al.* (1997).)

establishment of steady state in the sputtering process make the technique highly inaccurate, and so one of the methods for shallow composition gradients, such as ARXPS, should be used.

11.2.5 AES and XPS imaging

Surface microanalysis is used in order to unravel localized surface-related phenomena involving compositional variations, and it can be achieved by using focused beams to excite atoms in a restricted, small area of the sample. In such a mode of AES (or XPS), elemental maps of the surface can be created by scanning the excitation beam over the surface while simultaneously detecting the signal of each element (scanning Auger microscopy or microprobe – SAM). As with other types of microscopies, it is desirable to map the smallest feature of interest with the highest sensitivity. These two conflicting requirements, sensitivity and spatial resolution, cannot usually be optimized simultaneously, so a compromise in operating conditions has to be made. Reducing the electron-beam current can improve the lateral resolution (by reducing the beam size), but the emitted Auger electron signal intensity is consequently lowered.

Elemental imaging is highly important in quality control of modern microelectronic devices having deleterious tiny defects or surface contamination, for which AES (using state-of-the-art instrumentation) clearly surpasses the capability of other techniques. For larger-scale inhomogeneities, scanning Auger microprobe with poorer lateral resolution is sufficient. Such instrumentation (with ~5 µm beam size) was used to study elemental distributions and lateral correlations in a bronze archeological arrowhead (Polak *et al.* 1983), the first application of AES in archeological artifact analysis. First, in order to analyze the general composition of the material, an 'average', survey Auger spectrum was recorded while

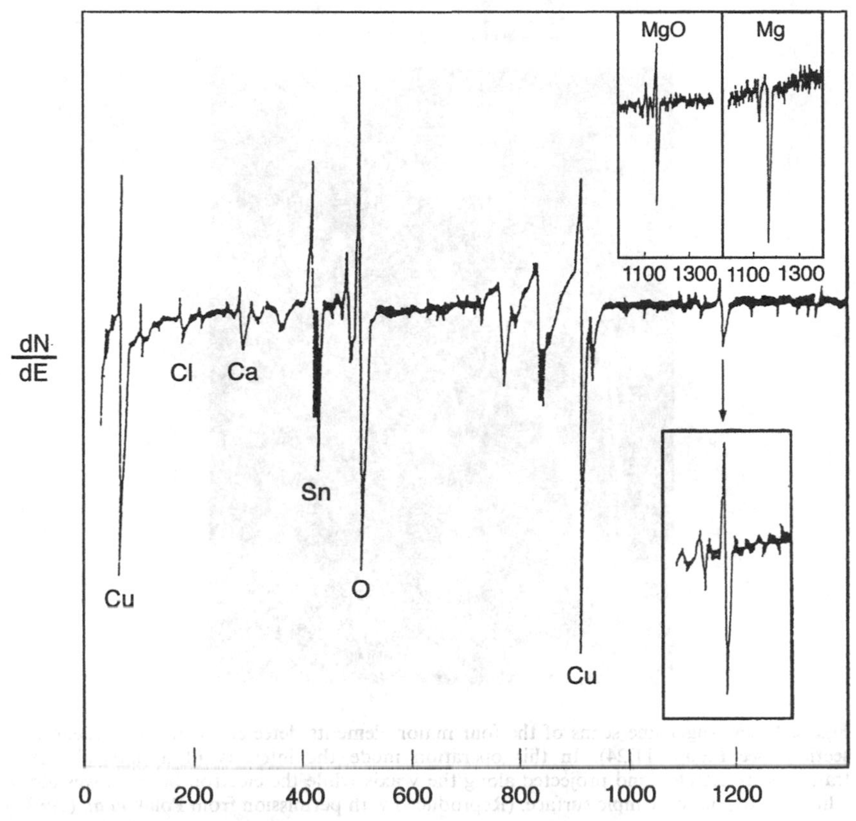

Figure 11.24 Auger survey spectrum of sputter-cleaned cross-section of an archaeological bronze arrowhead. This 'averaged' spectrum was obtained during fast scanning of the sample surface by the primary electron beam. The reference Auger spectra characteristic of metallic and oxidized Mg (upper inset) show that the bronze magnesium is in the oxidized state. (Reproduced with permission from Polak *et al.* (1983).)

scanning the beam over a large area of the sample cross-section (Figure 11.24). The striking feature of this bronze arrowhead analysis was the appearance of significant amounts of magnesium (as MgO; Figure 11.24 inset). Auger scans were then taken along a line over the cross-section (Figure 11.25), revealing general lateral correlations of Sn, Mg and O, with a complementary Cu distribution. The more informative and complete picture emerging from the elemental maps (Figure 11.26) indicates only partial Sn/Mg correlations, namely the appearance also of Sn-rich/Mg-deficient regions (and vice versa) in the microstructure. This demonstrates the application and capability of SAM in the analysis of bulk inhomogeneities. An example of a state-of-the-art application of SAM is the identification of submicrometer particulate contamination on semiconductor devices. Since such defects can result in significant yield loss,

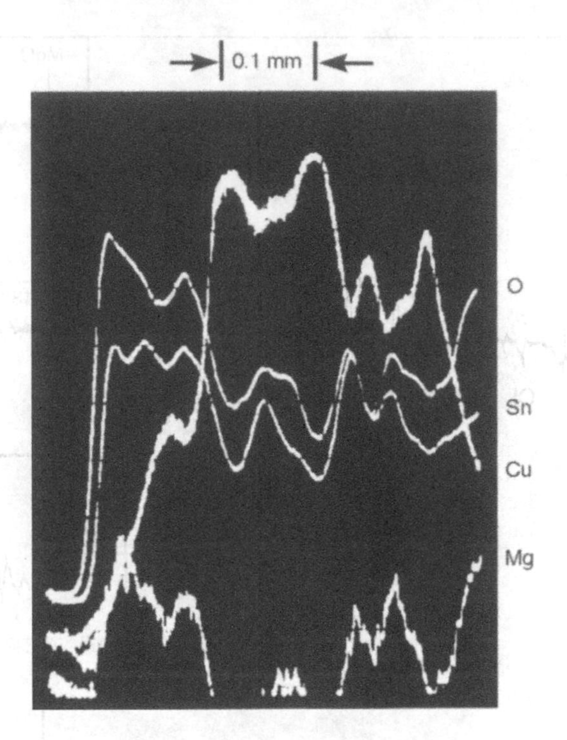

Figure 11.25 Auger line scans of the four major elements detected on the arrowhead cross-section (see Figure 11.24). In this operation mode the intensity of a particular Auger transition is detected and projected along the y axis while the electron beam moves along a line (x axis) on the sample surface. (Reproduced with permission from Polak *et al.* (1983).)

finding the source of particles via their composition and eliminating them from the fabrication process is essential. As the size of integrated-circuit design rules decreases for each new generation of devices, particle compositional analysis becomes more difficult. The performance of SAM analysis of submicrometer particles (which depends critically on the primary electron-beam energy) and its advantages over scanning electron microscopy (SEM)/EPMA or SIMS analysis were recently demonstrated with model Al particles deposited on Si (Figure 11.27).

XPS imaging techniques are instrumentally more complicated and the lateral resolution is lower. Unique to XPS imaging is the use of the

Figure 11.26 SAM images of the tin (top) and magnesium (bottom) distributions at the arrowhead cross-section. Similarly to line scanning, for each Auger map the spectrometer is tuned to the characteristic electron energy while the primary beam rasters a selected surface area. The corresponding relative intensity displayed for each point reflects qualitatively the relative amount of surface atoms (i.e. a brighter region corresponds to higher local concentration of the particular element detected). Polak *et al.* unpublished results.

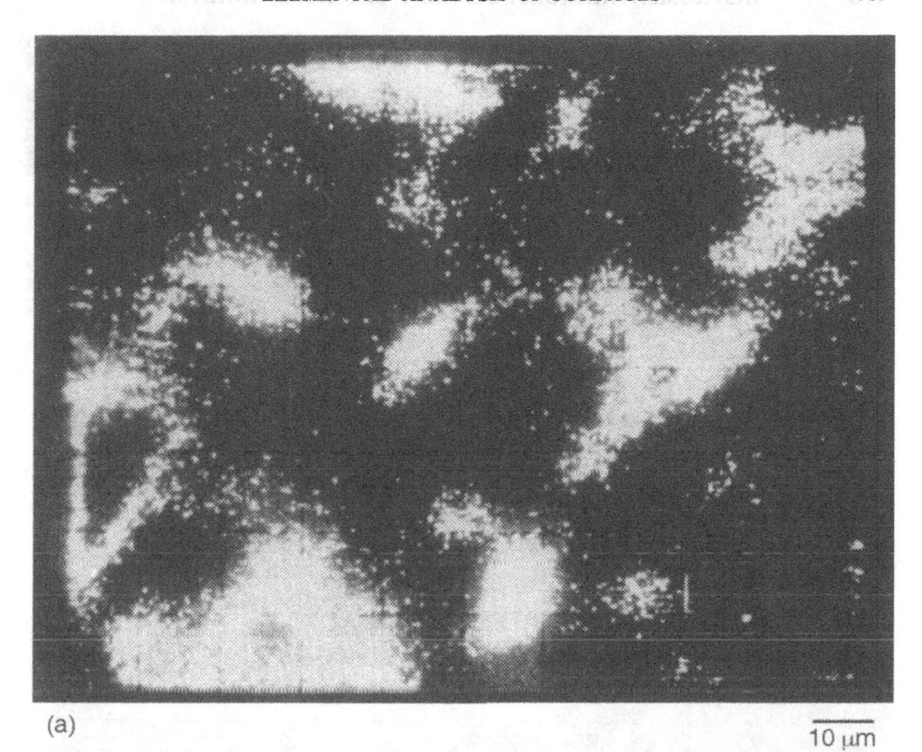
(a)
10 μm

(b)

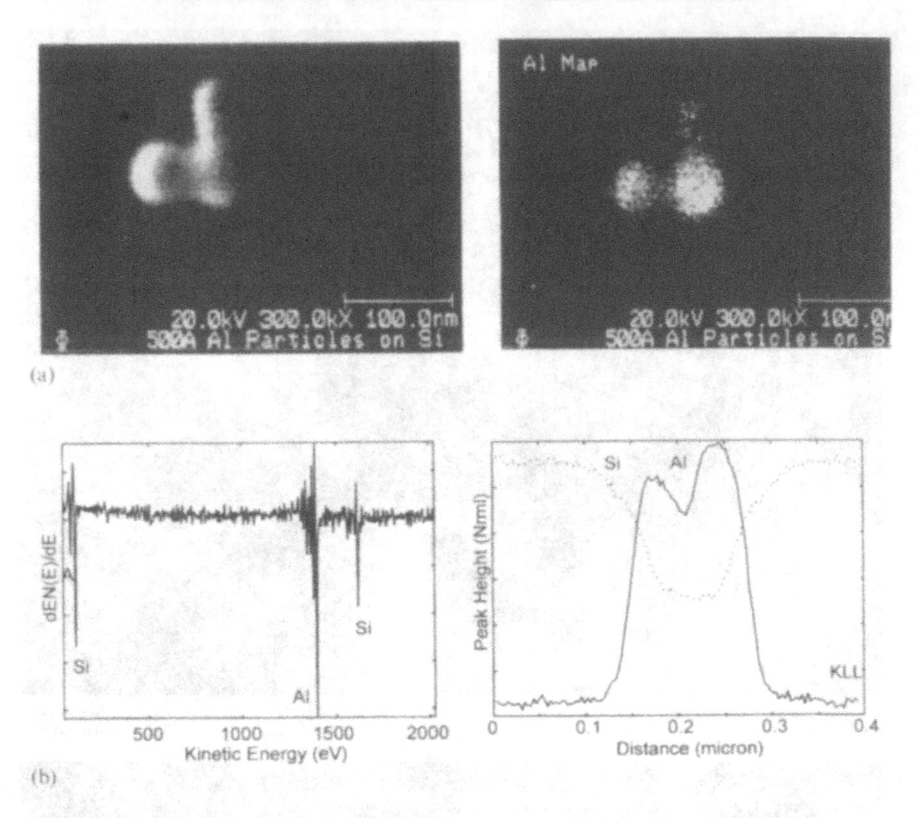

Figure 11.27 Submicrometer particle analysis by scanning Auger microprobe (~50 nm Al particles on Si, using a 20 kV, 10 nA primary electron beam): upper left – secondary electron image (SEM); upper right – SAM Al map; bottom left – Auger point analysis of the central particle acquired in 1.8 min; bottom right – Al and Si Auger line scans acquired horizontally across the bottom two particles (the Si signal is due to the scattering of primary electrons to the surrounding Si substrate). (Reprinted with permission from Childs *et al.* (1996), copyright 1996 American Vacuum Society.)

commonly distinct chemical shifts in binding energy to get mapping of different elemental chemical states distributed on the surface. In one mode of XPS imaging, a focused electron beam is scanned over the anode surface, generating X-rays that are focused by an ellipsoidal monochromator and produces a scanned, narrow X-ray beam ($<10\,\mu m$) on the sample surface. But, the best performance has been achieved so far in 'scanning photoemission microscopy' (SPEM) using a high-brightness (third-generation X-ray) synchrotron source (Casalis *et al.* 1997). The incident photon beam is focused to a submicrometer size with a zone-plate optical system, providing lateral resolution better than 200 nm, and the spectral resolution is good enough to distinguish between different elemental chemical states. Illustration of space-resolved composition

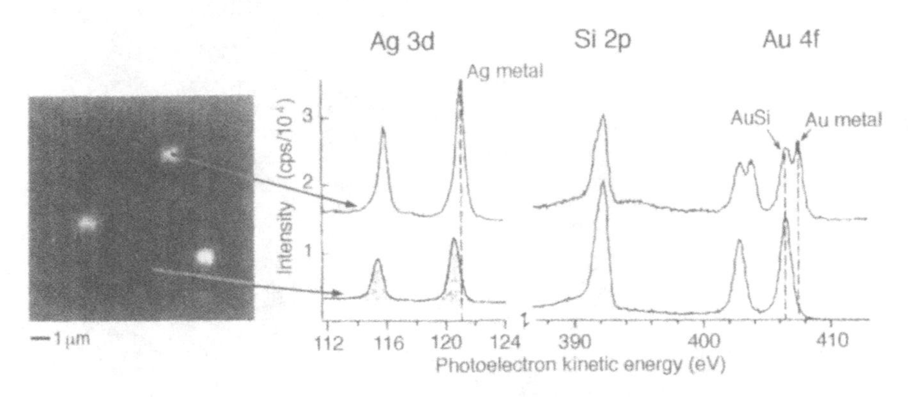

Figure 11.28 State-of-the-art (synchrotron-excited) XPS imaging – Ag+Au on Si(111): left – metallic silver image where the bright features correspond to 3D subphase particles; right – small-spot spectra taken from such particles (top) and from the 2D area in between (bottom): the particles contain metallic Ag and Au, whereas the 2D phase contains only reacted Ag and Au (with Si). (Reproduced with permission from Casalis *et al.* (1997).)

mapping done for heterogeneous metal/semiconductor interfaces with submicrometer phases of different composition and chemical bonding is given in Figure 11.28.

11.3 Secondary-ion mass spectrometry

11.3.1 General principles and operation

Surface and interface analysis by SIMS is based on the mass detection of charged atoms or molecular clusters ('secondary ions') ejected from the surface layer by a focused beam of 'primary ions' with appropriate energy dissipated into the near-surface region. Usually, 1–15 kV argon ions have been used, but Cs^+ or O^+ can enhance and stabilize the secondary-ion yield. All elements and isotopes, from hydrogen to uranium, can be mass-analyzed with very high sensitivity. Generally, there are three modes of SIMS operation: **surface-specific analysis** of the top one or two layers (static SIMS), **composition profiling** (including ppm–ppb trace elements) down to micrometer depths (dynamic SIMS), and **elemental mapping** of a selected surface area (imaging SIMS). Analogous modes of operation were described above for AES and XPS.

When an energetic ion beam impinges on a solid surface, electrons, photons and a variety of secondary heavy particles can be ejected (sputtered) from the target material, including neutral atoms in excited

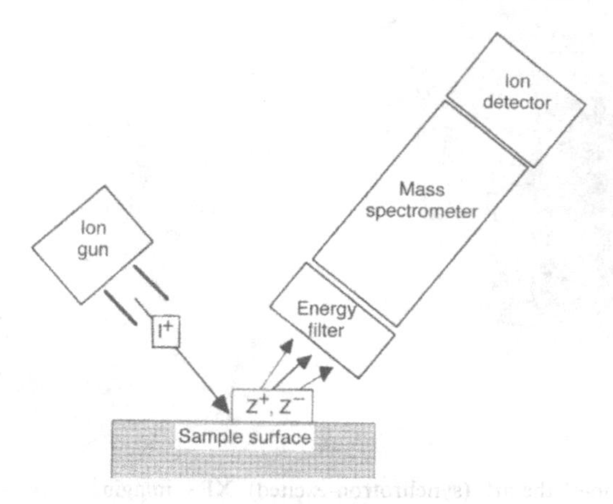

Figure 11.29 Schematic diagram of a SIMS instrument.

states, positive and negative monatomic ions (with different charges) and molecular clusters (neutral or charged). In conventional SIMS (see Figure 11.29) singly charged, positive (usually the most abundant) or negative secondary ions are first separated from the other particles, their energy is selected from a broad energy distribution, and then they are analyzed according to the mass/charge ratio by means of a mass spectrometer (the detector usually operates in the pulse-counting mode to detect single ions). A high-resolution partial spectrum taken from a contaminated video tape is shown in Figure 11.30. The full secondary-ion spectrum is usually quite complex, especially for organics and polymers, but the multitude of lines enables full identification of the molecular species comprising the surface.

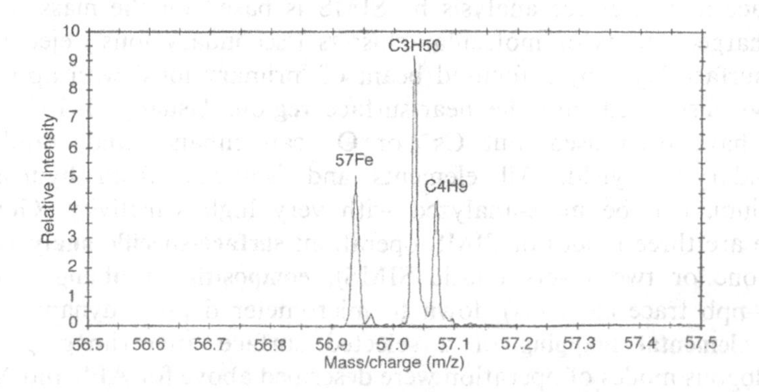

Figure 11.30 Partial high-resolution secondary-ion mass spectrum of a video tape showing the ^{57}Fe isotope and organic fragment peaks. (Reproduced with permission from Reichlmaier (1992).)

The use of a mass spectrometer (MS) into which secondary ions can be collected very efficiently, and which is relatively free of background noise, is a key factor in the high sensitivity and low detection limits obtainable in SIMS. Under favorable conditions, including the absence of mass interferences and relatively high secondary-ion yields (achieved with reactive, e.g. Cs, ion bombardment), the detectable atomic density can be as low as 10^{12} atoms/cm^3 (Figure 11.31).

Thus, a unique combination of merits establishes SIMS as one of the leading and most popular surface and thin-film analytical techniques. Actually, for certain applications, such as trace-element profiling, SIMS is the only suitable technique.

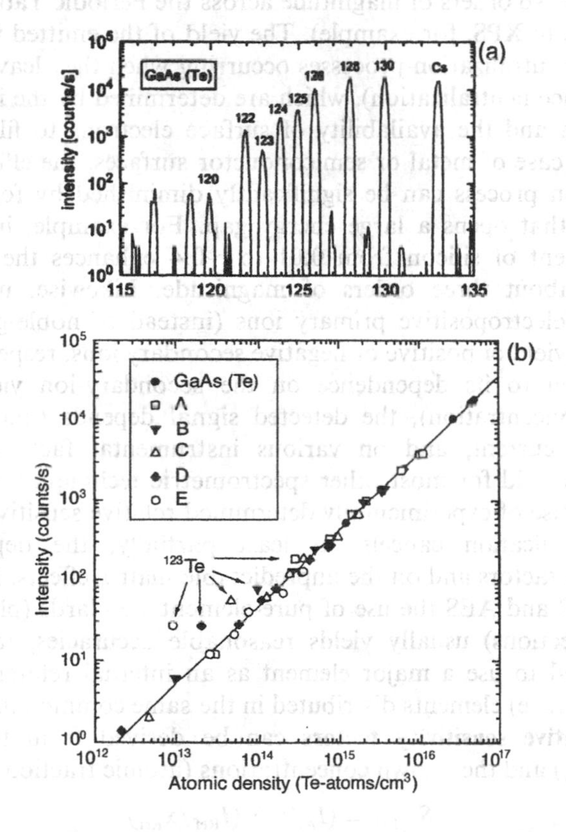

Figure 11.31 Recent demonstration of the extremely low elemental concentrations detectable by SIMS under favorable conditions: (a) The partial mass spectrum (in the range of the Te isotopes) obtained from the Te-doped GaAs standard sample used (1.4×10^{17} Te/cm^3). Cs^+ primary-ion beam: 14.5 keV, 30 nA. (b) The atomic-density dependence of the negative secondary-ion intensities measured for all the Te isotopes in the Te-doped GaAs standard (from the spectrum in (a)) and in undoped GaAs contaminated *in-situ* by Te from a sputtered HgCdTe sample (data labeled A to E). The straight line is a fit to the GaAs(Te) data, and the ^{123}Te data offset is attributed to mass interference). (Reproduced with permission from Gnaser (1997).)

11.3.2 Quantification

SIMS is commonly considered as a semi-quantitative technique mainly because of relatively strong and dominant matrix effects. Just like XPS or AES, quantitative SIMS does not constitute an *absolute* analytical method, but has to rely on spectral relative intensities measured on standard samples having accurately known compositions. The main difficulty in SIMS quantification stems from the large variability of the probability for ionizing surface atoms. Thus, the secondary-ion yield depends strongly on the target chemical/electronic character. For pure elements under the same conditions of the primary-ion beam and the secondary-ion detection, it varies over ~6 orders of magnitude across the Periodic Table (compared to ~2 orders in XPS, for example). The yield of the emitted ions is highly sensitive to neutralization processes occurring when they leave the surface (e.g. resonance neutralization), which are determined by the ion electronic energy levels and the availability of surface electrons to fill the ionized level. In the case of metal or semiconductor surfaces, the efficiency of the neutralization process can be significantly diminished by forming a thin oxide layer that opens a large energy gap. For example, increasing the oxygen content of silicon from 0.01 to ~0.4 enhances the Si^+ and Si^- signals by about three orders of magnitude. Likewise, using electronegative or electropositive primary ions (instead of noble-gas ions) can enhance the yield of positive or negative secondary ions, respectively.

In addition to its dependence on the secondary-ion yield (and the elemental concentration), the detected signal depends (linearly) on the primary-ion current, and on various instrumental factors (equivalent relationships hold for most other spectrometric techniques, e.g. equation (11.9)). The use of experimentally determined relative sensitivity factors in SIMS quantification cancels, at least partially, the dependence on instrumental factors and on the unpredictable matrix effects. In particular, while in XPS and AES the use of pure-element standards (plus calculated matrix corrections) usually yields reasonable accuracies, in SIMS it is recommended to use a major element as an internal reference for other (minor and trace) elements distributed in the same common matrix (Figure 11.32). **Relative sensitivity factors** can be derived from the measured intensities (I_i) and the known concentrations (atomic fractions, x_i):

$$S_{\mathrm{A/Ref}} = (I_{\mathrm{A}}/x_{\mathrm{A}})/(I_{\mathrm{Ref}}/x_{\mathrm{Ref}}) \tag{11.13}$$

Then, an element in a sample containing the reference element as a major constituent can be determined as usual from the relation

$$x_{\mathrm{A}}/x_{\mathrm{Ref}} = (I_{\mathrm{A}}/I_{\mathrm{Ref}})/S_{\mathrm{A/Ref}} \tag{11.14}$$

where, obviously,

$$\sum x_i + x_{\mathrm{Ref}} = 1 \tag{11.15}$$

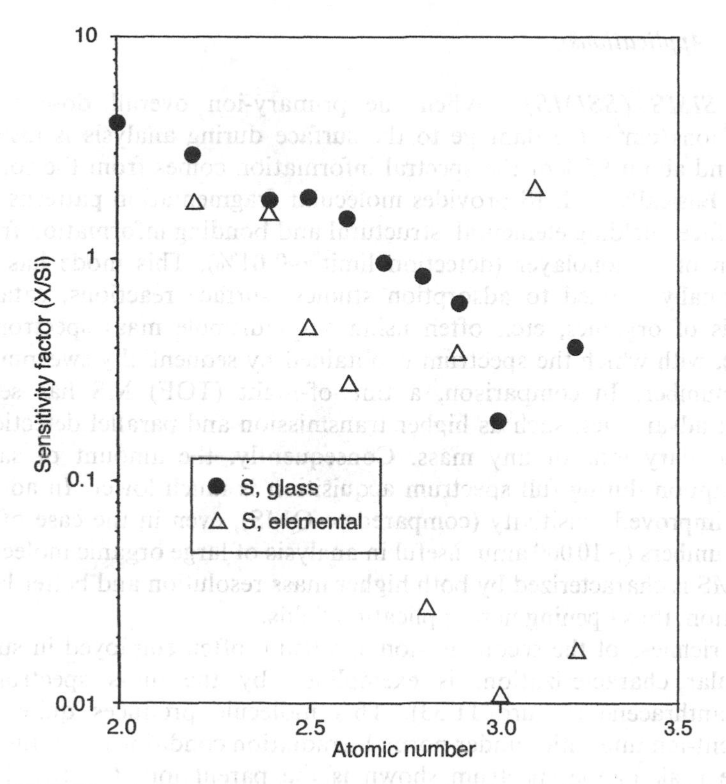

Figure 11.32 The significance of matrix effects in SIMS quantification: Relative (to Si) sensitivity factors ($S_{X/Si}$) measured for several minor and trace elements in the same silicate glass matrix, and as calculated from the pure element measured sensitivities. Note that the $S_{X/Si}$ range is strongly compressed in the case of the elements measured in the common matrix, which affects the various emitted secondary ions in a similar manner. (Reproduced with permission from Briggs and Seah (1992).)

The development of standard materials for SIMS is somewhat problematic since there is no other independent technique with such a combination of high spatial resolution and high sensitivity. Other complications in using multi-element standards in SIMS involve surface modifications due to the prior treatment and initial sputtering effects. Ion implantation is a common procedure for the preparation of standards, e.g. for semiconductor materials (Briggs and Seah 1992).

The usual limitations on SIMS quantification (especially matrix effects) can be largely resolved by the detection of MCs^+ molecular ions, emitted under Cs ion bombardment, and formed by the association of neutral atoms M and Cs^+, rather than detecting M^+ or M^- (Benninghoven *et al.* 1994). Thus, the flux of neutral atoms, which represents the actual M concentration in the sample, is probed, improving the accuracy of quantitative analysis of major components.

11.3.3 Applications

Static SIMS (SSIMS). When the primary-ion overall dose is low ($< 10^{13}$ ions/cm^2), the damage to the surface during analysis is insignificant, and about 95% of the spectral information comes from the top two layers. Basically, SSIMS provides molecular fragmentation patterns from the surface, yielding elemental, structural and bonding information from a fraction of a monolayer (detection limit ~0.01%). This mode has been traditionally applied to adsorption studies, surface reactions, catalysis, analysis of organics, etc., often using a quadrupole mass spectrometer (QMS), with which the spectrum is obtained by sequentially sweeping the mass number. In comparison, a time-of-flight (TOF) MS has several distinct advantages, such as higher transmission and parallel detection of all secondary ions of any mass. Consequently, the amount of sample consumption during full spectrum acquisition is much lower. In addition to the improved sensitivity (compared to QMS), even in the case of high mass numbers (>10 000 amu, useful in analysis of large organic molecules), TOF MS is characterized by both higher mass resolution and better lateral resolution, thus opening new application fields.

The richness of the secondary-ion spectrum, often employed in surface molecular characterization, is exemplified by the mass spectrum of dibenzanthracene (Figure 11.33). This molecule produces quite weak fragment-ion intensities under normal irradiation conditions, and the most intense peak in the spectrum shown is the parent ion ($C_{22}H_{14}^+$). Lower masses, $m/z = 276$, 265 and 279, for example, are due to losses of H_2 and CH, and to H capture, respectively. Higher masses, such as $m/z = 293$ and 294, are due to partial oxidation of the molecule. The identification of molecular species in SSIMS can be aided by the use of extensive databanks of fingerprint spectra available through printed (e.g. Figure 11.34) and electronic media.

Depending on the irradiated sample, the negative-secondary-ion spectrum can be more informative than the positive-ion spectrum, such as in the case of carbon nitride films (Figure 11.35). Among the many C_xN_y cluster fragments, the CN peak is the most intense. Fragment series characteristic of pyridine or graphite systems, such as six-atom aromatic rings, were not observed. Also, the spectrum shows that the film is not significantly hydrogenated, and that the C and N atoms are in configurations analogous to those in the theoretical β-C_3N_4 lattice (Lopez *et al.* 1997).

Dynamic SIMS. Probably the most important technological contribution and application of SIMS is the composition–depth profiling of thin films, in semiconductor structures for example, which includes dopant and impurity elements. In this mode the intensities of several selected mass

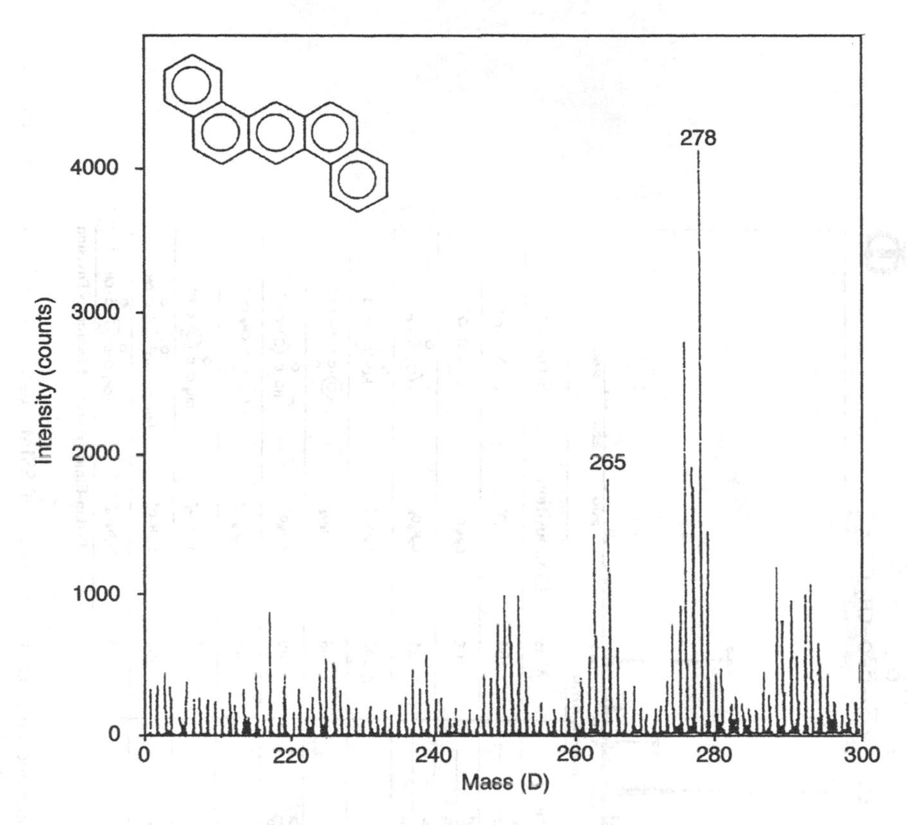

Figure 11.33 A partial spectrum of the positive secondary ions emitted from Dibenz-anthracene. (Reprinted from Delcorte *et al.* (1997), with kind permission of Elsevier Science-NL.)

lines are measured cyclically as a function of sputter time of a (raster scanning) ion gun with a small beam diameter (the term 'dynamic' refers to times of a few seconds or less to remove a monolayer). High sensitivity can be obtained with relatively high fluxes of the primary ions (using powerful guns giving also relatively fast erosion rates), and magnetic sector field MS or QMS (TOF MS has too slow erosion rates). Usually 10^4–10^5 sputtered atoms are sufficient to give statistically useful signals, so that very low detection limits can be obtained for a very small volume of material. This quite unique capability, already realized during the early days of SIMS as an analytical tool, has contributed greatly to its development. Typically, it can take less than 10 min to obtain a complete ~1 μm thick depth profile (10–50% precision) with 10 nm resolution and a detection limit of $<10^{17}$ atoms/cm^3. Under favourable conditions a much lower detection limit is accessible (Fig. 11.31). The dynamic range is also of importance, and it can be extended by certain modifications of the

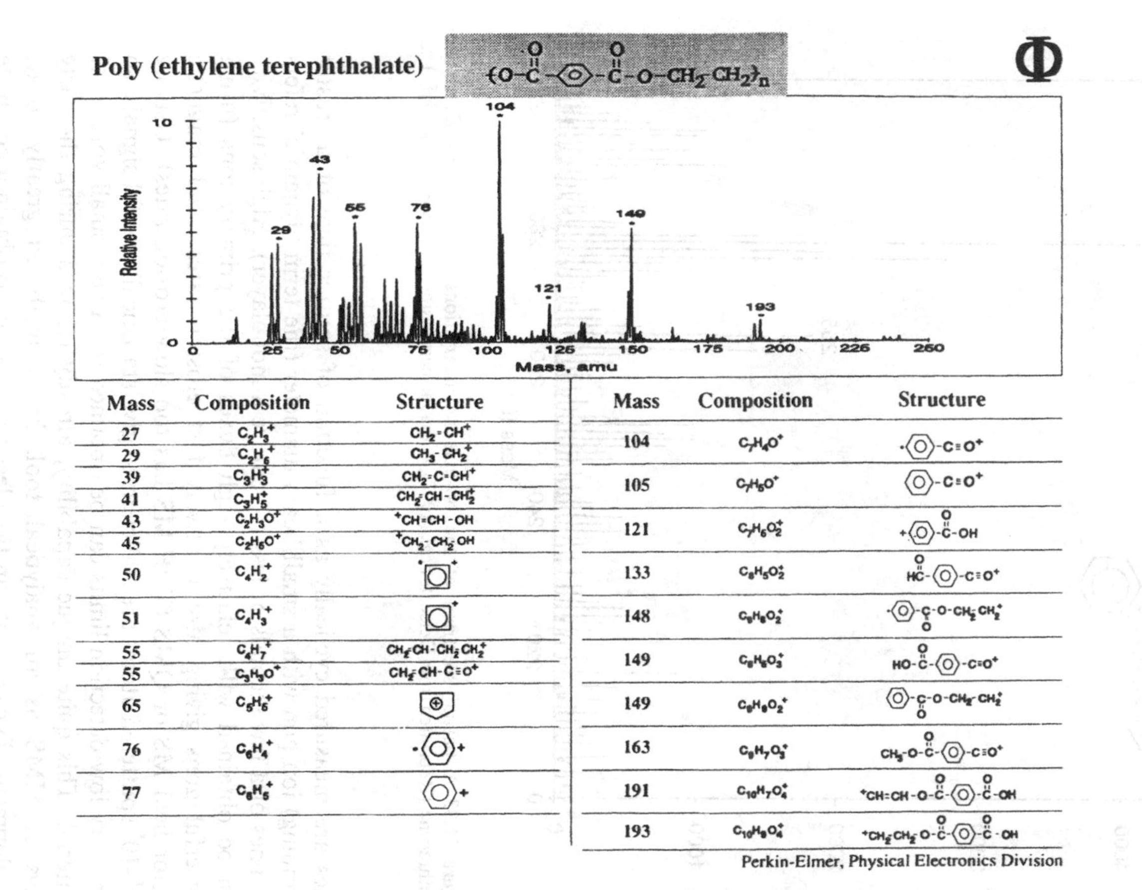

Mass	Composition	Structure
27	$C_2H_3^+$	$CH_2{=}CH^+$
29	$C_2H_5^+$	$CH_3{-}CH_2^+$
39	$C_3H_3^+$	$CH_2{=}C{=}CH^+$
41	$C_3H_5^+$	$CH_2{=}CH{-}CH_2^+$
43	$C_2H_3O^+$	$^+CH{=}CH{-}OH$
45	$C_2H_5O^+$	$^+CH_2{-}CH_2{-}OH$
50	$C_4H_2^+$	
51	$C_4H_3^+$	
55	$C_4H_7^+$	$CH_2{=}CH{-}CH_2CH_2^+$
55	$C_3H_3O^+$	$CH_2{=}CH{-}C{\equiv}O^+$
65	$C_5H_5^+$	
76	$C_6H_4^+$	
77	$C_6H_5^+$	

Mass	Composition	Structure
104	$C_7H_4O^+$	
105	$C_7H_5O^+$	
121	$C_7H_5O_2^+$	
133	$C_8H_5O_2^+$	
148	$C_9H_8O_2^+$	
149	$C_8H_5O_3^+$	
149	$C_9H_9O_2^+$	
163	$C_9H_7O_3^+$	
191	$C_{10}H_7O_4^+$	
193	$C_{10}H_9O_4^+$	

Figure 11.34 An example from a reference source used for identification and interpretation of SSIMS data. (Reproduced with permission from Newman *et al.*)

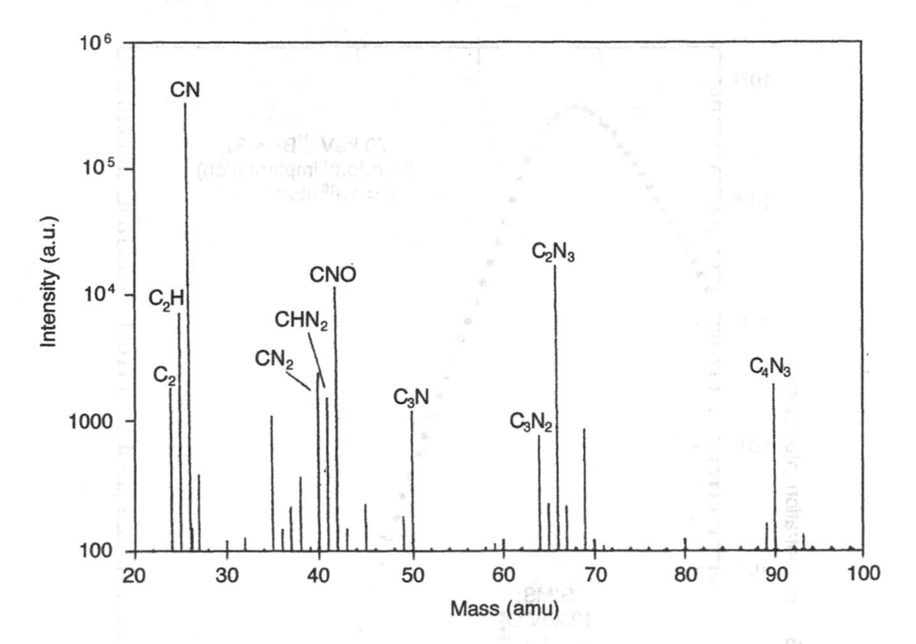

Figure 11.35 Time-of-flight SIMS negative ion spectrum for magnetron-sputtered carbon-nitride films. (The positive SIMS spectrum was dominated by superficial carbon contamination and by Cu sputtered from the r.f. coil.) (Reproduced with permission from Lopez *et al.* (1997).)

experimental set-up. For example, the depth distribution of boron implanted in silicon, shown in Figure 11.36, covers six orders of magnitude, down to the 10^{15} atoms/cm^3 range (Wittmaack and Clegg 1980). It should be noted that, owing to transient variations during the initial sputtering process of the surface region, special precautions have to be taken in order get accurate and reliable compositional gradients from the top ~50 nm. One possible procedure involves extrapolating the profiles measured with low primary energies (1–2 kV) to estimate a true, 'zero'-beam-energy profile (Clegg 1987).

Imaging SIMS. Elemental images in SIMS are obtained by detecting the lateral intensity distribution of the secondary-ion emission. A common application of secondary-ion imaging is in spatial characterization of localized contamination on a surface (in microelectronic devices, for example). In one type of instrument, a focused primary-ion beam is raster-scanned in a controlled way over the sample surface, and simultaneously the secondary-ion signal intensity of a particular mass number is detected (and stored) as a function of the beam position (**secondary-ion microprobe**). Quite a high lateral resolution (several tens of nanometers) can be achieved by using a microfocused liquid–metal (In, Ga) ion source (LMIS) in

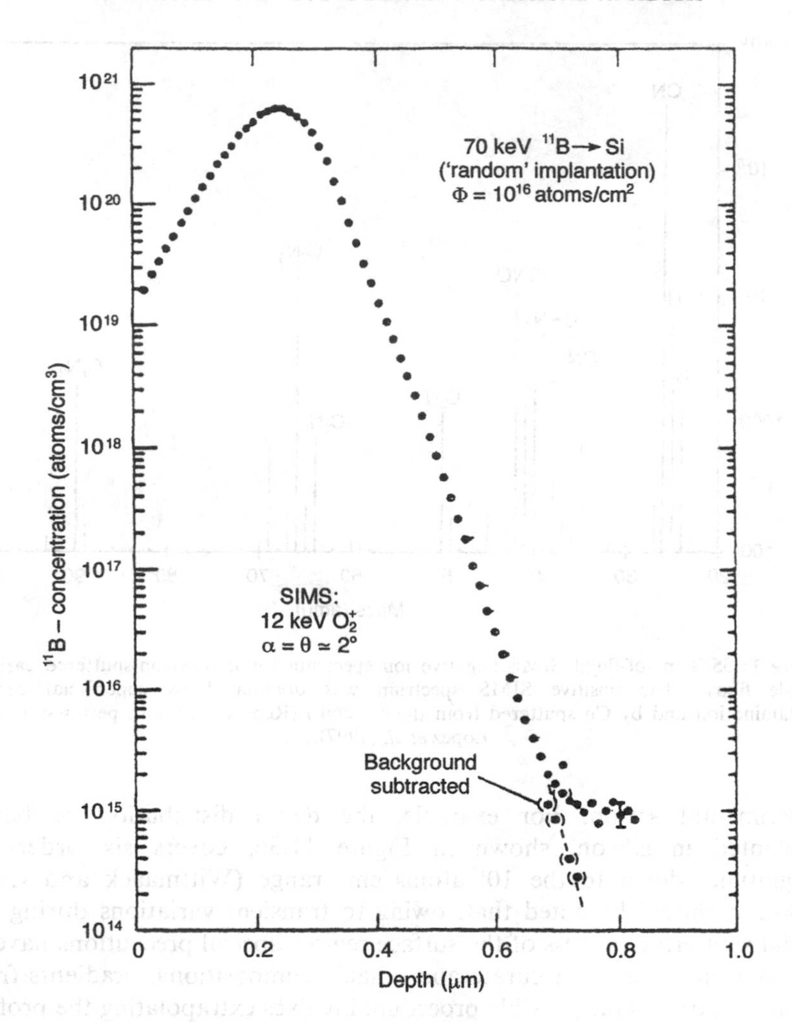

Figure 11.36 SIMS depth-profiling with extended dynamic range: the case of boron (70 keV) implanted in silicon (12 keV O_2^+ primary beam). (Reproduced with permission from Wittmaack and Clegg (1980).)

conjunction with TOF MS. In direct imaging mass spectrometers (using TOF or magnetic image field), the sample is irradiated by a large-diameter beam, and the entire image is simultaneously detected (**secondary-ion microscope**). In this case, micrographs can be obtained faster and at higher erosion rates. In both modes of operation, image processing can furnish correlation between elemental maps and thus firm identification of defects or contamination becomes possible. As an example of a practical application of imaging SIMS, results of TOF SIMS analysis of video head contamination, using a LMIS with ~0.3 μm beam diameter, are shown in

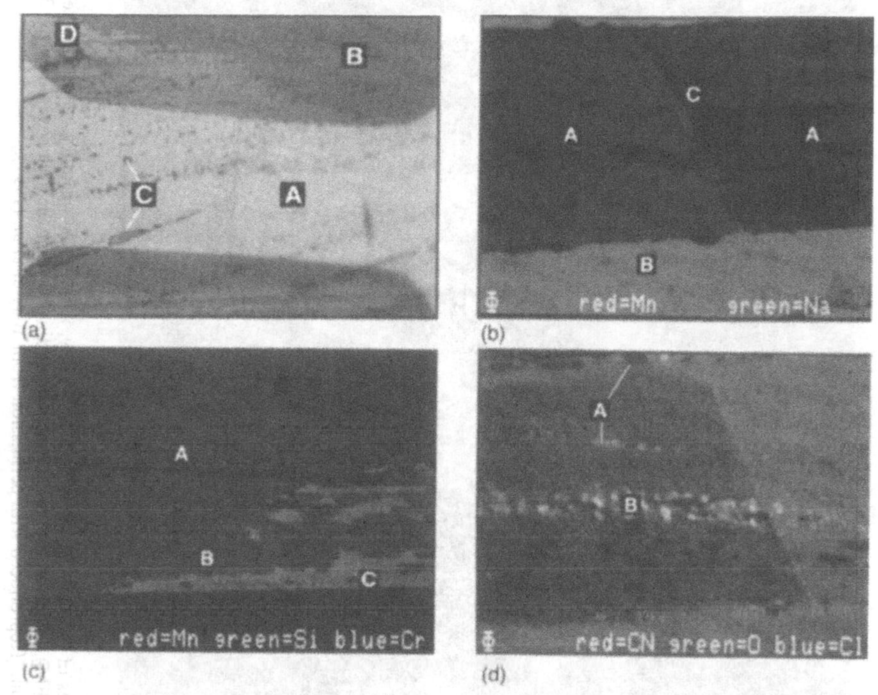

Figure 11.37 TOF-SIMS elemental maps of a contaminated used video head: (a) – optical microscope image (190 × 135 μm): A – magnetic poles, B – glass inlay, C,D – two types of residue; (b) positive SIMS images (40 × 80 μm) of Mn (A, from the ferrite) and Na (B, from the glass); (c) positive SIMS images (50 × 100 μm) of the second type of residue consisting of Cr (area B, blue) and Fe (not shown), at the front side of the glass. Both were transferred from the tape onto the head; (d) negative SIMS images (40 × 80 μm) of CN enriched area (A) that fill the dark spots seen in the positive map, and of Cl enriched regions (B). (Reproduced with permission from Reichlmaier (1992).)

Figure 11.37 (Reichlmaier 1992). Finding the composition and the origin of contamination on video heads can help to maximize their lifetime and optimize cleaning procedures.

When two-dimensional mapping is accomplished under relatively high primary-ion fluxes ('dynamic' erosion rates), three-dimensional (3D) elemental characterization of the complete sample volume becomes possible (**image depth profiling**). A review emphasizing the interplay of resolution, size of the analysis volume and detection limits in 3D SIMS analysis can be found in Clegg (1995). In order to avoid possible distortions due to topographic effects, the topography of the original surface and the relative sputter rates of the different structures within the volume have to be taken into account. For example, AFM topographic images of SIMS-analyzed area before and after depth profiling can be convoluted with the 3D SIMS images to produce the correct image of the compositional variations within the sample volume (Figure 11.38).

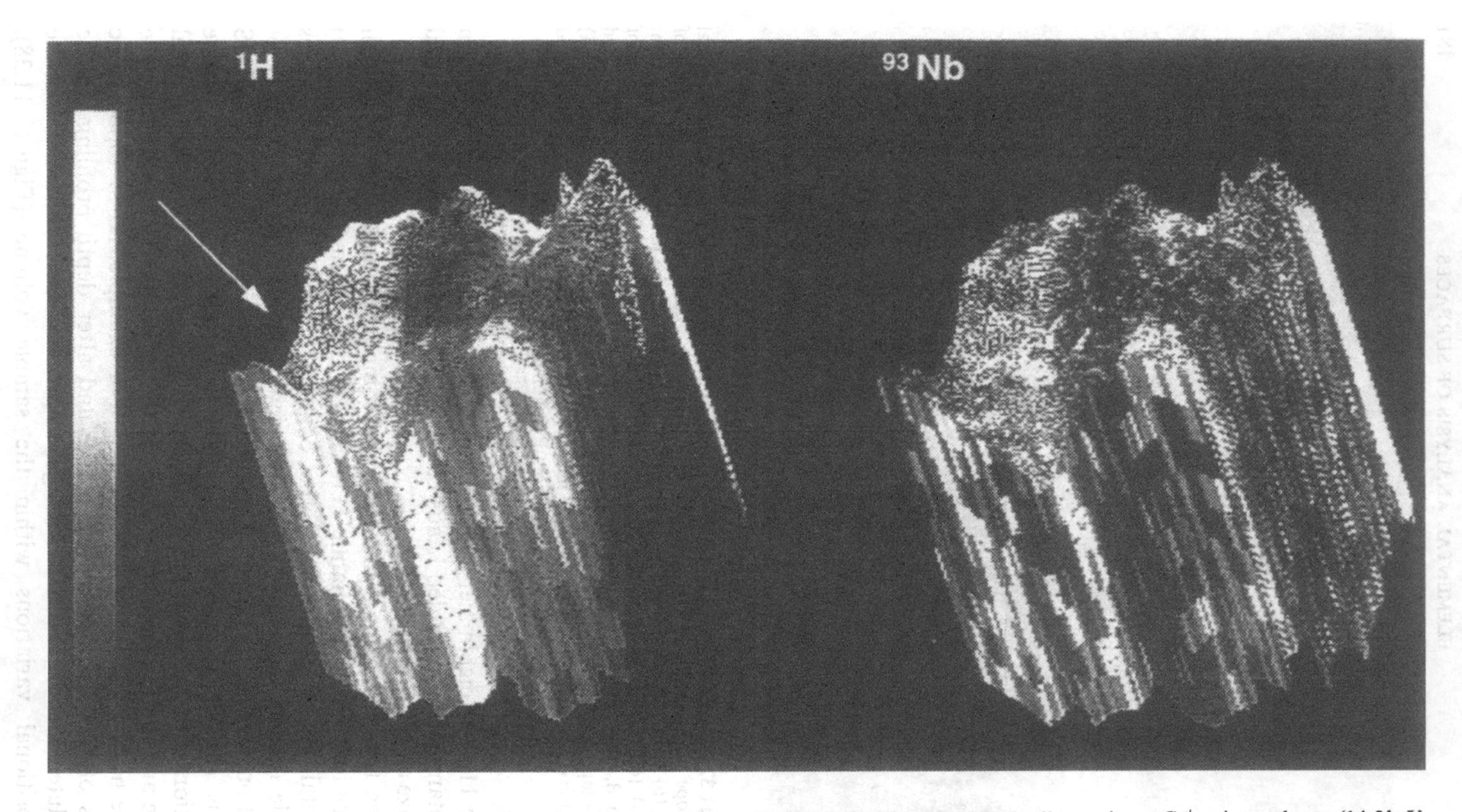

Figure 11.38 Topographically-corrected SIMS 3-dimensional depth-profiles of Nb^- and H^- in a Zr–2.5Nb alloy, using a Cs^+ primary beam (14.5 keV, 1000 nA) rastered over a 250 μm square area. Atomic force microscopy (AFM) images of the topography were adjusted to fit the area covered by the 3D SIMS profile. The arrow points at a mechanical depression. (Reproduced with permission from Wagter *et al.* (1997).)

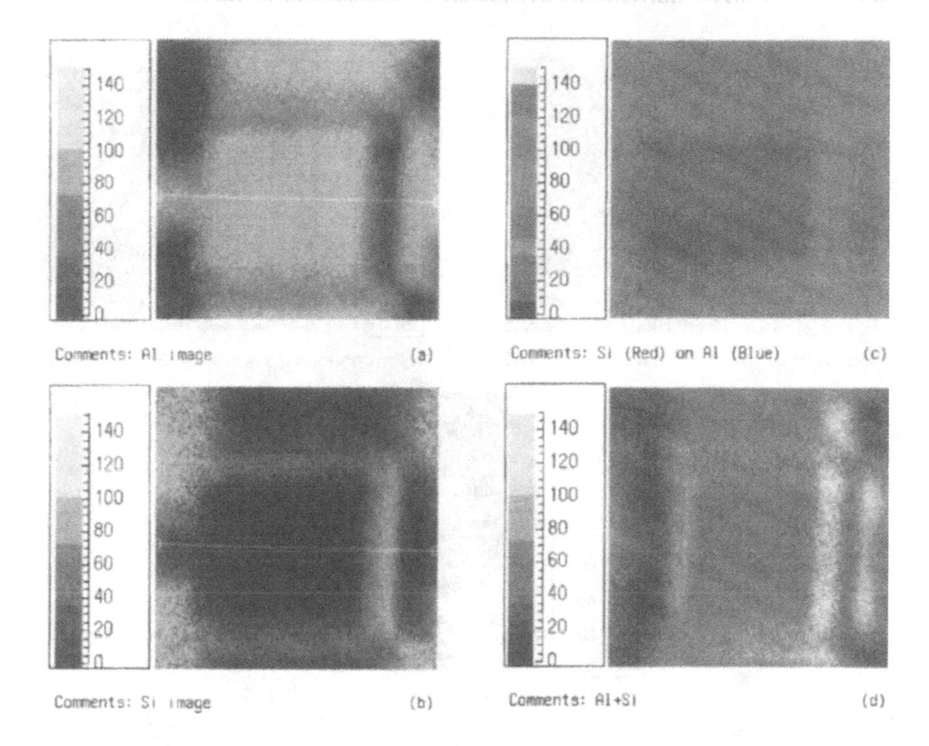

Figure 11.39 Quantitative SIMS imaging (75 × 75 μm) of a p-metal-oxide-Si transistor (7 keV, 1.5 nA beam, 3 mm C_s^+ beam): Elemental distribution maps of Al and Si corresponding to a sputtered depth halfway through the Al contact (the Ti concentration was low to produce an image). (a) Al image, (b) Si image, (c) overlay of the Al and Si images to emphasize the separate regions, (d) sum of the RSF (relative sensitivity factor)-weighted AlC_s^+ and SiC_s^+ images which is proportional to the local sputtering yield (the variation is largely due to the sample topography). (Reprinted with permission from Gnaser (1997), copyright 1997 American Vacuum Society.)

Correlations between SIMS images and topography can provide insight into the relations between microstructure and composition. Another study applied the MCs^+ molecular-ion detection procedure described above to the 3D quantification of a vertical p–metal oxide–semiconductor (p-MOS) transistor structure (Gnaser 1997a). Complete micrographs (750 images) of $AlCs^+$, $SiCs^+$ and $TiCs^+$ (e.g. Figure 11.39) were recorded during depth profiling, and cross-sectional images of the three elements (Figure 11.40) were reconstructed from the complete set of images along the line marked in Figure 11.39. To correct for the actual roughness of the surface, in this work the reconstruction of the sample volume was assisted by 2D profilometry of the surface topography before and after sputtering. Selected-area depth profiles were constructed by summing in each image the counts in all the pixels of the selected area. An example using the 250

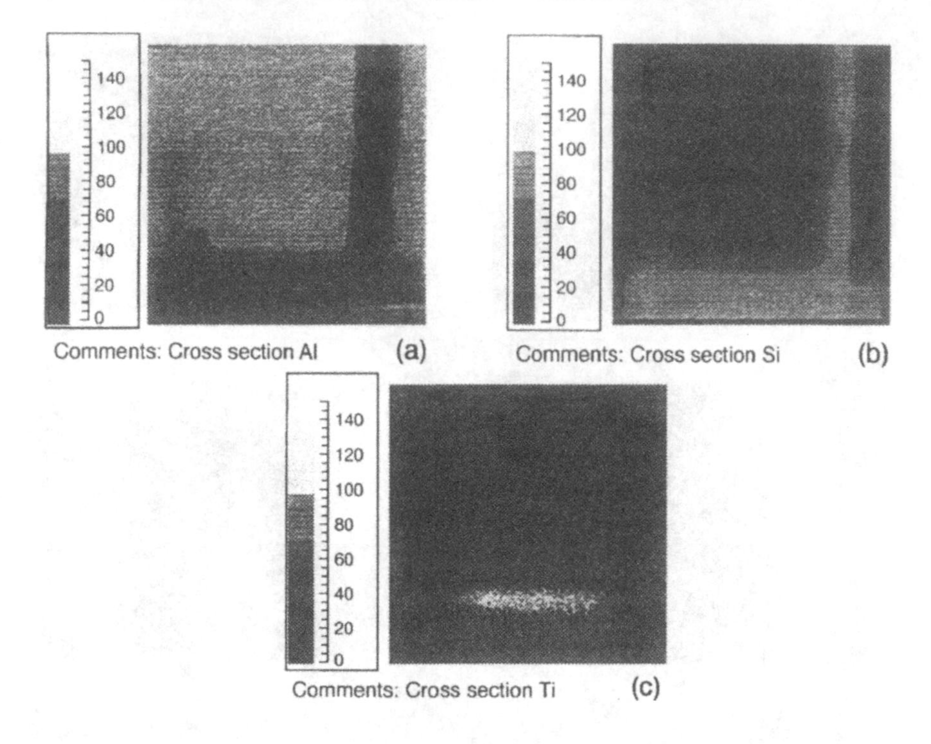

Figure 11.40 Cross-sectional images of Al, Si and Ti in the p-MOS transistor along the line marked in Figure 11.39. They were reconstructed from the complete set of images (250 for each element). The scales give the atomic concentration (at%). (Reprinted with permission from Gnaser (1997a), copyright 1997 American Vacuum Society.)

images for the elements monitored is presented in Figure 11.41. The detection of MCs^+ ions, employed in this work, represents a further step in achieving the goal of 3D quantification in SIMS microanalysis.

11.4 Comparative evaluation of the performance of the three techniques

As described in some detail in this chapter, each of the surface-specific techniques has distinct merits and certain drawbacks. Obviously, the overall performance dictates the optimal technique to be chosen and applied for a given surface-related issue.

Compared to AES, the most prominent advantages of XPS include the more direct chemical-state information usually obtainable, the ability to analyze all materials, including insulators, and the relatively easy interpretation of the spectrum. On the other hand, AES has a superior lateral resolution, reaching a few nanometers in the state-of-the-art instrumentation, and it offers relatively rapid data acquisition. Among the

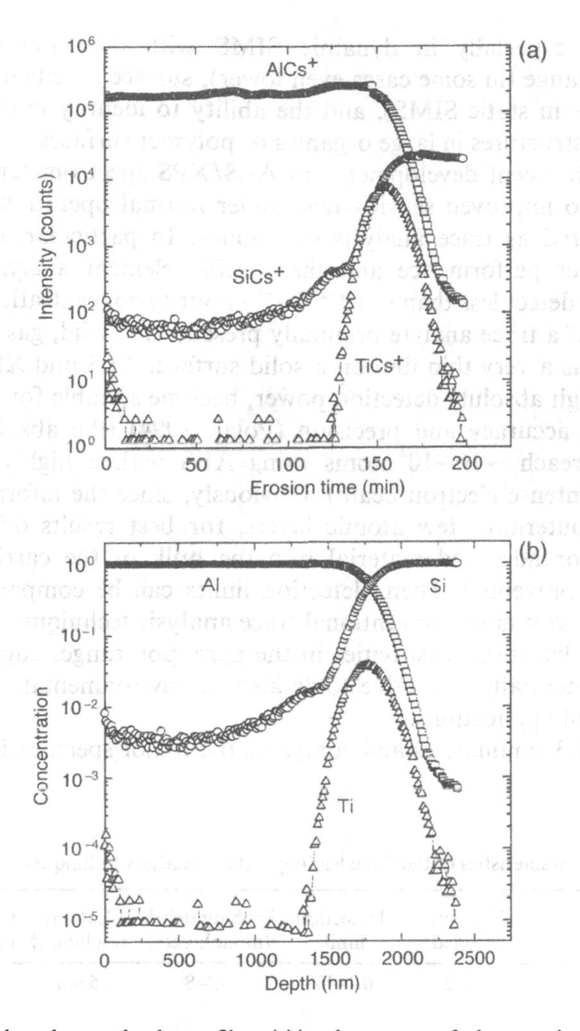

Figure 11.41 Selected-area depth profile within the center of the transistor Al contact (Figure 11.39), reconstructed from the image depth profile data depicted in Figures 11.39 and 11.40. The depth scale in (b) was determined from evaluated relative sputtering yields and the erosion rate, interpolating the atomic densities from those of the pure elements. (Reprinted with permission from Gnaser (1997a), copyright 1997 American Vacuum Society.)

three methods, it can provide compositional mapping at the highest magnification (also compared to SEM/EPMA), and it should be suitable to the technologically important analysis of sub 0.1 μm particulate contamination on semiconductor devices needed in the near future.

SIMS has several distinct advantages that make it unique among the existing multitude of surface analysis tools. These capabilities include the detection of hydrogen and helium (and isotopes of all elements), high sensitivity, especially in dynamic SIMS with detection limits in the ppm–ppb range (in some cases even lower), surface specificity (one or two monolayers in static SIMS), and the ability to identify in detail complex molecular structures in large organics or polymer surfaces.

Although recent developments in AES/XPS spectrometer technologies include also improved sensitivities, under normal operation both cannot be considered as trace analysis techniques. In particular, depending on spectrometer performance and the specific element analyzed, it is not possible to detect less than $\sim 10^{-2}$–10^{-4} impurity concentration in the bulk. However, if a trace analyte originally present in a solid, gas or solution is deposited as a very thin film on a solid surface, AES and XPS, with their intrinsic high absolute detection power, become suitable for trace analysis with good accuracy and precision (Polak 1994) (the absolute detection limit can reach $\sim 10^3$–10^4 atoms using AES with a highly focused and relatively intense electron beam). Obviously, since the information comes from the outermost few atomic layers, for best results diffusion of the deposited or adsorbed material into the bulk of the carrier should be absolutely prevented. Then, detection limits can be comparable to those achievable with more conventional trace analysis techniques. Thus, several reports of detection sensitivities in the ppm–ppb ranges suggest XPS and AES as potentially attractive tools also in environmental, biomedical or geochemical applications.

Table 11.3 summarizes and compares the major specifications and some

Table 11.3 Characteristics of the three leading surface analysis techniques

Technique	Elements detected	Detection limit*	Probing depth (monolayers)	Lateral resolution†	Chemical-state information
AES	$Z > 2$	0.1–1%	~2–8	5 nm	Mainly from CVV lineshapes
XPS	$Z > 2$	0.1–1%	~2–8	2 µm‡	Directly from 'chemical shifts' in core levels
SIMS (static)	all Z (incl. isotopes)	0.01%	1–2	1 µm	From molecular fragment mass numbers
SIMS (dynamic, depth profiling)	all Z (incl. isotopes)	ppm–ppb	~10	50 µm	
SIMS (dynamic, imaging)	all Z (incl. isotopes)	ppm–ppb	~10	20 nm	

* The range represents variations in elemental detection sensitivity. It can change depending on the spatial resolution and the spectrometer performance.
† Best performance.
‡ 0.2 µm in SPEM.

state-of-the-art capabilities of XPS, AES and SIMS. But, at the same time, one should be aware also of the limitations of each technique:

1. SIMS is a more destructive surface technique, compared to AES or XPS (the least destructive among the three), even under the relatively mild primary-ion fluxes used in the static mode.
2. Quantification is typically more difficult in SIMS, in spite of some recent progress described above.
3. The operation and use of SIMS instrumentation is typically more complicated compared to XPS/AES.
4. Depending on the electron-beam current density, surface charging effects and irradiation-induced chemical changes (e.g. electron-stimulated desorption, ESD) can be severe during AES analysis of adsorbates, insulators or even semiconductors.
5. In AES and XPS the detected signal intensity is always a superposition originating from a number of atomic layers, thus making the derived chemical information concerning the top atomic layer somewhat ambiguous. Determination of the actual (shallow) composition gradients by means of AES/XPS requires the quite elaborate procedures described in section 11.2.4.

Complementary techniques. When several surface-specific techniques are available, data of one method can complement another and thus yield a more comprehensive and unambiguous description of a surface phenomenon. For instance, the synergistic advantages of XPS and SIMS in organic or polymeric applications, such as characterization of intentionally modified surfaces (for better adhesion, for example), can be combined to improve the interpretation of surface chemical features. In the case of very complex organic molecules, the use of XPS alone for identification may be insufficient since it can provide only the elemental composition and chemical states which can be almost identical for different molecular structures. Static SIMS, on the other hand, is capable of identifying even subtle structural differences between surface molecular species of high complexity.

When, in addition to elemental/chemical characteristics of the surface, atomic structural features are of interest, low-energy electron diffraction (LEED) or one of the more direct, so-called ‘local probe microscopies’, such as scanning tunneling microscopy (STM), can be used. Their application can elucidate the information achievable by AES, XPS or SIMS. An instructive example for the integration of techniques sensitive to the surface chemistry and to the surface structure can be found in a recent study of the oxygen-induced reconstruction of a Pt–Rh(100) alloy surface (Matsumoto *et al.* 1997). While all that could be learned on this system from AES intensity measurements was a gradual process of Rh

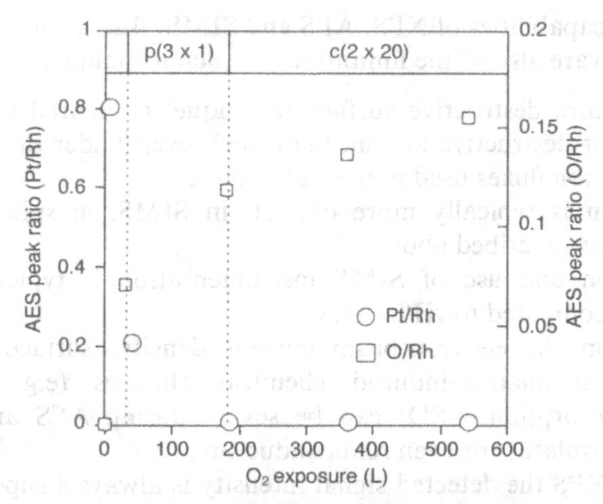

Figure 11.42 Variations of the AES Pt/Rh and O/Rh intensity ratios (reflecting relative concentrations) that occur upon exposing the Pt enriched, clean (100) surface of the 25Pt-75Rh random alloy to oxygen at 600 K. The corresponding LEED patterns observed are indicated at the top. (Reprinted from Matsumoto *et al.* (1997) with kind permission of Elsevier Science-NL.)

segregation by exposure to O_2 (Figure 11.42), LEED and STM revealed two distinct surface reconstructions, and enabled deduction of the corresponding atomic structure models (Figure 11.43).

It is worth while noting that local atomic structural information can be obtained also in AES or XPS by measuring the angular distributions of the emitted electrons, but it requires a special experimental set-up and relatively long acquisition times.

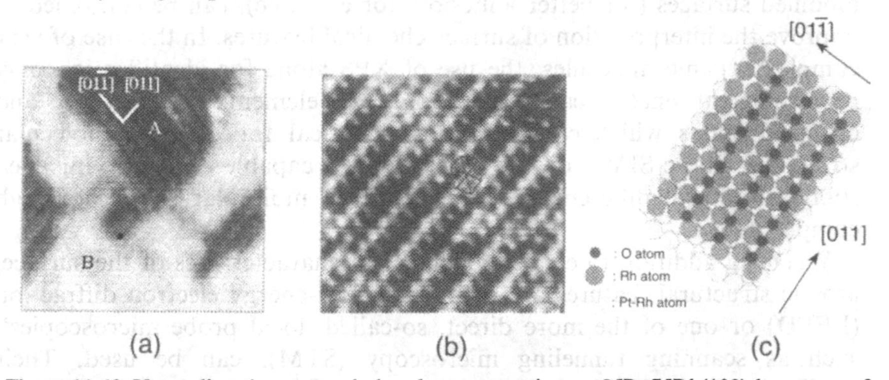

Figure 11.43 Unraveling the oxygen-induced reconstruction on 25Pt-75Rh(100) by means of scanning tunneling microscopy (STM): (a) STM image from a surface that gives a combined LEED pattern – p(3 × 1)-O (from domain A) and c(2 × 20)-O (from domain B). 13.2 nm × 13.2 nm. (b) A highly (atomic)-resolved STM image for the c(2 × 20)-O surface structure. The bright protrusions are arranged in a quasi-hexagonal lattice which is a quasi-Rh(111)-O overlayer on the (100) alloy surface. 4 nm × 4 nm. (c) The resultant structural model deduced for the c(2 × 20)-O reconstruction of the 25Pt-75Rh(100) surface. (Reprinted from Matsumoto *et al.* (1997) with kind permission of Elsevier Science-NL.)

11.5 Summary

This chapter has focused on the three most widely used methods for surface elemental/chemical analysis, AES, XPS and SIMS. The basic principles and capabilities of each were introduced together with selected applications and some state-of-the-art developments. In addition to fundamental research on well-defined samples, including mechanisms of gas adsorption or surface segregation and reactions on atomically clean single-crystal surfaces, the three techniques have been particularly useful in 'real-world' surface analysis. Thus, they are often applied to material issues and problems in metallurgy, thin-film semiconductors, ceramics and polymers. Hence, the widespread activity is not limited to basic research, but includes also trouble-shooting and quality control in the fabrication and processing of 'conventional' and 'high-tech' materials.

All this is definitely due to a combination of several common merits:

1. Relatively straightforward and fast multi-elemental identification (excluding H and He in AES/XPS).
2. Chemical-state information accessible via chemical shifts and lineshape features in XPS and AES, and by the mass numbers of molecular cluster fragments in SIMS.
3. Laterally resolved elemental microanalysis (including imaging).
4. Analysis of shallow or extended in-depth composition gradients (depth profiling). This capability is often of interest, since modifications or variations in the chemical composition in a multi-element solid material are usually not confined to the outermost surface layer. Furthermore, combining elemental imaging with controlled sputter depth profiling can furnish in all three techniques a detailed **three-dimensional microanalysis** of the surface region in bulk samples or of layered thin-film structures.
5. Relatively easy specimen preparation.
6. The availability of advanced, multi-technique equipment, and vast support data (this is the outcome of all the other advantages).

When selecting one of the three techniques for a specific surface issue, one should be aware also of merits (and limitations) unique to each. Thus, the extremely high lateral resolution achievable in AES (down to the nanoscale in state-of-the-art instrumentation), the elemental chemical bonding and ionic charge reflected directly by line shifts in XPS, and the identification of complex organic or polymeric structures in static SIMS as well as trace-element sensitivity in dynamic SIMS, are the most prominent advantages. When several surface-related techniques are available, data of one method can complement another and thus yield a more comprehensive and unambiguous description of surface phenomena. Further progress and instrumental developments are likely to improve their performance (lateral resolution, sensitivity, quantification, etc.). Nevertheless, the basic characteristics described in this chapter are not expected to change significantly for these well-established methods.

References

Benninghoven, A., Nihei, Y., Shimizu, R. and Werner, H.W. (eds) (1994), *Secondary Ion Mass Spectrometry SIMS IX*, John Wiley & Sons, Chichester, pp. 377–425.

Briggs, D. and Seah, M.P. (eds) (1990), *Practical Surface Analysis*, vol. 1, *Auger and X-ray Photoelectron Spectroscopy*, John Wiley & Sons, Chichester.

Briggs, D. and Seah, M.P. (eds) (1992), *Practical Surface Analysis*, vol. 2, *Ion and Neutral Spectroscopy*, John Wiley & Sons, Chichester.

Casalis, L., Gregoratti, L., Kishinova, M. *et al.* (1997), *Surf. Interface Anal.* **25**, 374.

Certified Reference Material (1983), BCR No. 261 (NPL No. S7B83).

Chatelier, R.C., St John, H.A.W., Gengenbach, T.R *et al.* (1997), *Surf. Interface Anal.* **25**, 741–746.

Childs, K.D., Narum, D., LaVanier, L.A. *et al.* (1996), *J. Vac. Sci. Technol.* **A14**(4), 2392–2404.

Clegg, J.B. (1987), *Surf. Interface Anal.* **10**, 332.

Clegg, J.B. (1995), *J. Vac. Sci. Technol.* **A13**, 143.

Cumpson, P.J. (1995), *J. Electron Spectrosc. Relat. Phenom.* **43**, 25–52.

Cumpson, P.J. and Seah, M.P. (1997), *Surf. Interface Anal.* **25**, 430–446.

Davis, L.E., MacDonald, N.C., Palmberg, P.W. *et al.* (1978), *Handbook of Auger Electron Spectroscopy*, Physical Electronics Inc., Eden Prairie, MN 55344, USA.

Feldman, L.C. and Mayer, J.W. (1986), *Fundamentals of Surface and Thin Film Analysis*, Elsevier Science-NL, Sara Burgerhartst. 25, 1055 KV Amsterdam, Netherlands, p. 223.

Gnaser, H. (1997a), *J. Vac. Sci. Technol.* **A15**(3), 445–450.

Gnaser, H. (1997b), *Surf. Interface Anal.* **25**, 737–740.

Grant, J.T. (1989), *Surf. Interface Anal.* **14** 271–283.

Gunter, P.L.J., Gijzeman O.L.J. and Niemantsverdriet, J.W. (1997), *Appl. Surf. Sci.* **115**, 342–346.

Guo, H., Maus-Friedrichs, W. and Kempter, V. (1997), *Surf. Interface Anal.* **25**, 390–396.

Guttierez, A., Lopez, M.F., Garcia, I. and Vazquez, A. (1997), *J. Vac. Sci. Technol.* **A15**(2), 296.

Hall, P.M. and Morabito, J.M. (1979), *Surf. Sci.* **83**, 391–405.

Jablonsky, A. and Powell, C.J. (1997), *J. Vac. Sci. Technol.* **A15**(4), 2095–2106.

Langeron, J.P. (1989), *Surf. Interface Anal.* **14**, 381–387.

Leveque, G. and Bonnet, J. (1995), *Appl. Surf. Sci.* **89**, 211–219.

Lopez, S., Donlop, H.M., Benmalek, M. *et al.* (1997), *Surf. Interface Anal.* **25**, 315–323.

Matsumoto, Y., Aibara, Y. Mukai, K. *et al.* (1997), *Surf. Sci.* **377/379**, 32–37.

Moon, D.W. and Kim, K.J. (1996), *J. Vac. Sci. Technol.* **A14**(5), 2744–2756.

Newman, J.G. and Carlson, B.A., Michael, R.S. *et al.* (undated), *Static SIMS Handbook of Polymer Analysis*, Physical Electronics Inc., 6509 Flying Cloud Dr, Eden Prairie, MN 55344, USA.

Polak, M. (1994), Determination of trace elements by electron spectroscopic methods, in *Determination of Trace Elements* (ed. Z.B. Alfassi), VCH, Weinheim, pp. 359–392.

Polak, M., Baram, J. and Pelleg, J. (1983), *Archaeometry*, **25**, 59–67.

Powell, C.J. and Seah, M.P. (1990), *J. Vac. Sci. Technol.* **A5**, 735.

Procop, M. and Weber, E.-H. (1990), *Surf. Interface Anal.* **15**, 583–584.

Reichlmaier, S. (1992), *PHI Interface*, **14**(2), 1–16, Physical Electronics Inc., 6509 Flying Cloud Dr, Eden Prairie, MN 55344, USA.

Satory, K., Haga, Y., Minatoya, R. *et al.* (1997), *J. Vac. Sci. Technol.* **A15**(3), 468–484.

Seah, M.P. and Dench, W.A. (1979), *Surf. Interface Anal.* **1**, 2–11.

Seah, M.P. and Gilmore, I.S. (1996), *J. Vac. Sci. Technol.* **14**, 1401.

Seal, S., Barr, T.L., Sobczak, N. *et al.* (1997), *J. Vac. Sci. Technol.* **A15**(3), 505–512.

Shiffman, B. and Polak, M. (1986), *Surf. Interface Anal.* **9**, 151–155.
Shirley, D.A. (1972), *Phys. Rev.* **B5**, 4709.
Silvain, J.F., Turner, M.R. and Lahaye, M. (1996), *Composites* **27A**, 793–798.
Tanuma, S., Powell, C.J. and Penn, D.R. (1991), *Surf. Interface Anal.* **17**, 911–926.
Tougaard, S. (1996a), *Appl. Surf. Sci.* **100/101**, 1–10.
Tougaard, S. (1996b), *J. Vac. Sci. Technol.* **A14**(3), 1415–1423.
Tougaard, S. and Jansson, C. (1992), *Surf. Interface Anal.* **19**, 171–174.
Tyler, B.J., Gastner, D.G. and Ratner, B.D. (1989), *Surf. Interface Anal.* **14**, 443–450.
Wagner, C.D. (1983), *J. Electron Spectrosc. Relat. Phenom.* **32**, 99–102.
Wagner, C.D. (1990), Photoelectron and Auger energies and the Auger parameter: a data set, in *Practical Surface Analysis – Vol. 1, Auger and x-ray photoelectron spectroscopy* (eds Brigg, D. and Seah, M.P.), John Wiley & Sons, Chichester, Appendix 5.
Wagner, C.D., Riggs, W.M., Davis, L.E. *et al.* (1979), *Handbook of X-ray Photoelectron Spectroscopy*, Physical Electronics Inc., Eden Prairie, MN 55344, USA.
Wagter, M.L., Clarke, A.H., Taylor, K.F. *et al.* (1997), *Surf. Interface Anal.* **25**, 788–789.
Werner, W.S.M., Gries, W.H. and Stori, H. (1991), *Surf. Interface Anal.* **17**, 693–704.
Wittenaack, K. and Clegg, J.B. (1980), *Appl. Phys. Lett.* **37**, 285.
Yoshitake, M. and Yoshihara, K. (1997), *Internet Database of AES and XPS Spectra*, National Research Institute for Metals, Japan. E-mail: michystk@nrim.go.jp
Zalar, A. (1986), *Surf. Interface Anal.* **9**, 41.

Index

Page numbers in *italics* refer to tables; page numbers in **bold** refer to figures

9780751404272